Organotransition Metal Chemistry

Organotransition Metal Chemistry

Fundamental Concepts and Applications

AKIO YAMAMOTO

Research Laboratory of Resources Utilization
Tokyo Institute of Technology
Yokohama, Japan

A Wiley-Interscience Publication

JOHN WILEY & SONS

New York • Chichester • Brisbane • Toronto • Singapore

Library of Congress Cataloging in Publication Data

Yamamoto, Akio, 1930–
Organotransition metal chemistry.

"A Wiley-Interscience publication."
Includes index.
1. Organometallic compounds. 2. Transition metal compounds. I. Title.

QD411.Y36 1986 547′.05 85-26349
ISBN 0-471-89171-1

Printed in the United States of America

10 9 8 7 6 5 4 3 2 1

Preface

Organotransition metal compounds are useful in many respects and show such a rich and diverse chemistry that researchers derive enormous pleasure from working with them. Yet the chemistry of organotransition compounds is reasonably well understood only by the specialists involved. Courses on organometallic chemistry are given in only a few universities, and lecturers feel the lack of suitable textbooks that can give students an idea of what organometallic chemistry is and how interesting it is to study.

The primary reason for this is that progress in the field has been so rapid. Stimulated by discoveries made in the 1950s, such as ferrocene, the Ziegler catalyst, and the Hoechst–Wacker process, an enormous expansion occurred in the chemistry of organotransition metal compounds. A variety of exciting discoveries was published in rapid succession, and the number of papers dealing with organotransition metal compounds increased almost exponentially for some time. Following this rapid progress, a number of excellent review articles and books have been published, culminating in the launching of *Comprehensive Organometallic Chemistry*, a nine-volume encyclopedia. Nevertheless, there are still relatively few readable texts that deal with organotransition metal chemistry and its fundamental concepts. The prototype of this book was written in Japanese and was based on my lectures given to graduate students at the Tokyo Institute of Technology. Fortunately, the book has been well received by the Japanese audience, and, somewhat recklessly, I decided to write an English version, despite the considerable language barrier.

Significant alterations have been made from the Japanese version. A chapter dealing with organometallic compounds of the main-group elements was entirely omitted. On the other hand, other chapters have been updated and expanded considerably. No attempt has been made to make this book a comprehensive account. Many reviews and reference books exist. It is my wish to provide beginners in this field with a readable book that surveys various aspects of organotransition metal chemistry, ranging from fundamental concepts to applications in catalysis and organic synthesis. In most chapters I avoided making the list of references too long but I included references to pertinent review books and articles so that the interested reader has access to the original literature.

In the hope of making this book as readable as possible the text has been interspersed with various anecdotes entitled Intermezzo. These were presented during my lectures to help students stay awake. I hope they work that way for the reader as well.

Some of my own group's results are included in the text, not from a wish to neglect others' work, but simply because one knows one's own work best. For those who contributed to the development of the work in our group I owe sincere gratitude. They have also helped me by critically reading and discussing the manuscript, in its English and Japanese versions. In particular, the help of Takakazu Yamamoto, Takashi Ito, Sanshiro Komiya, Fumiyuki Ozawa, Kotaro Osakada, Hiroshi Nakazawa, and Kenji Sano is gratefully acknowledged. Most of the manuscript was typed by Chiemi Ezawa and figures were drawn by Kenji Sano, whose effort and skill were indispensable in completing the book.

The last part of this book was written during my stay at the Australian National University and the University of Auckland, New Zealand. Martin Bennett and Warren Roper served as my hosts and helped me enormously by carefully reading the manuscript, giving advice in various aspects with proper criticisms, and brushing up my English text. Particular thanks are due to Martin Bennett, who read the entire manuscript and gave me invaluable advice. Various other chapters were read by other colleagues, including R. Hoffmann, C. A. Tolman, J. Latten, N. Roberts, B. Wild, R. Rickards, H. Yamazaki, and K. Tatsumi, to whom I owe my sincere thanks. Any errors and omissions are, of course, the author's responsibility, and I would welcome any criticism and suggestions from the reader. I am very grateful to Syōkabo Publishing Co. for the kind permission to publish this English version, part of which is based on the Japanese version.

The excellent service of John Wiley and Sons in the production of this book is gratefully acknowledged.

Last, but not least, I owe heartfelt thanks to my family for their help and encouragement.

AKIO YAMAMOTO

August 1985
Yokohama

Contents

CHAPTER 3
NATURE OF THE METAL-CARBON BOND AND ELECTRONIC CONFIGURATIONS OF TRANSITION METAL COMPLEXES 42

CHAPTER 4
SYNTHESES OF ORGANOTRANSITION METAL COMPLEXES 92

CHAPTER 5
EXPERIMENTAL TECHNIQUES IN ORGANOMETALLIC CHEMISTRY 155

CHAPTER 6 FUNDAMENTAL PROCESSES IN REACTIONS OF ORGANOTRANSITION METAL COMPLEXES 195

CHAPTER 7 APPLICATION OF TRANSITION METAL COMPLEXES TO CATALYSIS 305

List of Abbreviations

Ac	Acetyl group
acac	Acetylacetonato ligand (2,4-pentanedionato)
Ar	Aryl group
bipy	2,2′-Bipyridine
Bu	Butyl group
Bz	Benzyl group
COD	1,5-Cyclooctadiene (written in lowercase letters to represent ligand)
COT	Cyclooctatetraene
Cp	Cyclopentadienyl group, η^5-C_5H_5
Cy	Cyclohexyl group
Δ	Heating or thermolysis
DMF	*N,N*-Dimethylformamide
dmpe	1,2-Bis(dimethylphosphino)ethane
DMSO	Dimethyl sulfoxide
dppb	1,2-Bis(diphenylphosphino)butane
dppe	1,2-Bis(diphenylphosphino)ethane
dppm	1,2-Bis(diphenylphosphino)methane
dppp	1,2-Bis(diphenylphosphino)propane
ee	Percentage enantiomeric excess
en	1,2-Diaminoethane ($H_2NCH_2CH_2NH_2$)
E	Electrophile
EHMO	Extended Hückel molecular orbital
Et	Ethyl group
ESR	Electron spin resonance
fac	Facial
Fp	$FeCp(CO)_2$
HMPA	Hexamethylphosphoric triamide $(Me_2N)_3PO$

HOMO	Highest occupied molecular orbital
L	Any unidentate ligand
LUMO	Lowest unoccupied molecular orbital
M	The central metal in a complex or a metal atom on a surface
Me	Methyl group
MO	Molecular orbital
Nu	Nucleophile
NMR	Nuclear magnetic resonance
[O]	Oxidation
Oh	Octahedral
OAc	Acetoxyl group, acetato anion
Pc	Phthalocyanine
Ph	Phenyl group
phen	1,10-Phenanthroline
Pr	Propyl group
py	Pyridine (coordinated)
R	An alkyl or aryl group
S	Solvent
spl	Square planar
spy	Square pyramidal
salen	Bis(salicylaldehyde)ethylenediamine ligand
tbp	Trigonal bipyramidal
TCNE	Tetracyanoethylene
THF	Tetrahydrofuran
TMED	*N,N,N',N'*-Tetramethylethylenediamine
VB	Valence bond
X	Halogen atom or halide ion

Organotransition Metal Chemistry

CHAPTER ONE

Introduction

Few other fields in chemistry have developed as remarkably as organometallic chemistry in the past three decades. The number of papers published in international journals has grown rapidly, and we now have two journals dealing specifically with organometallic chemistry. One is the *Journal of Organometallic Chemistry*, issued now weekly, and the other is *Organometallics*, recently launched by the American Chemical Society, both bearing witness to the importance of organometallic chemistry and its relevance to other disciplines in chemistry.

Organometallic chemistry has attracted such tremendous attention because it is *interesting* and *useful*. Organometallic compounds are fun to study and useful in many respects. And many interesting discoveries and useful applications are still expected in the future. Organometallic compounds, particularly those of transition metals, are interesting to study because we encounter many examples of unexpected behavior and their chemistry is full of variety.

1.1. BACKGROUND OF ORGANOMETALLIC CHEMISTRY AND OVERVIEW

The history of organometallic chemistry may be described as one of unexpected discoveries (Table 1.1). The oldest compound in the history of organometallic chemistry was prepared by W. C. Zeise, a Danish chemist, in 1827 by the reaction of ethanol with a mixture of $PtCl_2$ and $PtCl_4$ in the presence of KCl.[1] This was about the same time as the first successful synthesis of urea in 1828 by Wöhler and about 40 years *prior* to the proposal of the Periodic Table by A. D. Mendeleev in 1869.

The compound[1b] prepared and formulated as $PtCl_2(C_2H_4)\cdot KCl\cdot H_2O$ by Zeise must have been regarded as quite bizarre at the time. How can ethylene,

Table 1.1 Development of Organometallic Chemistry[a]

1827	Discovery of Zeise's salt (W. C. Zeise)
1837	Synthesis of the first organoarsenic compound, cacodyl (tetramethyldiarsine) (R. Bunsen)
1849	Synthesis of diethylzinc (E. Frankland)
1859	Synthesis of the first organoaluminum compound (W. Hallwachs, A. Schaferik, A. Cahours)
1863	Synthesis of the first organosilicon compound (C. Friedel, J. M. Crafts)
1868	Synthesis of the first metal carbonyl $(PtCl_2CO)_2$ (M. P. Schützenberger)
1869	Proposal of the Periodic Table by A. D. Mendeleev
1890	Synthesis of $Ni(CO)_4$ (L. Mond et al.)
1893	Proposal of octahedral and square planar coordination compounds (A. Werner)
1900	Discovery of Grignard reagents (V. Grignard; P. A. Barbier, 1899)
1907	Synthesis of the first organoplatinum compound (W. J. Pope, S. J. Peachey)
1917	The first isolation of alkyllithium compounds (W. Schlenk, J. Holtz)
1919	Synthesis of Hein's polyphenyl chromium complexes, which were subsequently shown to be π-arene chromium complexes (M. Tsutsui, H. H. Zeiss, 1954)
1921	Synthesis of tetraethyllead for use as an antiknock agent in gasoline engines (T. Midgeley, T. A. Boyd)
1925	Discovery of the Fischer–Tropsch process; development of a direct preparative method for alkyllithiums (K. Ziegler, M. Colonius); the first synthesis of a butadiene complex, $Fe(C_4H_6)(CO)_3$ (H. Reihlen)
1931	The first synthesis of a metal carbonyl hydride, $H_2Fe(CO)_4$ (W. Hieber)
1938	Discovery of the oxo process (O. Roelen); discovery of Kharasch reaction (M. S. Kharasch); synthesis of silver–olefin complexes (S. Winstein, H. J. Lucas)
1938–1945	Development of Reppe reactions
1939	Discovery of homogeneous catalytic hydrogenation by rhodium complexes (M. Iguchi)
1944	Discovery of a direct synthetic method for organosilicon compounds (E. G. Rochow)
1951	Discovery of ferrocene (T. J. Kealy, P. L. Pauson; S. A. Miller, J. A. Tebboth, J. F. Tremaine, 1952); proposal of a theory for the olefin–metal π bond (M. J. S. Dewar, 1951; J. Chatt, L. A. Duncanson, 1953)
1952	Preparation of the first phenyltitanium compound (D. F. Herrman, W. K. Nelson)
1953	Discovery of the Ziegler process (K. Ziegler et al.); proposal of molecular orbital (MO) theory to explain the bonding in electron-deficient compounds (G. N. Lewis, R. E. Rundle); discovery of the Wittig reaction (G. Wittig)
1955	Discovery of fluxional behavior of $C_5H_5Fe(CO)_2C_5H_5$ (G. Wilkinson, T. S. Piper)
1956	Discovery of hydroboration (H. C. Brown)
1957	Discovery of hydrosilylation (J. L. Speier et al.); discovery of Wacker process (J. Smidt)

Table 1.1 Development of Organometallic Chemistry[a] *(Continued)*

1958	Discovery of butadiene oligomerization process (G. Wilke)
1961	Discovery of Vaska's complex, $IrCl(CO)(PPh_3)_2$ (L. Vaska); X-ray structural analysis of vitamin B_{12} coenzyme (D. Crowfoot-Hodgkin)
1963	Holding of the first International Conference on Organometallic Chemistry (Cincinnati, Ohio); publication of *Journal of Organometallic Chemistry*
1964	Discovery of nitrogen fixation by transition-metal-containing systems (M. E. Vol'pin); synthesis of the first carbene–metal complex (E. O. Fischer); discovery of olefin metathesis reaction (R. L. Banks)
1965	Discovery of Wilkinson's catalyst (G. Wilkinson, R. S. Coffey); discovery of the first dinitrogen complex (A. D. Allen, C. V. Senoff)

[a] Main source: J. S. Thayer, *Adv. Organometal. Chem.*, **13,** 1 (1975).

a gaseous compound under ordinary conditions, combine with platinum? When the synthesis of this compound, which is now called Zeise's salt, was first reported (in Latin at that time), Zeise was severely attacked by Liebig.[2] The compound was condemned as a fantasy by the renowned Liebig, the founder of Justus Liebig's *Annalen der Chemie*, the inventor of the Liebig condenser, and a celebrity regarded as the father of organic chemistry. However, the infallible Liebig was wrong on this occasion.

The analysis of this compound established the essential soundness of Zeise's formulation unequivocally. Later, structural determination revealed a structure having ethylene combined with platinum through its double bond (see structure 1-1).[3]

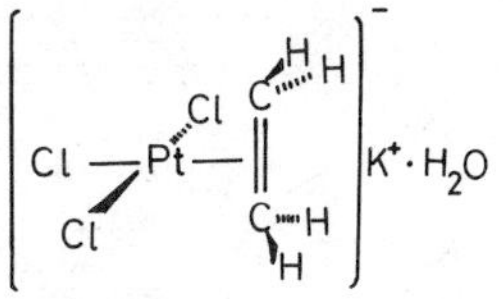

1-1

(Throughout this book, a bond on the surface of the paper is represented by —, a bond protruding toward the reader by ▬, and a bond away from the reader by ııııı or ---.)

Such a chemical bond between the platinum atom and the double bond of ethylene is now known as a π bond, the formation of which plays a key role in catalytic transformation of olefins by transition metal catalysts. It was not until the middle of the twentieth century that the first theory satisfactorily accounting for the π bond was proposed by M. J. S. Dewar[4] and later by Chatt and Duncanson.[5]

The first organometallic compound having a direct metal-to-alkyl σ bond was synthesized,[6] also accidentally, by E. Frankland, the discoverer of helium. In 1849, chemistry itself was in its infancy. What Frankland was trying to prove was the presence of organic radicals. Reasoning that abstraction of iodine from ethyl iodide by zinc should give an ethyl radical, he heated a mixture of C_2H_5I and Zn. He obtained a volatile, colorless liquid that analyzed roughly as C_2H_5 and at first thought that he had demonstrated the occurrence of a radical. Later molecular weight determination, however, revealed that it was not the ethyl radical but butane that was formed by decomposition of an ethylzinc compound generated by the reaction of zinc with ethyl iodide:

$$C_2H_5I + Zn \longrightarrow [C_2H_5ZnI] \longrightarrow \tfrac{1}{2}Zn(C_2H_5)_2 + \tfrac{1}{2}ZnI_2$$

This experiment, which was called "the most fruitful failure," led to a method for preparing ankylzinc compounds. Despite their inflammable nature, these were extensively used as alkylating agents before they were replaced by Grignard reagents, which are much easier to prepare and handle.

These particularly outstanding "accidental" discoveries of organometallic compounds were followed by the discoveries of nickel carbonyl by Mond[7] in 1890 and Grignard reagents in 1900.[8] After the steady growth of chemical knowledge through the first half of the century, the most exciting findings that spurred the tremendous growth of organometallic chemistry came in the early 1950s. These were the discovery of ferrocene in 1951[9-12] (see Section 4.1) and the invention of the Ziegler process in 1953[13] (see Section 7.2).

In a sense, the discovery of ferrocene added a new dimension to the then two-dimensional chemistry of organometallic compounds. Ferrocene was first prepared independently by two groups. Miller et al. obtained it by heating iron powder with dicyclopentadiene[10] and Pauson and Kealy prepared it in their effort to synthesize fulvalene by the reaction of cyclopentadienyl magnesium bromide with ferric chloride.[9] Pauson proposed a structure having an Fe–cyclopentadienyl σ bond with a canonical ionic formula as shown:

This discovery coincided with the period when modern analytical instruments were available in some laboratories. Immediately after the publication of the papers reporting the unusual, thermally stable orange complex of iron, Wilkinson and Fischer independently realized the problem of representing the structure of this complex as depicted above. They proposed instead a structure consisting of an iron sandwiched between two planar cyclopentadienyl ligands that symmetrically bind the central iron atom.[11,12] The correctness of this proposal was quickly and firmly established with the aid of physicochemical methods such as infrared and NMR spectroscopy and later by X-ray crystallographic analysis. This was the advent of a new era. The discovery of ferrocene together with the invention of the Ziegler process triggered the explosive growth of organotransition metal chemistry.

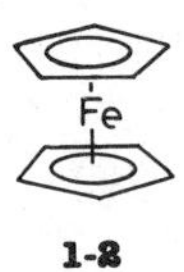

1-8

The revolutionary new process for converting ethylene into polyethylene by a catalyst composed of $TiCl_4$ and $AlEt_3$ was invented by Ziegler and his group at the Max-Planck Institute for Coal Research only 2 years after the discovery of ferrocene.[13] The story of how Ziegler discovered the process during his study of the reaction of ethylene with triethylaluminum is fascinating. It is even more fascinating to see how the process was transferred to Natta's group in Italy, where processes for the polymerization of propylene, butadiene, and isoprene were developed quite rapidly.[14] This is the most outstanding example of how basic research on organometallic compounds, which were regarded as curiosities at the time, suddenly produced a worldwide industrial process—an incident proving the *usefulness* of organometallic chemistry.

The intensive studies following the discovery of the Ziegler process on the mechanism of olefin polymerization with the mixed catalyst system composed of transition metal salts and main-group metal alkyls revealed the importance of organotransition metal complexes in various catalytic reactions.

The discoveries of ferrocene and Ziegler catalysts revealed a new field—organotransition metal chemistry—an unexpectedly fertile field full of potential for further development. Wilkinson and Fischer in 1973 and Ziegler and Natta in 1963 jointly received Nobel prizes.

The discovery of the Ziegler process was followed by the invention of the Wacker process for converting ethylene into acetaldehyde.[15] Both processes had a profound impact on the rapidly developing petrochemical industry at that time. On the other hand, a variety of novel, unprecedented complexes were discovered in the academic field in succession, for example, Vaska's complex[16] (Section 4.5.e), carbene complexes[17] (Section 4.3) and dinitrogen complexes[18] (Section 4.6), to name a few. The number of existing purely organic compounds is already tremendous. (In 1981 the registered number of compounds exceeded five million, and it is increasing at a rate of 5000 new compounds per week.) The combination of the organic moiety with metals, which dominate the Periodic Table, can be astronomical. Furthermore, transition metal atoms in complexes may exist in various oxidation states, and the complexes themselves can display various stereochemistries. One does not need to worry about the waning of research topics in organotransition metal chemistry.

Although less spectacular than the development of organotransition metal chemistry, but nevertheless as steady, the study of organometallic compounds of main-group elements has also continued. Organometallic compounds of alkali metals, alkaline earths, aluminum, boron, mercury, tin, lead, silicon, and other metals have been studied and utilized for the benefit of society. Silicones, for example, used in lubricants, rubbers, and surfactants, rank among the most useful commercial products. Organosilicon compounds, organoboron compounds, as well as organometallic compounds of other main-group metals such as aluminum and tin are attracting interest among preparative organic chemists. The work of Brown,[19] who received the Nobel prize for his study on hydroboration of olefins, together with Wittig, who discovered the Wittig reaction in his studies on organophosphorus compounds, may be mentioned to illustrate the importance of the field.

Chapters 2–5 are introductory, for inexperienced readers. Most of the basic concepts dealt with in Chapter 2 may be found in any elementary textbooks, but the minimum essentials of coordination chemistry have been included to help the reader understand later chapters. Chapter 3 also deals with basic concepts regarding the nature of metal–carbon bonds and introduces the 18-electron rule.

Chapter 4 will make the reader familiar with a variety or organometallic complexes. Chapter 5 lowers the barrier for beginners in organometallic chemistry by providing illustrations for handling air-sensitive compounds. The chapter also includes an account of NMR spectroscopy and other techniques applied to organometallic compounds.

Fundamental processes involving organotransition metal complexes are dealt with in Chapter 6. Although in other chapters extensive citation of references has been avoided, this chapter is somewhat exceptional. It is hoped that this chapter may be useful for organometallic chemists as well as for nonspecialists. An understanding of the basic principles in the reactions of organotransition metal compounds is of vital importance in the elucidation of the mechanisms of transition-metal-promoted catalytic and stoichiometric reactions as dealt with in Chapters 7 and 8. Various new synthetic methods utilizing transition metal complexes have been recently developed in rapid succession. Multistep conventional synthetic approaches can sometimes be drastically shortened by using transition metal complexes, which promote reactions in good yields, and with high regio- and stereochemical specificities.

As described in Chapter 9, organotransition metal complexes sometimes play particular roles in biological systems. The cobalt-containing coenzyme vitamin B_{12} provides a prominent example of the versatile roles of organotransition metal complexes. Further studies may well unveil yet-unknown organometallic compounds in biological systems.

In the same chapter we treat various recent topics, including supported transition metal catalysts, one-dimensional polymers containing transition metals, and cluster complexes.

1.2. DEFINITIONS

Before we proceed to the next chapter, it is necessary to clarify the definitions regarding organotransition metal compounds. There are some misunderstandings even among chemists concerning the definition of organometallic compounds. Organometallic compounds are *organic compounds containing direct metal–carbon bonds.* Thus, compounds where metals are bonded to organic moieties through heteroatoms such as oxygen, nitrogen, and sulfur should not be called organometallic compounds even though these metal-containing compounds appear organic. Thus, various transition metal alkoxides and coordination compounds such as acetylacetonates are not organometallic since the organic moieties are bonded to transition metals through oxygen atoms. Metal alkoxides such as $Ti(OR)_4$ are sometimes called alkyl titanates. This term arises by employing a conventional nomenclature regarding the alkoxide as the ester formed between orthotitanic acid $Ti(OH)_4$ and the alcohol, but it is somewhat misleading and can give the impression that the alkyl groups may be bonded to titanium directly. The terminology based on alkoxide is less confusing.

Although various coordination compounds containing acetylacetonato (2,4-pentanedionato) and analogous ligands possess metal–oxygen bonds and thus are not organometallic compounds, one should note the versatility of this type of complex. These are "noninnocent" complexes and are prone to be transformed into organometallic compounds containing a metal–carbon bond by simple reactions.[20] An example of this behavior is the addition of pyridine to platinum or palladium complexes containing two *O,O'*-bonded acetylacetonato ligands.[21]

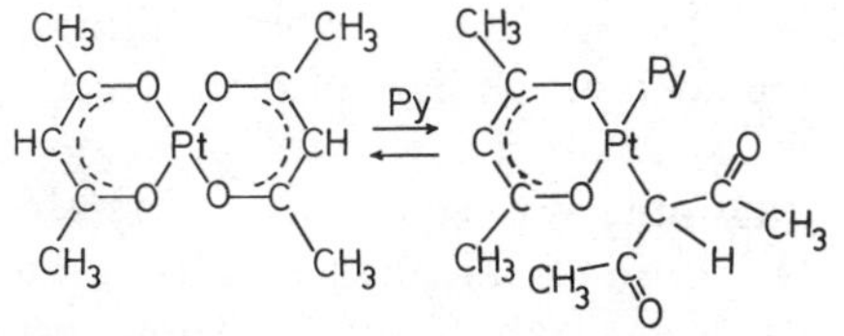

1-3 **1-4**

Square planar complexes such as **1-3** are sometimes called Werner complexes after A. Werner (1866–1919), who established the concepts of coordination chemistry by his studies on octahedral, square planar, and tetrahedral complexes and received the Nobel prize for this work in 1913. The many organotransition metal complexes having formulas not consistent with the conventional concept of Werner complexes are called non-Werner complexes. It is amusing that acetylacetonato complexes, which are usually regarded as typical Werner complexes, may take non-Werner configurations. The list of non-Werner acetylacetonato complexes is increasing all the time.[22] Even a complex of composition $K[Pt(acac)_2Cl]$ prepared by Werner himself that had been considered to have *O,O'*-bonded acetylacetonato ligands was later revealed to be a C-bonded organoplatinum complex (**1-5**). Complex **1-6** represents another example of non-Werner acetylacetonato complex of Pt(IV). Simply by addition of pyridine one of the *O,O'*-bonded acetylacetonato ligand in **1-3** flips to the C-bonded group in **1-4**.

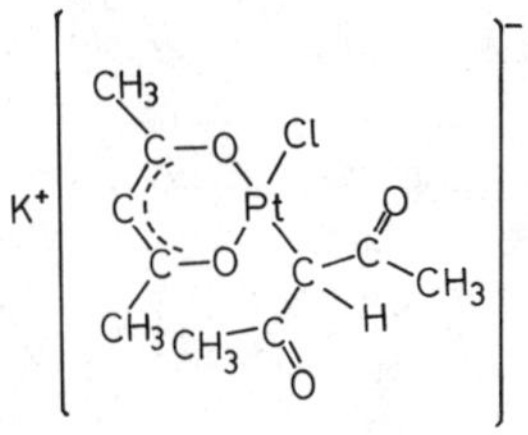

1-5

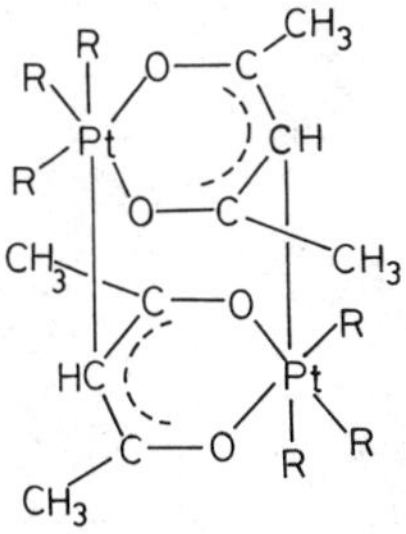

1-6

Purely inorganic compounds such as carbides and cyanides are not classified as organometallic compounds even though they contain metal–carbon bonds. The distinction, however, is somewhat arbitrary; metal carbonyls and hydrides are usually included in the field of organometallic chemistry because they show organic character in many cases. Recently found N_2-containing compounds are treated similarly. Distinction of metals from nonmetals also provides a problem. Boron and silicon, particularly the former, are usually regarded as nonmetals. But organoboron and organosilicon compounds are treated as organometallic compounds. The term *organoelement compounds* used in the USSR and the Eastern Bloc countries is free of this contradiction in terminology, but it has never been used in Western scientific literature.

Inclusion of organophosphorus compounds as organometallic compounds is disputable. In *Chemical Abstracts* and the *Journal of Organometallic Chemistry* they are treated as organometalloid compounds. Metal complexes having tertiary phosphines, PH_3 or PF_3, are often treated in organometallic chemistry because the inclusion of these complexes, particularly those of the low-valent transition metals, in the discussion of organometallic compounds is more convenient.

Transition metals may be strictly defined as those that have partly filled d shells as elements. It is more common to adopt a broader definition to include copper, silver, and gold as transition metals, since they have partly filled d shells in some of their oxidation states. The organozinc, cadmium, and mercury compounds are not treated here since these compounds contain no partly filled d shells in their most common oxidation state of +2 and they are more appropriately discussed as organometallic compounds of nontransition metals.

Organometallic compounds of lanthanoids and actinoids,† those elements having partly filled f shells, are treated only briefly in this book in terms of a comparison of their properties with those of the d-block elements. They are now attracting the belated attention of organometallic chemists, and it is difficult to discuss their chemistry when it is undergoing rapid expansion. Preliminary indications are that it may be fascinatingly different from that of the d-block elements.[23-26]

REFERENCES

1. (a) W. C. Zeise, *Pogg. Ann.*, **9**, 632 (1827); **21**, 497 (1831) (in German); (b) G. B. Kauffmann ed., *Classics in Coordination Chemistry,* Dover, New York, 1976.
2. J. Liebig, *Ann.*, **23,** 12 (1837).
3. J. A. Wunderlich and D. P. Mellor, *Acta Crystallogr.*, **7,** 130 (1954); J. A. J. Jarvis, B. T. K. Kilbourn and P. G. Owston, *Acta Crystallogr.*, **B27,** 366 (1971).
4. M. J. S. Dewar, *Bull. Soc. Chim. Fr.*, **18** C71 (1951).
5. J. Chatt and L. A. Duncanson, *J. Chem. Soc.*, 2939 (1953).
6. E. Frankland, *Ann.*, **71,** 171 (1849).
7. L. Mond, C. Langer, and F. Quincke, *J. Chem. Soc.*, **57,** 749 (1890).
8. V. Grignard, *Compt. Rend.*, **130,** 1322 (1900).
9. T. J. Kealy and P. L. Pauson, *Nature,* **168,** 1039 (1951).
10. S. A. Miller, J. A. Tebboth, and J. F. Tremaine, *J. Chem. Soc.*, 632 (1952).
11. G. Wilkinson, M. Rosenblum, M. C. Whiting, and R. B. Woodward, *J. Am. Chem. Soc.*, **74,** 2125 (1952).
12. E. O. Fischer and W. Pfab, *Z. Naturforsch.*, **7B,** 377 (1952).
13. K. Ziegler, E. Holtzkampf, H. Breil, and H. Martin, *Angew. Chem.*, **67,** 543 (1955).
14. J. Boor, Jr., *Ziegler–Natta Catalysts and Polymerizations,* Academic Press, New York, 1979.
15. J. Smidt, W. Hafner, R. Jira, J. Seldmeier, R. Sieber, R. Rutlinger, and H. Kojer, *Angew. Chem.*, **71,** 176 (1959).
16. L. Vaska and J. W. DiLuzio, *J. Am. Chem. Soc.*, **83,** 2784 (1961).
17. E. O. Fischer and A. Massböl, *Angew. Chem. Int. Ed. Engl.*, **3,** 580 (1964).
18. A. D. Allen and C. V. Senoff, *Chem. Commun.*, 621 (1965).
19. H. C. Brown and B. C. Subba Rao, *J. Am. Chem. Soc.*, **78,** 5694 (1956).
20. (a) D. Gibson, *Coord. Chem. Rev.*, **4,** 225 (1969); (b) D. W. Thompson, *Structure and Bonding,* **9,** 27 (1971).
21. (a) S. Baba, T. Ogura, and S. Kawaguchi, *Bull. Chem. Soc. Jpn.*, **47,** 665 (1974); (b) T. Ito, T. Kiriyama, Y. Nakamura, and A. Yamamoto, *Bull. Chem. Soc. Jpn.*, **49,** 3257 (1976).
22. S. Okeya, H. Sazaki, M. Ogita, T. Takemoto, Y. Onuki, Y. Nakamura, B. K. Mohapatra, and S. Kawaguchi, *Bull. Chem. Soc. Jpn.*, **54,** 1978 (1981).
23. (a) T. J. Marks, *Acc. Chem. Res.*, **9,** 223 (1976); (b) T. J. Marks, *Prog. Inorg. Chem.*, **24,** 51 (1978).
24. T. J. Marks and J. R. Kolb, *Chem. Rev.*, **77,** 263 (1977).

† We follow the IUPAC recommendation here. In English literature the terms lanthanides and actinides are usually used.

25. M. Tsutsui, N. Ely, and R. Dubois, *Acc. Chem. Res.*, **9**, 217 (1976).

26. T. J. Marks and R. D. Fischer, eds. *Organometallics of the f-Elements*, Reidel, Dordrecht, Holland, 1979.

ADDITIONAL REFERENCES

A. Newer books

G. Wilkinson, F. G. A. Stone, and E. W. Abel, eds., *Comprehensive Organometallic Chemistry*, Pergamon Press, Oxford, 1982.

F. A. Cotton and G. Wilkinson, *Advanced Inorganic Chemistry, A Comprehensive Text*, 4th ed., Wiley, New York, 1980.

J. P. Collman and L. S. Hegedus, *Principles and Applications of Organotransition Metal Chemistry*, University Science Books, Mill Valley, CA 1980.

J. K. Kochi, *Organometallic Mechanisms and Catalysis*, Academic Press, New York, 1978.

R. F. Heck, *Organotransition Metal Chemistry, a Mechanistic Approach*, Academic Press, New York, 1974.

F. R. Hartley, *The Elements of Organometallic Chemistry*, Monograph for Teachers, No. 26, The Chemical Society, 1974.

D. St. C. Black, W. R. Jackson, and J. M. Swan, eds., *Comprehensive Organic Chemistry*, Vol. 3, Pergamon Press, Oxford, 1979, p. 1127.

A. F. Trotman-Dickenson, ed., *Comprehensive Inorganic Chemistry*, Vols. 1–5, Pergamon Press, Oxford, 1973.

E. Negishi, *Organometallics in Organic Synthesis*, Vol. 1, Wiley, New York, 1980.

C. M. Lukehart, *Fundamental Transition Metal Organometallic Chemistry*, Brooks/Cole, Monterey, CA, 1985.

A. J. Pearson, *Metallo-organic Chemistry*, Wiley-Interscience, Chichester, 1985.

B. Older Books

H. Zeiss, ed., *Organometallic Chemistry*, Reinhold, New York, 1960.

P. L. Pauson, *Organometallic Chemistry*, Edward Arnold Publishing, London, 1967.

G. E. Coates, *Organometallic Compounds*, Methuen, London, 1960.

E. G. Rochow, D. J. Hurd, and R. N. Lewis, *The Chemistry of Organometallic Compounds*, Wiley, New York, 1957.

J. J. Eisch, *The Chemistry of Organometallic Compounds. The Main Group Elements*, Macmillan, New York. 1967.

G. E. Coates, M. L. H. Green, P. Powell, and K. Wade, *Principles of Organometallic Chemistry*, Methuen, London, 1968.

G. E. Coates and K. Wade, *Organometallic Compounds, Vol. 1, The Main Group Elements*, 3rd ed., Methuen, London, 1967.

M. L. H. Green, *Organometallic Compounds, Vol. 2, The Transition Elements*, 3rd ed. Methuen, London, 1968.

R. B. King, *Transition-Metal Organometallic Chemistry*, Academic Press, New York, 1969.

M. Dub, ed., *Organometallic Compounds, Methods of Synthesis, Physical Constants and Chemical Reactions*, Springer Verlag, Berlin, 1966.

M. Tsutsui, M. N. Levy, A. Nakamura, M. Ichikawa, and K. Mori, *Introduction to Metal π Complex Chemistry,* Plenum Press, New York, 1970.

E. O. Fischer and H. Werner, *Metal π-Complexes,* Vol. 1, Elsevier, Amsterdam, 1966.

M. Herberhold, *Metal π-Complexes,* Vol. 2, Elsevier, Amsterdam, 1972.

N. Hagihara, M. Kumada, and R. Okawara, *Handbook of Organometallic Compounds,* Benjamin, New York. 1968.

D. S. Matteson, *Organometallic Reaction Mechanisms,* Academic Press, New York. 1974.

J. P. Candlin, K. A. Taylor, and D. T. Thompson, *Reactions of Transition-Metal Complexes,* Elsevier, Amsterdam, 1968.

CHAPTER TWO

Fundamentals of Coordination Chemistry

Before discussing the chemistry of organotransition metal compounds, it may be necessary to review briefly the fundamental principles of coordination chemistry. Readers who already have a sound knowledge of coordination chemistry may skip this chapter and proceed to the next chapter.

Coordination compounds (or simply complexes) are molecules containing a central metal atom or atoms bound to organic and/or inorganic moieties called ligands. The term ligand stems from the Latin word *ligare* (join). In organometallic complexes, alkyl groups, aryl groups, and cyclopentadienyl groups serve as ligands in the same way as unsaturated compounds such as olefins, acetylenes, dienes, and a variety of organic groups. Thus, the variety in organotransition metal complexes far exceeds the classical Werner-type complexes containing ligands such as halide ions and organic groups bonded through heteroatoms as ligands. Some ligands, such as acetylacetonato and bipyridine, are attached to the central metal atom through two atoms. These are called bidentate ligands, and the formation of complexes containing bidentate or polydentate ligands gives chelate complexes (chelate, from Greek *chele,* meaning claw).

The total number of atoms attached to the central atom is called the coordination number. The coordination number is determined by the size and nature of the ligands as well as by the size, nature, and oxidation state of the central metal atom. Larger metal atoms can bind more ligands around them than smaller ones, and the number of ligands capable of being coordinated with the metal is limited by the relative size of the ligand. Thus, a wide range of transition metal complexes with two, three, four, five, six, and even higher coordination numbers is known. A complex with the coordination number 2 may be formed by a metal coordinated with very bulky ligands[1] or when a linear structure is favored for electronic reasons, as in some complexes of Cu(I), Ag(I), Au(I), and Hg(II). The Roman letter in parentheses represents the oxidation number of the metal. When the size of the ligand is small and the central metal is relatively large, a complex with a higher coordination number may be formed. This situation often occurs when the smallest ligand, hydride, is present and a coordination number of 9 is found in complexes $MoH_6(PPr_3)_3$ and K_2ReH_9.[2] The most common types of transition metal complexes are octahedral ones having six ligands around the central metal atom. Following the octahedral ones in number are complexes having the square planar and tetrahedral configurations with four ligands surrounding the central metal. Let us first briefly review the basic bonding theories describing these transition metal complexes.

2.1. CRYSTAL FIELD THEORY

The crystal field theory was originally proposed by Bethe[3] and Van Vleck.[4] This simple theory explains the interaction between the central metal orbitals and the surrounding ligands by electrostatic interaction. Here the negative ions are regarded as point charges and neutral molecules as dipoles with their negative ends directed toward the metal. Covalent bonding is completely neglected. This is a rather naive theory and of limited use in discussing the behavior of real transition metal complexes but is of some value in helping us visualize the effects of ligands on the metal orbitals.

The s, p, and d orbitals (wave functions) in an atom have the spatial arrangements shown in Figure 2.1. The spatial arrangement of each orbital is represented by lobes within which most of the total electron density is included. The s orbital has spherical symmetry, whereas the p and d orbitals have angular dependence and have lobes with positive and negative signs representing the wave functions ψ. The electron density is given by ψ^2, which represents the probability of finding the electron in a unit volume. The p and d orbitals have nodal planes where ψ^2 is zero.

In a discussion of organic chemistry one need not bother about the d orbitals, but in a discussion of transition metal complexes the d orbitals play important roles in bonding with ligands. The d orbitals (3d, 4d, and 5d orbitals depending on the transition series) are composed of five orthogonal sets of independent orbitals with lobes, as shown in Figure 2.1. It should be noted that the plus and minus signs on the lobes along each axis are different from those in the p orbitals. The lobes along the axis or in the direction bisecting two Cartesian axes have the same sign in d orbitals. A consideration of the signs is important in understanding the bonding of a transitional metal with unsaturated ligands such as olefins and carbon monoxide.

The five d orbitals are equivalent and have the same energy when the atom is present in the absence of ligands, for example, as a gas in a vacuum. In other words, these orbitals are degenerate. Each orbital can accommodate two

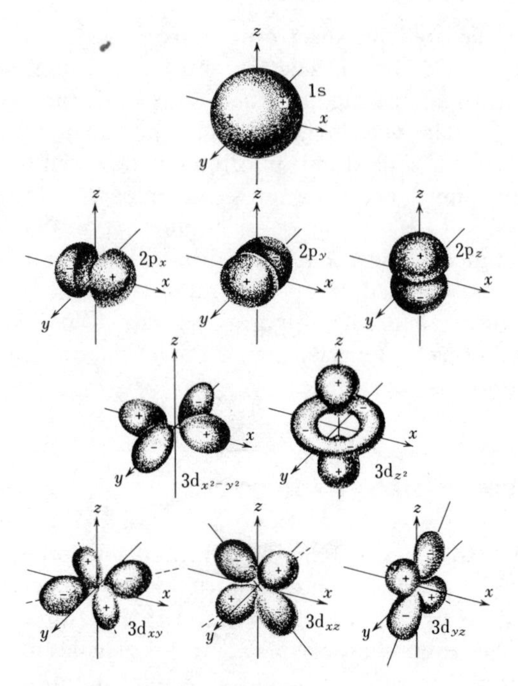

FIGURE 2.1. Shape of s, p, and d orbitals. (From F. A. Cotton and G. Wilkinson, *Basic Inorganic Chemistry,* Wiley, New York, 1976, with permission.)

electrons with their spins paired. Thus, five d orbitals can accommodate a maximum number of 10 electrons, the s orbital 2 electrons, and the three p orbitals 6 electrons. When discussing the bonding in complexes of lanthanoids and actinoids, one has to consider further the *f* orbitals. The organometallic chemistry of f-elements is now in the process of rapid development[5] but will not be discussed here. In transition metal complexes the presence of the d orbitals is the important factor that gives the diversity of complexes described in this book. When the transition metal is placed in an environment surrounded by ligands, namely in the ligand field, the orbitals become nonequivalent and they are split into groups with different energies. The crystal field theory describes how the splitting of the d orbitals takes place, as shown below.[6]

a. Octahedral Complexes

Let us first treat an octahedral complex (Figure 2.2). Electrons in the d orbitals of a central metal surrounded by six ligands are affected differently by the negative charges depending on whether the d orbital is distributed along the axes connecting the metal and ligands or in the direction bisecting the axes. Figures 2.2*b,c* compare the effect of the negatively charged ligands in the *xy*

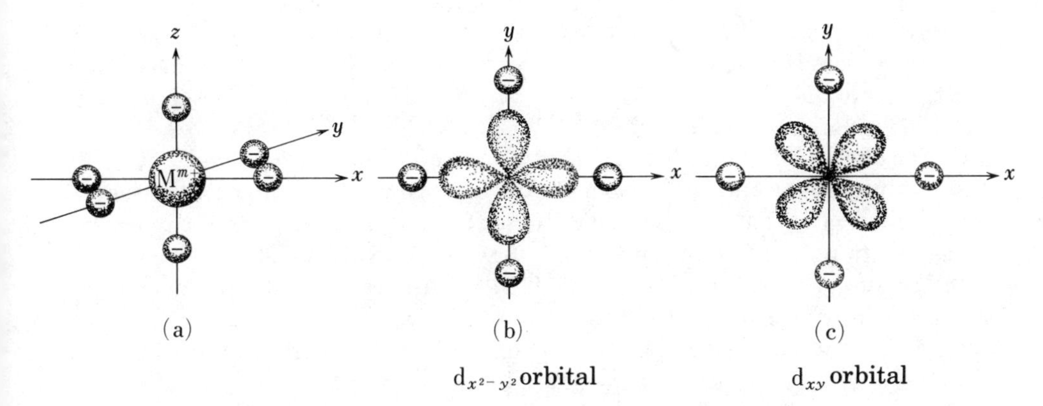

FIGURE 2.2. Concept of electrostatic crystal field theory. (*a*) Arrangement of ligands around the central ion in an octahedral complex. (*b,c*) An electron in the $d_{x^2-y^2}$ orbital is repelled more strongly by the ligands than an electron in the d_{xy} orbital.

plane on electrons in the $d_{x^2-y^2}$ and d_{xy} orbitals. An electron in the $d_{x^2-y^2}$ orbital experiences electrostatic repulsion by the negative charge of the ligands and the $d_{x^2-y^2}$ orbital is destabilized relative to its energy in the absence of ligands.

On the other hand, the repulsion by the ligands along the x and y axes on an electron in the d_{xy} orbital oriented between the x and y axes is less pronounced than that on the electron in the $d_{x^2-y^2}$ orbital. The effect of the ligands along the z axis on the electron in the $d_{x^2-y^2}$ and d_{xy} orbitals is exactly the same.

Next we consider the d_{xz} and d_{yz} orbitals. These have the same spatial orientation relative to the ligands in the xz and yz planes as the d_{xy} orbital has to the ligands in the xy plane. Therefore, the energy of the d_{xz} and d_{yz} orbitals should be the same as that of the d_{xy} orbital. The d_{z^2} orbital experiences repulsion primarily by the ligands along the z axis and the d_{z^2} orbital is destabilized to the same extent as the $d_{x^2-y^2}$ orbital is relative to the d_{xy}, d_{yz}, and d_{xz} orbitals.

Therefore, the five d orbitals that originally had identical energy in the absence of ligands are now split in the octahedral environment into two sets, one triply degenerate set of d_{xy}, d_{yz}, and d_{xz} and the other less stable, doubly degenerate set, $d_{x^2-y^2}$ and d_{z^2}. In terms of group theory the former set is called t_{2g} and the latter e_g. The symbols t and e mean that the orbitals are triply and doubly degenerate, respectively, while nondegenerate orbitals are denoted by a or b. The subscript g means centrosymmetric (*gerade* in German), the antisymmetric being represented by u (*ungerade*).

Figure 2.3 shows how the splitting of the d orbitals takes place when the metal is placed in the octahedral field. The energy level of the d orbitals is raised by the repulsion between d electrons and the negative charges of the ligands in the octahedral environment, although the total energy of the positively charged metal ion is decreased on coordination of the negatively charged ligands. The degeneracy of the five orbitals is now lifted to give two sets of orbitals, the t_{2g} set of d_{xy} , d_{yz} , and d_{xz} orbitals and the e_g set of $d_{x^2-y^2}$ and d_{z^2} orbitals.

The energy difference between the two sets of orbitals is denoted by Δ_o. The subscript o represents the splitting in the octahedral environment. It is due to this difference that the characteristics of the transition metal complexes emerge. In place of Δ_o the symbol of 10 Dq is sometimes used, which is called the crystal field splitting or ligand field splitting parameter. Its magnitude depends on the nature of the metal ion and ligands, their charges, and the distance between the metal and the ligands. The value of Δ_o is in the range of 100–400 kJ mol^{-1} (1 cal = 4.184 J, 1 kJ mol^{-1} = 83.54 cm^{-1}).

Imagine a hypothetical situation where the central metal with 10 electrons in the d orbitals is placed in a spherical field (Figure 2.3*b*), and then lift the orbital degeneracy by converting the spherical field into an octahedral field without changing the total energy of the set of d orbitals so as to maintain the barycenter. It is necessary then for the two e_g orbitals to be raised in energy by $6Dq$ to balance the stabilization of the three t_{2g} orbitals to the extent of $4Dq$, as shown in Figure 2.3*c*. If an electron is placed in one of the d_{xy}, d_{yz}, and d_{xz} orbitals, the stabilization energy gained is $-4Dq$ for one electron put in the t_{2g} set if one compares with the case without the splitting. This is called the crystal field stabilization energy (CFSE). If four electrons are accommodated in the d orbitals, three in each of the t_{2g} orbitals and one in either d_{z^2} or $d_{x^2-y^2}$, the crystal field stabilization energy is $3(-4Dq) + 6Dq = -6Dq$. By this model some of the fundamental properties of transition metal complexes, such as optical and magnetic properties, can be explained reasonably well. Many transition metal complexes having electrons in d orbitals are colored. For example, an aqueous solution of Ti(III) containing the hydrated ion $[Ti(OH_2)_6]^{3+}$ shows a violet color due to the absorption at 490 nm (20,400 cm^{-1}). This is called d–d absorption since it is caused by an electronic excita-

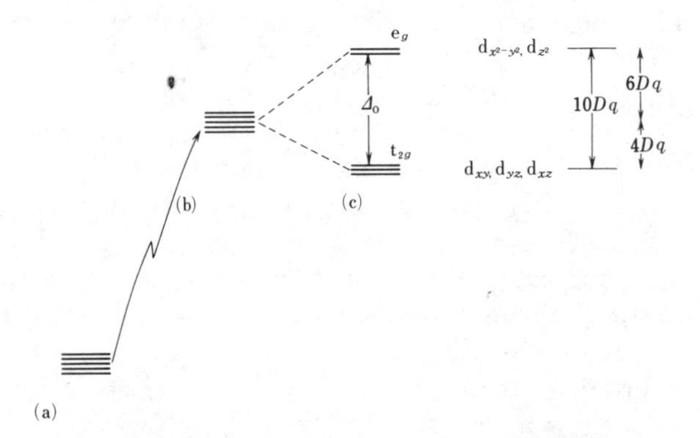

FIGURE 2.3. Splitting of d orbitals in an octahedral ligand field. (*a*) Free ion. (*b*) Hypothetical ion in a spherically symmetric field. (*c*) Hypothetical ion in an octahedral field.

tion between the two sets of d orbitals. The octahedral complexes of Ti(III) coordinated with ligands other than water molecules show absorptions at different wavelengths. Since the absorption maximum corresponds to the energy gap Δ_o, the magnitude of the ligand field splitting can be determined by measuring the d–d absorption. For a fixed cation the magnitude of the ligand field splitting increases in the order

$$I^- < Br^- < Cl^- < OH^- < RCO_2^- < F^- < H_2O < SCN^- < NH_3 < en < bipy < CN^-$$

where bipy is 2,2′-bipyridine and en is ethylenediamine. The order is called the spectrochemical series. As the value of Δ_o increases, the energy required for the electronic excitation is increased, and hence the absorption band is shifted toward the shorter wavelength.

The magnitude of the ligand field splitting is affected by the ligands as well as by the nature of the metal ion involved. Within a given transition series the differences are not great, but noticeable changes occur between the congeners in a given group in progressing 3d < 4d < 5d. As a consequence, the heavier metal complexes tend to be low spin, whereas complexes of the first transition series can be either high or low spin. The other important characteristics of a transition metal complex is its magnetic behavior. By experimental techniques, most directly by magnetic susceptibility measurement, we can determine the number of unpaired electrons in the d orbitals of the complex. When usual organic compounds without unpaired electrons are placed in a magnetic field, they are repelled by the magnetic field and are thus called *diamagnetic*. A compound with unpaired electrons can be regarded as a tiny magnet because the unpaired electron behaves as if it were spinning about an axis. Therefore, a compound that has an unpaired electron is attracted to the applied magnetic field, and such a compound is called *paramagnetic*. The magnetic moment of a complex can be estimated by measuring the magnetic susceptibility of the complex, for example, by measuring the weight change of a complex before and after application of a magnetic field.

The effective magnetic moment μ_{eff} derived from the magnetic susceptibility measurement can be expressed as

$$\mu_{eff} = \sqrt{n(n+2)}\ \mu_0 \tag{2.1}$$

where n stands for the number of unpaired electrons and μ_0 is a constant called the Bohr magneton. Knowing the value of μ_{eff} from measurement of the magnetic susceptibility, one can find the number of unpaired electrons in the complex.

The five d orbitals can accommodate various numbers of electrons, ranging from 1 to the maximum number of 10. How these electrons are distributed between the higher-energy e_g and the lower-energy t_{2g} orbitals depends on the number of d electrons and the magnitude of the splitting between the two sets of d orbitals. Let us consider what happens when an increasing number of electrons is put in these orbitals (Figure 2.4). For a complex having one elec-

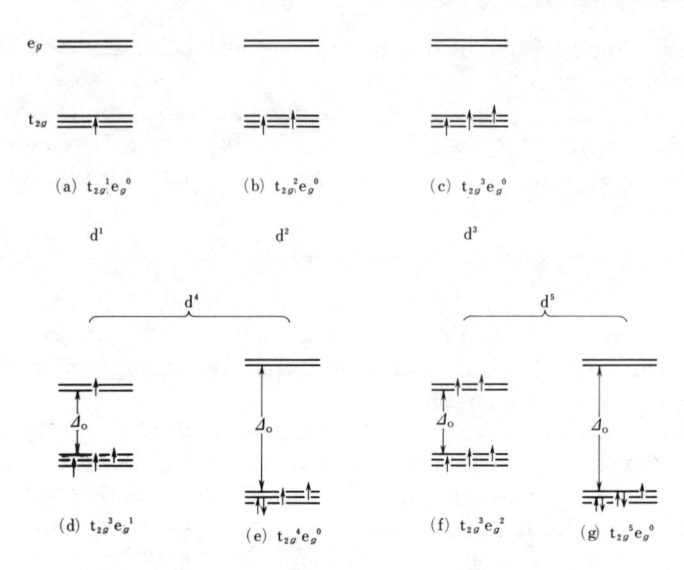

FIGURE 2.4. Electron configurations in octahedral complexes.

tron, such as a Ti(III) complex, the electron is placed in the more stable t_{2g} orbital. The electron configuration in this case is expressed as $t_{2g}^1e_g^0$, which signifies that there is one electron in the t_{2g} orbital and none in the e_g orbital (Figure 2.4*a*). When one places two electrons (Figure 2.4*b*), the electrons are put in two of the three t_{2g} orbitals with their spins parallel by Hund's rule, because putting two electrons with the spins paired in the same orbital requires a pairing energy *P*. The same can be said for placing three electrons (Figure 2.4*c*) as is seen in Cr(III) complexes.

To place four electrons in d orbitals in complexes such as Mn(III), two possibilities exist, depending on the magnitudes of Δ_o and *P*. Figure 2.4*d* represents the case where the fourth electron occupies the upper e_g level, whereas Figure 2.4*e* shows the case where the fourth electron enters the lower t_{2g} orbital pairing with the electron already placed in one of the three orbitals. When the ligand field is large (strong field), the fourth electron enters the lower t_{2g} level, resulting in two unpaired electrons. This is called the low-spin configuration. If the ligand field splitting Δ_o is small (weak field), the fourth electron would prefer to enter the upper level rather than the lower level, overcoming the pairing energy *P* (Figure 2.4*d*). This is called the high-spin configuration with four unpaired electrons. In the case of Mn(III) complexes, Mn(III) surrounded by strong-field ligands such as CN^- takes the electron configuration as shown in Figure 2.4*e*, whereas Mn(III) in $Mn(acac)_3$ containing the weak-field acetylacetonate forms a high-spin complex possessing a magnetic moment corresponding to four unpaired electrons.

Figures 2.4*f,g* show two possibilities of electron configurations for d^5 complexes such as Fe(III). Figure 2.4*f* indicates the high-spin configuration where five electrons are accommodated in each of t_{2g} and e_g orbitals, whereas in the low-spin configuration (Figure 2.4*g*) all five electrons are placed in the lower t_{2g} orbitals, leaving one electron unpaired. Ligands producing a strong field such as CN^- give a low-spin complex as $[Fe(CN)_6]^{3-}$, whereas $[Fe(OH_2)_6]^{3+}$ with the weak-field ligand H_2O is a high-spin complex. A similar argument can be extended to d^6 complexes such as those of Co(III), which give high-spin complexes of configuration $t_{2g}^4e_g^2$ and low-spin (in this case diamagnetic) complexes of t_{2g}^6 types. The possible configurations for d^7 complexes are high-spin type of $t_{2g}^5e_g^2$ and low-spin type of $t_{2g}^6e_g^1$. For d^8 and d^9 complexes only one sort of configuration is possible regardless of the ligand field strength.

Table 2.1 shows all possible electron configurations together with the numbers of unpaired electrons and magnetic moments.

Although the crystal field theory assumes point charges for ligands and regards d orbitals as nonbonding, it can be seen from the spectrochemical series that many ligands that produce a strong field are also those expected to form covalent bonds to the metal rather than purely electrostatic bonds. This reveals a shortcoming of the simple crystal field theory and suggests the desirability of a theory taking into account covalency, as will be discussed later.

b. Tetrahedral Complexes

By an analogous line of reasoning we can understand how the d orbitals in a tetrahedral environment are split. The spatial arrangement of the d orbitals in the tetrahedral field is somewhat less straightforward to visualize than in the octahedral field. In Figure 2.5*a* we place a metal in the center of a cube with

Table 2.1. Electron Configurations in Octahedral Complexes[a]

Number of d electrons	Electron configurations in a weak ligand field		n	μ	Electron configurations in a strong ligand field		n	μ
	t	e			t	e		
1	(↑)( )( )	( )( )	1	1.73	(↑)( )( )	( )( )	1	1.73
2	(↑)(↑)( )	( )( )	2	2.83	(↑)(↑)( )	( )( )	2	2.83
3	(↑)(↑)(↑)	( )( )	3	3.87	(↑)(↑)(↑)	( )( )	3	3.87
4	(↑)(↑)(↑)	(↑)( )	4	4.90	(↑↓)(↑)(↑)	( )( )	2	2.83
5	(↑)(↑)(↑)	(↑)(↑)	5	5.92	(↑↓)(↑↓)(↑)	( )( )	1	1.73
6	(↑↓)(↑)(↑)	(↑)(↑)	4	4.90	(↑↓)(↑↓)(↑↓)	( )( )	0	0
7	(↑↓)(↑↓)(↑)	(↑)(↑)	3	3.87	(↑↓)(↑↓)(↑↓)	(↑)( )	1	1.73
8	(↑↓)(↑↓)(↑↓)	(↑)(↑)	2	2.83	(↑↓)(↑↓)(↑↓)	(↑)(↑)	2	2.83
9	(↑↓)(↑↓)(↑↓)	(↑↓)(↑)	1	1.73	(↑↓)(↑↓)(↑↓)	(↑↓)(↑)	1	1.73
10	(↑↓)(↑↓)(↑↓)	(↑↓)(↑↓)	0	0	(↑↓)(↑↓)(↑↓)	(↑↓)(↑↓)	0	0

[a] n = number of unpaired spins; μ = predicted spin-only moment in Bohr magnetons.

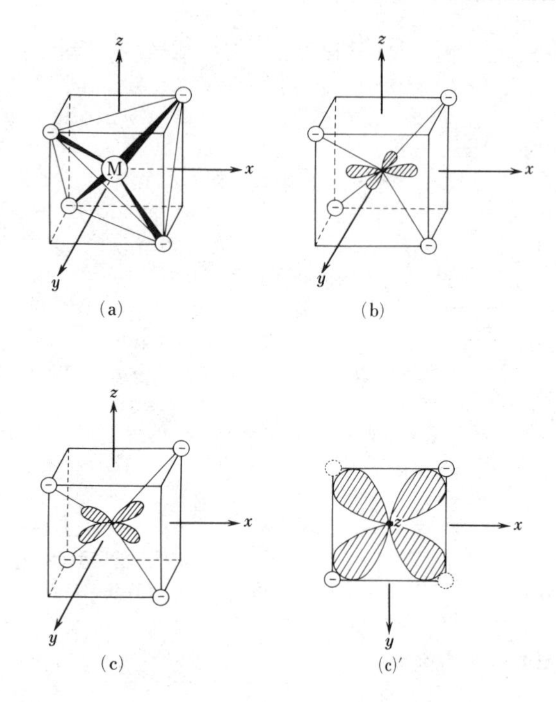

FIGURE 2.5. Relationship between coordinate axes and d orbitals in tetrahedral complexes. (*a*) Tetrahedral complex; (*b*) the lobes of $d_{x^2-y^2}$ orbital are oriented to directions away from ligands; (*c*) the d_{xy} orbital is placed in the cube: (*c*)′ top view of the d_{xy} orbital in (*c*) observed along the *z* axis (each lobe of the d_{xy} orbital comes close to ligand).

the four ligands sitting at the vertices forming a tetrahedron. If we take the Cartesian axes as shown in Figure 2.5*b*, it can be seen that the $d_{x^2-y^2}$ orbital is oriented in the direction furthest removed from the lines connecting the metal and the ligands, whereas each lobe of the d_{xy} orbital comes close to one ligand, as seen in Figure 2.5*c*. The situation is clearly understood if we look at the d_{xy} and $d_{x^2-y^2}$ orbitals along the *z* axis (see Figure 2.5*c*′).

Thus we can see that the degenerate d_{xy}, d_{yz}, and d_{xz} orbital set (referred to as t_2 orbitals) suffers a stronger repulsion than the $d_{x^2-y^2}$ and d_{z^2} set (e orbitals) in the tetrahedral environment. The order of the energy levels is inverted relative to that for octahedral environments, the t_2 level being less stable than the e level, as shown in Figure 2.6*a*. It can be shown by calculations and was confirmed by experiment that the energy gap, or ligand field, in a tetrahedral complex is $\frac{4}{9}$ of Δ_o for comparable octahedral complexes with ligands at an identical distance from the metal.

The way to place electrons in the d orbitals of a tetrahedral complex is the same as in the octahedral complex with the difference that the ligand field Δ_t is smaller than Δ_o. Figure 2.6*b* illustrates the two possibilities of electron configurations in strong and weak fields.

c. Square Planar and Other Types of Complexes

If two trans ligands in an octahedral complex are moved away from the metal, for example, along the *z* axis, the resulting complex is said to be tetragonally distorted. If this process continues and two ligands are completely removed, we have a square planar arrangement of the four remaining ligands. During this process of moving the two trans ligands away from the metal, the repulsion between the d_{z^2} orbital and the two axial ligands will diminish and the d_{z^2} orbital is more stabilized than the $d_{x^2-y^2}$ orbital. The further the axial ligands are removed, the more strongly the ligands on the *x* and *y* axes are attracted with the resultant destabilization of the $d_{x^2-y^2}$ orbital. The d_{xy} orbital is also destabilized relative to the d_{yz} and d_{xz} orbitals. Figure 2.7 is a schematic representation of the change in energy levels of the d orbitals upon increasing the tetragonal distortion. In the figure is shown the energy level diagram of the square planar complex formed by removal of the axial ligands. Metal ions such as Ni(II), Pd(II), Pt(II), and Au(III), which have a d^8 configuration and ligands high in the spectrochemical series, favor a square planar configuration. This combination gives low-spin, diamagnetic complexes with two elec-

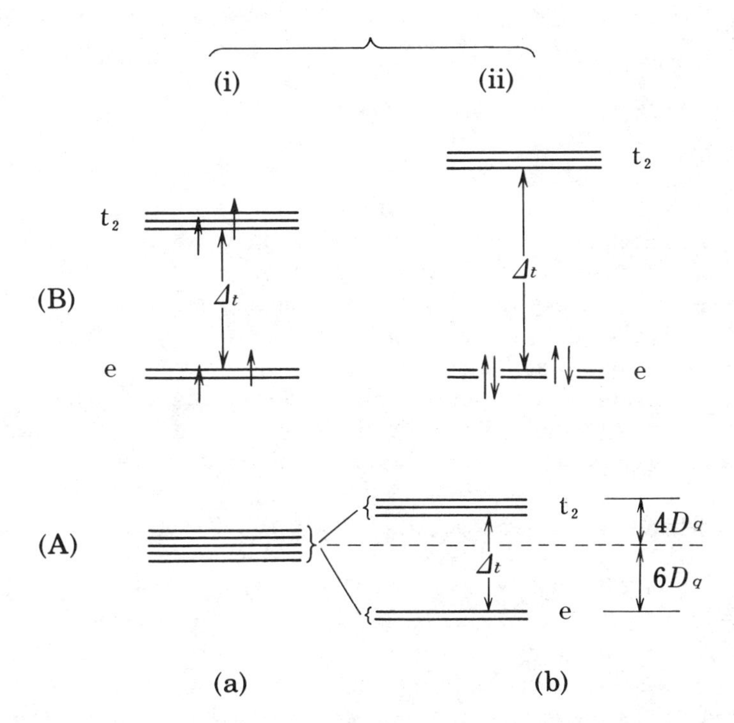

FIGURE 2.6. (A) Splitting of energy levels of d orbitals in a tetrahedral ligand field: (*a*) degenerate d orbital, (*b*) split d orbitals. (B) Occupation of d orbitals by electrons for d^4 configurations, (i) the number of unpaired electron is 4 in a weak field, (ii) the number of unpaired electrons is zero in a strong field.

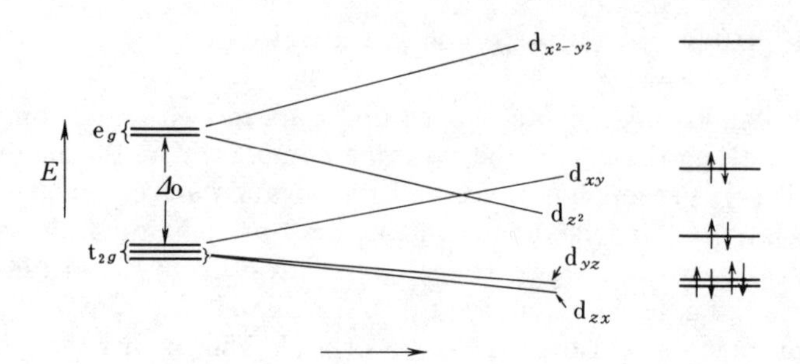

FIGURE 2.7. Energy diagram showing the splitting of d orbitals on removing the trans ligands from octahedral complexes.

trons occupying each of the low-energy d_{yz}, d_{xz}, and d_{z^2} and d_{xy} orbitals, whereas the high-energy $d_{x^2-y^2}$ orbital remains unoccupied. Examples of this type of complexes are $Pt(acac)_2$ and $PdMe_2(PR_3)_2$. It is noted that the energy gap between $d_{x^2-y^2}$ and d_{xy} corresponds to the ligand field Δ_o in the octahedral complex. Thus, strong-field ligands tend to give low-spin complexes, and heavier metals such as Pt(II) tend to form square planar complexes more readily than lighter metals such as Ni(II), as is observed experimentally.

2.2. MOLECULAR ORBITAL THEORY

Although the crystal field theory is effective in accounting for some properties of transition metal complexes, such as electronic spectra and magnetism, it is seriously inadequate in explaining various properties of many transition metal complexes, particularly those of organotransition metal complexes. On the other hand, although its fundamental concepts are less easy to grasp, molecular orbital theory provides a powerful means of accounting for the structures, spectra, and chemical behavior of a variety of transition metal complexes consistently. In this section we provide an explanation of the fundamental concepts without employing the sophisticated mathematical equations.

a. Fundamental Concepts of Molecular Orbital Theory

Let us first consider the simplest molecule H_2^+. This molecule is comprised of two protons and one electron. In the absence of the electron, the two positively charged protons are repelled from each other, as shown in Figure 2.8*a*[7]. When the electron is present in the neighborhood of the two protons, it can serve to connect the two protons. We cannot specify the position of an electron, but we can describe the behavior of an electron by a wave function ψ, the square of which gives the relative probability of its being at any given point. The electron traveling in the space near the protons may be at a posi-

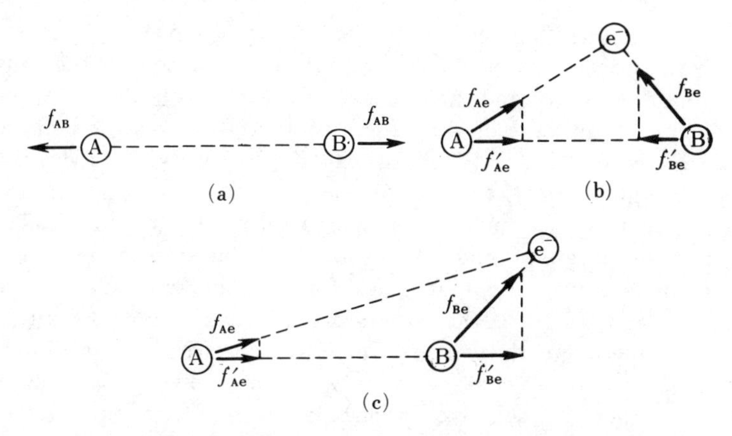

FIGURE 2.8. Forces in H_2^+. (*a*) Internuclei repulsion always hinders binding the nuclei together. (*b*) A possible electron position that contributes to binding the nuclei together. (*c*) A possible electron position that works against binding. (From G. C. Pimentel and R. D. Spratley, *Chemical Bonding Clarified through Quantum Mechanics,* Holden-Day, Oakland, CA, 1969, with permission.)

tion shown in Figure 2.8*b* at one instant and at a position given in Figure 2.8*c* at another. In the position of an electron shown in Figure 2.8*b* the nuclei A and B experience the attractive forces f_{Ae} and f_{Be} of an electron. The magnitude of the forces increases as the electron approaches the nuclei. It can be seen that the components of the attractive forces f'_{Ae} and f'_{Be} along the axis connecting the A and B nuclei contribute to pull the nuclei toward each other. On the other hand, when the electron is present in a position shown in Figure 2.8*c*, the electron serves to pull both nuclei to the right, but the net influence in this case is to move B away from A since f'_{Be} is greater than f'_{Ae}. In other words, the electron in this position serves to disrupt the binding. Figure 2.9 illustrates the regions where the electron serves to bind the two nuclei and to disrupt the binding. In the binding region the electron works as a kind of "adhesive" to bind the nuclei. The bonding results when the electron is present near both nuclei at the same time. To separate the nuclei in the H_2^+ molecule, an energy of 270 kJ mol^{-1} is required.

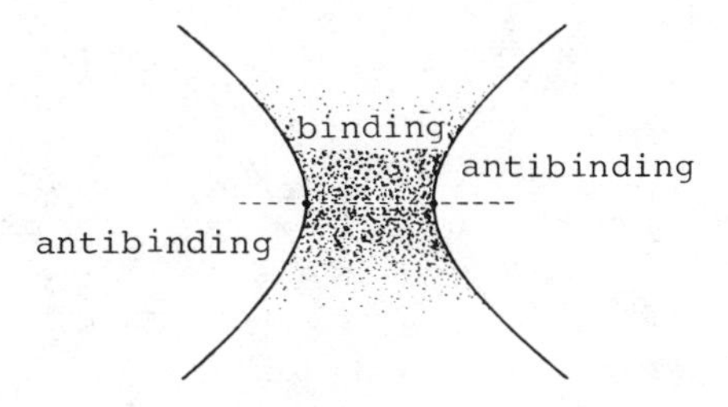

FIGURE 2.9. Binding and antibinding regions in a homonuclear diatomic molecule. (From G. C. Pimentel and R. D. Spratley, *Chemical Bonding Clarified through Quantum Mechanics,* Holden-Day, Oakland, CA, 1969, with permission.)

When a neutral H_2 molecule is formed from two hydrogen atoms, two protons are tied together by means of two electrons. Since two electrons serve as a stronger adhesive to bind the nuclei than one electron, greater stability of an H_2 molecule results, the binding energy for the H_2 molecule being 450 kJ mol^{-1}, almost twice, that for H_2^+. The reason it is not exactly twice is due to the Coulomb repulsion between the two electrons.

When two hydrogen atoms are bonded to form a hydrogen molecule, the electrons in the separate atomic orbitals are now placed in the new molecular orbitals. There are two possibilities for forming the molecular orbitals, as illustrated in Figure 2.10. If the two 1s orbitals are combined with positive overlap, a bonding interaction results. The positively overlapping combination, $\phi_1 + \phi_2$, forms a new molecular orbital ψ_b, where the subscript b stands for bonding. The other is a negative overlap, $\phi_1 - \phi_2$, forming an antibonding orbital ψ_a, where a stands for antibonding. In the bonding molecular orbital the electron distribution is along the line connecting the two nuclei and the electron serves to bind the nuclei. On the other hand, in the antibonding molecular orbital a nodal plane, where the electron density is zero, is formed between the nuclei. In the antibonding molecular orbital, as shown in Figure 2.10, the electron is repelled from the binding region to the peripheral antibinding region, as shown in Figure 2.9. Thus, when an electron is present in the antibonding orbital, it will exert a force to separate the two nuclei. The bonding orbital is symmetric, while the antibonding orbital is antisymmetric. Although the two molecular orbitals differ in the nature of the nodal plane, the electron distribution is along the direction connecting the two nuclei. These types of molecular orbitals are called the σ orbitals. The antibonding orbital is denoted by σ^*, with an asterisk. As in the atomic orbitals each molecular orbital can accommodate two electrons with spins antiparallel. Figure 2.11*a* shows the energy levels of the bonding σ and antibonding σ^* orbitals formed on interaction of the two hydrogen atomic orbitals. In the ground

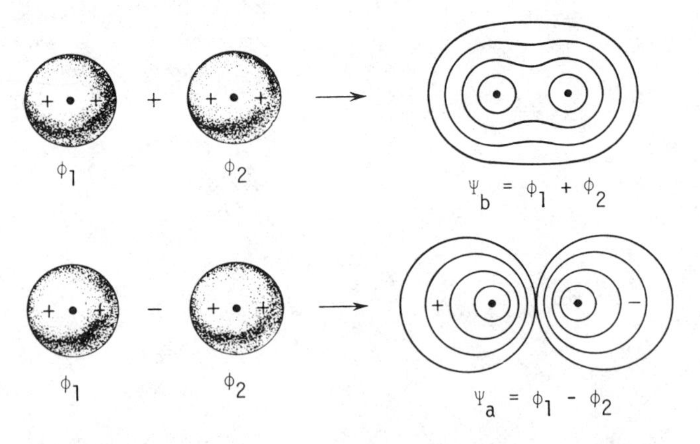

FIGURE 2.10. Formation of a bonding orbital ψ_b and an antibonding orbital ψ_a from two atomic orbitals ϕ_1 and ϕ_2. (From F. A. Cotton and G. Wilkinson, *Basic Inorganic Chemistry*, Wiley, New York, 1976, with permission.)

ψ_a σ^* orbital
ϕ_A ϕ_B
ψ_b σ orbital
(a)

ψ_a σ^* orbital
ϕ_A ϕ_B
ψ_b σ orbital
(b)

FIGURE 2.11. Formation of a bonding molecular orbital (ψ_b) and an antibonding molecular orbital (ψ_a) from two atomic orbitals (ϕ_A, ϕ_B) of two atoms. (*a*) When energy of each atomic orbital is identical. (*b*) When each atomic orbital has a different energy.

state the bonding orbital is occupied by two electrons and the molecule is stabilized.

Figure 2.11*b* shows the energy level diagram of a heteronuclear molecule. In this case the initial energies ϕ_A and ϕ_B of the two atomic orbitals are different. The energy level of the newly formed bonding MO is closer to that of the lower-energy atomic orbital, and the energy level of the antibonding MO is closer to that of the higher-energy atomic orbital. Quantum mechanical theory tells us that the closer the two original energy levels are, the more strongly they interact and the more stable is the resulting molecular orbital. When the initial atomic orbitals ϕ_A and ϕ_B are different in energy, the bonding MO ψ_b has more contribution from the lower atomic orbital ϕ_B, whereas, conversely, the antibonding MO ψ_a (σ^* MO) has a preponderance of ϕ_A character. When ϕ_A and ϕ_B differ very greatly in energy, the interaction becomes so small that ψ_a is quite close in its shape and energy to ϕ_A and ψ_B is similar to ϕ_B.

A similar line of discussion can be extended to the construction of molecular orbitals from atoms containing more electrons. Figure 2.12 shows the schematic energy level diagram of molecular orbitals of a nitrogen molecule. Each 1s and 2s atomic orbital forms new bonding and antibonding MOs where two electrons are accommodated for each MO from the lower level.

When the MO is constructed from 2p orbitals, one has to consider the matching between the atomic orbitals because the p orbitals have a dumbbell shape, with positive and negative signs in each lobe of the orbital. For overlap of the orbitals the following three cases should be considered: (a) bonding, (b) antibonding, and (c) nonbonding. When the atomic orbitals overlap positively with matching of the symmetry, they form a bonding orbital; on negative overlap they form an antibonding orbital. There are other cases where no overlap of the atomic orbitals is possible. For example, if p_x and p_y orbitals are orthogonal, no overlap results between them.

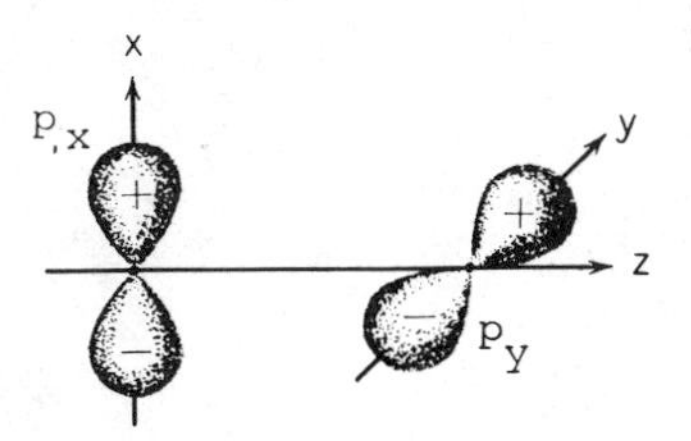

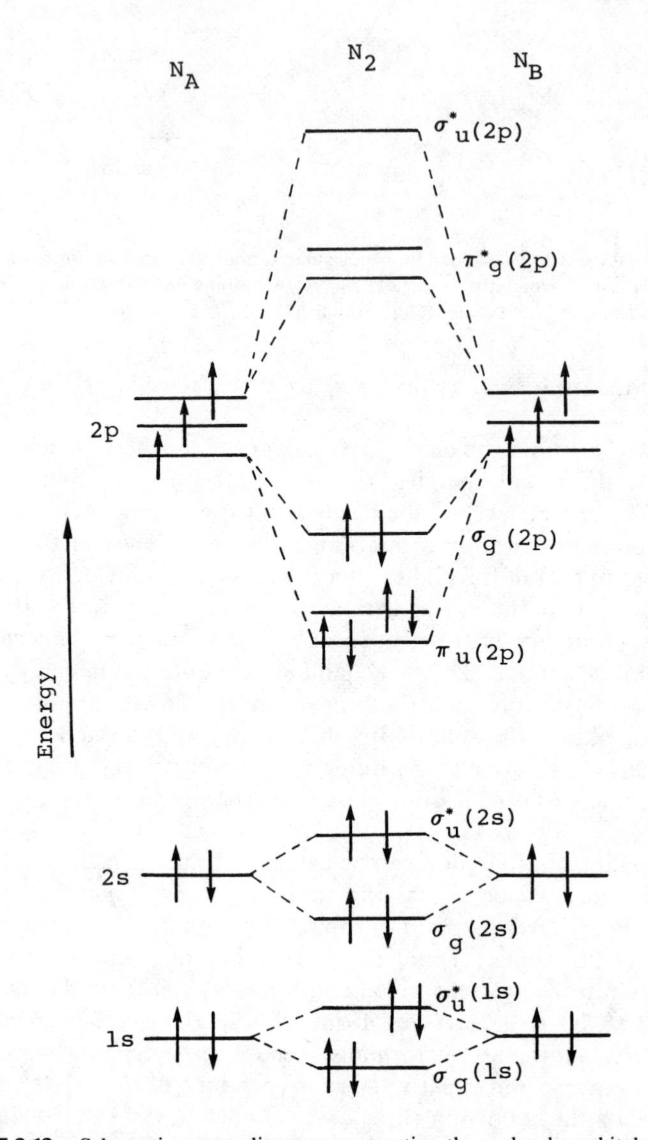

FIGURE 2.12. Schematic energy diagram representing the molecular orbitals of N_2.

In the combination of an s orbital with a p_x or p_y orbital or of an s orbital with a d_{xy} orbital, for example, a positive overlap between the wave functions of the same signs is canceled by a negative overlap.

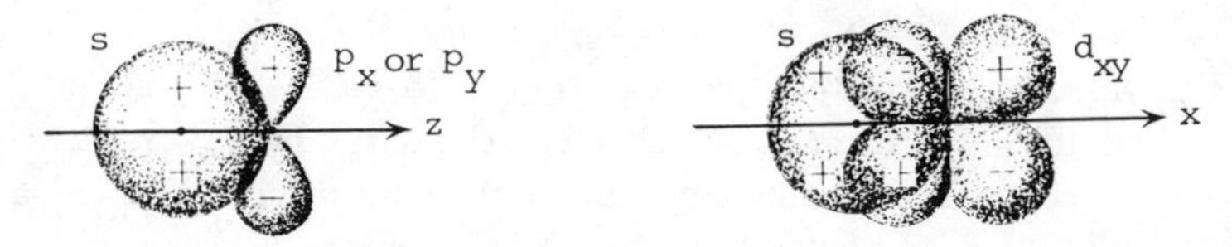

In this case there is neither bonding nor repulsion, and hence this situation is nonbonding. Various combinations between atomic orbitals are shown in Figure 2.13.

The σ_s and σ_s^* orbitals correspond to the molecular orbitals for the hydrogen molecule. The atomic p orbitals can form a σ bond from direct ("head-on") overlap of the p_z orbital, giving σ_p (or σ_g) and σ_p^* (or σ_u^*) orbitals, whereas parallel overlap of the p_x and p_y orbitals gives bonding π_p (or π_u) and antibonding π_p^* (or π_g^*) orbitals. We can now fill the N_2 orbitals in Figure 2.12 from the lower levels, placing two electrons per orbital. Orbitals to the σ_g (2p) level are occupied with electrons, leaving higher MOs unoccupied. The σ_g (2p) orbitals are the highest occupied molecular orbitals (HOMO) and the π_g^* (2p) orbitals are the lowest unoccupied molecular orbitals (LUMO); these are

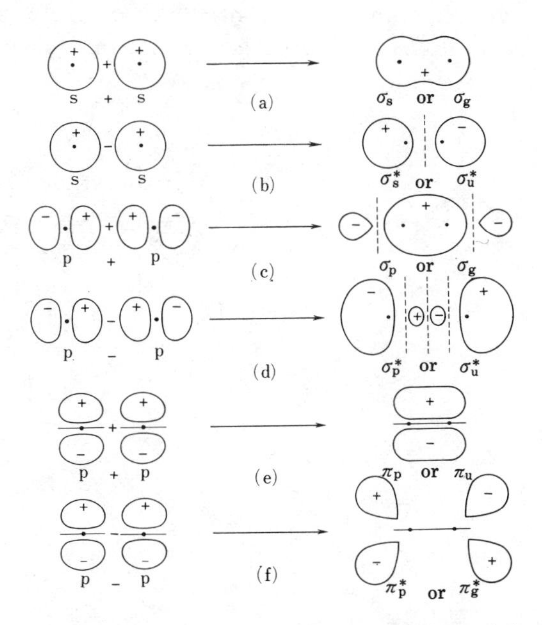

FIGURE 2.13. Various combinations of atomic orbitals and shapes of the resultant molecular orbitals. Combinations (*a*)–(*d*) form σ orbitals whereas combinations (*e*) and (*f*) give π orbitals. (From J. E. Huheey, *Inorganic Chemistry, Principles of Structures and Reactivity*, 3rd ed. Copyright © 1983 by James E. Huheey. Reprinted by permission of Harper & Row, Publishers, Inc.)

called frontier orbitals and play the most important roles in determining the behavior of the molecule. The energy level of the HOMO can be estimated from the ionization energy, which corresponds to the energy required for removing an electron from the HOMO level; the energy generated by putting an electron into the LUMO is the electron affinity. From the excitation energy required to transfer an electron from a HOMO to a LUMO, the energy level of the LUMO may be estimated if the HOMO energy is known. A molecule having a high HOMO level can serve as an electron donor, whereas a molecule having a low LUMO level tends to act as an electron acceptor.

b. Molecular Orbital Description of Transition Metal Complexes

Having reviewed the fundamental concepts in molecular orbital theory, we now proceed to treat transition metal complexes by a simple MO theory.[6]

We first consider a linear molecule MH_2, made up of a transition metal and two hydrogen atoms. As shown in Figure 2.14*a*, the metal s orbital having a spherical symmetry can combine with the s orbitals in the H atoms of the same symmetry in phase. The p_z orbital of the metal can combine with the s orbitals of the hydrogen atoms, as shown in Figure 2.14*b*, whereas the p_x orbital cannot overlap with the s orbitals of the hydrogen atoms regardless of their phase, as shown in Figure 2.14*c*. Among the d orbitals, only the d_{z^2} orbital has the correct symmetry to interact with the hydrogen atoms.

Next we consider the combinations of the atomic orbitals between the metal and hydrogen atoms in a hypothetical square planar metal hydride

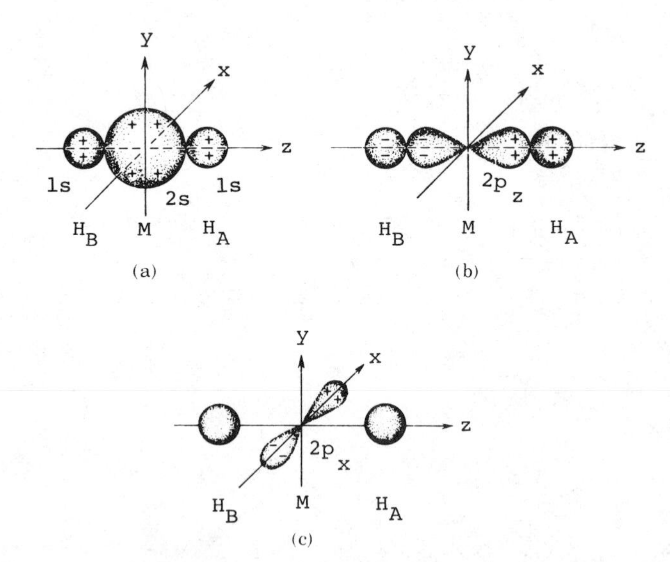

FIGURE 2.14. Molecular orbitals of a linear metal complex, MH_2. (*a*) Bonding, (*b*) bonding, (*c*) Nonbonding. (Adapted from L. E. Orgel, *An Introduction to Transition Metal Chemistry: Ligand Field Theory,* Methuen, London, 1960.)

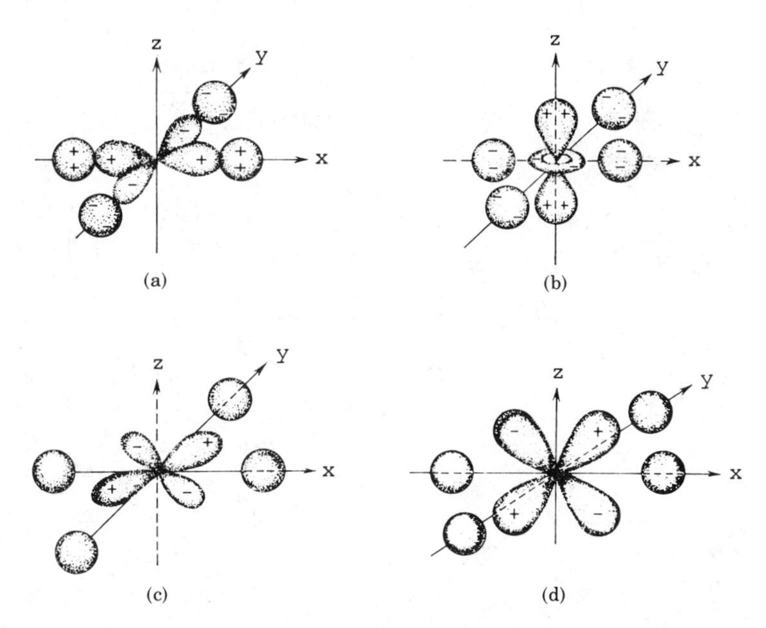

FIGURE 2.15. Molecular orbitals of a square planar complex, MH_4. (*a*) Combination of the $d_{x^2-y^2}$ orbital with the ligand 1s orbitals. (*b*) Weak combination between the d_{z^2} orbital with the ligand 1s orbitals. (*c*) The d_{xy} orbital has no net overlap with the ligand 1s orbitals. (*d*) The d_{xz} also has no overlap with the ligand 1s orbitals. (Adapted from L. E. Orgel, *An Introduction to Transition Metal Chemistry: Ligand Field Theory,* Methuen, London, 1960.)

MH_4. We consider only the d orbitals of the central metal, and we choose the bond directions as the x and y axes, as shown in Figure 2.15. We can see that the $d_{x^2-y^2}$ and d_{z^2} orbitals of the central metal atom can combine with the hydrogen s orbitals, as shown in Figures 2.15*a,b,* where interaction in (*a*) is stronger than in (*b*). On the other hand, none of the d_{xy}, d_{yz}, and d_{xz} orbitals can combine with the hydrogen s orbitals.

We now proceed to the case of an octahedral complex where the central metal atom having ns, np, and $(n-1)$d orbitals bonds six ligand molecules with σ orbitals, having roughly axial symmetry about the bond directions, as in Figure 2.16. It can be seen that the s orbital and the p_x, p_y, and p_z orbitals of the metal can combine with the ligand σ orbitals. Among the metal d orbitals, only the $d_{x^2-y^2}$ and d_{z^2} orbitals can combine with the ligand orbitals, whereas the d_{xy}, d_{yz}, and d_{xz} orbitals cannot combine with the ligand σ orbital and thus remain nonbonding.

From these considerations we can draw a qualitative energy level diagram for an octahedral complex of the first transition metal series having 4s, 4p, and 3d orbitals, as shown in Figure 2.17. In constructing the molecular orbitals, we follow the procedure employed for constructing the molecular orbitals shown in Figures 2.11 and 2.12. The metal 4s orbital combines with the ligand a_{1g} orbital to form a bonding orbital a_{1g} and an antibonding orbital a_{1g}^*, and

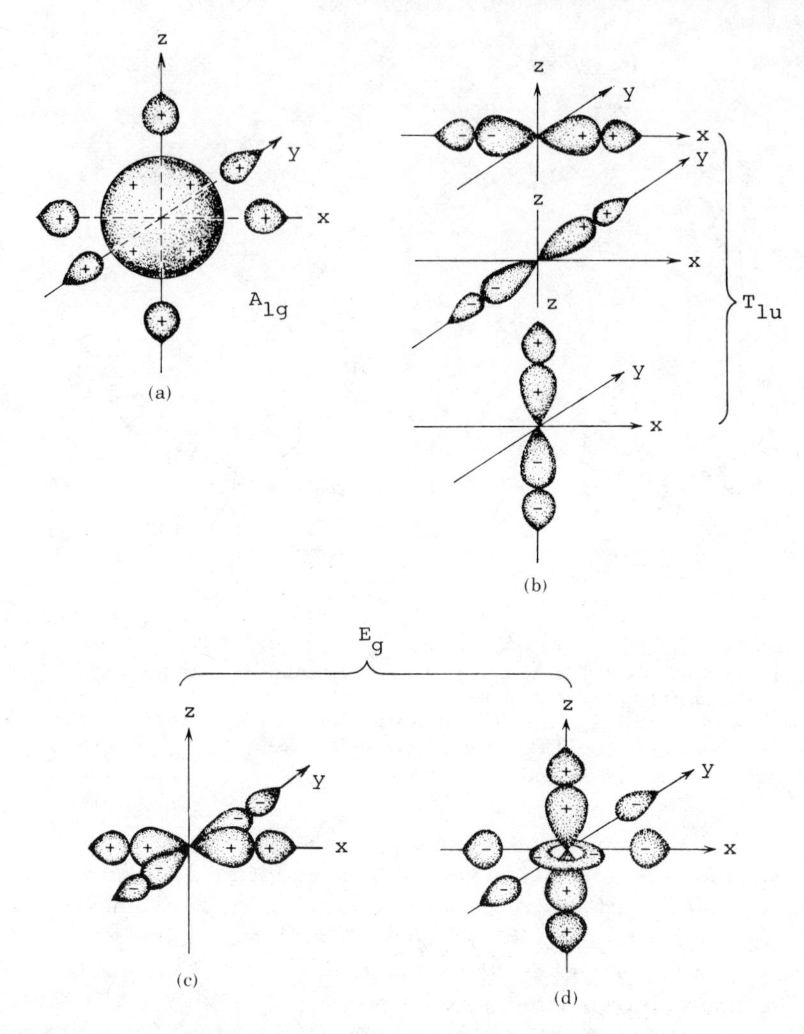

FIGURE 2.16. The σ bonding orbitals of an octahedral complex. Each ligand has a σ orbital directed toward the metal (s, p, sp, sp^2, or sp^3). (*a*) Combination with the metal s orbital. (*b*) Combination with the metal p_x (or p_y or p_z) orbital. (*c*) Combination with the $d_{x^2-y^2}$ orbital. (*d*) Combination with the d_{z^2} orbital.

the metal 4p orbital combines with the ligand t_{1u} orbitals, giving bonding t_{1u} and antiboding t_{1u}^* orbitals. Among the five 3d orbitals of the metal, the e_g orbitals (i.e., $d_{x^2-y^2}$ and d_{z^2}) can overlap with the corresponding e_g orbitals of the ligands, giving the bonding e_g and antibonding e_g^* orbitals. Since the three t_{2g} orbitals, that is, d_{xy}, d_{yz}, and d_{xz}, have no ligand orbitals of matching symmetry, their energy remains unaffected.

Among the molecular orbitals thus formed, the low-energy a_{1g}, t_{1u}, and e_g orbitals are mainly ligand orbitals, and the antibonding e_g^* orbitals are mainly

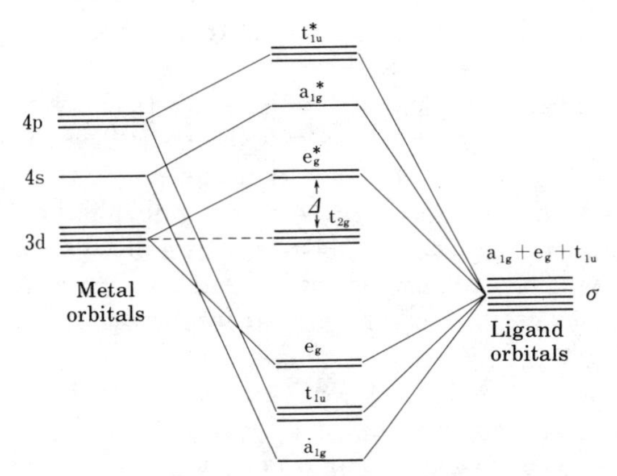

FIGURE 2.17. Schematic representation of the energy diagram in molecular orbitals of an octahedral complex.

metal orbitals, provided the original ligand orbitals have lower energies than the metal atomic orbitals. Electrons may now be added to the molecular orbitals of the complex in order of increasing energy. If each of the ligand orbitals originally contained an electron pair, the t_{1u}, a_{1g}, and e_g molecular orbitals will be filled with 12 electrons originating from the six ligand orbitals. The d electrons of the metal are now placed in the t_{2g} and e_g^* orbitals from the lower levels. Thus, the essential feature of the result derived from molecular orbital theory is quite similar to that of crystal field theory with regard to the metal d orbitals (compare Figure 2.3): the d orbitals are split into a set of two e_g^* and a set of three t_{2g}, with the former having a higher energy than the latter. The energy difference between the t_{2g} and e_g^* orbitals corresponds to Δ_o in Figure 2.3*c*. The way of putting d electrons in t_{2g} and e_g^* orbitals is exactly the same as that in the crystal field theory. The main difference between the molecular orbital and crystal field treatments is that the e_g^* orbitals obtained in the MO treatment are not pure d orbitals. The magnitude of the energy difference between e_g^* and t_{2g} orbitals now results from the strength of the ligand–metal bond rather than from electrostatic effects only. For strongly bonding ligands, the a_{1g}, t_{1u}, and e_g levels will be lowered and, correspondingly, the antibonding e_g^*, t_{1u}^*, and a_{1g}^* orbitals are raised. As the ligands are moved away from the metal, the overlap of the orbitals decreases and the antibonding levels are lowered.

In the crystal field theory the idea of double bonding cannot be accommodated because the ligands are treated simply as point charges. In the molecular orbital theory, on the other hand, the bonding of a transition metal to ligands having π symmetry can be treated generally. Since in organotransition metal chemistry we are often concerned with complexes having ligands with π orbitals, molecular orbital theory is far more useful than the simple crystal field approach in understanding structures and chemical reactivities of organotransition metal complexes.

In a transition metal complex having ligands with π orbitals, the metal d_{xy}, d_{yz}, and d_{xz} orbitals participate in bonding with these ligands, as illustrated for the d_{xy} orbital in Figure 2.18. Ligands that can participate in bonding with metals include Cl^- having atomic p_π orbitals, phosphines or arsines having vacant d_π orbitals, and polyatomic ligands with molecular orbitals of π symmetry such as CO, CN^-, olefins, pyridines, and N_2. Figure 2.18 demonstrates the simplest case where the metal d_π orbital combines with the ligand p_π orbitals having the proper symmetry.

We now consider the effects of π bonding via molecular orbitals of the t_{2g} type on the energy levels. The metal t_{2g} level, which remained nonbonding when the ligands are engaged only in σ bonding, is now affected by the interaction with the ligand π orbitals. There are two cases depending on the relative energies of the metal t_{2g} and the ligand π orbitals and on whether the ligand orbitals are filled or empty.

Figure 2.19 shows the case where the metal t_{2g} orbitals interact with empty π orbitals of higher energy than the metal t_{2g} orbitals. In the molecular orbitals of the transition metal complex formed by the interaction of the metal t_{2g} orbitals with the ligand orbitals, the newly formed t_{2g} orbitals are stabilized relative to the original, nonbonding t_{2g} orbitals, thus resulting in an increase of Δ_o. This situation is found in complexes having π-bonding ligands such as phosphines and arsines. In the other case the metal t_{2g} orbitals interact with filled π orbitals of ligands of lower energy than that of the metal t_{2g} orbitals, as depicted in Figure 2.20. The interaction here destabilizes the metal t_{2g} orbitals, causing a decrease in Δ_o. Many ligands encountered in organotransition metal chemistry such as CO, olefins, and dienes have filled π and vacant π^*

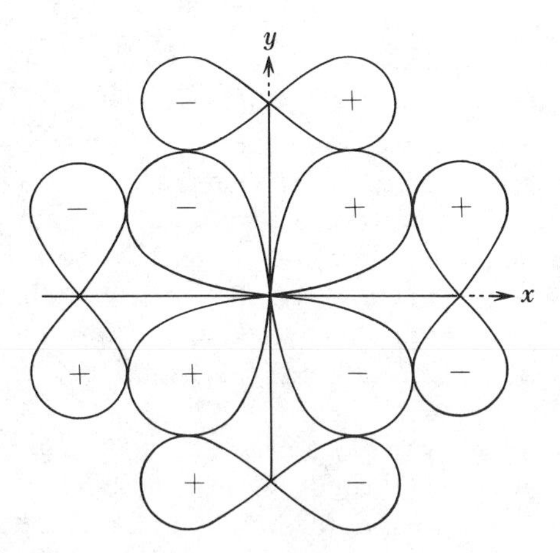

FIGURE 2.18. π Bonding between the metal d_{xy} orbital and ligand p orbitals. (Adapted from L. E. Orgel, *An Introduction to Transition Metal Chemistry: Ligand Field Theory,* Methuen, London, 1960.)

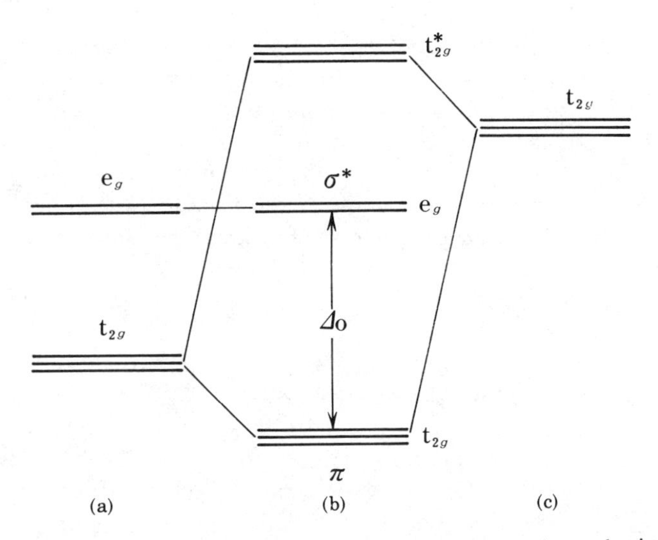

FIGURE 2.19. Effect of π bonding on the energy of the MO of a metal complex in combination of the filled d orbitals with vacant ligand orbitals. Note that the π bonding increases Δ_o in this case. (*a*) Filled metal d orbitals, (*b*) MO of the complex, and (*c*) vacant ligand orbitals.

orbitals that are capable of interacting with the metal d orbitals. Such interactions will be treated in Section 3.2.

2.3. VALENCE BOND THEORY

The concept of the valence bond theory applied to coordination compounds has been developed primarily by Linus Pauling (Nobel laureate in chemistry,

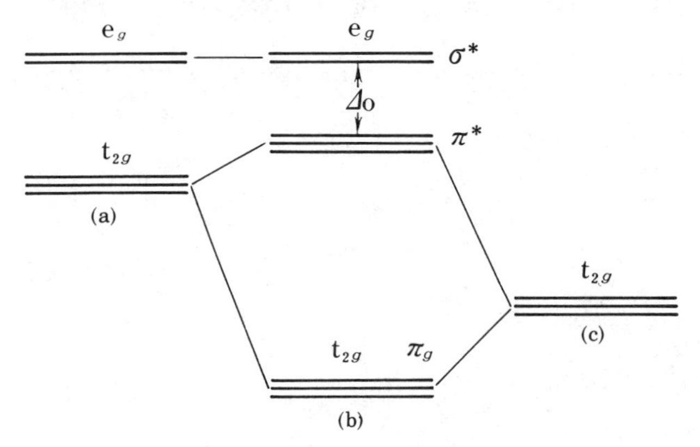

FIGURE 2.20. Energy levels in combination of vacant metal d orbitals with electron-donating ligand orbitals. (*a*) Vacant or partially filled metal d orbitals, (*b*) molecular orbitals of the metal complex, and (*c*) ligand orbitals (filled). Note the decrease in Δ_o on complexation.

1954). This theory has the advantage of interpreting the directionality of chemical bonds by regarding the bonding in polyatomic molecules as a collection of localized bonds between adjacent pairs of atoms. Let us begin an account of the valence bond theory with a simple compound.

A beryllium atom in its ground state has an electron configuration of $1s^2 2s^2$, which is not suitable for making chemical bonds with other atoms. For making a chemical bond with other atoms, it is necessary to promote one electron in the 2s orbital to the 2p orbital of the next highest energy so that it can form an electron pair with an electron of the other atom. When an atom is in this state and ready to form a chemical bond by electron pairing with other atoms, the atom is said to be in a valence state. For beryllium to form a linear compound BeH_2, the promotion energy for raising an electron from the 2s orbital to the 2p orbital is about 323 kJ mol^{-1}. Now mixing of the 2s and 2p orbitals can occur to form two hybridized orbitals, which are disposed for better overlap with the hydrogen 1s orbitals. The result of mixing the 2s and 2p orbitals to give two hybrid sp orbitals is illustrated in Figure 2.21.

Each of the hybrid orbitals has a large positive lobe concentrated along the z axis as a consequence of mixing the 2s and 2p orbitals. The two new hybrid orbitals are suitable for maximum overlap with the two hydrogen 1s orbitals to form a linear BeH_2 molecule. The bonding in the Be–H bonds in the linear BH_2 molecule formed as a result of sp hybridization is much stronger than the bonding formed between the two beryllium 2s electrons and the hydrogen 1s electron without hybridization, and this bonding more than compensates for the energy required for promoting the 2s electron to the 2p orbital (Fig. 2.22).

A similar procedure can be extended to form a planar, triangular compound and a tetrahedral compound. For boron the ground-state electron configuration is $1s^2 2s^2 2p$. In this case an electron in the 2s orbital is promoted to one of the 2p orbitals to give a valence state configuration of $2s2p_x2p_y$, where the choice of the $2p_x$ and $2p_y$ is arbitrary. Upon mixing of the s and two p orbitals, sp^2 hybrid orbitals can be formed. The three sp^2 hybrid orbitals lie on the xy plane along the lines that are 120° apart. Thus, the BH_3 molecule formed by electron pairing with three hydrogen atoms has a trigonal planar

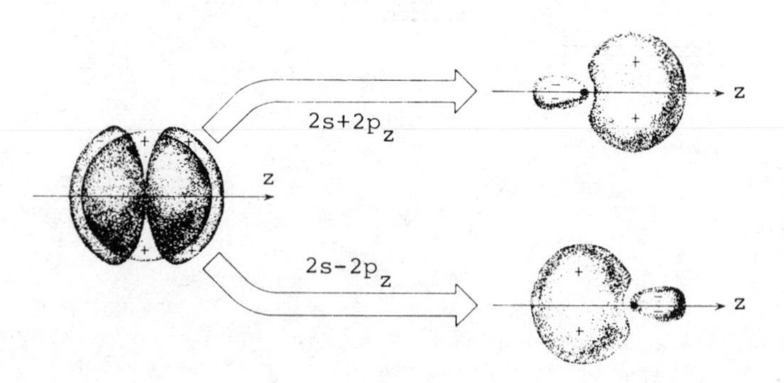

FIGURE 2.21. Formation of sp orbital by hybridization of 2s and $2p_z$ orbitals.

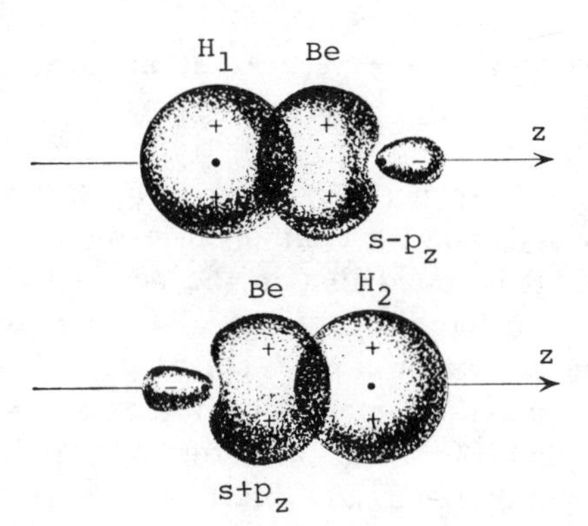

FIGURE 2.22. Formation of BeH_2 molecule by overlap of beryllium sp hybrid orbitals with 1s orbitals of hydrogen atoms.

configuration with a vacant $2p_z$ valence orbital. Having the low-lying vacant $2p_z$ orbital, the BH_3 molecule has a residual bonding ability, so the molecule is unstable in its monomeric form and tends to form a dimer B_2H_6.

The sp^3 hybridization can be achieved similarly by promoting an electron in the s orbital to one of the p orbitals. For carbon atom the valence state of $2s2p_x2p_y2p_z$ can be generated by an electron promotion from the ground-state electron configuration of $1s^22s^22p^2$. By mixing the one s and the three p orbitals, the familiar sp^3 hybrid orbitals are obtained. The four equivalent sp^3 hybrid orbitals are directed toward the vertices of a tetrahedron, and thus CH_4 has a tetrahedral configuration.

Hybridization is not limited to s and p orbitals. When d orbitals as well as s and p orbitals are available, a variety of hybrid orbitals can arise. First let us see how the octahedral hybrid is constructed from s, p, and d orbitals. When the $d_{x^2-y^2}$ and d_{z^2} hybridize with one s and a set of three p orbitals, a d^2sp^3 hybrid is formed. The six hybrid orbitals thus formed are equivalent, with

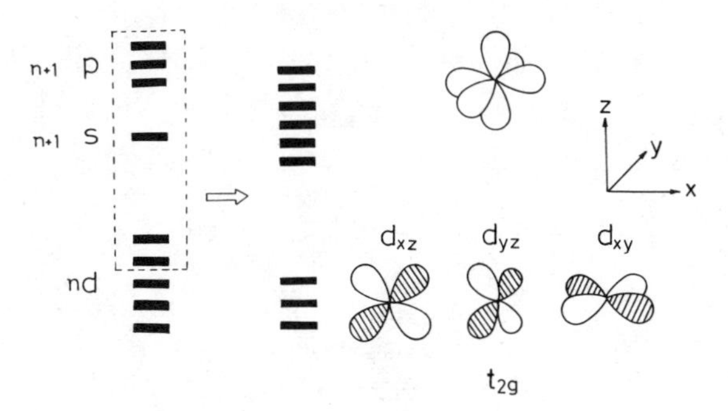

FIGURE 2.23. Formation of d^2sp^3 hybrid orbitals from two d, one s, and three p orbitals. (From R. Hoffmann, *Angew. Chem. Int. Ed. Engl.*, **21,** 711 (1982), with permission. © The Nobel Foundation, 1982.)

their lobes pointing toward the vertices of an octahedron. In octahedral hybridization the d_{xy}, d_{yz}, and d_{xz} orbitals remain unused, as shown in Figure 2.23.

Other hybridizations are performed similarly. Figure 2.24 illustrates the shapes of hybrid orbitals and the component orbitals from which the hybrid orbitals are formed. It is noted that in the octahedral (Fig. 2.24*a*), square planar (Fig. 2.24*b*), and tetrahedral (Fig. 2.24*c*) hybridizations, each hybrid orbital is equivalent in composition (percentage of s and percentage of p character) and in its spatial orientation with respect to each other, whereas in the trigonal bipyramidal (Fig. 2.24*d*) and square pyramidal hybridizations the hybrid equatorial orbitals and axial orbitals are different in bond lengths and strengths even when the ligands are identical.

The valence bond approach has the advantages of simplicity and of providing a pictorial concept of the chemical bonds in various metal complexes, but it has become somewhat less popular because it is less amenable to quantita-

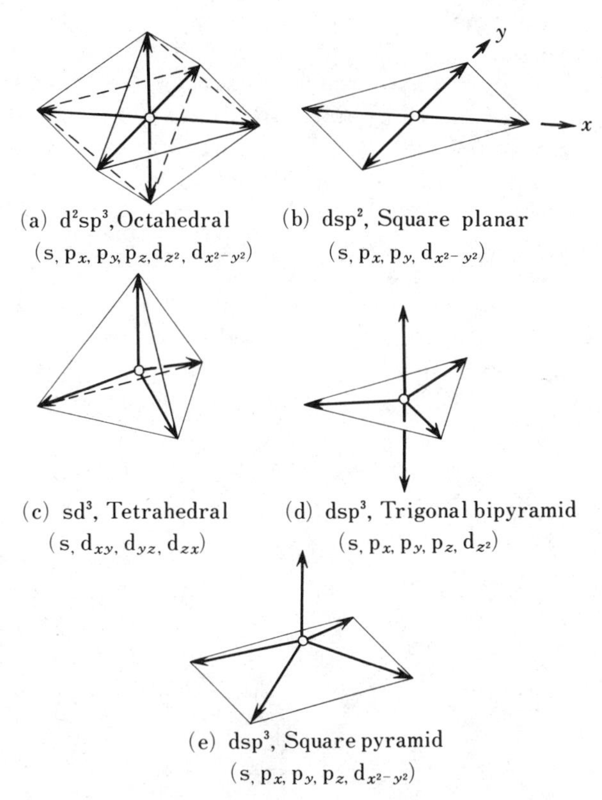

FIGURE 2.24. Five important hybridization schemes involving d orbitals. Heavy arrows show the directions in which the lobes point. (From F. A. Cotton and G. Wilkinson, *Basic Inorganic Chemistry,* Wiley, New York, 1976, p. 76, with permission.)

tive calculations than molecular orbital theory. With the progress of fast digital computers we are now entering an era where quite complete nonempirical (ab initio) calculations can be carried out for some simple transition metal complexes. As we shall occasionally mention in later discussions of the structures and chemical behavior of certain types of organotransition metal complexes, ab initio calculations are now producing results that show reasonable agreements with experimental findings.

On the other hand, it is evident that the sophisticated and costly nonempirical approach cannot be applied to too many problems of transition metal complexes. Thus, less complete but simpler theoretical approaches will retain their value for elucidating structures and chemical behavior of a variety of transition metal complexes. In one such approach Roald Hoffmann (Nobel laureate in chemistry, 1981) introduced the concept of the valence bond approach into the molecular orbital approach and achieved considerable success in accounting for the structures and behavior of some important organotransition metal complexes and, in some cases, in predicting the synthesis of

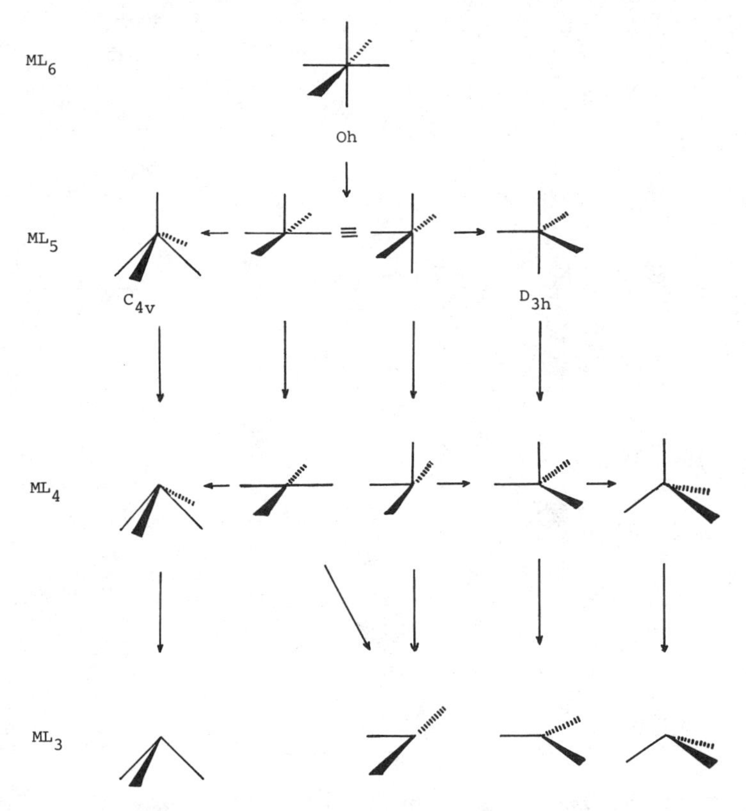

FIGURE 2.25. Derivation of three-, four-, and five-coordinate complexes by successive removal of a ligand from an octahedron.

new compounds and novel chemical reactivities of organotransition metal complexes. The following is a brief account of his approach.[8]

In this approach a molecule is conceptually broken up into a metal fragment ML_n and a ligand. The orbitals of these simple fragments are easily visualized. One then "reconstructs" the complex by examining the interaction of the ligand orbitals with the orbitals of the ML_n fragment. Figure 2.25 shows how the fragment geometries are derived from an octahedron by successive stripping away of one ligand, followed by a sequence of geometrical rearrangements of the remaining ML_n fragment. We can see in the procedure shown in Figure 2.25 how the square pyramidal and trigonal bipyramidal structures are related to the octahedral structure and further how the square planar and tetrahedral structures are derived from the five coordinate structures. In the valence bond approach we have mixed two d orbitals, an s orbital, and three p orbitals to give the octahedral d^2sp^3 hybridization (Fig. 2.23). The square pyramidal fragment can now be formed by interaction of five ligand orbitals in an octahedral field with five of the six hybrid orbitals, as shown in Figure 2.26. In the fragment orbitals, as shown in the center of Figure 2.26, five bonding and five antibonding orbitals are formed by the interaction of the five ligand orbitals with five hybrid orbitals. One hybrid orbital remains unused, along with the three unhybridized d orbitals. These four orbitals are the valence or frontier levels of the ML_5 fragment. They

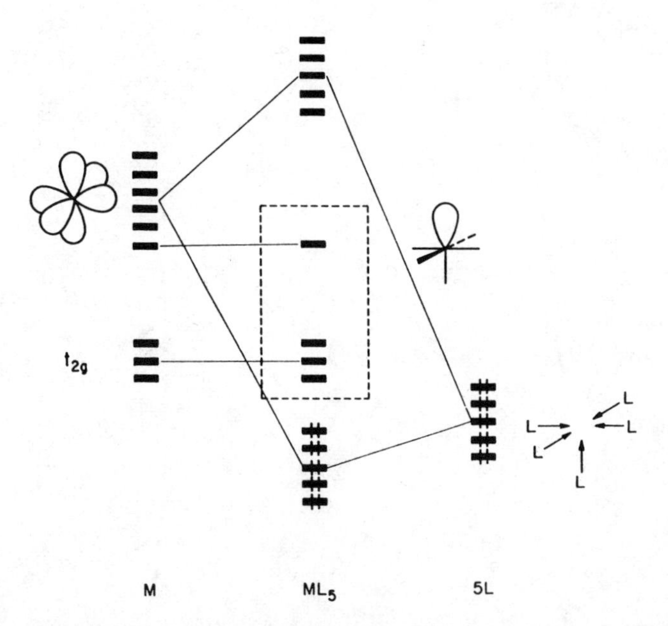

FIGURE 2.26. Fragment orbitals of a square pyramidal complex. [Figures 2.26–2.30 are reproduced from R. Hoffmann, *Angew. Chem. Int. Ed. Engl.*, **21,** 711 (1982), with permission. © The Nobel Foundation, 1982.)

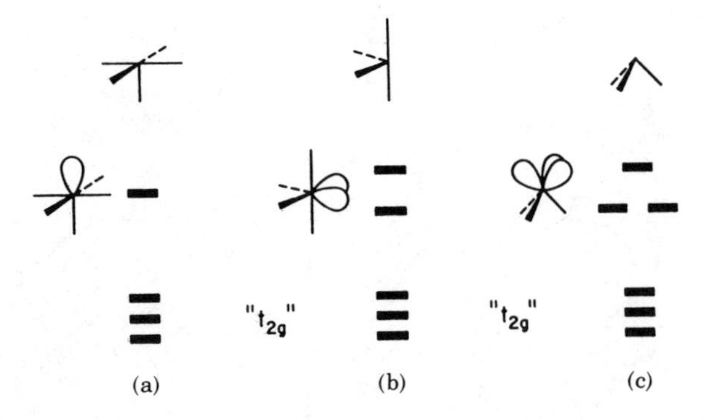

FIGURE 2.27. Fragment orbitals of (*a*) ML_5, (*b*) ML_4, and (*c*) ML_3.

carry all the metal d electrons, and they will interact with any other ligands that come near.

The patterns of interaction for ML_4 and ML_3 are no different. Four or three ligands interact with four or three of the octahedral hybrids, leaving two or three hybrid orbitals, respectively, untouched. The conceptual pictures of ML_5, ML_4, and ML_3 fragments are given in Figure 2.27. In each ML_n fragment there are three orbitals of lower energy, essentially d_{xy}, d_{yz}, and d_{xz}, and $6-n$ hybrids of higher energy pointing toward the octahedral sites.

Let us examine a specific example. The complex $CH_3Mn(CO)_5$ is known as a stable complex having a metal–carbon σ bond. If we split the complex into a methyl radical and an $Mn(CO)_5$ fragment containing seven d electrons, the seven d electrons may be accommodated in the fragment orbitals as shown in Figure 2.28, which compares the ML_5 fragment with the methyl radical. A similarity between the MnL_5 fragment orbital and the methyl radical orbital is seen.

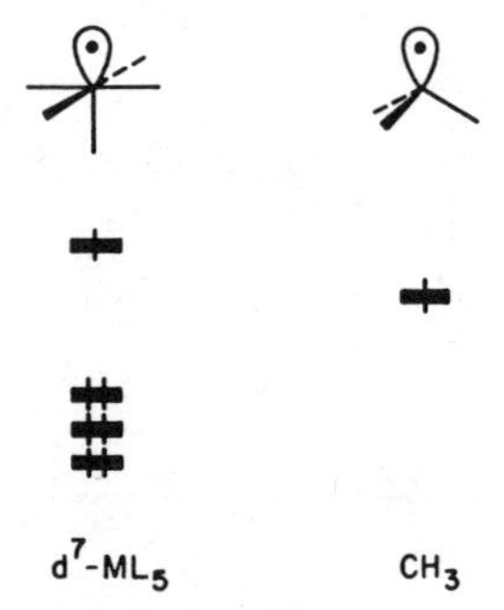

FIGURE 2.28. Comparison of the electron configurations in the d^7 fragment and the methyl radical.

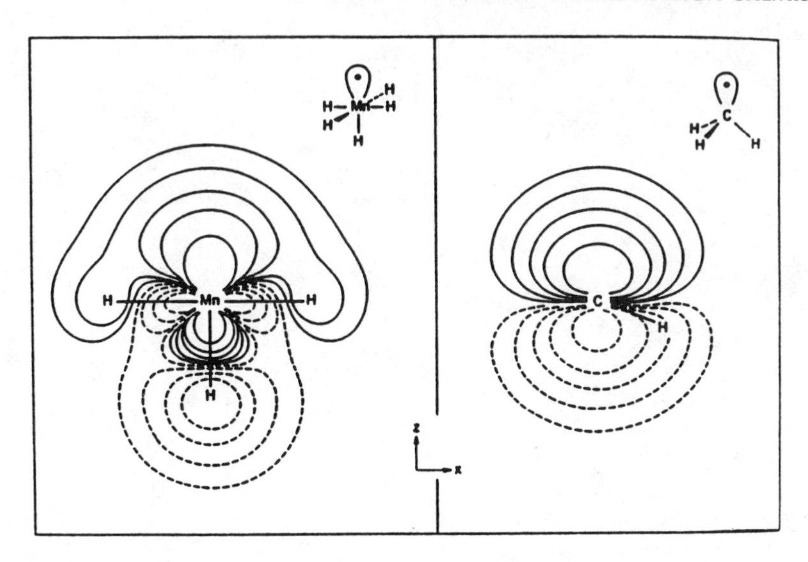

FIGURE 2.29. Comparison of contour diagrams of MnH_5^{5-} (left) and CH_3 (right) a_1 orbitals, as computed by the extended Hückel method. The contours of ψ, plotted in a plane passing through manganese and three hydrogen molecules (left) and carbon and one hydrogen (right) are ±0.2, ±0.1, ±0.055, and ±0.01.

For further comparison, contour plots of the frontier orbitals of MnH_5 and of the CH_3 radical computed by the extended Hückel method are given in Figure 2.29. Noting the similarity between the two orbitals, Hoffmann developed a concept of *isolobal* analogy and called the CH_3 radical and the $Mn(CO)_5$ fragment isolobal. Figure 2.30 illustrates the concept. As two methyl radicals can combine to give ethane, the $Mn(CO)_5$ fragment can dimerize to a known complex with a direct Mn–Mn bond, $Mn_2(CO)_{10}$. It is also shown that by overlap of the orbitals of $Mn(CO)_5$ and CH_3 entities, each housing an unpaired electron, the known methylmanganese complex $CH_3Mn(CO)_5$ having a covalent Mn–C bond can be generated.

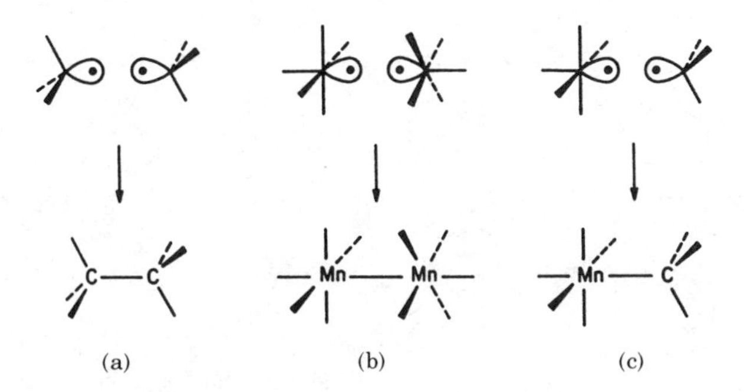

FIGURE 2.30. Isolobal analogy between the methyl radical and the ML_5 fragment.

The analogy can be extended to describe the bonding of a variety of organotransition metal complexes, which will be discussed in later chapters.

REFERENCES

1. T. Yoshida, M. Matsuda, T. Okano, T. Kitami, and S. Otsuka, *J. Am. Chem. Soc.*, **101,** 2027 (1979).
2. (a) S. C. Abrahams, A. P. Ginsberg, and K. Knox, *Inorg. Chem.*, **3,** 558 (1964); (b) R. H. Crabtree and G. G. Heatky, *J. Organometal. Chem.*, **238,** C21 (1982).
3. H. Bethe, *Ann. Physik,* **3**(5), 133 (1929).
4. J. H. Van Vleck, *Theory of Electric and Magnetic Susceptibilities,* Oxford University Press, London, 1932.
5. (a) T. J. Marks and R. D. Fischer, eds., *Organometallics of the f-Elements,* D. Reidel, Dordrecht, 1979; (b) T. J. Marks, *Prog. Inorg. Chem.*, **24,** 51 (1978); (c) M. Tsutsui, N. Ely, and R. Dubois, *Acc. Chem. Res.*, **9,** 217 (1976); (d) T. J. Marks, *Acc. Chem. Res.*, **9,** 223 (1976).
6. L. E. Orgel, *An Introduction to Transition-Metal Chemistry, Ligand Field Theory,* 2nd ed., Methuen, London, 1966.
7. (a) G. C. Pimentel and R. D. Spratley, *Chemical Bonding Clarified through Quantum Mechanics,* Holden-Day, San Francisco, 1969; (b) E. Heilbronner and H. Bock, *The HMO Model and Its Applications,* Wiley, New York, 1975; (c) R. Hoffmann, *Acc. Chem. Rev.*, **4,** 1 (1971).
8. (a) R. Hoffmann, *Science,* **211,** 995 (1982); (b) Nobel Lectures, The Nobel Foundation, 1982; *Angew. Chem. Int. Ed. Engl.*, **21,** 711 (1982); (c) M. Elian and R. Hoffmann, *Inorg. Chem.*, **14,** 1058 (1975); (d) M. Elian, M. M.-L. Chen, D. M. P. Mingos, and R. Hoffmann, *Inorg. Chem.*, **15,** 1148 (1976).

ADDITIONAL REFERENCES

J. E. Huheey, *Inorganic Chemistry. Principles of Structure and Reactivity,* 3rd ed., Harper & Row, New York, 1983.

F. A. Cotton and G. Wilkinson, *Basic Inorganic Chemistry,* Wiley, New York, 1976.

B. E. Douglas, D. H. McDaniel, and J. J. Alexander, *Concepts and Models of Inorganic Chemistry,* 2nd ed., Wiley, New York, 1983.

K. F. Purcell and J. C. Kotz, *An Introduction to Inorganic Chemistry,* Holt-Saunders, Philadelphia, 1980.

R. B. Heslop and K. Jones, *Inorganic Chemistry. A Guide to Advanced Study,* Elsevier, Amsterdam, 1976.

F. Basolo and R. G. Pearson, *Mechanisms of Inorganic Reactions. A Study of Metal Complexes in Solution,* 2nd ed., Wiley, New York, 1967.

C. J. Ballhausen, *Introduction to Ligand Field Theory,* McGraw-Hill, New York, 1962.

F. A. Cotton, *Chemical Applications of Group Theory,* 2nd ed., Wiley, New York, 1971.

A. F. Williams, *A Theoretical Approach to Inorganic Chemistry,* Springer-Verlag, Berlin, 1979.

CHAPTER THREE

Nature of the Metal–Carbon Bond and Electronic Configurations of Transition Metal Complexes

In this chapter we consider the nature of the metal–carbon bonds in organometallic compounds in general and the classification of ligands in organotransition metal complexes and the 18-electron rule in particular.

Organotransition metal compounds may be broadly classified as those having either M–C σ bonds or M–C π bonds. The former have counterpart in the nontransition elements, but the latter are unique to transition metal complexes. Let us first discuss the organometallic compounds having M–C σ bonds in comparison with nontransition metal alkyls.†

In nontransition metal alkyls some M–C bonds are more ionic and some are covalent. The M–C bond in organometallic compounds of electropositive metals has considerable ionic character, as seen in alkyl compounds of alkali metals (compare the electronegativities of nontransition metals with those of transition metals as shown in the Periodic Table on the back cover.[1] Carbon has an electronegativity value of 2.5, and metals have smaller values. The electronegativities of transition metals range from 1.2 to 1.8, whereas the alkali and alkaline earth metals have smaller values. Thus, organic groups attached to alkali and alkaline earth elements show carbanionic properties.

† A compound having a hydrocarbon entity bonded to a metal through a σ bond is sometimes called a hydrocarbyl, a term including aryl compounds as well as alkyl compounds. In this book the simple term *alkyl* is used to describe all metal compounds with an M–C σ bond including aryl compounds as well when the differentiation is not required.

However, the nature of M–C bonds is strongly influenced by the ligands attached to the metal. Electron-withdrawing ligands coordinated to the metal cause an increase in the electronegativity of the metal fragment, making the other organic ligand less carbanionic. Electron-donating ligands exert an opposite effect. The nature of the M–C bond is also influenced by the nature of entities attached to carbon. For example, the nature of the M–CH_3 bond is considerably different from that of the M–CF_3 bond. This is due to the greater electronegativity of fluorine, having an electronegativity of 4.1, compared to that of hydrogen (2.1). The enhanced stability of trifluoromethyl compounds of transition metals in comparison to that of transition metal methyl compounds is a consequence of the change of the nature in the M–C bond.

3.1. COMPOUNDS WITH METAL–CARBON σ BONDS AND THE METAL–CARBON BOND STRENGTH

In contrast to organometallic compounds of nontransition elements such as alkylzinc, alkylmagnesium, and alkylaluminum compounds, which have been long known as thermally stable compounds, the number of isolated alkyl compounds of transition metals was quite limited until recently. An exception to this generalization is provided by the platinum alkyls, which have been known since 1907. In a review article[2] published in 1955 only a few examples of chromium and titanium were cited as isolated transition metal alkyls characterized unequivocally. And in 1953 a paper theoretically explaining the reason for the *intrinsic* instability of transition metal to carbon bonds was reported.[3]

However, the number of isolated transition metal alkyls has increased rapidly since the sixties, and now transition metal alkyls are known for virtually all transition metals. Why then had transition metal alkyls been considered intrinsically unstable? What factors determine the stability of transition metal alkyls? Before we proceed with our discussion, it is necessary to define the term *stability,* which is often used rather loosely. When one says that an organometallic compound is unstable, it is sometimes not clear whether the compound is thermally unstable or the compound is so air-sensitive that it is readily decomposed by attack of oxygen or water in air. Thus, it is necessary to ask a somewhat awkward question: Against what is the compound unstable?

If the compound is prone to be decomposed by air, one should be able to isolate the compound by handling it in the absence of air, by working in a vacuum or in the atmosphere of a protecting gas such as argon or nitrogen. (In chapter 5 we deal with the techniques for handling organometallic compounds.) One of the reasons for the previous failure to isolate transition metal alkyls was the inadequacy of the available experimental techniques in handling air-sensitive compounds. However, some organometallic compounds are thermally unstable and may be decomposed spontaneously, even in the absence of air, by raising the temperature. These compounds should be kept and handled at low temperatures. We call these complexes thermally *unstable* and avoid using the term stability in relation to air-sensitivity.

In discussing the stability of transition metal alkyls, it is necessary to consider thermodynamic as well as kinetic factors. A transition metal alkyl, which is thermodynamically unstable and has a tendency to be transformed to other forms of products, may still exist under certain conditions if a suitable energy barrier for that transformation process is provided to block the decomposition pathway. For this purpose certain special alkyl groups or stabilizing ligands such as CO, tertiary phosphines, and cyclopentadienyl ligands have been used successfully. Thus, we now have a considerable range of transition metal alkyls available for study. For a discussion of the stability of transition metal alkyls information on the M–C bond strength is required. However, we have had very limited data until recently. According to Skinner,[4] this is partly due to "the diffidence of thermochemists to tackle organometallic compounds until recently, and the failings of conventional thermochemical techniques when applied to these compounds." Such a situation is being remedied, albeit slowly, and we are obtaining an increasing body of data concerning the M–C bond dissociation energies of transition metal alkyls.[5-9]

Table 3.1 summarizes mean bond dissociation energies $\bar{D}$ of nontransition metal alkyls of the type MR_n.[4] This type of binary metal alkyl is sometimes

Table 3.1. $\bar{D}$(M–C) Values in MR_n Compounds of Nontransition Metals[a]

M–C Bond	$\bar{D}$ (kJ mol^{-1})	M–C Bond	$\bar{D}$ (kJ mol^{-1})
Li–Et	209	Ge–Pr	238
Li–Bu	248	Sn–Me	218
Zn–Me	176	Sn–Et	193
Zn–Et	145	Sn–Pr	197
Cd–Me	139	Sn–Bu	195
Cd–Et	109	Sn–Ph	257
Hg–Me	122	Pb–Me	153
Hg–Et	101	Pb–Et	129
Hg–Pr	103	P–Me	275
Hg–*i*–Pr	89	P–Et	258
Hg–Ph	136	P–Ph	301
B–Me	363	As–Me	229
B–Et	342	As–Ph	267
B–Bu	344	Sb–Me	215
Al–Me	276	Sb–Ph	244
Al–Et	242	Bi–Me	143
Ga–Me	247	Bi–Ph	177
Ge–Et	237	Se–Et	242

[a] From ref. 4.

called a homoleptic alkyl.[5] One should note the difference between the mean bond dissociation energy and the bond dissociation energy for cleaving an individual M–C bond. A metal alkyl MR_n may be homolytically decomposed, that is, the M–R bonds may be cleaved to give radicals, as shown by Eq. (3.1) and the standard heat of disruption, ΔH°, of an MR_n molecule may be given by Eq. (3.2).

$$MR_n(g) \longrightarrow M(g) + nR\cdot(g) \tag{3.1}$$

$$\Delta H^\circ = \Delta H_f^\circ(M, g) + \Delta H_f^\circ(R, g) - \Delta H_f^\circ(MR_n, g) \tag{3.2}$$

where ΔH_f° (M, g), ΔH_f° (R, g), and ΔH_f° (MR_n, g) stand for the standard heats of formation of gaseous metal, the alkyl radical, and the metal alkyl. The mean bond dissociation energy $\bar{D}$(M–R) of MR_n molecule is given by

$$\bar{D}(M\text{–}R) = \Delta H^\circ/n \tag{3.3}$$

As an example, the standard heat of reaction ΔH° for decomposition of $HgMe_2$ is 244 ± 9 kJ mol^{-1}. Thus, D(Hg–Me) is 122 kJ mol^{-1} (29.2 kcal mol^{-1}). The value implies that the Hg–Me bond is a weak bond. However, dimethyl-mercury is more stable than implied by the $\bar{D}$(Hg–Me) value since more energy than indicated by the $\bar{D}$ value is required to initiate disruption of $HgMe_2$.

For the first step of thermolysis,

$$\text{Me-Hg-Me} \longrightarrow \text{Me-Hg} + \text{Me}\cdot$$

the activation energy of the reaction determined by a kinetic method was found to be 215.5 ± 8 kJ mol^{-1}; this leaves only 29 ± 12 kJ mol^{-1} as the bond dissociation energy for the second step of cleavage of the Hg–Me bond. This example clearly shows that one cannot estimate the ease of the first M–R bond scission from the mean bond dissociation energy $\bar{D}$(M–R).

Furthermore, the mechanism of thermolysis may vary depending on the metal alkyl. For example, dipropylmercury is thermolyzed by stimultaneous rupture of *both* Hg–C bonds,

$$\text{Pr-Hg-Pr} \longrightarrow \text{Pr}\cdot + \text{Hg} + \text{Pr}\cdot$$

so that the activation energy for thermolysis corresponds to twice the value of $\bar{D}$(Hg–Pr). It should also be noted that the dissociation energies $\bar{D}$(Me–HgX) (X = Cl, Br, I) are in the range of 247–269 kJ mol^{-1}. These values are much larger than the $\bar{D}$(Hg–Me) value of 122 kJ mol^{-1} in $HgMe_2$ and are also appreciably larger than the $\bar{D}$(Me–HgMe) value of 215 kJ mol^{-1} in the first Hg–Me rupture.

With these cautions concerning the implication of the $\bar{D}$ value in mind, one can still use $\bar{D}$(M–R) as a relative measure for estimating ease of M–R bond scission for related groups of organometallic compounds, which decompose by a similar mechanism.

The *mean* bond dissociation energies of $M(CH_3)_n$ compounds of some non-transition metals[4] are shown in Figure 3.1. For trimethyl compounds, MMe_3,

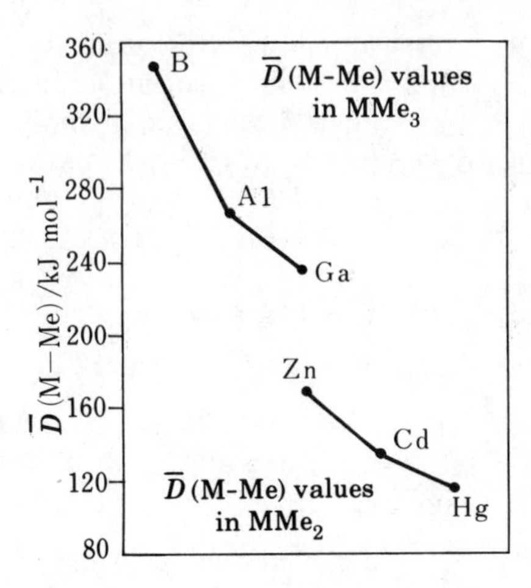

FIGURE 3.1. $\bar{D}$(M–Me) values for group 12 and 13 metals. (From H. A. Skinner, *Adv. Organometal. Chem.*, **2**, 99 (1964), with permission.)

the mean dissociation energy D(M–Me) is shown to decrease in the order B > Al > Ga, and for dimethyl compounds, MMe_2, in the order Zn > Cd > Hg; thus, $\bar{D}$(M–Me) decreases in both series as one descends in the same group in the periodic table.

Also, for MR_4 compounds $\bar{D}$(M–R) decreases with increasing atomic number in group 14,† as shown in Figure 3.2. It can be seen from the same figure that the mean bond dissociation energy of the M–H, M–Me, and M–Et bonds decreases in the order $\bar{D}$(M–H) > $\bar{D}$(M–Me) > $\bar{D}$(M–Et). Also, $\bar{D}$(M–Ph) is generally larger than $\bar{D}$(M–Me) by ~40 kJ mol^{-1} and smaller than $\bar{D}$(M–H).

Although bond dissociation energy data for transition metal alkyls are still limited, we now have just enough for comparison with those of nontransition metal alkyls. Table 3.2 summarizes mean bond dissociation energies (in some cases dissociation energies for particular bonds) of various transition metal alkyls and aryls. The footnote to the table also shows the C–C and C–H bond dissociation energies of representative hydrocarbons.

Table 3.2 shows that the mean bond dissociation energies of transition metal-alkyl bonds are in the range of 125–260 kJ mol^{-1}, whereas those of metal-aryl bonds are somewhat higher and range from 250 to 350 kJ mol^{-1}.

† In this book the periodic group notation is in accord with recent recommendation by IUPAC and ACS committees. A and B notation is eliminated. Groups IA and IIA become groups 1 and 2. The d-transition elements comprise groups 3 through 12, and the p-block elements comprise groups 13 through 18. Note that the former Roman number designation is preserved in the last digit of the new numbering: e.g., III → 3 and 13.

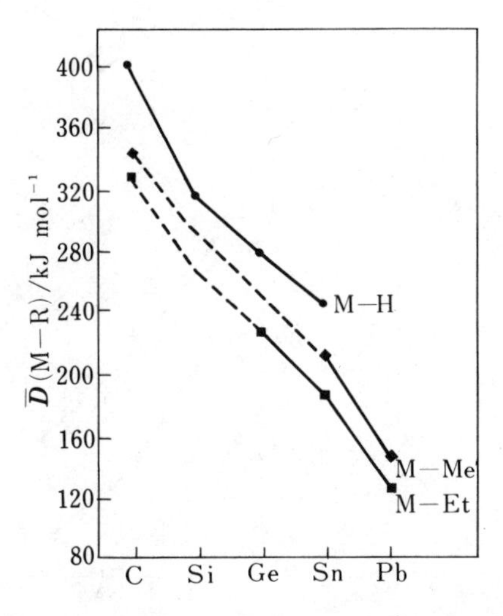

FIGURE 3.2. $\bar{D}$(M–R) values for group 14 metals. (From H. A. Skinner, *Adv. Organometal. Chem.*, **2**, 99 (1964), with permission.)

These values are not so different from those of nontransition metal alkyl or aryl compounds summarized in Table 3.1.

For some transition metals the mean bond dissociation energies for other types of bonds such as metal–halogen, metal–alkoxide, and metal–amide as

Table 3.2. $\bar{D}$(M–R) for Organotransition Metal Compounds[a]

Compound	Bond	$\bar{D}$	Reference
$Ti[CH_2C(CH_3)_3]_4$	$Ti\text{-}CH_2R$	170	7
$Ti(CH_2C_6H_5)_4$	$Ti\text{-}CH_2R$	240	7
$Ti[CH_2Si(CH_3)_3]_4$	$Ti\text{-}Ch_2R$	250	7
$Cp_2Ti(CH_3)_2$	$Ti\text{-}CH_3$	250	6
$Cp_2Ti(C_6H_5)_2$	$Ti\text{-}C_6H_5$	350	6
$(OC)_3MnCH_3$	$Mn\text{-}CH_3$	150	11
$(OC)_5ReCH_3$	$Re\text{-}CH_3$	220	10b
$Zr[CH_2C(CH_3)_3]_4$	$Zr\text{-}CH_2R$	220	7
$Zr(CH_2C_6H_5)_4$	$Zr\text{-}CH_2R$	380	7
$Zr[CH_2Si(CH_3)_3]_4$	$Zr\text{-}CH_2R$	225	7
$CpPt(CH_3)_3$	$Pt\text{-}CH_3$	165	6
$(Et_3P)_2Pt(C_6H_5)_2$	$Pt\text{-}C_6H_5$	250	6
$Ta(CH_3)_5$	$Ta\text{-}CH_3$	260	8
$W(CH_3)_6$	$W\text{-}CH_3$	160	8

[a] Bond energies in organic compounds (kJ mol^{-1}) from ref. 6.
$CH_3—CH_3$ ($C_{sp^3}—C_{sp^3}$), 370 — $H_5C_2—H$ ($C_{sp^3}—H$), 410
$C_6H_5—C_6H_5$ ($C_{sp^2}—C_{sp^2}$), 420 — $H_5C_6—H$ ($C_{sp^2}—H$), 430
$HC{\equiv}C—C{\equiv}CH$ ($C_{sp}—C_{sp}$), 630 — $HC{\equiv}C—H$($C_{sp}—H$), 525

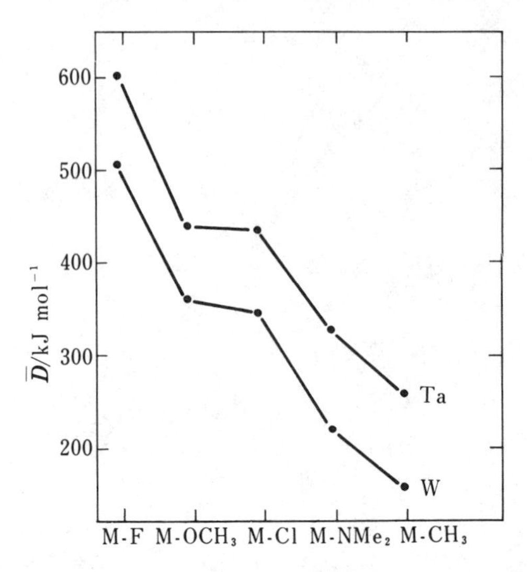

FIGURE 3.3. Mean bond dissociation energies for TaX_5 and WX_6 complexes (X = F, OCH_3, Cl, NMe_2, and CH_3). (From H. A. Skinner, *J. Chem. Thermodyn.*, **10**, 314 (1978), with permission.)

well as metal–alkyl have been measured. Figure 3.3 compares such $\bar{D}$ values for tantalum and tungsten compounds. The mean bond dissociation energies are seen to decrease in the order F > OMe > Cl > NMe_2 > Me.[8,9]

Exactly similar trends have been found for Ti, Zr, and Hf. If one assumes that this trend holds for other transition metal compounds, one can roughly estimate an approximate $\bar{D}$ value of a transition metal–alkyl compound, even when its value has not been measured, by extrapolation from the known $\bar{D}$ values. For example, the mean bond dissociation energy of V–F was determined as 468 kJ mol^{-1}, which is considerably smaller than the $\bar{D}$ value of 600 kJ mol^{-1} for the Ta–F bond. Since the mean bond dissociation energy of the Ta–Me bond was determined as 261 kJ mol^{-1}, the bond dissociation energy for the V–Me bond is expected to be smaller than that for the Ta–Me bond. Judging from the behavior of the compound $TaMe_5$, one may anticipate that VMe_5 may be thermally unstable and will decompose at room temperature. This may explain why VMe_5 has not yet been prepared. From similiar considerations, $MoMe_6$ is also expected to be thermally less stable than the tungsten congener.

Figure 3.4 shows the comparison of the mean bond dissociation energies of metal–methyl bonds in transition metal methyls and nontransition metal methyls. For nontransition elements the $\bar{D}$(M–CH_3) value decreases with an increase in the atomic number *Z*, whereas the $\bar{D}$ value in transition metal methyls increases with an increase in *Z*.

This apparent discrepancy is removed if one plots the $\bar{D}$ value against ΔH_f° (M, g), not against *Z*, as in Figure 3.5. It can be seen that $\bar{D}$ increases in both cases with an increase in ΔH_f° (M, g). The results indicate the existence

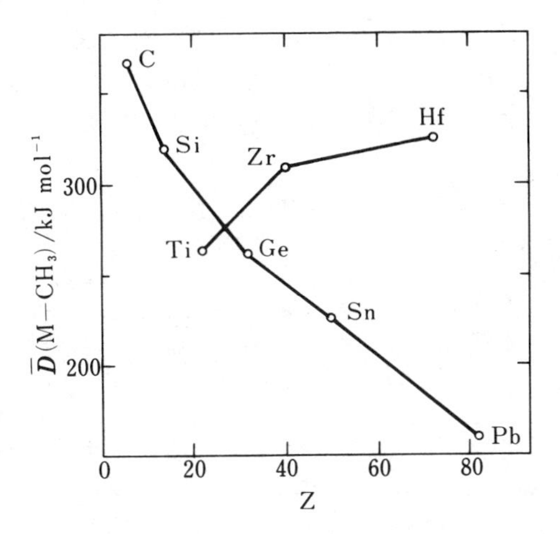

FIGURE 3.4. Plot of bond dissociation energy against atomic number *Z*. (From H. A. Skinner, *J. Chem. Thermodyn.*, **10**, 314 (1978), with permission.)

of a parallel relationship between the M–C bond strength and the M–M bond strength.

Results of thermochemical studies also indicate that there is not much difference between the bond energy of the M–CH_3 bond and the M–CO bond. For example, the D(W–CH_3) value for $W(CH_3)_6$ was found to be 159 ± 7 kJ mol^{-1} and is somewhat smaller than the $\bar{D}$(W–CO) value of 178 ± 2 kJ mol^{-1} for $W(CO)_6$. However, because the methyl groups are sterically bulkier than the CO ligand, the bond energy of the W–CH_3 bond is probably about the same as that of the W–CO bond.

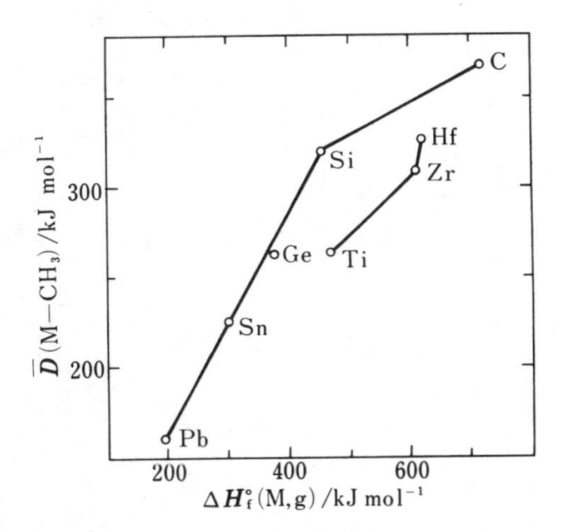

FIGURE 3.5. Plot of bond dissociation energy against enthalpy of formation (From H. A. Skinner, *J. Chem. Thermodyn.*, **10**, 314 (1978), with permission.)

Very recently bond dissociation energies for a series of organomanganese[11,15] and organocobalt[12-14,16] complexes have been measured. Table 3.3 shows Mn–C bond dissociation energies estimated from the following heats of reactions[11]:

$$[\mathrm{R{-}Mn(CO)_5}](g) \longrightarrow \mathrm{R}\cdot(g) + [\mathrm{Mn(CO)_5}](g) \tag{3.4}$$

$$D(\mathrm{R{-}Mn(CO)_5}) = \Delta H_f^\circ(\mathrm{R}, g) + \Delta H_f^\circ([\mathrm{Mn(CO)_5}], g) - \Delta H_f^\circ([(\mathrm{R{-}Mn(CO)_5}], g) \tag{3.5}$$

It is seen that the R–Mn bond dissociation energy decreases in the order H $> CF_3 \gtrsim C_6H_5 > CH_3 > CH_2C_6H_5$. Table 3.3 also shows that the bond dissociation energies of Mn–acyl bonds are somewhat smaller than the values for the corresponding alkyl or aryl complexes. The *D*(Mn–R) value for the trifluoromethyl compound, which has the strongest alkyl-metal bond, is only slightly smaller than that of the Mn–I bond, whereas the Mn–H bond strength exceeds that of the Mn–I bond.

Indirect data obtained from measurement of heats of reaction for oxidative additions of R–I and I_2 (for oxidative addition reaction see Section 6.2) indicate that the Ir–CH_3 bond dissociation energy is as large as the Ir–I bond dissociation energy.[17]

The *D*(Co–R) values for alkyl groups bonded to planar cobalt complexes having dimethylglyoximato and corrin (cf. Section 9.1) ligands[12-16] are in the range 72–100 kJ mol^{-1}. The M–C bond dissociation energy appears to be affected by electronic as well as by steric factors.

Although the available *D*(M–R) data for transition metal alkyls are still extremely limited, these data would allow one to make the following statements.

Table 3.3. Bond Dissociation Energies $D[R\text{–}Mn(CO)_5]$

Complex	$D[R\text{–}Mn(CO)_5]^a$ (kJ mol^{-1})
$[Mn(CO)_5CF_3]$	172 ± 7
$[Mn(CO_5C_6H_5]$	170 ± 11
$[Mn(CO)_5CH_2C_6H_5]$	87 ± 12
$[Mn(CO)_5CH_3]$	153 ± 5
$[Mn(CO)_5COCH_3]$	129 ± 12
$[Mn(CO)_5COC_6H_5]$	89 ± 10
$[Mn(CO)_5COCF_3]$	147 ± 11
$[Mn(CO)_5H]$	213 ± 10
$[Mn(CO)_5I]$	195 ± 6
$[Mn(CO)_5Br]$	242 ± 6
$[Mn(CO)_5Cl]$	294 ± 10
$[Mn_2(CO)_{10}]$	94

[a] Relative to $D[(CO)_5Mn\text{–}Mn(CO)_5] = 94$ kJ mol^{-1}

1. The M–C bond in transition metal alkyls is not as weak as was previously thought. There is no inherent instability attached to transition metal alkyls.

2. The D(M–R) value increases as the atomic number increases among the congeners in the same group in an opposite trend to nontransition metal alkyls.

3. The $D(M-CH_3)$ value is as strong as the D(M–I) value or slightly smaller than that.

4. The D(M–R) value decreases in the order $D(M-H) > D(M-CF_3) > D(M-C_6H_5) > D(M-CH_3) > D(M-C_2H_5) > D(M-CH_2C_6H_5)$.

As will be discussed in Section 6.5, these data suggest that the instability of transition metal alkyls is of kinetic origin. Thus, if one can block low-energy decomposition pathways such as β-elimination and reductive elimination by proper means, many transition metal alkyls may be isolated as thermally stable compounds. Conversely, even if a transition metal alkyl is not too prone to homolysis, it may be decomposed by some other pathways having low kinetic barriers.

3.2. BONDING SCHEME IN METAL ALKYLS

The bonding scheme for an M–R σ bond may be accounted for by sharing of electrons between the transition metal and the carbon atom as depicted in Figure 3.6.

Figure 3.6*a* shows an MO diagram for M–C σ bond formation between a metal fragment having n ligands L and an alkyl radical $R\cdot$, each contributing an electron to form a covalent bond. The figure is a reproduction of Figure 2.30*c*. A metal–alkyl bond of exactly the same nature and energy may be constructed as in Figure 3.6*b* by combination of the L_nM^+ entity with an anionic alkyl group R^-. In this case the carbanionic R group is considered to donate two electrons to the metal. The initial energy of R^- is higher than that of the $R\cdot$ radical in Figure 3.6*a* and the energy of L_nM^+ is lower than that of

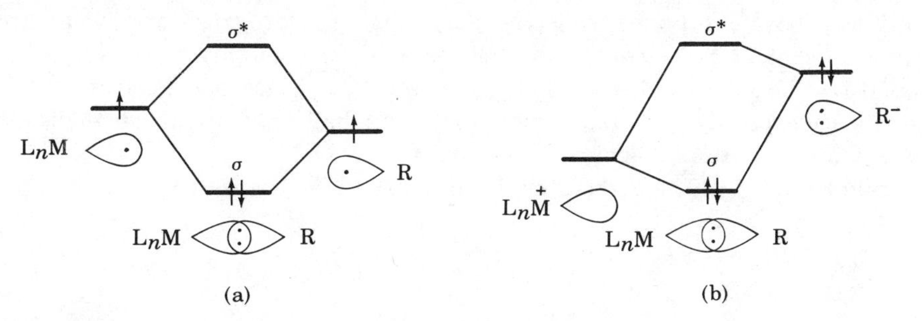

FIGURE 3.6. Formation of molecular orbitals (*a*) by interaction of L_nM fragment orbital and R orbital and (*b*) by interaction of L_nM^+ entity with R^- anion.

L_nM, but the energy of the resultant L_nMR σ and σ^* orbitals should be identical with those in Figure 3.6*a*. Figures 3.6*a,b* are MO descriptions of different formalisms.

In valence bond terms the metal–alkyl σ bond may be represented by two resonance forms,

$$M\text{-}R \longleftrightarrow M^+R^-$$

where the contribution of each canonical form depends on the nature of the alkyl group as well as that of the metal entity.

In addition to these simple metal alkyls, there are cases where the available electrons are insufficient for making normal electron pair covalent bonds, and the alkyl groups serve as bridging ligands between two or more metal atoms. The bridges may span transitional metal atoms **(3-1)**[18a] or atoms of a transition metal and a nontransition metal **(3-2)**.[18b]

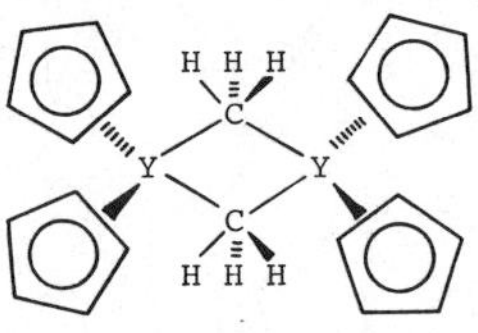

3-1 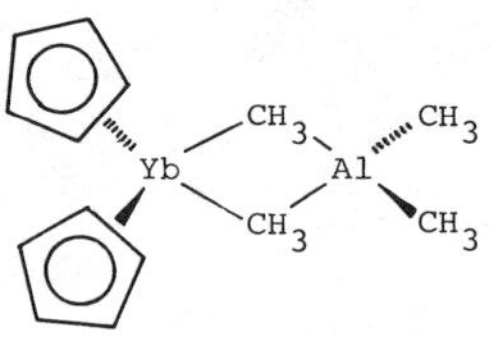 **3-2**

The bonding scheme in these complexes may be similar to that in nontransition metal compounds such as B_2H_6 and Al_2Me_6. Let us briefly examine these cases. As described in Section 2.3, the BH_3 molecule has one vacant 2p orbital and has a strong tendency to form the dimer **3-3**, shown below.

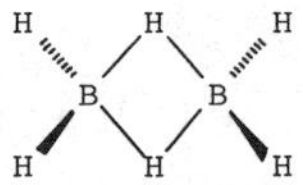

3-3

In this molecule there are eight adjacent pairs of atoms, but there are only six pairs of electrons available. Four terminal B–H bonds can be accounted for as ordinary covalent bonds, whereas the bridging hydrogens form what are called 3-center, 2-electron bonds, abbreviated 3c-2e bonds. In this type of bond two electrons bind one B–H–B set together. Since one electron pair is shared between three atoms, the B–H bond has only about one-half the strength of a normal 2-center, 2-electron bond.

Similarly, Al_2Me_6 has the dimeric structure shown in **3-4**.

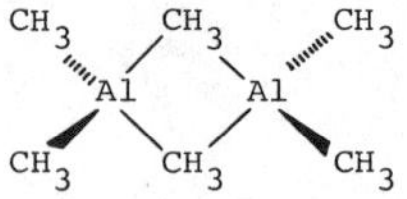

3-4

In this case the bridging CH_3 group contributes one electron for forming an Al–Me–Al bridge, utilizing its sp^3 orbital to form a similar 3c-2e bond. Again, the bond strength for each bridging Al–Me group is about half that for the terminal Al–Me bond. Thus, the bridging Al–C bonds are more susceptible to cleavage than the terminal Al–Me bonds.

The bonding scheme in transition metal analogs of these electron-deficient molecules may be interpreted similarly. In complexes such as **3-1** and **3-2** the methyl groups serve as the bridging ligands. The electron-deficient bonds may be formed between the transition metal and the nontransition metal, as illustrated by **3-2**. It should be noted that these complexes are different from the following complexes, which have bridging ligands but are not electron deficient.

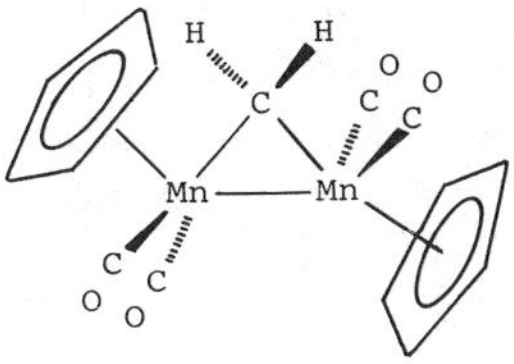

3-5 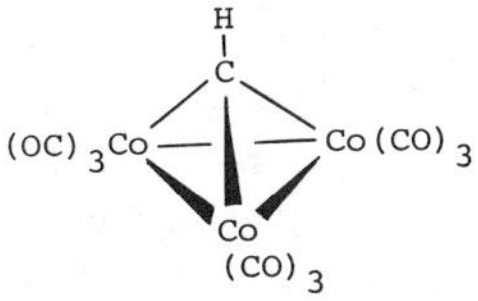 **3-6**

The electron-deficient bonds are also formed with two hydrogen atoms between a transition metal atom and a nontransition metal atom as in **3-7**.[19]

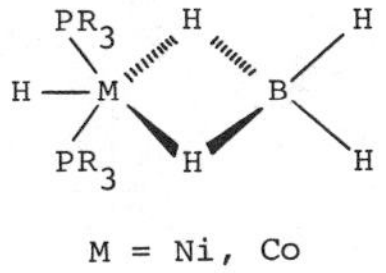

M = Ni, Co

3-7

As in borane compounds, hydrogen acts as a bridging ligand. There are various ways in which it does this, as shown in **3-8 – 3-12**.

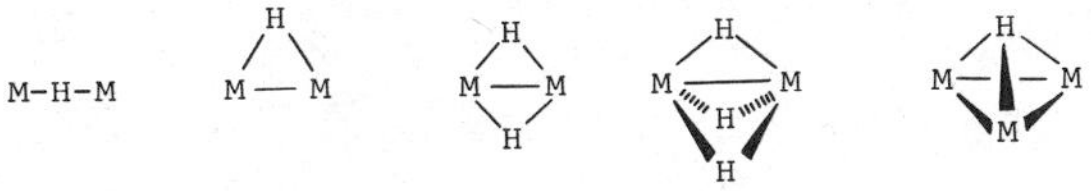

M–H–M

3-8 **3-9** **3-10** **3-11** **3-12**

Ate Complexes. Among the nontransition metal hydrides and alkyls there are anionic compounds, sometimes called ate complexes.[20a] Typical examples are $NaBH_4$, $LiAlH_4$, and $LiAlEt_4$. Analogous transition metalate complexes are known. Particularly useful for organic synthesis are the lithium alkyl cuprates, usually formulated as $LiCuR_2$. The precise structures of most cuprates have not been established, but for $LiCuMe_2$ the following dimeric structure has been proposed.[20b]

$$\begin{array}{ccc} CH_3 & — Li — & CH_3 \\ | & & | \\ Cu & & Cu \\ | & & | \\ CH_3 & — Li — & CH_3 \end{array}$$

Among group 8-10 transition metal complexes various alkyl and aryl ate complexes with the composition $Li_2[MR_4]$ are known.[20c]

3.3. COMPLEXES WITH METAL-CARBON π BONDS

The bonding situations in the Pt–C_2H_4 bond of Zeise's salt, in the Fe–C_5H_5 bonds of ferrocene, and in the metal–CO bond of metal carbonyls are unique to organotransition metal complexes. The interpretation of the bonding scheme in these cases provided a considerable challenge for theoretical chemists. These complexes are collectively called π complexes and show quite intriguing chemical properties that are usually not found among organometallic compounds of nontransition metals.

(1) *Metal Carbonyl Complexes*

Carbon monoxide has a nonbonding σ–electron pair localized on carbon available for donation to the metal. Its resonance forms are as shown below.

$$:\overset{-}{C}\equiv\overset{+}{O}: \longleftrightarrow :C=\ddot{O}$$

However, carbon monoxide is known to be a very poor electron donor to most Lewis acids, and for many years most inorganic chemists found it puzzling that metal atoms should form strong adducts with CO.

The metal–CO bonding may be accounted for by considering the following resonance forms:

$$\underset{A}{\overset{-}{M}\leftarrow C\equiv\overset{+}{O}:} \longleftrightarrow \underset{B}{\overset{-}{M}\leftarrow C=\overset{+}{\ddot{O}}} \longleftrightarrow \underset{C}{M\rightleftarrows C=\ddot{O}} \qquad (3.6)$$

The bonding may be more appropriately described in MO terms as follows. Figure 3.7*a* shows the formation of a σ bond by overlap of a metal d orbital with the filled carbon orbital. The electron flow from CO to the metal would lead to enhanced electron accumulation on the metal, which attempts to reduce the charge by pushing electrons back to the ligand. The metal d orbital has the right symmetry to overlap with an antibonding $2\pi^*$ orbital of the CO ligand, as shown in Figure 3.7*b* and forms a π bond.† By bond formation the electron density accumulated on the metal by σ donation from the unshared electron pair may be reduced by *back donation,* or *back bonding,* to the

† Interaction of the filled metal d orbital with one π orbital of CO is not negligible, albeit small.[21]

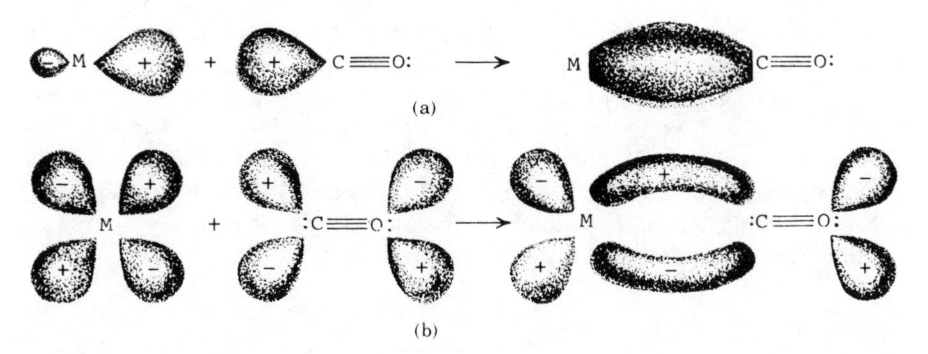

FIGURE 3.7. (*a*) Formation of C → metal σ bond. (*b*) Formation of M → C π bond by back bonding. (From F. A. Cotton and G. Wilkinson, *Basic Inorganic Chemistry*, Wiley, New York, 1976, with permission.) Other orbitals are omitted for clarity.

empty π* (and π) orbital of the CO ligand. The effect in the donation and back donation is synergic; in carbon-to-metal donation the CO becomes electropositive and susceptible to receiving electrons from the metal. At the same time the drift of metal electrons into the CO orbitals will make the CO ligand more negative and makes donation from the CO ligand to the metal orbitals easier.

Because of back bonding, the C≡O bond will be weakened and will approach double-bond character, whereas the metal–carbon bond will be strengthened, likewise approaching a double bond, which corresponds to the canonical form C in the valence bond expression [Eq. (3.6)]. Because the CO ligand acts as an electron acceptor from the filled metal d orbitals to the empty orbitals that have π symmetry with respect to the M–CO bond axis, it is called a π acid.

The multiplicity in the M–CO bond is reflected in the bond lengths and vibrational spectra. The bond length in the free CO molecule is 112.8 pm, whereas the reported bond length values in the metal carbonyls are somewhat lengthened, to 115 pm. On the other hand, considerable shortening in M–C bond distance is observed, reflecting the M–C double-bond character. The bond length of the Mn–CO bond in $MeMn(CO)_5$ was observed as 186.0 pm in comparison with the Mn–Me bond of 218.5 pm.[22]

Although the elongation of the CO bond distance on coordination is small, the IR absorption spectra of the metal carbonyl compounds reflect the back bonding much more sensitively. Thus, the CO bond stretching frequency of the free CO at 2143 cm^{-1} is shifted to lower frequencies in the range of 2125–1850 cm^{-1} in metal carbonyls. The CO frequency is further decreased when the metal carbonyl complex contains a ligand of stronger electron donor ability or when the metal complex takes an anionic character. On the other hand, if the carbonyl complex has a cationic character, an increase in the CO stretching frequency from that of the neutral complex is observed.

Besides mononuclear carbonyl complexes such as $Ni(CO)_4$, binuclear complexes such as $Co_2(CO)_8$ and $Fe_2(CO)_9$ and a variety of polynuclear complexes are known. In some of these polynuclear complexes (details to be described later) the CO ligand acts as a bridge connecting two or more metal atoms. Figure 3.8 shows two examples of such bonding modes. The bridging carbonyl is written as μ-CO. Metal carbonyls having μ-CO ligands show CO stretching frequency at lower values (1750–1850 cm^{-1}) than that of terminal carbonyl ligands. Thus, the bridging CO unit may be formally considered to resemble the ketonic group in organic compounds.

FIGURE 3.8. Two bonding modes of a bridging carbonyl ligand: (*a*) between two metal atoms and (*b*) among three metal atoms.

The bonding in Figure 3.8*b* where CO connects three metal atoms is a form of electron-deficient bond, where one orbital on each atom is directed toward the carbon lone pair orbital and overlaps with it to form a 4c–2e bond.

(2) *π-Bond Formation with Ligands Resembling Carbon Monoxide*

Isocyanides (RN≡C), dinitrogen (N_2), thiocarbonyl (CS), and nitric oxide (NO) somewhat resemble carbon monoxide in their behavior toward transition metals and form complexes containing these ligands.

Isocyanides generally appear to be stronger donors than CO and typically form complexes with metal cations such as $[Ag(CNR)_4]^+$ and $[Fe(CNR)_6]^{2+}$, but they are also capable of accepting electrons from low-valent metals to form complexes such as $Cr(CNR)_6$ and $Ni(CNR)_4$. The CN stretching frequency of the terminally bonded isocyanide ligand is usually lowered, reflecting this back bonding. The N–N stretching frequency of the coordinated dinitrogen is also lowered, as described in Section 4.6.

(3) *π Bonds Formed with Ligands of Group 15 Elements*

Compounds of trivalent phosphorus, arsenic, and antimony form complexes with transition metals. These compounds are able to act as electron donors and combine with Lewis acids such as BR_3. More importantly, these compounds have vacant d orbitals suitable for accepting electrons to form π bonds. Figure 3.9 illustrates the π bonding formed between a metal d orbital and an empty phosphorus d orbital.

Tertiary phosphines PR_3 and phosphites $P(OR)_3$ are particularly well utilized ligands and are known to stabilize low-valent transition metal complexes and transition metal alkyls.

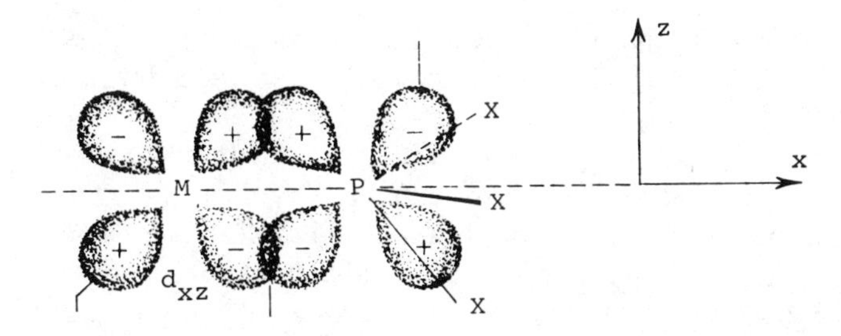

FIGURE 3.9. Bonding formed between a filled metal orbital and an empty phosphorus 3d orbital in the PX_3 ligand. The internuclear bond is taken as the x axis. (From F. A. Cotton and G. Wilkinson, *Basic Inorganic Chemistry,* Wiley, New York, 1976, with permission.)

(4) π Bonding with Ligands Having Delocalized Conjugate Systems

Ligands such as 2,2′-bipyridine(bipy) and 1,10-phenanthroline(phen) containing two or more nitrogen atoms are often used for syntheses of various transition metal complexes.

(bipy) (phen)

These ligands serve as both bidentate electron donors and electron acceptors through bonds formed between filled metal d orbitals and extended π^* orbitals of the ligands.

Among other ligands having delocalized π systems, 1,2-dithiolene may be mentioned.[23] With dithiolene ligands the later transition metals form the complexes shown below.

The ligand has a delocalized π system capable of accepting electrons by back bonding with transition metals.

(5) π Bonding with Olefins

The concept of π bonding was once described as a "panacea to cure the ills of coordination compounds."[24] The idea of synergism by electron giving and taking describes bonding with a variety of ligands, including CO, isocyanides, dinitrogen, bipyridine, and dithiolene, as we have seen. The concept works again for interpretation of the bonding between transition metals and unsaturated organic compounds,[25-29] the oldest example being Zeise's salt, which remained a puzzle for a long time to theoretical chemists trying to explain the bonding.

The first interpretation of the bonding was made by Dewar[30a] describing the silver–olefin bond and was later extended by Chatt and Duncanson[30b] to other transition metals such as platinum. The bonding consists of two components, as shown in Figure 3.10 for metal–ethylene bonding.

Figure 3.10*a* illustrates that the filled orbital of the coordinated ethylene overlaps with a σ-type acceptor orbital on the metal atom (note that the ethylene $\rightarrow$ metal dative bond has σ symmetry); in Figure 3.10*b* the π^* orbital on ethylene overlaps with a filled metal d orbital. As in the π bonding with CO, the d orbital of the metal, as shown by the plus and minus signs, has the right symmetry to overlap with the acceptor π^* orbital, forming a metal–olefin π bond. The result is a synergism that leads to relatively strong bonding. When the metal–olefin bond is formed, the electron density on the π orbital of the coordinated olefin is withdrawn by donation to the metal atom, but at the same time the electrons are returned to the antibonding π^* orbital of the olefin. Thus, the C=C bond of the coordinated olefin is weakened. The C=C bond length of 134 pm in free ethylene is generally lengthened to 139–148 pm on coordination to a transition metal. The C=C bond stretching frequency is likewise lowered by 140–160 cm^{-1} on olefin coordination. Furthermore, in the proton NMR spectrum a shift of the olefin protons toward a higher magnetic field is observed due to increased shielding of the protons caused by back bonding from the metal atom.

On coordination to a metal the chemical properties of the olefin are influenced depending on the extent of olefin-to-metal electron donation and metal-to-olefin back donation. When the olefin is coordinated to a charged metal

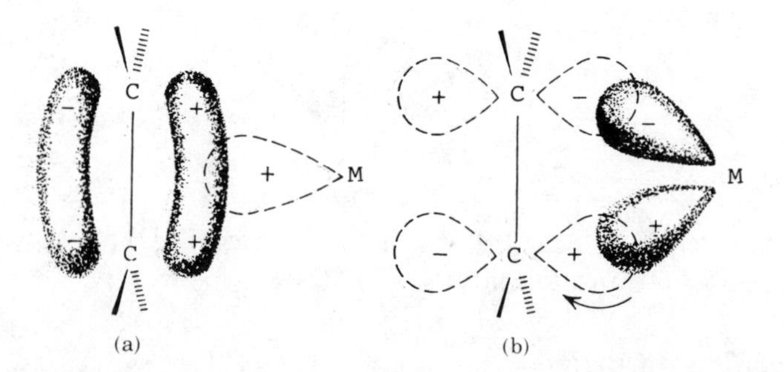

FIGURE 3.10. The molecular orbital views describing the bonding between a transition metal and an olefin. (*a*) Overlap of the olefin π orbital with an empty metal d orbital. (*b*) Overlap of a filled metal d orbital with the empty olefin π^* orbital (electron back donation from the metal to the olefin). (From F. A. Cotton and G. Wilkinson, *Basic Inorganic Chemistry,* Wiley, New York, 1976, with permission.)

ion and/or the metal atom is in a high positive oxidation state, as in Zeise's salt, there is an electron drift from olefin to metal, thus making the coordinated ethylene susceptible to attack by nucleophiles. In the Hoechst–Wacker process,[31] in which Pd(II) catalyzes the conversion of ethylene into acetaldehyde, the reaction is initiated by nucleophilic attack by OH^- on the ethylene coordinated to palladium (see Section 7.4). When the olefin is coordinated to a low-valent transition metal, the contribution of back bonding is more pronounced and the C=C bond of the coordinated olefin is lengthened. The effect is particularly evident when the coordinated olefin has electronegative substituents such as CN groups, and in the case of tetracyanoethylene (TCNE) metal complexes, complete metal-to-olefin charge transfer is observed.

Who Proposed the π-Bond Theory First?

The π-bond theory is commonly referred to as the Dewar–Chatt–Duncanson theory. However, Dewar is unhappy to include the names of Chatt and Duncanson and claims in his paper[32] that "the theory was not due in any way or any part to Chatt and Duncanson, in the interests of historical accuracy." Certainly it is true that Chatt and Duncanson developed their idea on the basis of Dewar's original theory. Furthermore, the *Journal of the Chemical Society,* where Chatt and Duncanson published their paper, is more readily accessible than the *Bulletin de la Societé Chimique de France,* which reported Dewar's idea. (Incidentally, his idea was presented in a question and answer format.)

Possibly a more important reason linking the names of Chatt and Duncanson to the π-bond theory was that they used it for elucidation of the Pt–C_2H_4 bond. The effect of the π bond on the modification of the nature of the coordinated ethylene was utilized to successfully account for the mechanism of an important industrial process, the Hoechst–Wacker oxidation of ethylene to acetaldehyde.[31]

Implications of the ethylene–metal π bond formation, however, were not exploited at the Imperial Chemical Industries, where Chatt and Duncanson worked, for development of the ethylene oxidation process, although ICI did later develop a commercial process converting ethylene into vinyl acetate, which was unfortunately abandoned after several years' operation because of corrosion trouble.

In coordination of the olefin to a transition metal complex, two coordination modes are possible: one is the coordination with the C=C bond perpendicular to the molecular plane and the other with the C=C bond in the molecular plane.

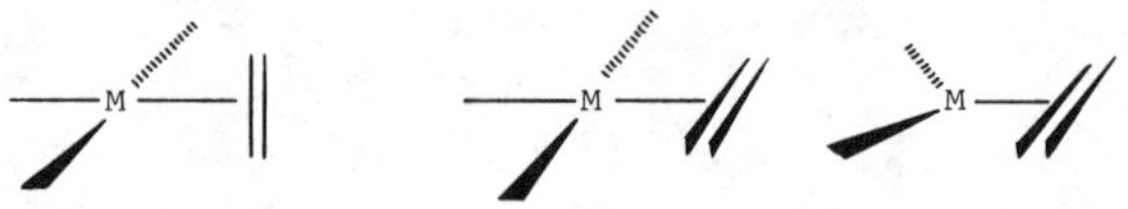

(a) perpendicular coordination (b) in-plane coordination

Which coordination mode is adopted is determined by the electronic and steric properties of the transition metal fragment and the coordinated olefin. In Zeise's salt, where ethylene is coordinated to d^8 Pt(II), perpendicular coordination is preferred, whereas in a d^{10} complex of Ni(0) such as $Ni(C_2H_4)_3$ in-plane coordination is favored. Rationales for the preferred coordination mode in these cases have been provided by semiempirical as well as ab initio MO calculations.[32-38]

In some cases the coordinated olefin can rotate about the metal–olefin bond axis. A barrier of about 60 kJ mol^{-1} has been calculated theoretically for changing from in-plane coordination to perpendicular coordination in d^8 and d^{10} olefin complexes,[21,36-38] this value being in reasonable agreement with the experimental value for the olefin rotation barrier.[39,40]

As the back bonding from the metal to the olefin increases, the C=C bond is weakened and approaches a single bond, as can be seen in tetracyanoethylene (TCNE) complexes. In addition, the originally planar olefin becomes nonplanar, with the substituents bending away from the metal atom and other ligands. In extreme cases the carbon atoms bound to the metal approach tetrahedral sp^3 hybridization, as established by the X-ray structural determination.[27] The bond can be thus regarded as a metal-containing cyclic compound called a metallacycle:

(3.7)

(a) side-on coordination (b) metallacycle

In going from side-on coordination to a metallacycle formation, the metal may be regarded as being oxidized by two units. However, as a convention, we regard the sideways bonded molecules as neutral ligands that do not alter the formal oxidation state of the metal, since the transition from (*a*) to (*b*) is a question of degree, with a smooth gradation from one description to another, and there is no stepwise change of the formal oxidation state.

(6) π Bonding with Acetylenes

Since acetylene has two mutually orthogonal orbitals, it is capable of bonding to two metal atoms through the π orbitals.[41,42] Figure 3.11 shows the molecu-

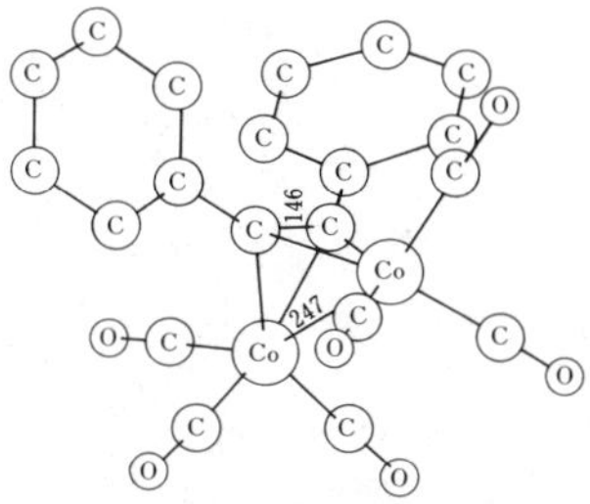

3-13

FIGURE 3.11. The molecular structure of (diphenylacetylene)hexacarbonyldicobalt. Numbers are in picometers. (Reprinted with permission from W. G. Sly, *J. Am. Chem. Soc.*, **81**, 18 (1959). Copyright 1976 American Chemical Society.)

lar structure of an acetylene complex, $(C_6H_5C{\equiv}CC_6H_5)Co_2(CO)_6$ **(3-13)**, where diphenylacetylene is bound to two cobalt atoms.[43] The two carbon atoms of the acetylene molecule have a distorted tetrahedral configuration, with the phenyl rings bending away from the cobalt atoms.

The acetylene molecule need not always coordinate with two metal atoms and can bond to a single metal atom using only one pair of π electrons in the same way as do olefins. Figure 3.12 shows the molecular structure of a diphenylacetylene complex of this type **(3-14)**.[44]

In this complex two phosphorus atoms of the triphenylphosphine ligands, the platinum atom, and two carbon atoms of the acetylene molecule are in a plane with the two phenyl groups bending away from platinum. The C≡C bond length is lengthened from 119 pm of the original, uncoordinated diphenylacetylene to 132 pm, which is close to the normal C=C double bond. The C≡C stretching frequency is also greatly lowered, to 1750 cm^{-1}, from that of free diphenylacetylene, at 2223 cm^{-1}. For acetylene complexes of earlier transition metal complexes, however, acetylene–metal interaction utilizing the two orthogonal sets of π and π* orbitals involving four-electron donation has been noted.[21,42]

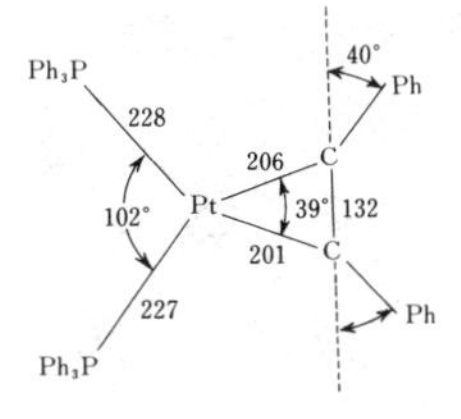

3-14

FIGURE 3.12. Molecular structure of bis(triphenylphosphine)diphenylacetyleneplatinum(0). (From J. O. Glanville et al., *J. Organometal. Chem.*, **7**, 7 (1967), with permission.)

(7) π-Allyl Bonding

An allyl group, $CH_2{=}CH{-}CH_2{-}$, can form a σ bond with a transition metal, as in (*a*) below.

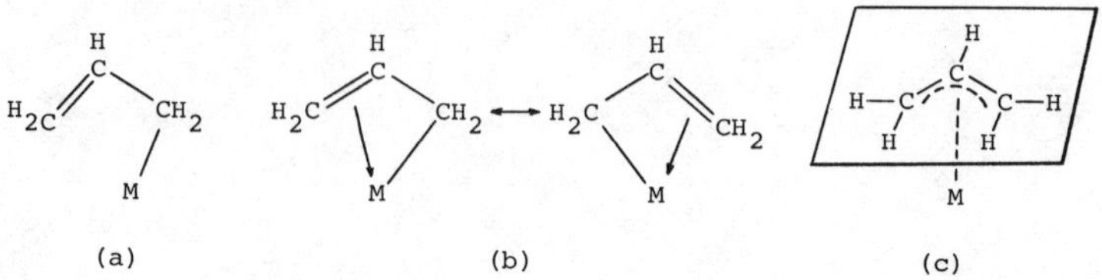

(a) (b) (c)

However, the remaining double bond of the allyl group is still capable of forming a further π bond with the metal. If such a π bond is formed in addition to the σ bond between the metal and the allyl group, the resultant π-allyl–metal bonding can be expressed in resonance forms as in (b) or in MO terms as in (c) as interactions between the transition metal orbitals and the orbitals of the planar allyl group where the electrons are completely delocalized over three carbon atoms.[45]

For example, bis(π-(2-methyl)allyl)nickel (**3-15**), prepared by the reaction of nickel dichloride and methylallylmagnesium bromide, has a symmetrical structure with the central nickel atom attached to two planar methylallyl groups, the distances from nickel to the terminal CH_2 groups being equal (Fig. 3.13).

The bonding between the π-allylic groups and the metal can be expressed in a manner similar to that applied for interpreting metal–olefin bonding. Figure 3.14 illustrates sets of the π-allyl group orbitals and the corresponding metal orbitals, which are able to overlap most effectively with the ligand group orbitals.

The three pπ orbitals of the allyl group combine to form three MOs (ligand group orbitals). ψ_1 (bonding), ψ_2 (nonbonding), and ψ_3 (antibonding). In Figure 3.14 the metal orbitals, which can overlap with these ligand group orbitals, are shown, respectively, below the ψ_1, ψ_2, and ψ_3 orbitals. The ψ_1 orbital overlaps with the metal s, d_{z^2} and p_z orbitals or hybrid orbitals constructed

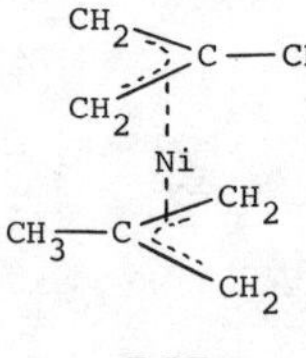

3-15

FIGURE 3.13. Molecular structure of bis(2-methylallyl)nickel. Methyl groups substituted on the allyl group are slightly deviated out of the plane of the allyl groups.

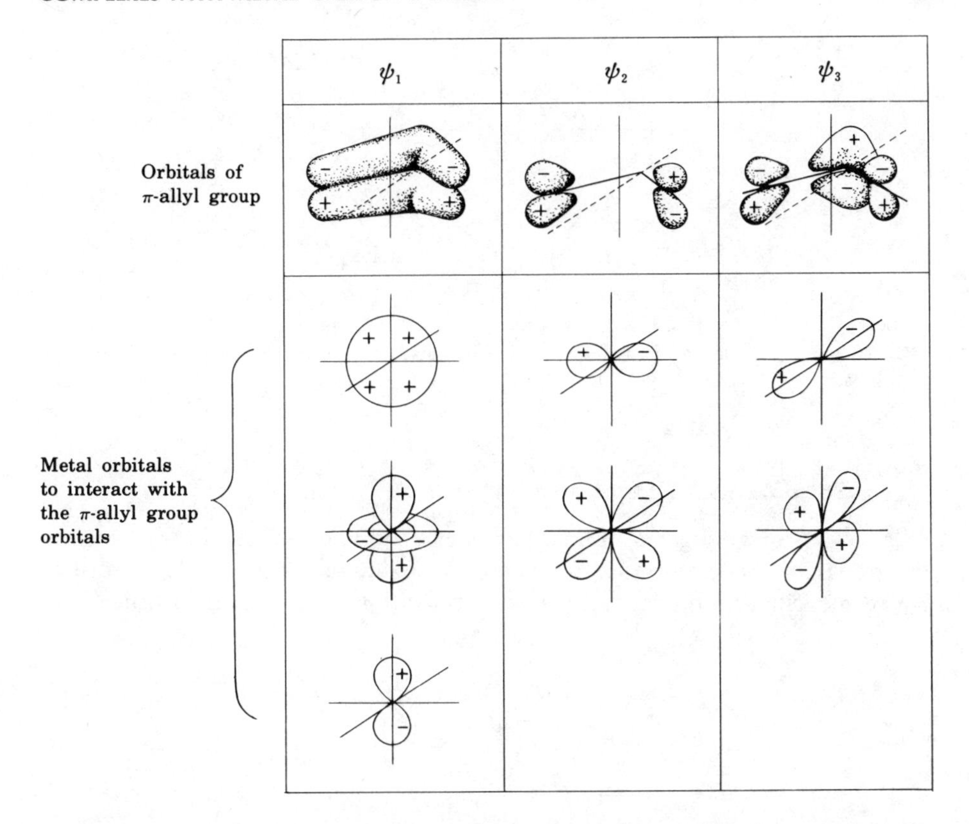

FIGURE 3.14. π-Allyl group orbitals ψ_1, ψ_2, and ψ_3 and the corresponding metal orbitals, which overlap most effectively with ligand group orbitals. The metal lies below the allyl group along the vertical axis. (Adapted from M. L. H. Green, *Organometallic Compounds,* Vol. 2, *The Transition Elements,* Methuen, London, 1968, p. 45.)

from them. Similarly, the nonbonding ψ_2 orbital is suitable for overlap with the metal p_y and d_{yz} orbitals or with their hybrid orbitals, whereas the antibonding orbital ψ_3 has a right symmetry to overlap with the p_x and d_{xz} orbitals or their hybrids to form metal-to-ligand back bonding.

Some allylic groups form unsymmetrically substituted π-allylic complexes. For example, the reaction of butadiene-1,3 with hydridotetra(carbonyl)cobalt gives a mixture of isomeric methylallyl cobalt complexes, as shown in Figure 3.15.

The methylallyl cobalt complexes **3-16a** and **3-16b** are called *syn* and *anti* forms, respectively. The one that has the methyl group on the same side of the methine proton is called the syn form and the one with the methyl group situated on the other side of the methine proton is called the anti form. Regarding the mutual relationship between the methyl group and the terminal methylene group, the syn form corresponds to a trans form and the anti form to a cis form.

Syn form **3-16a** Anti form **3-16b**

FIGURE 3.15. Formation of two types of π-allyl cobalt complexes on reaction of $CoH(CO)_4$ with butadiene.

(8) π Bonding with Dienes and Polyenes

Dienes with unconjugated double bonds can form complexes through respective double bonds with transition metals.[46,47] Examples of π complexes formed between cyclic unconjugated dienes and transition metals are shown below.

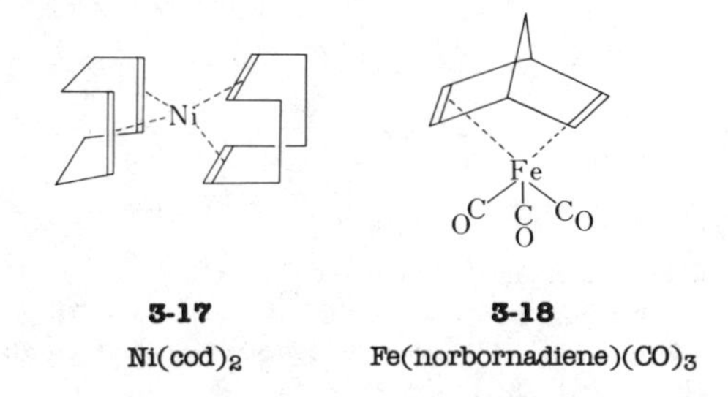

3-17 Ni(cod)$_2$ **3-18** Fe(norbornadiene)$(CO)_3$

Ni(cod)$_2$ (**3-17**) is a relatively stable Ni(0) complex, having only the cyclooctadiene ligands, and provides a convenient starting material for the preparation of various zero-valent nickel complexes.

The bonding between transition metal atoms and conjugated dienes may be accounted for in a manner similar to the bonding in π-allyl complexes. Butadiene forms a π complex of composition $Fe(CO)_3(C_4H_6)$ (**3-19**) whose molecular structure is shown in Figure 3.16*a*. The butadiene ligand in **3-19** has a cisoid form (s-cis form), and the four carbon atoms are approximately equidistant from iron (210 pm). The molecular orbitals of the butadiene molecule, ψ_1, ψ_2, ψ_3, and ψ_4, and their corresponding metal orbitals are shown in Figure 3.16*b*. The ψ_1 and ψ_2 orbitals are each occupied by two electrons, whereas ψ_3 and ψ_4 MOs are unoccupied. It is seen that the HOMO ψ_2 has the right symmetry to overlap with p_x and d_{xz} orbitals, whereas the LUMO ψ_3 can interact with p_y and d_{yz} orbitals, which are available for metal-to-ligand back bonding.

Figure 3.17 shows another example of a butadiene-coordinated complex, (butadiene)(1-methylallyl)(triphenylphosphine)cobalt. It has been shown that

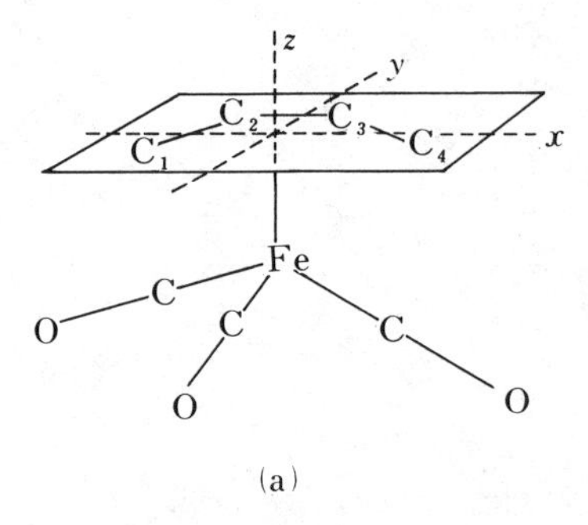

Molecular structure of (butadiene) tricarbonyliron(0) **(3-19)**

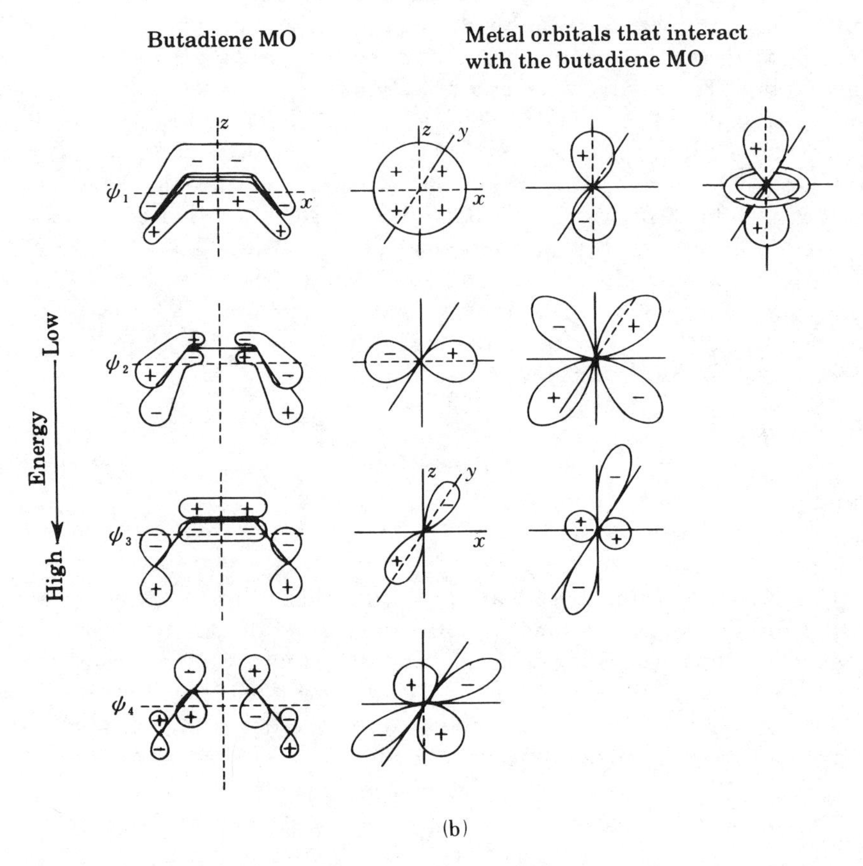

FIGURE 3.16. (*a*) Molecular structure of (butadiene)tricarbonyliron(0) and (*b*) molecular orbitals constituting the diene–iron bonding. (Adapted from M. L. H. Green, *Organometallic Compounds,* Vol. 2, *The Transition Elements,* Methuen, London, 1968.)

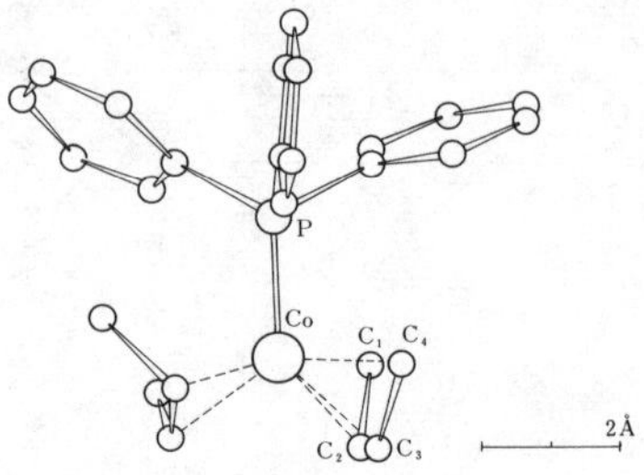

FIGURE 3.17. Molecular structure of a cobalt complex containing triphenylphosphine, butadiene and *anti*-1-methylallyl ligands. Bond distances are in picometers; C_1–C_2 146.7 (31), C_2–C_3 140.5 (24), and C_3–C_4 146.2 (27). [From L. Porri et al. *J. Chem. Soc. Chem. Commun.*, 276 (1969).]

the C_1–C_2 and C_3–C_4 bond distances in the coordinated butadiene ligand are longer than the internal C_2–C_3 bond distance; this is in contrast to the ground-state bond distances of butadiene, which has a short–long–short bond order alternation. This may be taken as a reflection of the enhanced contribution of the LUMO ψ_3 by back bonding. On the other hand, complex **3-19**, in which butadiene is coordinated to the $Fe(CO)_3$ fragment, has almost equal bond distances of 145 pm for C_1–C_2 and 146 pm for C_2–C_3. This may be caused by weaker back bonding from the $Fe(CO)_3$ entity, which is less electron releasing due to the ligation of three strong π-acid CO ligands.

With an increase of the metal-to-butadiene back donation, the C–C bond order alternation approaches the long–short–long pattern. This corresponds to one form of the resonance structure with two σ-Fe–CH_2 bonds and one π-Fe–olefin bond shown below.

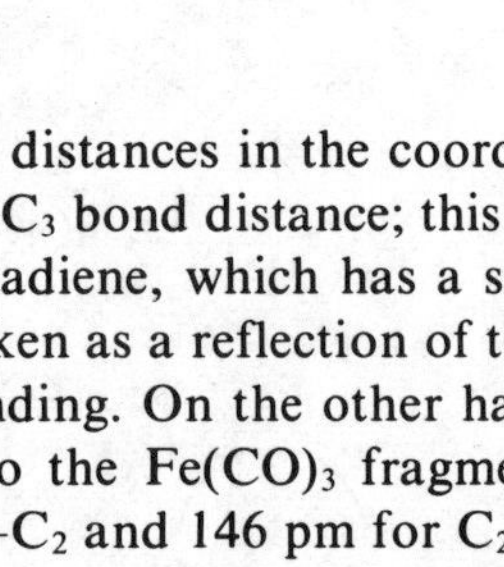

This situation is similar to that discussed earlier for metal–ethylene bonding, where extensive back bonding to the ethylene π^* orbital in the MO model leads to a metallacyclopropane structure. NMR studies provide some support for the presence of the σ-bonded species for some butadiene zirconium complexes in solution.[48,49] X-ray crystallographic analysis of $Cp_2Zr(C_4H_6)$ also established the involvement of such a bonding.[49]

Intermezzo — *A Prophecy Proving the Utility of MO Theory*

A prophecy made at the right time makes a strong impression when it is quickly fulfilled. Dimitri Mendeleev made a bold prophecy in 1871 about the properties of the new elements ekaaluminum and ekasilicon (as he named them) according to the periodic law. The discovery and confirmation of the predicted properties of the new element gallium (ekaaluminum) in 1875 by Boisbaudran led to the almost immediate acceptance of the periodic law. Prophecy of a stable cyclobutadiene–transition metal complex provides another such example.

Cyclobutadiene is a very unstable compound, having a very short life time at room temperature, but its instability was difficult to explain in terms of the valence bond theory. In 1956 Longuet-Higgins and Orgel proposed, on the basis of MO considerations, that a square planar cyclobutadiene molecule would be stabilized when it is attached to a transition metal such as Ni, Pd and Pt.[50] The molecular orbitals of cyclobutadiene are illustrated in Figure 3.18*a*. The nonbonding ψ_2 and ψ_3 orbitals are degenerate and have radical character, so this molecule will be extremely reactive and ephemeral. The ψ_2 and ψ_3 orbitals, however, can overlap with the corresponding metal orbitals B and C, as in Figure 3.18*b*, to form two covalent bonds, resulting in a stable complex.

The prophecy was realized three years later, in 1959, when Criegee isolated the first π-tetramethylcyclobutadiene complex **3-20** by the following reaction,[51] thus proving the remarkable usefulness of molecular orbital theory.

3-20

Cyclobutadiene complexes can be prepared by other routes, for example, by the reaction of diphenylacetylene and metal carbonyls.[52,53]

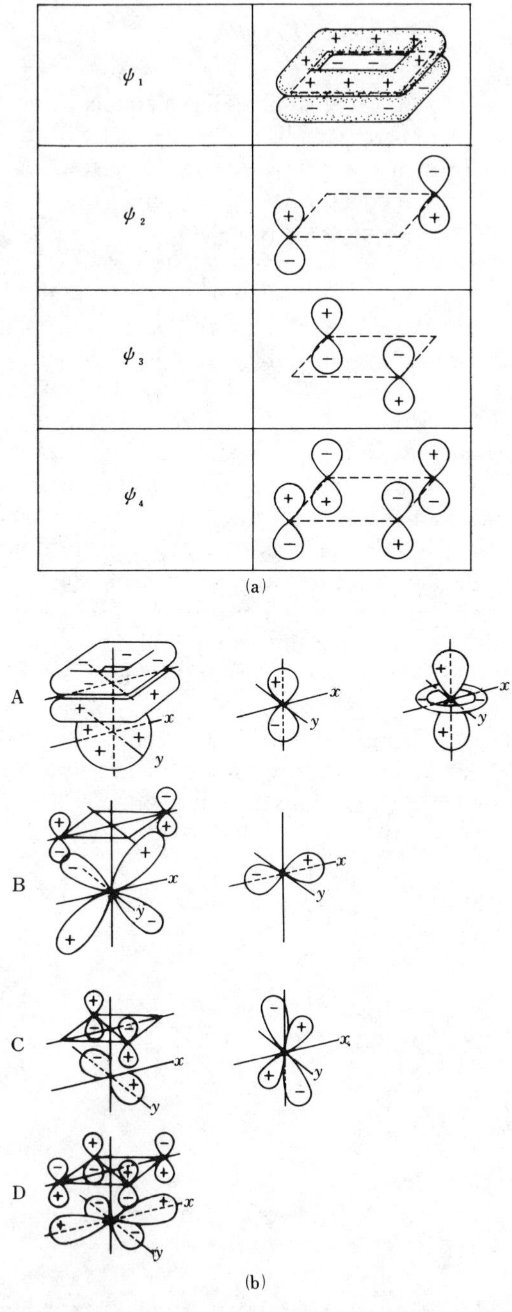

FIGURE 3.18. Molecular orbitals of a metal–cyclobutadiene complex. (*a*) Molecular orbitals of cyclobutadiene. (*b*) The corresponding metal orbitals A–D to interact with ψ_1–ψ_4 orbitals of cyclobutadiene in (*a*). (Adapted from M. L. H. Green, *Organometallic Compounds,* Vol. 2, *The Transition Elements,* Methuen, London, 1968.)

$$Fe(CO)_5 + 2\,PhC{\equiv}CPh \longrightarrow (Ph_4C_4)Fe(CO)_3 \qquad (3.8)$$

In this reaction two diphenylacetylene molecules are considered first to coordinate with the iron atom and then to form cyclobutadiene by rearrangement. The stability of the cyclobutadiene–iron complex probably serves to drive the ring-forming reaction.

(9) Bonding in Metal Cyclopentadienyl Complexes

The most typical bonding pattern of the cyclopentadienyl-type complexes can be seen in ferrocene and its related sandwich-type complexes. In addition to these bis-cyclopentadienyl-type complexes, many other complexes having cyclopentadienyl or substituted cyclopentadienyl ligands are known. They include mono-cyclopentadienyl complexes of half-sandwich type. The first of these cyclopentadienyl complexes, ferrocene, was found to have surprising stability, and it attracted the attention of theoretical chemists who tried to explain the nature of the bonding between the cyclopentadienyl rings and the iron atom. The molecular orbital theory proved to be the most useful in elucidation of the bonding,[21,54–58] whereas the conventional valence bond approach revealed its shortcomings in dealing with this type of complex.

The molecular orbitals of ferrocene can be considered either as a combination of the atomic orbitals of Fe(0) with ligand group orbitals (LGO) of the two cyclopentadienyl radicals $C_5H_5\cdot$ or as a combination of Fe(II) atomic orbitals with LGOs of cyclopentadienide anions $C_5H_5^-$. Both approaches lead to the same conclusion, but we shall give a brief explanation of the ferrocene MOs as a combination of Fe(II) and two $C_5H_5^-$ ligand orbitals.

Each carbon in the cyclopentadienyl ring $C_5H_5^-$ of a regular pentagon has a p_z orbital perpendicular to the plane of the cyclopentadienyl (Cp) ring. The five p_z orbitals of each C_5H_5 ring form five delocalized ligand group orbitals, thus 2 C_5H_5 rings carry as many as 10 such LGOs.

The lowest-energy LGOs are orbitals having doughnut-shaped electron clouds above and below the two pentagons. Above these orbitals are doubly degenerate e_1 orbitals, each having a nodal plane passing through the main axis of the pentagon. The highest-energy sets are e_2 orbitals having two nodal planes. When six electrons in each $C_5H_5^-$ unit are put into these LGOs (altogether 12 electrons in two Cp LGOs), the lowest a_1 and a_2 orbitals and the next lowest, two degenerate e_1 orbitals, are occupied, leaving the highest e_2 orbitals unoccupied. Figure 3.19 shows the cyclopentadienyl LGOs on the left side and the iron atomic orbitals of matching symmetries on the right side. The molecular orbitals of ferrocene constructed from the cyclopentadienyl

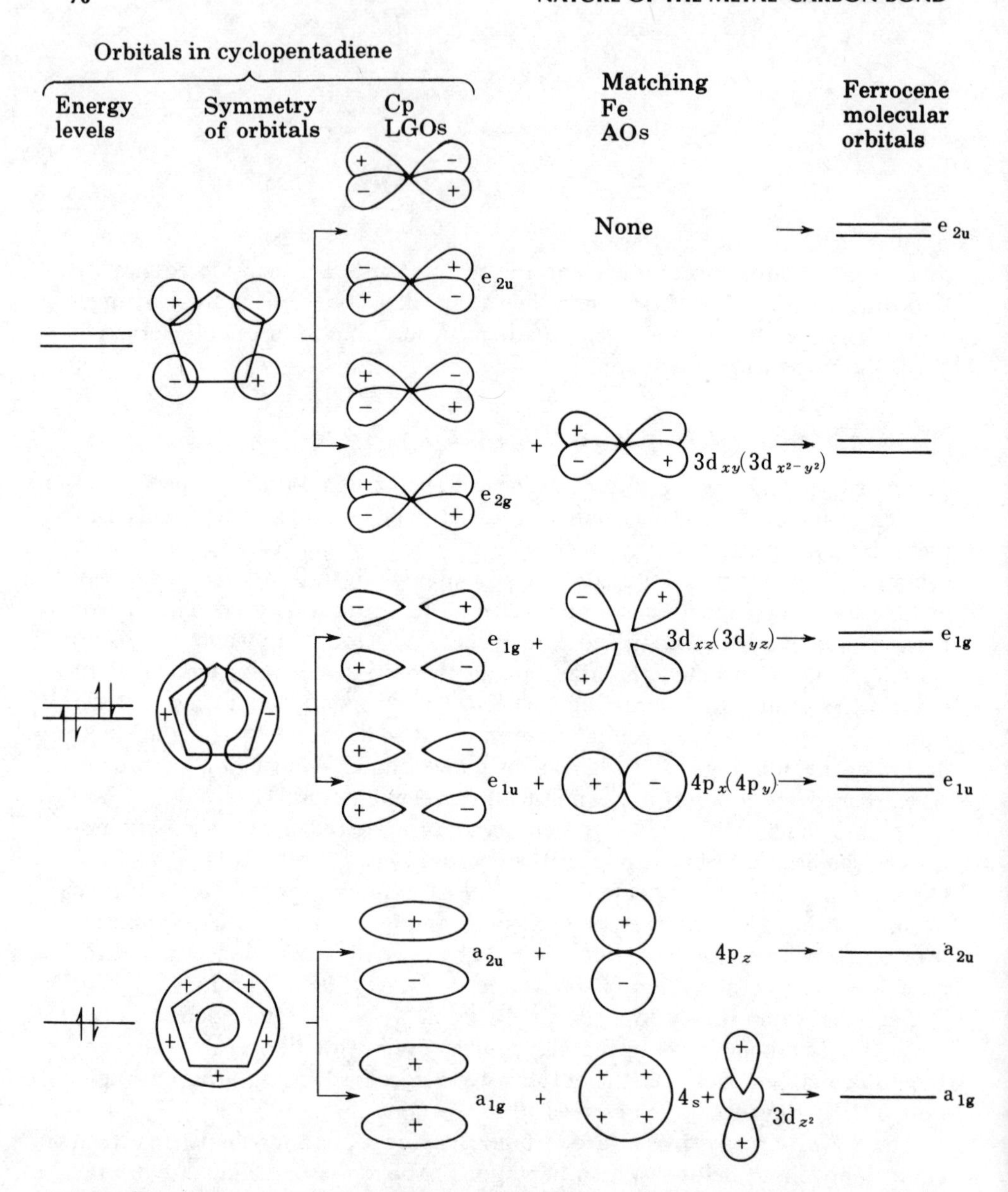

FIGURE 3.19. The cyclopentadienyl ligand orbitals and the iron AOs of the matching symmetry to form the ferrocene MOs. (From J. E. Huheey, *Inorganic Chemistry. Principles of Structures and Reactivity,* 3rd ed. Copyright © 1983 by James E. Huheey. Reprinted by permission of Harper & Row, Publishers, Inc.)

LGOs and iron AOs are shown on the extreme right in the order of decreasing energies.

In constructing the ferrocene MOs from the two Cp LGOs and iron AOs, two combinations are generated for the D_{5d} point group: one is centrosymmetric (with the symbol g), and the other antisymmetric (u), those combinations depending on the phase of the π orbitals of the cyclopentadienyl rings. The a_{1g} combination is the one with both positive (or both negative) signs on the sides of the Cp rings interacting with the iron AOs, whereas the a_{2u} is the combination of Cp π orbitals with different signs on the sides attached to the iron atom. The iron AOs of the right symmetry a_{1g} are 4s and $3d_{z^2}$ orbitals, whereas the iron AO with a_{2u} symmetry is the p_z orbital. The lowest-energy set of the ferrocene orbitals having a_{1g} and a_{2u} symmetry are thus formed. Similarly formed are the e_{1u} and e_{1g} ferrocene MOs. The unoccupied e_{2g} orbitals of the Cp rings, on the other hand, have the right symmetry to interact with the occupied iron $3d_{xy}$ or $3d_{x^2-y^2}$ orbital, the situation being similar to the back bonding in the olefin–metal complexes. There is some disagreement with the exact ordering of the ferrocene MOs, but the general pattern is widely accepted.

Figure 3.20 shows the energy level diagrams for metallocenes computed by Lauher and Hoffmann.[55] In the center of this energy diagram is shown how the new metallocene MOs are formed by interaction of the cyclopentadienyl e_{1g} and e_{2g} orbitals with the metal d orbitals. For ferrocene, formally regarded in

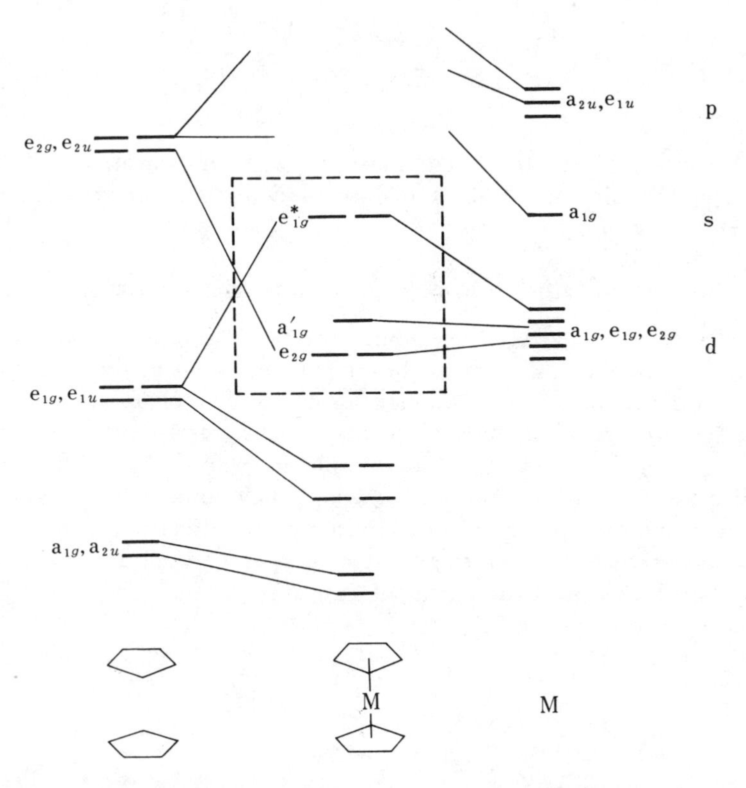

FIGURE 3.20. A qualitative energy level diagram of metallocene MOs. The metal d electrons are accommodated in the box of broken lines. (Reprinted with permission from J. W. Lauher and R. Hoffmann, *J. Am. Chem. Soc.*, **98,** 1729. Copyright 1976 American Chemical Society.)

the Fe(II) oxidation state as having six d electrons, the low-lying e_{2g} and a'_{1g} orbitals are occupied by the six d electrons. Because of the presence of the energy gap between the a'_{1g} orbital and the high-lying e_{1g}^* orbitals, all six electrons will enter the low-lying a'_{1g} and e_{2g} orbitals, leaving the e_{1g}^* orbital unoccupied and giving a diamagnetic complex. For nickelocene, however, which has eight d electrons in the Ni(II) state, two electrons still remain after the six electrons occupy the low-lying a'_{1g} and e_{2g} orbitals. These two electrons will enter the high-lying, degenerate e_{1g}^* orbitals, one electron occupying each orbital; this accounts for the fact that nickelocene has two unpaired electrons. Since the electrons in the high-lying e_{1g}^* orbitals tend to be easily removed, nickelocene is readily oxidized. Vanadocene (d^3) and chromocene (d^4), on the other hand, are electron deficient and tend to interact with additional ligands that can contribute more electrons. For example, $V(C_5H_5)_2$, reacts with CO to give $V(C_5H_5)_2(CO)$ (**3-21**), in which the Cp rings are bent back from their parallel positions to accommodate the CO ligand.

Cp_2V + CO ⟶ V–CO

3-21

Other bis-cyclopentadienyl complexes of earlier transition metal complexes take structures similar to **3-21** and show individual characteristics depending on the number of d electrons.

(10) Bonding of Other Carbocyclic Ligands with Transition Metals

Arenes such as benzene form similar sandwich complexes with transition metals. The first of this kind to be prepared was bis(benzene)chromium(0). The bonding pattern in the complex formed by combination of the benzene orbitals and the d^6 Cr(0) AOs is similar to that for ferrocene.

The tropylium ligand $C_7H_7^+$ and dianionic cyclobutadienyl ligand $C_4H_4^{2-}$ are also able to donate 6 π electrons, and such complexes are known. The cyclopropenium cation also forms π complexes donating 2 electrons. The cyclooctatetraene dianion forms complexes by donating up to 10 π electrons. Complexes **3-22 – 3-26** are an array of "mixed sandwich" complexes containing various carbocyclic ligands.[59]

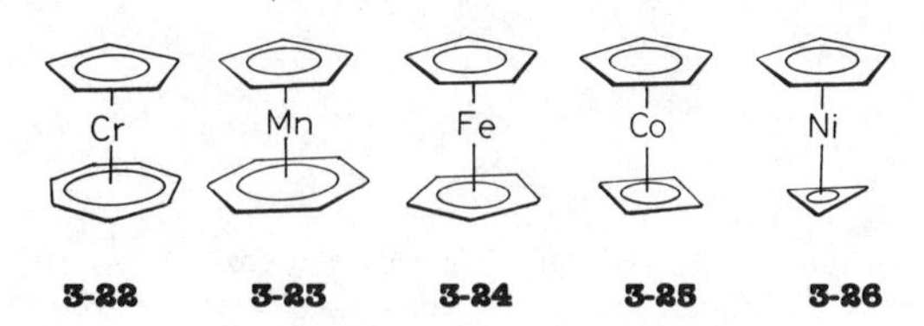

(11) Bonding in Carbene and Carbyne Complexes

Carbenes ($:CR_2$) are short-lived divalent carbon compounds that are considered to play important roles in many reactions. As in the case of the cyclobutadiene ligand, carbenes are known to be stabilized by forming complexes with transition metals. Numerous carbene–metal complexes are now known.[60,61] Some of these are important in connection with catalytic olefin metathesis reactions, as will be discussed in Section 7.5.

Carbene, a molecule having a lone pair of electrons, may be regarded as isoelectronic with carbon monoxide, and the bonding between the transition metal and the terminal carbon of the carbene ligand has some similarity with that in metal carbonyls.

Figure 3.21 shows the molecular structure of a tantalum complex bonded with the simplest carbene (alkylidene) ligand, CH_2.[62] Note that the sp^2 CH_2 plane is perpendicular to the plane containing the Ta atom and the carbon atoms of the CH_3 and CH_2 groups, and the Ta–CH_2 bond is considerably shorter than the Ta–CH_3 bond. Thus, the bond between the CH_2 entity and the Ta atom is considered to have double-bond character. The barrier for rotation of the CH_2 group around the Ta–C bond was found to be about 100 kJ mol^{-1}. The value may be taken as an approximate bond strength of the metal–alkylidene π bond. Since the bond strength of the Ta–C single bond is in the range of 160–240 kJ mol^{-1}, the bond strength of the metal–alkylidene bond (the ΔH value for dissociation of the carbene ligand from the metal) is estimated to be on the order of 300 kJ mol^{-1}.

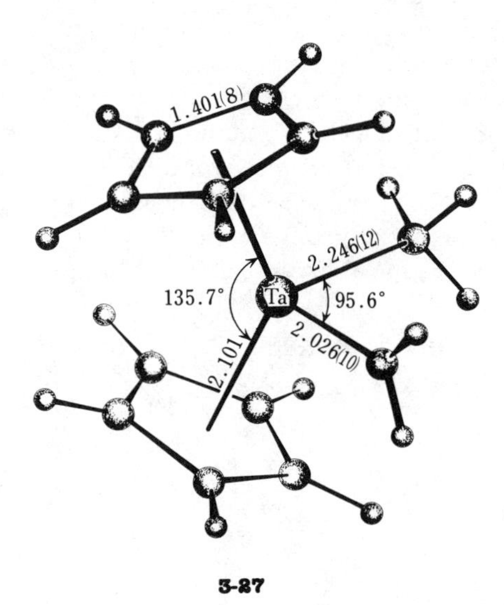

FIGURE 3.21. Molecular structure of a carbene complex $Cp_2Ta(CH_3)(CH_2)$. Numbers in the figure are in angstroms. (Reprinted with permission from L. H. Guggenberger and R. R. Schrock, *J. Am. Chem. Soc.*, **97,** 6578. Copyright 1975 American Chemical Society.)

Simple alkylidene complexes having only a hydrocarbon entity bound to a transition metal atom are sometimes called Schrock-type complexes, after their discoverer.

On the other hand, a variety of metal–carbene complexes have a heteroatom substituent attached to the carbene carbon, as shown below.

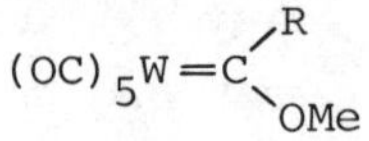

$(OC)_5W{=}C(R)(OMe)$

3-28

$(OC)_5Mo{=}C$ (cyclic, N–R, N–R)

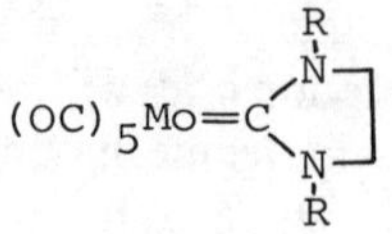

3-29

These carbene complexes were first prepared by Fischer and co-workers, before the Schrock-type complexes had been discovered, and these carbene complexes are sometimes called Fischer-type complexes.[61]

Complexes with a carbyne ligand, such as **3-30**,[63] (Fig. 3.22) are also known.[64] In this case the metal–carbon bond length is quite short, suggesting a metal–carbon triple bond. In forming the metal–carbyne bond an sp hybridized carbyne carbon is considered to donate a pair of electrons to the metal, and the metal may in turn back-donate electrons from two orthogonal π-type orbitals (e.g., d_{xz} and d_{yz}).

(12) *Complexes with Metal–*Metal Multiple Bonds

Until recently, examples of complexes with multiple metal–metal bonds were rare. We now have quite an array of transition metal complexes containing metal–metal double bonds, triple bonds, and even quadruple bonds.[65]

Figure 3.23 shows five possible overlaps between the sets of d orbitals on two metal atoms. The relative values of the overlaps are such that the σ interaction should be very strong, the π interactions of intermediate strength, and the δ interaction rather weak. Recent theoretical calculations provide support for the bonding scheme. Complexes with quadruple metal–metal bonds have been found among metals of groups 6 and 7, which have relatively high formal oxidation states. A typical organometallic example is $[Me_4Cr{\equiv}CrMe_4]^{4-}$.

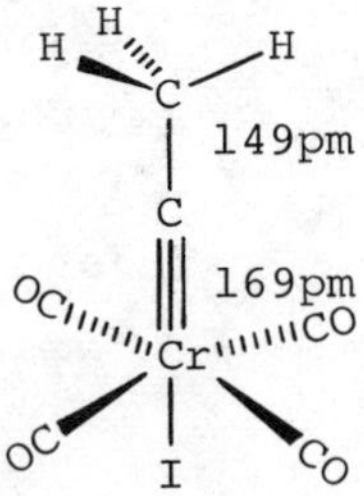

3-30

FIGURE 3.22. Molecular structure of a metal–carbyne complex $Cr({\equiv}CMe)I(CO)_4$ (from ref. 63).

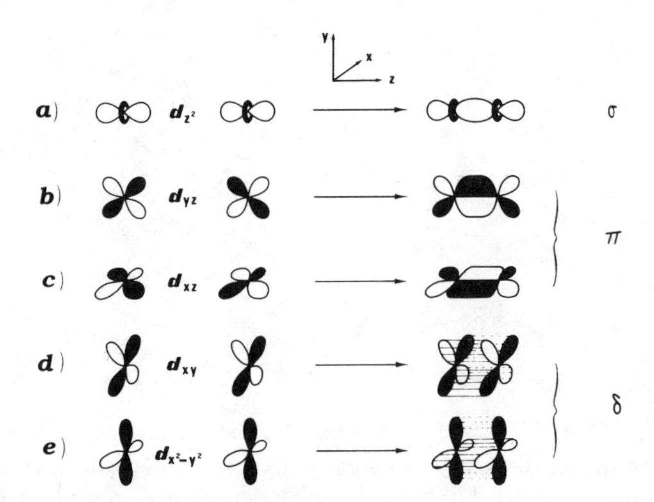

FIGURE 3.23. Possible combinations between two sets of d orbitals on adjacent metal atoms with the internuclear axis as the z axis. (Reproduced with permission from F. A. Cotton and G. Wilkinson, *Advanced Inorganic Chemistry*, 4th ed., Wiley, New York, 1980.)

3.4. CLASSIFICATION OF ORGANOTRANSITION METAL COMPLEXES AND THE EIGHTEEN-ELECTRON RULE

As we have seen so far, transition metals form a great variety of complexes with organic entities, and ways of classifying them are required.

The first approach is to classify the organic ligands according to the number of carbon atoms with which the metal is directly bonded. The prefix *hapto* (from the greek *haptein*, to fasten) was proposed by Cotton to designate the number of carbon atoms bonded to the metal. In describing the number of ligand atoms (n) bound to the metal, a shorthand notation, η, is now employed in the IUPAC nomenclature. Thus, ferrocene is formally named bis(η^5-cyclopentadienyl)iron(II), where η^5- is called *pentahapto*- or eta five-. Olefins bound to transition metals are η^2 ligands, π-allyl ligands are η^3 ligands, and cyclobutadiene ligands are η^4 ligands. The series of mixed-sandwich complexes **3-22 – 3-26** demonstrate examples where η^7, η^6, η^5, η^4, and η^3 ligands are attached to the metals that are bonded to η^5-cyclopentadienyl ligands on the other side. Although the prefix η was originally defined in the IUPAC nomenclature to describe the bonding of unsaturated ligands, the prefix η^1 is now widely used for describing the metal–carbon σ bond because of convenience.

Although the cyclopentadienyl ligand usually serves as a pentahapto ligand, it sometimes acts as a monohapto or trihapto ligand. Complex **3-31** is an exceptional case that contains η^5-, η^3-, and η^1-cyclopentadienyl ligands.

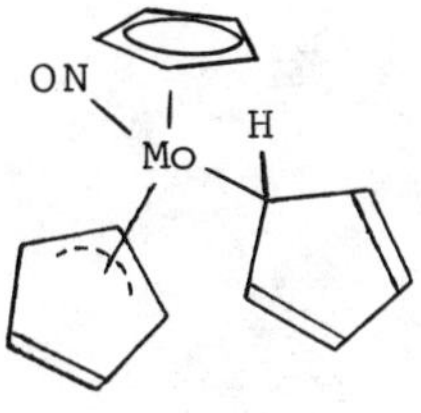

3-31 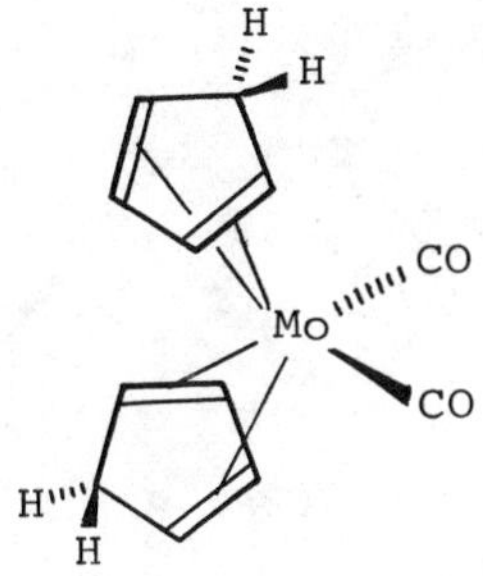 **3-32**

Cyclopentadiene C_5H_6, on the other hand, behaves as an η^4 ligand as in **3-32**.

It is sometimes necessary to explicitly designate the carbon atoms to which the metal atom is bonded. The cycloheptatriene ligand can bond with metals through two or three double bonds, as in **3-33** and **3-34**.

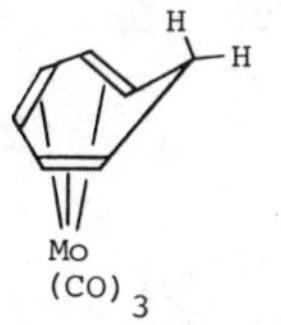

3-33 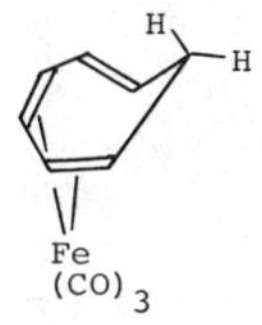 **3-34**

Complex **3-33** is named tricarbonyl(1-6-η^6-cycloheptatriene)molybdenum(0), whereas **3-34** is tricarbonyl(1-4-η^4-cycloheptatriene)iron(0). We shall later discuss the reason why cycloheptatriene sometimes bonds to metal atoms through two double bonds and at other times through three double bonds in relation to the 18-electron rule. If the cycloheptatriene ligand loses a hydride H^- (for example, on reaction with Ph_3C^+), a complex (**3-35**) with a planar tropylium ion $C_7H_7^+$ is formed.

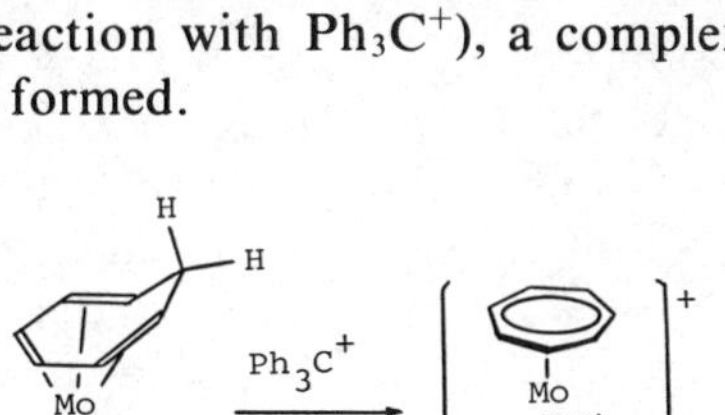

3-35

The heptahapto-$C_7H_7^+$ ion is considered to be bonded through seven equivalent carbon atoms involving six electrons.

An alternative approach to ligand classification is to specify the number of electrons contributed by the ligands in forming a bond with the metal. Table 3.4 summarizes the classification of ligands according to the number of electrons contributed.

Table 3.4. Classification of Ligands According to Number of Electrons Contributed to the Bonding with Transition Metal

Electrons Contributed	Ligand
1	H, alkyl, aryl, acyl, η^1-allyl groups
2	Olefins, CO, simple Lewis bases, PR_3, carbenes ($:CR_2$)
3	η^3-Allyl, NO (linear), carbynes ($\dot{:}CR$)
4	Diene
5	η^5-Cyclopentadienyl
6	Triene
7	Cycloheptatrienyl

In this convention the alkyl group is regarded as a one-electron ligand when it forms a covalent metal–carbon σ bond as in Figure 3.6*a*, where both the alkyl group and the metal contribute one electron each. The olefin is formally classified as a two-electron ligand, taking into account only the dative olefin-to-metal bond to be consistent with the convention for classical Lewis bases. The π-allyl ligand is considered to contribute three electrons to form the π-allyl bond, which is equivalent to the combination of one σ bond and one metal–alkene π bond.

Although it is not organic, NO is included in Table 3.4 as a three-electron ligand; the free NO molecule is a free radical having one nitrogen lone pair, and it is often found in organotransition metal complexes. The η^4-diene ligands are, of course, four-electron donors and the η^5-cyclopentadienyl ligand is a five-electron contributor. As illustrated in these examples, an advantage of this classificaiton is that the number of electrons contributed by the hydrocarbon ligand is in most cases identical with the number of carbon atoms involved in the formation of the metal–ligand bond. It is also consistent with naming the ligands as radicals such as alkyl ($R\cdot$), allyl ($C_3H_5\cdot$), and cyclopentadienyl ($C_5H_5\cdot$). In using this convention, however, it is necessary to avoid confusion regarding the formal oxidation number. The oxidation number (or oxidation state) for metal atoms in covalent compounds is defined as the number of electrons lost in forming covalent bonds where the ligand is more electronegative than the metal. Thus, the iron atom in ferrocene is regarded as being in oxidation state +2 since the cyclopentadienyl ligands are more electronegative than the iron atom. Thus, in ferrocene the cyclopentadienyl ligand is assigned a formal charge of −1 and is regarded as a carbanion. Similarly, an alkyl group in an alkyl metal complex is considered to have a formal negative charge. Hence, the oxidation number of manganese in $CH_3Mn(CO)_5$ is +1 and the methyl group is regarded as a carbanion. By extending this convention, we regard the H ligand in $HMn(CO)_5$ formally as a hydride anion and assign the formal oxidation number of +1 to the manganese. It should be noted that the chemical properties of hydride complexes may be different from what one would expect from the hydride notation. The five carbonyl groups, which serve as π acids withdrawing electrons from manganese by back donation, cause the hydride ligand in $MnH(CO)_5$ to be readily ionized as a proton in polar solvents.

To classify the alkyl ligand as a one-electron contributor and to regard it after the bond formation as an anion (for the purpose of bookkeeping with the oxidation number of the metal) may appear somewhat confusing. Therefore, some people prefer to regard the H and alkyl ligands as two-electron donors. Similarly, the allyl and cyclopentadienyl ligands are treated as four- and six-electron donor ligands, respectively.

However, another kind of discrepancy arises in this convention by assigning the electron count of 2 to the alkyl, alkene, and CO ligand. For example, the CO ligand often acts as a ligand bridging two metal atoms:

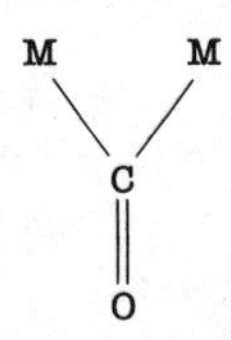

If we assign the electron count of 2 to the terminal acyl group, as for the alkyl group, then it would be more natural for the ketonelike bridging carbonyl group to have two electrons assigned for each bond connecting the metal and the carbonyl group. For the carbonyl group bridging the two metal atoms this would lead to a four-electron count and make the electron counting different from the terminal carbonyl group that donates the two electrons. To avoid this self-inconsistency, one should assign a two-electron count for the carbonyl group, disregarding the differences in nature between the terminal and bridging groups. This discrepancy does not exist in the electron counting system shown in Table 3.4.

Another discrepancy arises regarding the electron counting for an olefin ligand. When enhanced back bonding takes place in an olefin-coordinated complex, as in Eq. (3.7), the complex approaches a metallacyclopropane structure, and the olefin ligand is then better represented as a dianionic, four-electron donor following the convention by which the alkyl group is regarded as a two-electron donor. Since the gradation from the side-on bonded olefin to the metallacyclopropane representation is smooth, it is somewhat too artificial to call the olefin ligand a two-electron donor at one time and a four-electron donor at another. For practical electron counting purpose, it appears to be more natural to regard the coordinated olefin ligand as a two-electron donor at all times.

Unfortunately, no agreement has been reached regarding the electron counting of ligands, and authors of inorganic and organometallic chemistry textbooks employ different conventions regarding these ligands. The employment of either formalism regarding these noninnocent ligands is more or less a matter of taste. Jørgensen once used the terms *innocent* and *noninnocent* to describe ligands forming complexes in which the assignment of oxidation states is rigorous and ambiguous, respectively.[66] The ligands contributing odd numbers of electrons in Table 3.4 are noninnocent ligands, and the reader should be careful in electron counting with them.

Having discussed electron counting, we are now ready to proceed to the 18-electron rule. Many organotransition metal complexes, including most metal carbonyls and cyclopentadienyl complexes, obey the 18-electron rule or the effective atomic number (EAN) rule. The EAN rule was originally proposed by Sidgwick in 1927 as an expansion of the octet rule of Lewis to explain the stability of transition metal complexes.[67] The rule requires that the number of electrons possessed by the metal plus the number of electrons contributed by the ligands be equal to the number of electrons in the succeeding noble gas atom. The EAN rule proposed by Sidgwick was naturally not directed to organotransition metal complexes, but the usefulness of the rule when applied to organotransition metal complexes was later recognized. For example, in the case of $[Co(NH_3)_6]^{3+}$ the total number of electrons for the cobalt(III) ion is 24, and addition of 12 electrons donated by six lone pairs of ammonia gives a sum of 36, which corresponds to the electron configuration of krypton. Similarly in $[Ag(NH_3)_4]^+$ the total electron number is $46 + 2 \times 4 = 54$, which is equivalent to the electron configuration of xenon. In utilizing the EAN, the usual practice is to count the total number of electrons contained in the complex, and one has to remember the electron numbers of the noble gases. Furthermore, for heavier transition metals one has to count the f electrons by the EAN rule. In the 18-electron rule we disregard the number of core electrons in the closed shells and are concerned only with valence electrons about the metal; thus, the counting procedure is simpler. That a complex obeys the 18-electron rule means that all the valence orbitals, namely the five nd, the $(n+1)$s, and the three $(n+1)$p orbitals are fully occupied, and all their bonding capacity is used when the 18-electron configuration is reached.

Although the EAN rule has many exceptions when applied to inorganic coordination compounds in general, it is much more useful as a rule of thumb in organotransition metal chemistry. The majority of organotransition metal complexes are diamagnetic and obey the 18-electron rule, with certain well-defined exceptions for some square planar and trigonal complexes to which the 16-electron rule can be applied. For some complexes of Cu, Ag, and Au a number of complexes containing fewer electrons exist (14 or 12 are known), whereas the examples of complexes exceeding the 18 electrons are limited. Thus, the 18- or 16-electron rule is helpful in predicting and understanding the stoichiometries and even molecular structures of stable organotransition metal complexes.

The 18-electron rule also provides a useful guide in understanding reaction mechanisms of certain transition metal complexes.[68,69] Complexes that obey the 18-electron rule are called *coordinatively saturated.* In reactions involving organotransition metal complexes, coordinatively unsaturated complexes often become coordinatively saturated and vice versa. One should be cautious about accepting any mechanism that proposes chemical intermediates having more than 18 electrons.

a. Electron Counting

In counting electrons in organotransition metal complexes to see if the complex obeys the 18-electron rule, one should remember the number of electrons for all the transition metals. For quick reference Table 3.5 shows the part of the periodic table regarding transition metals and numbers of d electrons in various oxidation states of the metals. Note that the number of d electrons is not that in the ground state of the metal but that in its valence state. For example, the electron configuration of nickel in its ground state is $1s^22s^22p^63s^23p^63d^84s^2$; the 3d orbitals contain eight electrons and the 4s orbital contains two. However, it is energetically more favorable to promote the two electrons in the 4s orbital to the 3d orbitals for constructing chemical bonds with ligands. Therefore, the number of d electrons in nickel complexes in its zero oxidation state is 10.

In tetracarbonylnickle, $Ni(CO)_4$, the nickel atom having 10 d electrons receives 8 electrons from four CO ligands to satisfy the 18-electron rule.

The coinage metals Cu, Ag, and Au (group 11) have one s electron in their outer shells. Loss of the s electron gives the d^{10} ions Cu(I), Ag(I), and Au(I). Loss of two electrons gives divalent d^9 complexes. The group 12 elements Zn, Cd, and Hg are stable in their oxidation state of 2, having the d^{10} electron configuration, and these elements resist adopting electron configurations of less than d^{10} (and thus are usually not classified as transition elements).

Table 3.5. Numbers of d Electrons of Transition Metals in Various Oxidation States

	3	4	5	6	7	8	9	10	11	12
	Sc	Ti	V	Cr	Mn	Fe	Co	Ni	Cu	Zn
	Y	Zr	Nb	Mo	Tc	Ru	Rh	Pd	Ag	Cd
	(La)	Hf	Ta	W	Re	Os	Ir	Pt	Au	Hg
	(Ac)									
Oxidation Number	Number of d Electrons									
0	3	4	(5)	6	7	8	9	(10)		
I	2	3	4	5	6	7	8	9	(10)	
II	1	2	3	4	5	6	7	8	9	(10)
III	0	1	2	3	4	5	6	7	8	9
IV		(0)	1	2	3	4	5	6	7	8

In Table 3.5 the d^{10} electron configuration is circled. In going from Ni(0), having the d^{10} configuration to the left in the periodic table, the number of d electrons for each transition element decreases by one unit for the same oxidation number. One must remember the number of d electrons shown in the box of Table 3.5 for each transition element. Fortunately, the number of d electrons in each zero-valent metal atom for groups 3–10 coincides with the group notation according to the new IUPAC group nomenclature. To count the number of d electrons in the higher oxidation states, one should go down each column of Table 3.5. As a memory aid concerning the number of d electrons in the higher oxidation state, it is quicker to remember a reference number of d electrons. Ti(IV) serves as a good index for memorizing the d electrons because it is a d^0 metal ion. $TiCl_4$ is a colorless, diamagnetic liquid, whereas $TiCl_3$ has a violet color because of the presence of one d electron. In Table 3.5 the V(0) atom is also circled, indicating a d^5 metal. Examples of the application of the 18-electron rule to neutral complexes are given below.

(1) Metal Carbonyls

Binary metal carbonyls obey the 18-electron rule, the only exception being $V(CO)_6$. Let us first see the simplest cases of mononuclear complexes, as shown in Figure 3.24. $Cr(CO)_6$ has an octahedral structure, $Fe(CO)_5$ is trigonal bipyramidal, and $Ni(CO)_4$ is tetrahedral, all satisfying the 18-electron rule. Electron counting is illustrated below each structure of the metal carbonyls.

Metals with odd electron numbers cannot satisfy the 18-electron rule by addition of the CO ligands since the resultant moiety will have an odd number of electrons no matter how many carbonyls are added. One way to satisfy the 18-electron rule is that the two moieties, each with the odd electrons, dimerize, forming a metal–metal bond and sharing the contributed two electrons. The metal atom satisfies the 18-electron rule by regarding the electron of its counterpart as its own electron. Metal–metal bonding occurs in many dinuclear and polynuclear metal carbonyls. In some polynuclear metal carbonyls the carbonyl ligands serve as bridging ligands as well as terminal ligands. Figure 3.25 illustrates the structures of a variety of di- and polynuclear metal carbonyls having metal–metal bonds.

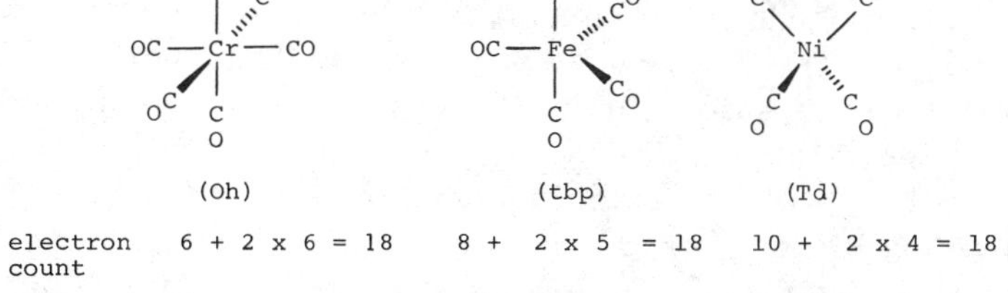

	(Oh)	(tbp)	(Td)
electron count	6 + 2 x 6 = 18	8 + 2 x 5 = 18	10 + 2 x 4 = 18

FIGURE 3.24. Structures of simple mononuclear metal carbonyls and electron counting.

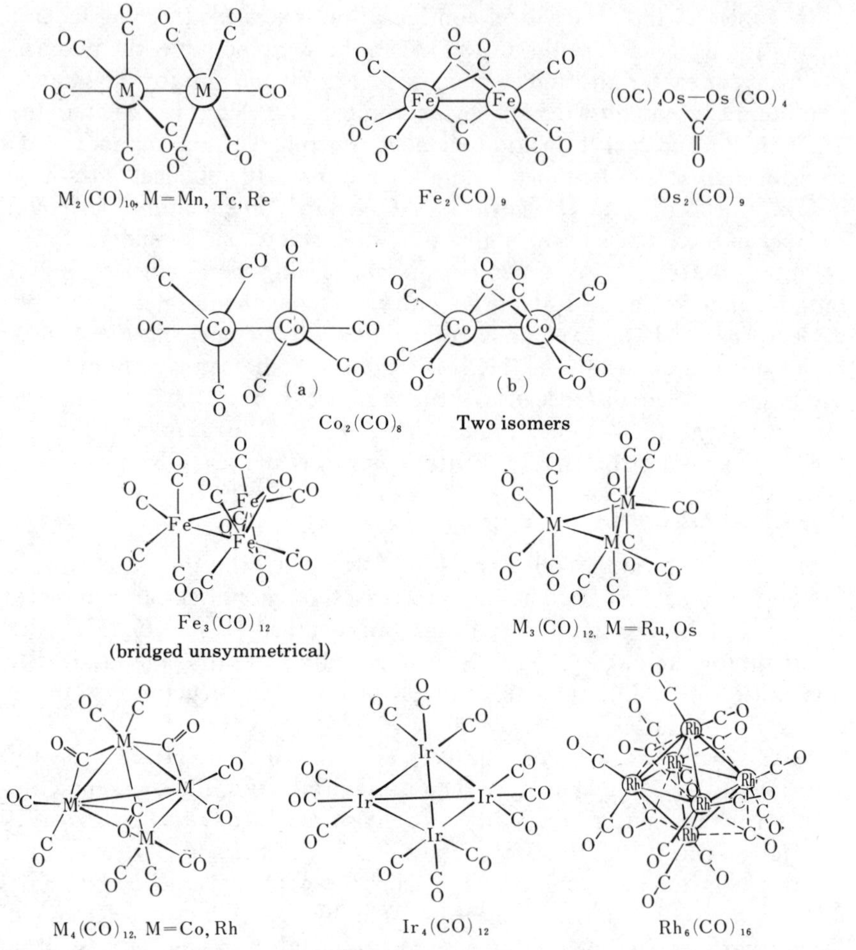

$M_2(CO)_{10}$, M=Mn, Tc, Re

$Fe_2(CO)_9$

$Os_2(CO)_9$

$Co_2(CO)_8$ Two isomers

$Fe_3(CO)_{12}$ (bridged unsymmetrical)

$M_3(CO)_{12}$, M=Ru, Os

$M_4(CO)_{12}$, M=Co, Rh

$Ir_4(CO)_{12}$

$Rh_6(CO)_{16}$

FIGURE 3.25. Structures of some di- and polynuclear metal carbonyls. (Reproduced with permission from F. A. Cotton and G. Wilkinson, *Basic Inorganic Chemistry,* Wiley, New York, 1976, p. 476.)

It can be seen that the 18-electron rule is satisfied in all of the metal carbonyls shown in Figure 3.25. For example, manganese, technetium, and rhenium, having d^7 electron configurations, all form a dimer in which an electron pair is shared between the metal atoms. Donation of 10 electrons from five terminal CO ligands to the d^7 manganese atom plus another electron provided by the metal–metal bonding gives the 18-electron complex. For the dimeric $Co_2(CO)_8$, which is an important catalyst precursor for hydroformylation of olefins, two isomers have been characterized, one without the bridging CO ligands (Fig. 3.25*a*) and the other with two bridging CO ligands (Fig. 3.25*b*). Both isomers differ by only a few kilojoules per mole in free energy, and they coexist in equilibrium mixtures at room temperature, the CO bridging isomer being predominant in solution at low temperature. Both isomers satisfy the 18-electron rule. In isomer *b* each of the two bridging CO ligands provides one electron to each cobalt atom. The 9 electrons from Co(0) plus the 2 electrons from the two bridging CO ligands together with 1 electron from the other cobalt and 6 from the three terminal CO ligands give 18 electron.

$Fe(CO)_5$ itself is a stable complex that has the 18-electron configuration. It releases one CO ligand upon UV irradiation and is converted into a golden yellow dimer, $Fe_2(CO)_9$, that also satisfies the 18-electron rule. Further removal of CO gives a CO-deficient moiety that achieves coordinative saturation by forming a trinuclear, metal–metal bonded species, triirondodecacarbonyl, $Fe_3(CO)_{12}$, which has an unsymmetrical structure having two bridging carbonyl (μ_2-) ligands. The formula that conveys the structural information is $Fe_3(\mu_2\text{-}CO)_2(CO)_{10}$. Electrons in $Fe_3(\mu_2\text{-}CO)_2(CO)_{10}$ are counted as follows.

Two bridged Fe: Fe(0) + 3CO + 2 × μ_2-CO + 2Fe-Fe bonds
8e + 3 × 2e + 2 × 1e + 2 × 1e = 18e

Unique Fe: Fe(0) + 4CO + 2Fe-Fe bonds
8e + 4 × 2e + 2 × 1e = 18e

It should be noted that the μ_2 bridge structure is not the only possibility for the $M_3(CO)_{12}$ type of complex. The higher congeners $Ru_3(CO)_{12}$ and $Os_3(CO)_{12}$ form trinuclear complexes without bridging CO ligands. This may be due to the larger sizes of the heavier atoms, which may make the effective bridge-bond formation difficult. In addition to the homonuclear complexes, heteronuclear complexes such as $FeRu_2(CO)_{12}$ and $Fe_2Os(CO)_{12}$ are also known. All of these satisfy the 18-electron rule.

The black octahedral complex $V(CO)_6$ is a conspicuous exception to the 18-electron rule, since it has one unpaired electron. The $V(CO)_6$ molecule, with 17 electrons, should dimerize, but if this happened, the vanadium atoms would become seven-coordinate. Presumably because of the strong repulsion of the CO ligands in the seven-coordinate moiety, the vanadium complex prefers to remain as a monomer without obeying the 18-electron rule. Vanadium carbonyl is less stable than other metal carbonyls, decomposing at 70 °C. It tends to become coordinatively saturated if given the opportunity. Thus, $V(CO)_6$ readily accepts an electron to form a hexacarbonyl vanadate ion:

$$Na + V(CO)_6 \longrightarrow Na^+ + [V(CO)_6]^-$$

In addition to these simple metal carbonyls there are an enormous number of other metal carbonyls that contain hydrido, alkyl, cyclopentadienyl, and other Lewis bases. The majority of these carbonyl-containing complexes also satisfy the 18-electron rule.

(2) Low-Valent Complexes Containing Phosphines and Related Ligands

With the recognition of tertiary phosphines and related soft ligands (cf. Section 3.3) as stabilizing ligands for low-valent complexes, we now have quite a number of phosphine-coordinated low-valent transition metal complexes. $Ni(PPh_3)_4$ and $Pd(PPh_3)_4$ are well known examples of zero-valent metal complexes that are frequently used in organic synthesis. These and other phosphine or phosphite-coordinated complexes together with those that have hydrido, olefin, or CO ligands generally obey the 18- or 16-electron rule.

(3) Metallocenes, Arene Complexes, and Related Complexes That Have Hydrocarbon Ligands

Ferrocene, the first prepared metallocene that has two cyclopentadienyl ligands, is a remarkably stable compound with the 18-electron configuration. The bis-arene sandwich complex, $Cr(C_6H_6)_2$, with two 6-electron ligands and a d^6 metal atom, also obeys the 18-electron rule and is known as a very stable complex. There are, however, other complexes that contain hydrocarbon ligands but do not conform to the 18-electron rule, such as paramagnetic $Ni(\eta^5\text{-}C_5H_5)_2$ and $Co(\eta^5\text{-}C_5H_5)_2$. These examples indicate that the 18-electron rule is not infallible even among organotransition metal complexes.

Despite the occasional exceptions, the 18-electron rule is quite useful in predicting the stoichiometry and even the molecular structures of some organotransition metal complexes. The following examples illustrate this point. The reaction of iron carbonyl with cyclooctatetraene gives three different complexes of the following compositions:[70] $Fe(CO)_3(C_8H_8)$ (**3-36**), $Fe_2(CO)_6(C_8H_8)$ (**3-37**), and $Fe_2(CO)_5(C_8H_8)$ (**3-38**). In complex **3-36** the sum of 8 electrons in Fe(0) and 6 electrons from three CO ligands gives 14 electrons. Thus, the prediction based on the 18-electron rule is that the cyclooctatetraene be coordinated with the iron moiety through two double bonds, leaving the other two double bonds uncomplexed. By a similar reasoning, with one cyclooctatetraene per two iron atoms, complex **3-37** is expected to have the cyclooctatetraene ligand bridging two iron atoms through two double bonds for each iron atom. The results of an X-ray analysis completely confirm this prediction, as shown in Figure 3.26*a,b*.

The ^1H NMR spectrum of **3-36** in solution at room temperature shows a sharp singlet, apparently indicating the equivalence of all protons and suggesting the presence of a planar, 8e donor cyclooctatetraene ring. However, this is incompatible with the 18-electron rule, and an X-ray analysis established that the ring has a boat conformation and is bound to iron as an η^4-1,3-diene, compatible with the 18-electron rule. The mechanism by which the protons are made equivalent as shown by variable-temperature NMR spectroscopy was deduced to be a 1,2-shift of the $Fe(CO)_3$ group about the cyclooctatetraene ring. This type of complex, which rearranges from one configuration into another, is called a fluxional molecule (see Section 5.3).

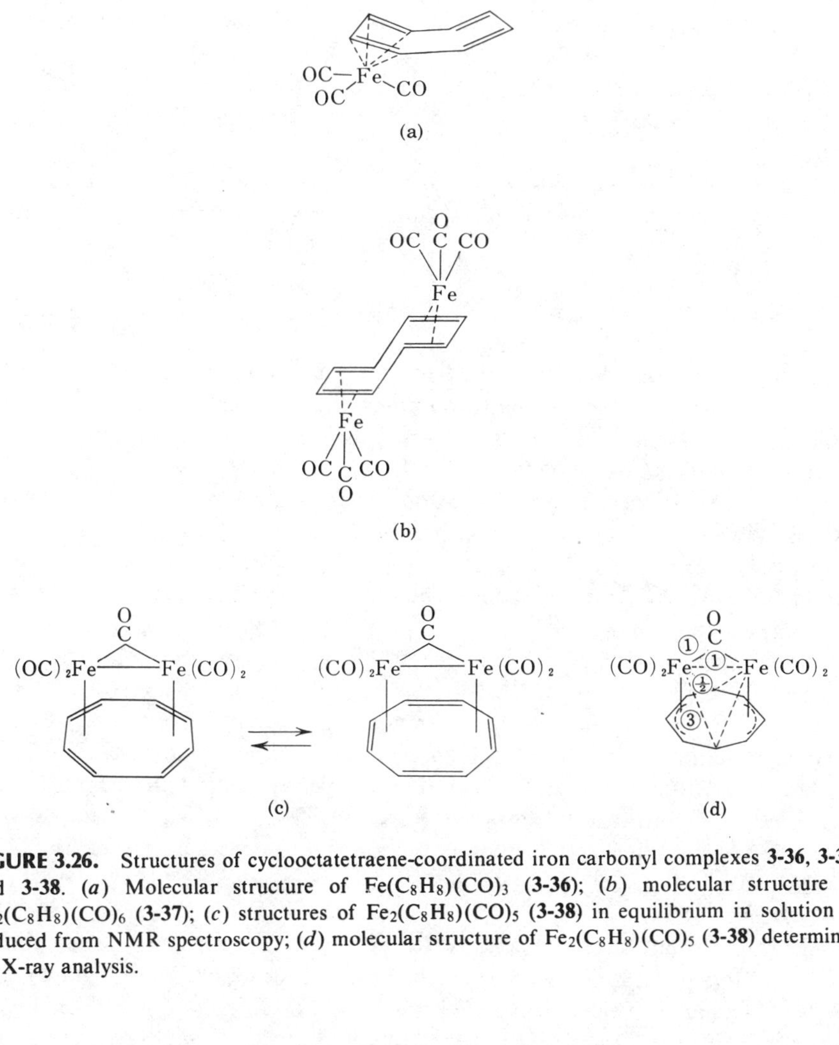

(a)

(b)

(c)

(d)

FIGURE 3.26. Structures of cyclooctatetraene-coordinated iron carbonyl complexes **3-36**, **3-37**, and **3-38**. (*a*) Molecular structure of $Fe(C_8H_8)(CO)_3$ (**3-36**); (*b*) molecular structure of $Fe_2(C_8H_8)(CO)_6$ (**3-37**); (*c*) structures of $Fe_2(C_8H_8)(CO)_5$ (**3-38**) in equilibrium in solution as deduced from NMR spectroscopy; (*d*) molecular structure of $Fe_2(C_8H_8)(CO)_5$ (**3-38**) determined by X-ray analysis.

The IR spectrum of complex **3-38** indicates the presence of a bridging carbonyl ligand, and a structure with an Fe–Fe bond is deduced from this information. If the 18-electron rule is obeyed, the electron count around the iron atom, except for the C_8H_8 moiety, is $8 + 1 + 1 + 2 \times 2 = 14$, leaving room for the iron to coordinate with two double bonds out of the four in the cyclooctatetraene entity. A structure consistent with this argument, which is also supported by NMR spectroscopy, is shown in Figure 3.26*c*. This complex is again believed to be fluxional, with the cyclooctatetraene double bonds migrating by a 1,2-shift mechanism. Results of X-ray analysis, however, revealed a more complicated structure for the solid sample, as shown in Figure 3.26*d*. The cyclooctatetraene ring is bent in the middle, and each half plane of the cyclooctatetraene ring is coordinated to an iron atom. In this case each carbon atom where the ring is bent is considered to form electron-deficient (3c-1e) bonds with the two iron atoms, utilizing one-half electron for each C–Fe bond. The electron count around each iron atom is 8e[Fe(0)] + 2 × 2e(CO) + 3e(π-allyl portion) + 1e(Fe–Fe) + 1e(μ-CO) + 2 × $\frac{1}{2}$ (Fe-bridging carbons) = 18.

Complexes **3-31 – 3-35,** which we have seen earlier, obey the 18-electron rule. The electron count in these cases should be self-evident from the foregoing explanations.

b. Rationale for the Eighteen-Electron Rule

As we have seen, the 18-electron rule is an empirical but quite useful rule for a variety of organotransition metal complexes, although it does have exceptions. The following is an attempt to rationalize the 18-electron rule and its exceptions in terms of molecular orbital theory.[71]

Let us begin the explanation with octahedral complexes ML_6. The energy level diagram of an octahedral complex is shown in Figure 3.27*a*, which is a reproduction of Figure 2.17, but for the sake of simplification, the metal–ligand bonding MOs are blocked at the bottom part of the energy diagram of the complex molecule.

The number of valence orbitals available for MO construction of the ML_6 complex is 9, namely five *n*d orbitals, one (*n*+1)s orbital, and three (*n*+1)p orbitals. When these orbitals interact with six ligand orbitals according to symmetry matching, the bonding orbitals a_{1g}, e_g, and t_{1u} and the antibonding ortibals t_{1u}^*, a_{1g}^*, and e_g^* are formed. The 12 electrons contributed by the six ligands are put into the bonding orbitals a_{1g}, e_g, and t_{1u}. When they do not have ligand orbitals of the proper symmetry with which to interact, the three t_{2g} orbitals of the five metal d orbitals remain nonbonding, as shown in Figure 3.27*a*.

As already mentioned, the energy difference Δ_o between the t_{2g} and e_g^* orbitals is the ligand field energy. When the complex has π-acceptor ligands such as CO, the energy of the t_{2g} orbitals is decreased on interaction with the ligand π orbitals, with a resultant increase in Δ_o as shown in Figure 2.19. Thus, six electrons can be accommodated in the t_{2g} orbitals to form stable ML_6 complexes. Hence the sum of d electrons and electrons contributed by the ligands is 12 + 3 × 2 = 18. If the Δ_o value is small, d electrons may enter the e_g^* orbitals, and a maximum number of 22 electrons can be accommodated.

A similar reasoning can be applied to complexes other than octahedral configurations. Let us examine the cases of ML_n complexes having *n* ligands ($n > 4$). Of nine metal orbitals, *n* orbitals can be employed for forming bonds with ligands, giving *n* bonding and *n* antibonding MOs. Hence the number of nonbonding d orbitals will be $9 - n$. These orbitals and the low-energy antibonding orbitals can be used for accommodating the metal d electrons.

Figure 3.27*b* illustrates the electron configuration for a tetrahedral complex ML_4. In this case the ligand field energy Δ_t is smaller than Δ_o for an octa-

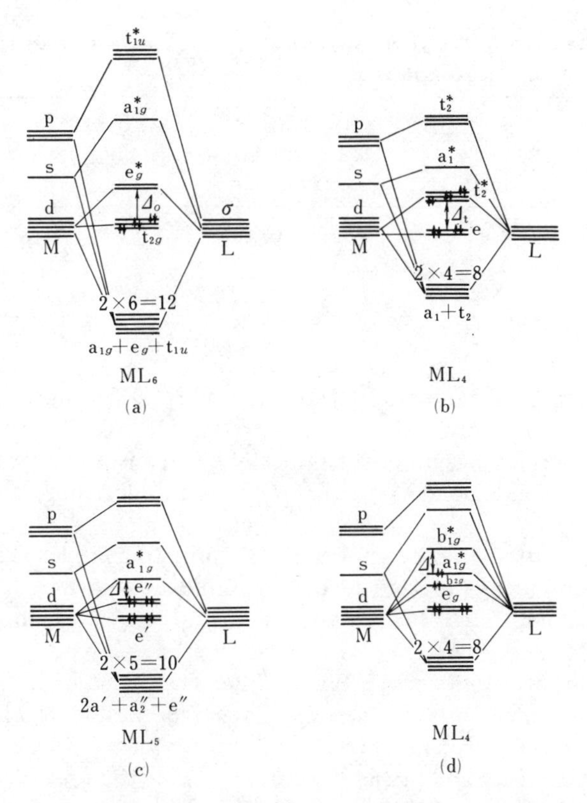

FIGURE 3.27. Rationale for the 18- and 16-electron rule for various transition metal complexes: (*a*) Octahedral, (*b*) tetrahedral, (*c*) trigonal bipyramidal, and (*d*) square planar (16e).

hedral complex. The d electrons can therefore enter not only the e orbitals but also the t_2^* orbitals. The sum of 10 d electrons and 8 electrons donated by the four ligands is 18.

For a trigonal bipyramidal complex ML_5 (Fig. 3.26*c*) orbitals up to e′ and e″ levels can be readily filled with d electrons, if the ligand field energy between e″ and a_{1g}^* is large, to give the total number of d electrons and ligand electrons as 18. For a square planar complex ML_4(d), five d orbitals are available, as in the tetrahedral case, to accommodate 18 electrons altogether if all the e_g, b_{2g}, a_{1g}^*, and b_{1g}^* orbitals are filled with d electrons. However, when the energy gap between the a_{1g}^* and b_{1g}^* orbitals is large, all orbitals up to the level of the a_{1g}^* orbital will be occupied by the d electrons giving relatively stable 16-electron complexes (Fig. 3.27*d*). Many examples of stable square planar complexes are known for Pt(II), Pd(II), and Ni(II) complexes that have acceptor ligands.

For metal carbonyls satisfying the 18-electron rule the number of ligands able to bind the metal is limited by the number of d electrons. As we have

Table 3.6. Relationship Between the Number of d Electrons and the Configuration of Metal Complexes

Metal d Electrons	Metal	Coordination Number	Number of Orbitals that Accommodate the d Electrons
6	Cr	6	3
8	Fe	5	4
10	Ni	4	5

seen with the simple metal carbonyls of Figure 3.24, the following relationship holds (Table 3.6).

When the complex has an odd number of electrons, it shows a tendency to form a metal–metal bond by pairing the odd electrons, thus achieving stabilization.

These arguments should be applicable not only to complexes having simple mono or bidentate ligands but also to complexes that are not easily categorized as octahedral, tetrahedral, and so on. In those cases a detailed estimation of the MO energy diagram is needed. We have already seen the energy level diagram of a metallocene in Figure 3.20. The orbitals that accommodate the d electrons are enclosed in the box. For ferrocene the e_{2g} and a'_{1g} orbitals are filled with 6 d electrons of Fe(II), and the rest of the six cyclopentadienyl orbitals are occupied by 12 ligand electrons, making the total electron number 18. However, for nickelocene the extra d electrons will enter the upper two e^*_{1g} orbitals to give a 20-electron complex. This complex shows paramagnetism since the extra 2 electrons will singly occupy the degenerate e^*_{1g} orbitals.

For the metallocenes of earlier transition metals such as Ti, V, and Cr situated on the left side of the periodic table, the electron count will be less than 18 provided the cyclopentadienyl ligands are mutually parallel. However, in these cases there is a tendency for the cyclopentadienyl planes to be tipped so as to accommodate more ligands, thus allowing the 18-electron rule to be obeyed. The cyclopentadienyl tantalum methylene complex shown in Figure 3.21 is a case in point.

REFERENCES

1. A. L. Allred and E. G. Rochow, *J. Inorg. Nucl. Chem.*, **5**, 264 (1958).
2. F. A. Cotton, *Chem. Rev.*, **55**, 551 (1955).
3. H. H. Jaffé, *J. Chem. Phys.*, **21**, 196 (1953).
4. H. A. Skinner, *Adv. Organometal. Chem.*, **2**, 49 (1964).
5. P. J. Davidson, M. F. Lappert, and R. Pearce, *Chem. Rev.*, **76**, 219 (1976).
6. R. Taube, H. Drevs, and D. Steinborn, *Z. Chem.*, **18**, 425 (1978).
7. M. F. Lappert, D. S. Patil, and J. B. Pedley, *J. Chem. Soc. Chem. Commun.*, 830 (1975).
8. J. A. Connor, *Top. Cur. Chem.*, **71**, 71 (1977).

9. (a) H. A. Skinner, *J. Chem. Thermodyn.*, **10**, 309 (1978); (b) G. Pilcher and H. A. Skinner, in *The Chemistry of the Metal–Carbon Bond*, F. R. Hartley and S. Patai, eds., Wiley, New York, 1982.
10. (a) F. A. Adedeji, J. A. Connor, H. Skinner, L. Gayer, and G. Wilkinson, *J. C. S. Chem. Commun.*, 159 (1976); (b) D. L. S. Brown, J. A. Connor, and H. A. Skinner, *J. Organometal. Chem.*, **81**, 403 (1974).
11. J. A. Connor, M. T. Zofurani-Moattar, J. Bickerton, N. I. ElSaied, S. Suradi, R. Carson, G. AlTakkin, and H. A. Skinner, *Organometallics*, **1**, 1166 (1982).
12. J. Halpern, *Acc. Chem. Res.*, **15**, 238 (1982).
13. F. T. T. Ng, G. L. Rampel, and J. Halpern, *J. Am. Chem. Soc.*, **104**, 621 (1982).
14. T.-T. Tsou, M. Loots, and J. Halpern, *J. Am. Chem. Soc.*, **104**, 623 (1982).
15. N. J. Nappa, R. Santi, S. P. Diefenbach, and J. Halpern, *J. Am. Chem. Soc.*, **104**, 619 (1982).
16. G. N. Schrauzer and J. H. Grate, *J. Am. Chem. Soc.*, **103**, 541 (1981).
17. G. Yoneda and D. M. Blake, *Inorg. Chem.*, **20**, 67 (1981).
18. (a) J. Holton, M. F. Lappert, D. G. H. Ballard, R. Pearce, J. L. Atwood, and W. E. Hunter, *J. Chem. Soc., Dalton Trans.*, 54 (1979); (b) J. Holton, M. F. Lappert, D. G. H. Ballard, R. Pearce, J. L. Atwood, and W. E. Hunter, *J. Chem. Soc. Dalton Trans.*, 45 (1979); (c) G. R. Scollary, *Aust. J. Chem.*, **31**, 411 (1978).
19. (a) M. Nakajima, T. Saito, A. Kobayashi, and Y. Sasaki, *J. Chem. Soc., Dalton Trans.*, 385 (1977); (b) T. Saito, M. Nakajima, A. Kobayashi, and Y. Sasaki, *J. Chem. Soc., Dalton Trans.*, 482 (1978).
20. (a) W. Tochtermann, *Angew. Chem.*, **78**, 355 (1966); (b) G. H. Posner, *An Introduction to Synthesis Using Organocopper Reagents*, Wiley, New York, 1980; (c) R. Taube, *Pure Appl. Chem.*, **55**, 165 (1983).
21. D. M. P. Mingos, in *Comprehensive Organometallic Chemistry*, Vol. 3, Pergamon Press, Oxford, 1982, p. 1.
22. H. M. Seip and R. Seip, *Acta Chim. Scand.*, **24**, 3431 (1970).
23. (a) J. A. McCleverty, *Prog. Inorg. Chem.*, **10**, 49 (1968); (b) R. P. Burns and C. A. McAuliffe, *Adv. Inorg. Chem. Radiochem.*, **22**, 303 (1979).
24. R. G. Wilkins, *Quart. Rev.*, **16**, 316 (1962).
25. E. O. Fischer and H. Werner, *Metal π-Complexes*, Vol. I, Elsevier, Amsterdam, 1966.
26. M. Herberhold, *Metal π-Complexes*, 2 vols., Elsevier, Amsterdam, 1972, 1974.
27. S. D. Ittel and J. A. Ibers, *Adv. Organometal. Chem.*, **14**, 33 (1976).
28. S. G. Davies, M. L. H. Green, and D. M. P. Mingos, *Tetrahedron*, **34**, 3047 (1978).
29. F. Hartley, *Angew. Chem.*, **84**, 657 (1972).
30. (a) M. J. S. Dewar, *Bull. Soc. Chim. Fr.*, C71 (1951); (b) J. Chatt and L. A. Duncanson, *J. Chem. Soc.*, 2939 (1953).
31. (a) J. Smidt, W. Hafner, R. Jira, J. Seldmeier, R. Sieber, R. Ruttinger, and H. Kojer, *Angew. Chem.*, **71**, 176 (1959); (b) *ibid.*, **74**, 93 (1962); (c) J. Smidt, *Chem. Ind. (London)*, 54 (1962).
32. M. J. S. Dewar and G. P. Ford, *J. Am. Chem. Soc.*, **101**, 783 (1979).
33. (a) T. A. Albright, R. Hoffmann, J. C. Thibeault, and D. Thorn, *J. Am. Chem. Soc.*, **101**, 3801 (1979); (b) N. Rosch and R. Hoffmann, *Inorg. Chem.*, **13**, 2656 (1974).
34. T. Ziegler and A. Rauk, *Inorg. Chem.*, **18**, 1558 (1979).
35. R. M. Pitzer and H. F. Schaeffer III, *J. Am. Chem. Soc.*, **101**, 7176 (1979).
36. B. Åkermak, M. Almemark, J. Almlof, J.-E. Bäckvall, B. Roos, and A. Stogard, *J. Am. Chem. Soc.*, **99**, 4617 (1977).

37. (a) P. J. Hay, *J. Am. Chem. Soc.*, **103,** 1390 (1981); (b) K. Kitaura, S. Sakaki, and K. Morokuma, *Inorg. Chem.*, **20,** 2292 (1981).
38. D. M. P. Mingos, *Adv. Organometal. Chem.*, **15,** 1 (1977).
39. J. A. Segal and B. F. Johnson, *J. Chem. Soc., Dalton Trans.*, 677, 1990 (1975).
40. B. E. Mann, in *Comprehensive Organometallic Chemistry,* Vol. 3, Pergamon Press, Oxford, 1982, p. 89.
41. F. L. Bowden and A. B. P. Lever, *Organometal. Chem. Rev.*, **3,** 227 (1968).
42. (a) S. Otsuka and A. Nakamura, *Adv. Organometal. Chem.*, **14,** 245 (1970); (b) J. L. Templeton, P. B. Winston, and B. C. Ward, *J. Am. Chem. Soc.*, **103,** 7713 (1981); (c) K. Tatsumi, R. Hoffmann, and J. L. Templeton, *Inorg. Chem.*, **21,** 466 (1982) and references therein.
43. W. G. Sly, *J. Am. Chem. Soc.*, **81,** 18 (1959).
44. J. O. Glanville, J. M. Stewart, and S. O. Grim, *J. Organometal. Chem.*, **7,** 7 (1967).
45. M. I. Loback, B. D. Babitskii, and V. A. Kormer, *Russ. Chem. Rev.*, **36,** 476 (1967).
46. G. Deganello, *Transition Metal Complexes of Cyclic Polyolefins,* Academic Press, New York, 1979.
47. R. Pettit and G. F. Emerson, *Adv. Organometal. Chem.*, **1,** 1 (1964).
48. H. Yasuda, Y. Kajihara, K. Mashima, K. Nagasuna, K. Lee, and A. Nakamura, *Organometallics,* **1,** 388 (1982).
49. G. Erker, J. Wicher, K. Engel, F. Rosenfeldt, W. Dietrich, and C. Krüger, *J. Am. Chem. Soc.*, **102,** 6344 (1980).
50. H. C. Longuet-Higgins and L. E. Orgel, *J. Chem. Soc.*, 1969 (1956).
51. R. Criegee and G. Schroder, *Ann.*, **623,** 1 (1959).
52. W. Hubel and E. H. Brye, *J. Inorg. Nucl. Chem.*, **10,** 250 (1959).
53. A. Efraty, *Chem. Rev.*, **77,** 691 (1977).
54. J. Dunitz and L. E. Orgel, *Nature,* **171,** 121 (1953).
55. J. W. Lauher and R. Hoffmann, *J. Am. Chem. Soc.*, **98,** 1729 (1976).
56. A. Haaland, *Acc. Chem. Res.*, **12,** 415 (1979).
57. M.-M. Rohmer and A. Veillard, *J. Chem. Phys.*, **11,** 349 (1975).
58. N. Rosch and K. H. Johnson, *Chem. Phys. Lett.*, **24,** 178 (1974).
59. M. D. Rausch, *Pure and Applied Chem.*, **30,** 523 (1972).
60. R. R. Schrock, *Acc. Chem. Res.*, **12,** 98 (1979).
61. (a) E. O. Fischer, *Adv. Organometal. Chem.*, **14,** 1 (1976); (b) D. J. Cardin, B. Cetinkaya, M. J. Doyle, and M. F. Lappert, *Chem. Soc. Rev.*, **72,** 545 (1972); (c) D. J. Cardin, B. Cetinkaya, and M. F. Lappert, *Chem. Rev.*, **72,** 545 (1972); (d) F. A. Cotton and C. M. Lukehart, *Prog. Inorg. Chem.*, **16,** 487 (1972).
62. L. H. Guggenberger and R. R. Schrock, *J. Am. Chem. Soc.*, **97,** 6578 (1975).
63. G. Huttner, H. Lorenz, and W. Gartzke, *Angew. Chem., Int. Ed. Engl.*, **13,** 609 (1974).
64. (a) E. O. Fischer, *J. Organometal. Chem.*, **100,** 59 (1975); (b) U. Schubert, in *The Chemistry of the Metal-Carbon Bond,* F. R. Hartley and S. Patai, eds., Wiley, New York, 1982.
65. (a) M. H. Chisholm and F. A. Cotton, *Acc. Chem. Res.*, **11,** 356 (1978); (b) F. A. Cotton, *Acc. Chem. Res.*, **11,** 225 (1978); (c) F. A. Cotton, *Chem. Soc. Rev.*, **4,** 27 (1975); (d) J. L. Templeton, *Prog. Inorg. Chem.*, **26,** 211 (1979); (e) W. C. Trogler and H. B. Gray, *Acc. Chem. Res.*, **11,** 232 (1978); (f) F. A. Cotton and R. A. Walton, *Multiple Bonds between Metal Atoms,* Wiley-Interscience, New York, 1982; (g) M. H. Chisholm, *Polyhedron,* **2,** 681 (1983).
66. C. K. Jørgensen, *Structure and Bonding,* **1,** 234 (1966).

67. N. V. Sidgwick, *The Electronic Theory of Valency,* Cornell University Press, Ithaca, N.Y., 1927.

68. J. E. Ellis, *J. Chem. Educ.,* **53,** 2 (1976).

69. C. A. Tolman, *Chem. Soc. Rev.,* **1,** 337 (1972).

70. (a) F. A. Cotton, *Acc. Chem. Res.,* **1,** 257 (1968); (b) F. A. Cotton, A. Davison, T. J. Marks, and A. Musco, *J. Am. Chem. Soc.,* **91,** 6598 (1969) and references therein.

71. P. R. Mitchell and R. V. Parish, *J. Chem. Educ.,* **46,** 811 (1969).

72. E. Elian and R. Hoffmann, *Inorg. Chem.,* **14,** 1058 (1975).

CHAPTER FOUR

Syntheses of Organotransition Metal Complexes

Organotransition metal complexes are full of variety, and there are so many different preparative methods that it is difficult to provide a systematic account of them. Two main approaches can be adopted, one based on the ligands and the other on the metals. Fortunately, we now have *Comprehensive Organometallic Chemistry* and other comprehensive reference books that cover most of the synthetic methods in detail. Therefore, we leave the detailed description of synthetic methods to these reference books and concentrate on describing the essential features of the syntheses.

Organotransition metal complexes may be broadly classified as those having metal–carbon σ bonds and metal–carbon π bonds. In this chapter we first treat the syntheses of representative σ- and π-bonded organotransition metal complexes and in the later part various types of complexes with hydrido, carbonyl, N_2, CO_2, CS_2, and O_2 ligands. Although some of them are not organometallic complexes in the strict sense of the term, the development of their chemistry has been closely associated with that of organotransition metal chemistry and they may be more conveniently dealt with here.

4.1. π COMPLEXES

a. η^2-Bonded Olefin Complexes

Since the metal-to-olefin back bonding plays a very important role in olefin–metal π bonding and back bonding is favored when the metal is electron rich, the majority of the olefin π complexes are found with metals in lower oxidation states.[1] However, because olefin-to-metal donation also plays an important part in metal–olefin bonding, there are some π complexes formed between olefins and transition metals in a relatively high formal oxidation state. Pt(II) complexes are among this limited number.

The first, historically important metal–olefin complex, $K[Pt(C_2H_4)Cl_3]H_2O$ (**1-1**), now called Zeise's salt, was prepared by heating a mixture of Pt(II) and Pt(IV) chlorides in ethanol accompanied by treatment with KCl.[2] This complex can now be prepared more conveniently by bubbling ethylene into an aqueous K_2PtCl_4 solution in the presence of $SnCl_2$ as a catalyst[3]:

$$K_2PtCl_4 + C_2H_4 \xrightarrow{SnCl_2} \underset{\textbf{1-1}}{K[Pt(C_2H_4)Cl_3]H_2O} + KCl \tag{4.1}$$

In this reaction complex **1-1** can be readily obtained since it precipitates as yellow crystals from the reaction mixture. Similar monoolefin complexes have been prepared with d^8 [Pd(II), Rh(I), and Ir(I)] as well as d^{10} [Cu(I) and Ag(I)] metals.

Olefins with two or more nonconjugated double bonds can bind metals through each double bond, forming chelate complexes whose stabilities are generally higher than those of the singly bonded monoolefin complexes. Typical examples are complexes of cycloocta-1,5-diene (1.5-COD) and bicyclo[2.2.1]diene(norbornadiene).[4] Because of the high stability of these chelate complexes, chelating diolefins such as COD can displace weakly bound ligands such as benzonitrile.

$$PdCl_2(PhCN)_2 + \underset{\textbf{4-1}}{\text{(COD)}} \xrightarrow{-2PhCN} \underset{\textbf{4-2}}{(cod)PdCl_2} \tag{4.2}$$

$PdCl_2(PhCN)_2$, Kharasch's complex, is a useful starting material for the preparation of other palladium complexes.

A great number of olefins form π complexes with metals in lower oxidation states. Therefore, the most common method for the preparation of olefin complexes is to reduce the metal salts in the presence of olefins or to displace other ligands bound to low-valent metal complexes with olefins. For example, a synthetically useful Ni(0) complex, $Ni(cod)_2$ (**4-3**)† can be prepared by reducing nickel acetylacetonate, $Ni(acac)_2$, with an alkylaluminum compound in the presence of COD.[5]

$$Ni(acac)_2 + AlEt_2(OEt) + COD \longrightarrow \underset{\textbf{4-3}}{Ni(cod)_2} \tag{4.3}$$

†The coordinated ligands are written in lowercase letters.

In the process of reduction of Ni(II) to Ni(0), an ethylnickel(II) complex is probably formed as an intermediate, which is then thermally decomposed to give an Ni(0) species to which COD molecules coordinate to give $Ni(cod)_2$. When the reaction is carried out without COD but in the presence of triphenylphosphine, PPh_3, an ethylene-coordinated Ni(0) complex is obtained. This result indicates that the intermediate ethylnickel(II) complex decomposes with liberation of ethylene, which coordinates to the Ni(0) moiety.

$$Ni(acac)_2 + AlEt_2(OEt) + PPh_3 \longrightarrow \underset{\textbf{4-4}}{Ni(PPh_3)_2(C_2H_4)} \tag{4.4}$$

A similar approach works for iron, molybdenum, and palladium acetylacetonates, and the method is called reductive olefination.

Weakly bound olefins can be displaced by more strongly binding olefins as follows.

$$Ni(PPh_3)_2(C_2H_4) + CH_2{=}CHCN \longrightarrow \underset{\textbf{4-5}}{Ni(PPh_3)_2(CH_2{=}CHCN)} \tag{4.5}$$

In certain cases the starting metal compounds are spontaneously reduced during reaction with the olefin; this occurs in solvents that can serve as reducing agents.

$$2IrCl_3 + 1,5\text{-COD} \xrightarrow{EtOH} \underset{\textbf{4-6}}{(\text{cod})Ir(\mu\text{-}Cl)_2Ir(\text{cod})} \tag{4.6}$$

Some ligands can be displaced thermally by olefins, but for displacement of strongly bound ligands, irradiation with UV light is often effective.[6]

$$Fe(CO)_5 + \begin{array}{c}CHCOOEt\\ \Vert \\ CHCOOEt\end{array} \xrightarrow[-CO]{UV} \underset{\textbf{4-7}}{Fe(CO)_4(\eta^2\text{-}EtO_2C\text{-}CH{=}CH\text{-}CO_2Et)} \tag{4.7}$$

Hydridotransition metal complexes can react with olefins to give olefin-coordinated complexes with liberation of their hydrogenated products.[7]

$$RuH_2(PPh_3)_4 + \text{olefin} \longrightarrow Ru(\text{olefin})(PPh_3)_3 \quad \text{olefin} = \text{ethylene, styrene} \tag{4.8}$$

In a rare case even benzene can coordinate with a transition metal through two carbon atoms.[8]

Re HX NEt3 Re+ X− OC N O R

4-8 (4.9)

R = H, Me, CF_3; X = BF_4, SO_3F

Olefin-coordinated low-valent transition metal complexes are useful as precursors to other low-valent metal complexes that are employed in various stoichiometric as well as catalytic reactions, since the coordinated olefins can be readily released from the complexes. This generates reactive, coordinatively unsaturated complexes whose chemical properties can be modified by addition of an appropriate ligand such as a tertiary phosphine. The COD-coordinated complexes of low-valent metals represent typical examples.[9]

Synthesis using metal vapors sometimes provides a useful approach to olefin complexes that are unavailable by other methods.[10] Metal vapors are cocondensed with olefins at low temperatures to give the olefin-coordinated complexes; for example,

$$\text{Fe} + \text{1,5-COD} \longrightarrow \underset{\textbf{4-9}}{\text{Fe(cod)}_2} \tag{4.10}$$

The technique of matrix isolation in which metal atoms and olefins are condensed in an argon matrix at low temperatures allows the observation of unstable olefin-coordinated species such as $Ni(C_2H_4)$ and $Ni(C_2H_4)_2$ by IR spectroscopy.[11]

b. Acetylene Complexes

Complexes containing coordinated acetylene or its derivatives can be similarly prepared by direct interaction of acetylenes with transition metal complexes of low oxidation state.[12] The difference from olefin complexes is that acetylenes can act not only as two-electron donors but also as four-electron donors, utilizing their orthogonal two pairs of orbitals either to bridge two metal atoms as in complex **3-13** or to coordinate to one metal atom, as in $W(CO)(PhC{\equiv}CPh)_3$ and $Mo(HC{\equiv}CH)(CO)(S_2CNEt_2)_2$.

Representative preparative routes are shown below:

$$Pt(PPh_3)_3 + RC{\equiv}CR \longrightarrow \underset{\textbf{4-10}}{Pt(PPh_3)_2(RC{\equiv}CR)} + PPh_3 \tag{4.11}$$

$$Co_2(CO)_8 + RC{\equiv}CR \longrightarrow \underset{\textbf{4-11}}{Co_2(CO)_6(\mu\text{-}RC{\equiv}CR)} + 2CO \quad (4.12)$$

It should be noted that the coordination of two acetylene molecules to a transition metal sometimes leads to metallacyclopentadiene complexes (which will be discussed later). After coordination of two acetylene molecules to a transition metal, further reactions also may take place, subsequently giving a variety of complexes. For example, diphenylacetylene reacts with $Fe(CO)_5$, giving at least eight compounds, of which three are shown below.

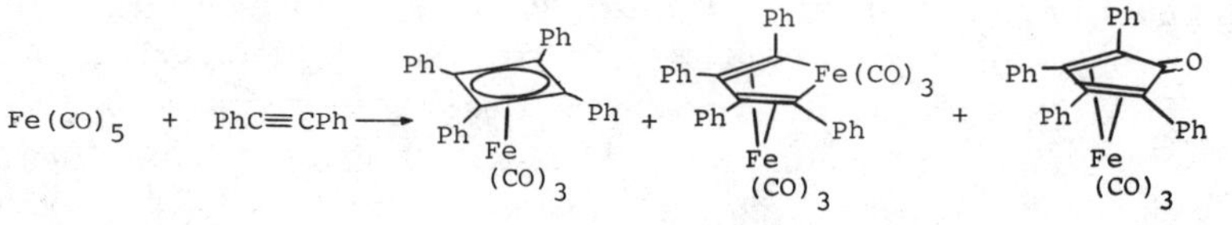

$$Fe(CO)_5 + PhC{\equiv}CPh \longrightarrow \textbf{4-12} + \textbf{4-13} + \textbf{4-14} \quad (4.13)$$

Complex **4-12** is an η^4-cyclobutadiene complex formed by cyclization of two acetylene molecules, and the upper half of **4-13** is a metallacyclopentadiene entity that acts as an η^4-diene for the lower $Fe(CO)_3$ moiety. The cyclopentadienone moiety in complex **4-14** may be formed by insertion of the CO ligand into the Fe–C bond in the metallacyclopentadiene entity (in **4-13**) followed by ring closure.

c. η^4- and η^6-Bonded Polyene Complexes

Diene- and triene-coordinated complexes are prepared by interaction of these polyenes with low-valent transition metal complexes.[13] Metal carbonyls are conveniently used as zero-valent metal complexes from which part of the carbonyl ligands is displaced on heating or UV irradiation.

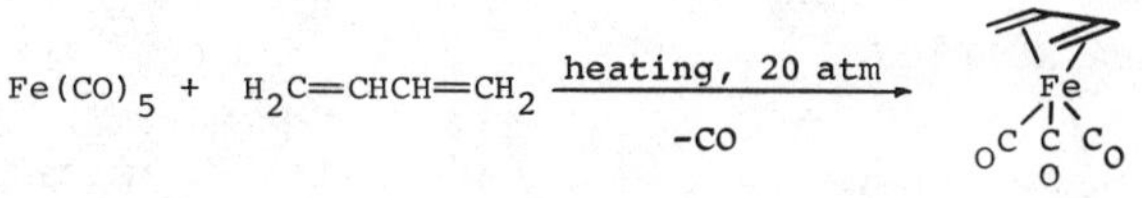

$$Fe(CO)_5 + H_2C{=}CHCH{=}CH_2 \xrightarrow[-CO]{\text{heating, 20 atm}} \textbf{4-15}\ (Fe(CO)_3) \quad (4.14)$$

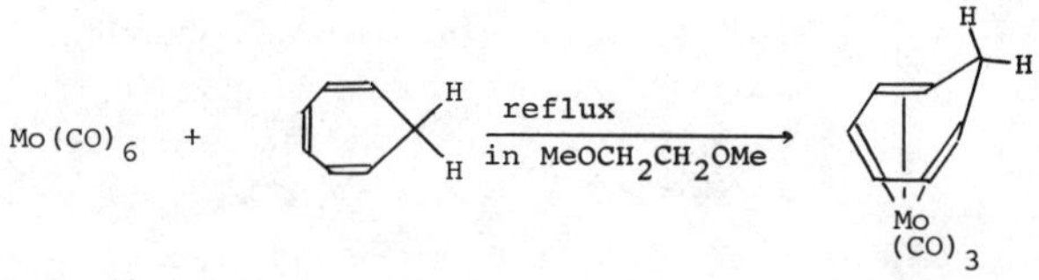

$$Mo(CO)_6 + C_8H_8 \xrightarrow[\text{in } MeOCH_2CH_2OMe]{\text{reflux}} \textbf{4-16}\ (Mo(CO)_3) \quad (4.15)$$

As discussed in Section 3.4, the numbers of double bonds that interact with metals in these and related complexes are restricted by the 18-electron rule. Various metal–diene complexes can be prepared by metal vapor synthesis.[10]

A new route to butadiene complexes of early transition metals containing cyclopentadienyl ligands that use a butadiene–magnesium adduct has been recently developed.[14,15]

$$Cp_2MCl_2 + (\text{—}Mg\text{—}CH_2\text{—}HC\text{=}CH\text{—}CH_2\text{—})_n \quad (M = Zr, Hf) \longrightarrow [Cp_2M\text{--}(\eta^2\text{-butadiene})] \rightleftharpoons \text{s-trans } \mathbf{4\text{-}17} \text{ or s-cis } \mathbf{4\text{-}18} \quad (4.16)$$

Trimethylenemethane complexes represent a special class of η^4-C_4 coordination in which the hypothetical trimethylenemethane radical is stabilized on complexation with a transition metal atom. Representative synthetic methods are shown below.

$$CH_2\text{=}C(CH_2Cl)_2 + Fe_2(CO)_9 \longrightarrow \underset{\mathbf{4\text{-}19}}{(\eta^4\text{-}C(CH_2)_3)Fe(CO)_3} \xleftarrow{\text{heat}} (\eta^3\text{-}C_4H_7)FeCl(CO)_3 \quad (4.17)$$

There are two main synthetic routes to η^4-cyclobutadiene complexes.[16] One uses reagents containing four-membered carbocyclic rings such as dichlorocyclobutenes:

$$\text{3,4-dichlorocyclobutene} \longrightarrow \begin{cases} \xrightarrow{M(CO)_6,\ Na/Hg\ (M = Mo,\ W)} (\eta^4\text{-}C_4H_4)M(CO)_4 \quad \mathbf{4\text{-}20} \\ \xrightarrow{Ru_3(CO)_{12},\ Na/Hg} (\eta^4\text{-}C_4H_4)Ru(CO)_3 \quad \mathbf{4\text{-}21} \\ \xrightarrow{Na_2Fe(CO)_4} (\eta^4\text{-}C_4H_4)Fe(CO)_3 \quad \mathbf{4\text{-}22} \end{cases} \quad (4.18)$$

$$\text{(benzocyclobutene-}X_2\text{)} \xrightarrow{Fe_2(CO)_9} \text{4-23} \qquad (4.19)$$

4-23

The other route is to cyclodimerize acetylenes and acetylene derivatives. The preparation of $Fe(\eta^4\text{-}C_4Ph_4)(CO)_3$ (**4-12**) in Eq. (4.13) is an example. An η^4-cyclobutadiene–cobalt complex containing a cyclopentadienyl ligand (**4-24**) is available from several starting complexes.

$$\left.\begin{array}{l} CoCp_2 \\ CpCo(CO)_2 \\ CpCo(cod) \\ CpCo(PPh_3)(PhC{\equiv}CPh) \end{array}\right\} \xrightarrow{PhC\equiv CPh} \text{4-24} \qquad (4.20)$$

4-24

The cyclodimerization of acetylenes is considered to proceed stepwise via a metallacyclopentadiene complex, which can sometimes be isolated. For example, **4-25** decomposes on heating to give **4-24**.

$$CpCo(PPh_3)(PhC{\equiv}CPh) \xrightarrow[80\ ^\circ C]{PhC\equiv CPh} \text{4-25} \xrightarrow{heat} \text{4-24} \qquad (4.21)$$

4-25

d. η^3-Allylic Complexes

Various η^3-allylic complexes[17] can be prepared by treatment of transition metal salts with allylic Grignard reagents.[18] For example, bis(η^3-allyl)nickel can be synthesized as follows:

$$2C_3H_5MgBr + NiCl_2 \xrightarrow[-10^\circ C]{ether} Ni(\eta^3\text{-}C_3H_5)_2 + 2MgBrCl \qquad (4.22)$$

4-26

Homoleptic η^3-allyl complexes (binary complexes containing only η^3-allyl groups) have been prepared for most of the transition metals, and their chemistry has been extensively studied, mainly by Wilke's group, in connection with their roles in butadiene oligomerization. They are summarized in Table 4.1.[17,18]

Table 4.1. Various Homoleptic η^3-Allyl Complexes[a]

	VA_3	CrA_3[b]	FeA_3	CoA_3	NiA_2
ZrA_4	NbA_3	MoA_4[b]		RhA_3	PdA_2
HfA_4	TaA_4	WA_4		IrA_3	PtA_2

[a] A: η^3-C_3H_5.
[b] For Cr and Mo dinuclear complexes Cr_2A_4 and Mo_2A_4 are also known.

The Grignard method can also be used to introduce allyl groups in complexes bearing other ligands. In the reaction of allylmagnesium halides with $TiCp_2Cl_2$, reduction of Ti(IV) to Ti(III) is also involved.

$$TiCp_2Cl_2 + allylMgX \longrightarrow Cp_2Ti(\eta^3\text{-allyl}) \quad (4.23)$$

Like zero-valent metal complexes with COD, these homoleptic η^3-allyl complexes can serve as reactive chemical species that can be used for various stoichiometric and catalytic reactions. This is so even though some of the homoleptic η^3-allyl complexes are thermally unstable. The η^3-allyl groups originally attached to the metal can be readily displaced by incoming substrates, and the chemical reactivities can be modified by addition of external ligands such as tertiary phosphines.

Intermezzo — *The Man Who Brought Sex into Chemistry*

In the beginning of the 1960s organotransition metal chemistry was just "taking off." The controversy about whether the active site of Ziegler-type catalysts was the transition metal or the aluminum (see Section 7.2) was not settled. It was Wilke and his group who prepared transition metal complexes free from the aluminum component and proved that these transition metal complexes in fact act as very active catalysts for butadiene and propylene oligomerization. With their excellent techniques using Schlenk tubes (see Section 5.1) they prepared a number of extremely air sensitive and sometimes thermally unstable transition metal complexes free of supporting ligands such as tertiary phosphines, CO, and cyclopentadienyl groups. In his lectures Wilke called the nickel complexes without supporting ligands "naked nickel." An American nicknamed him "a man who brought sex into chemistry." Wilke seemed to enjoy this name.

The work of Wilke's group removed much of the mystery attached to catalysis and demonstrated that the specific activity of the catalysts can be tuned by clothing the "naked metal" in various ligand additives.

It may be added that the most productive part of this work was accomplished in a one-story, temporary wooden building that Wilke's group called a barrack, not in the supermodern, impressive multistory building in which Wilke's group now works.

η^3-Allyl complexes having carbonyl groups can be prepared by reactions of hydrido- or alkyltransition metal complexes with butadiene[19] or by protonation of butadiene complexes with HCl or other strong acids.[20]

$$CoH(CO)_4 + H_2C{=}CHCH{=}CH_2 \xrightarrow{-CO} \text{4-27} + \text{4-28} \quad (4.24)$$

$$\text{(butadiene)}Fe(CO)_3 + HCl \longrightarrow \text{4-29} + \text{4-30} \quad (4.25)$$

For η^3-allylic complexes the form with a substituent (in this case the methyl group) on the same side as the central methine portion of the η^3 entity is called *syn* and the other form, with the substituent on the other side of the methine proton, is called *anti*. Generally the syn form is thermodynamically more stable than the anti form. In reaction (4.25) the anti form **(4-29)** is formed initially and then isomerizes to the syn form **(4-30)**.

η^3-Allylic complexes can also be synthesized conveniently by reactions of allylic halides with low-valent transition metal complexes.

$$CH_2{=}CHCH_2X \xrightarrow{Ni(CO)_4} \text{4-31}$$
$$CH_2{=}CHCH_2X \xrightarrow{Pt(cod)_2} \text{(allyl)}Pt(cod)X \xrightarrow{AgPF_6} [(\eta^3\text{-allyl})Pt(cod)]^+PF_6^- \;\; \text{4-32}$$
$$CH_2{=}CHCH_2X \xrightarrow{Fe_2(CO)_9} (\eta^3\text{-allyl})Fe(CO)_3X \;\; \text{4-33}$$
$$CH_2{=}CHCH_2X \xrightarrow{Na^+CpMo(CO)_3^-} Cp(\text{allyl})Mo(CO)_3 \xrightarrow[-CO]{UV} (\eta^3\text{-allyl})CpMo(CO)_2 \;\; \text{4-34} \quad (4.26)$$

The reaction of an allyl halide with a low-valent transition metal complex involves the cleavage of the C–X bond and an increase in the formal oxidation state of the metal. This is an example of oxidative addition (see Section 6.2). Oxidative addition reaction of allylic compounds to transition metal complexes can take place not only with the carbon–halogen bond cleavage but also with the cleavage of other bonds such as C–O and C–H. A few examples involving C–O bond cleavage are shown below.[21,22]

$$Ni(cod)_2 + L + CH_3COOCH_2CH{=}CH_2 \longrightarrow (\eta^3\text{-allyl})Ni(O_2CCH_3)(L) \quad (4.27)$$

$L = PPh_3, PEtPh_2, PCy_3$

4-35

$$Pd(PCy_3)_2 + CH_3COOCH_2CH{=}CH_2 \longrightarrow (\eta^3\text{-allyl})Pd(O_2CCH_3)(L) \quad (4.28)$$

PCy_3 = tricyclohexylphosphine

4-36

The allylic proton of coordinated propylene can be removed as shown in the following reactions.

$$2CH_2{=}CHCH_3 + 2PdCl_2 \longrightarrow [(CH_2{=}CHCH_3)PdCl_2]_2 \xrightarrow{-2HCl} [(\eta^3\text{-allyl})Pd(\mu\text{-Cl})]_2 \quad (4.29)$$

4-37

$$RuH_2(PPh_3)_4 + CH_3CH{=}CH_2 \longrightarrow (\eta^3\text{-}C_3H_5)_2Ru(PPh_3)_2 \quad (4.30)$$

4-38

It should be noted that the allyl group can behave as a bridging ligand in certain cases. For example, in reaction (4.28), in addition to the mononuclear complex **4-36**, a dinuclear complex (**4-39**) having a Pd–Pd bond and a bridging allyl and acetato ligands is also produced. The same complex can be obtained by heating **4-36** in benzene[22] and also by the reaction of an acetato-bridged dimer (**4-40**) with PCy_3.[23]

$$(\eta^3\text{-allyl})Pd(OAc)(L) \xrightarrow[\text{in benzene}]{70\ °C} L\text{-}Pd(\mu\text{-allyl})(\mu\text{-}O_2CCH_3)Pd\text{-}L \longleftarrow [(\eta^3\text{-allyl})Pd(\mu\text{-}O_2CCH_3)]_2 \quad (4.31)$$

4-36 **4-39** **4-40**

The bridging halogen ligands in the $[(\eta^3\text{-allyl})MX]_2$ dimers (M = Ni, Pd) can be cleaved by addition of Lewis bases, such as PPh_3, into mononuclear η^3-allyl complexes.[24]

$$[(\eta^3\text{-allyl})Ni(\mu\text{-Cl})]_2 + 2L \longrightarrow (\eta^3\text{-allyl})NiCl(L) \qquad (4.32)$$

4-41

The roles played by these η^3-allyl complexes as intermediates in catalytic reactions will be discussed in Section 7.2 and 8.2.

e. η^5-Cyclopentadienyl and Related Complexes

Since the first accidental discovery of ferrocene,[25] an enormous number of complexes having η^5-cyclopentadienyl (abbreviated Cp) ligands have been prepared.[26] The most general method for the preparation of metallocenes employs the sodium salt of cyclopentadiene. Cyclopentadiene, being a weak acid ($pK_a \sim 20$), reacts with a strong base such as Na or NaH in tetrahydrofuran (THF) and other ethers to form a salt $Na^+C_5H_5^-$. The subsequent reactions of the resultant solution with metal halides give metallocenes.

$$C_5H_6 + Na \xrightarrow{THF} Na^+C_5H_5^- + \frac{1}{2}H_2 \quad \text{(main reaction)} \qquad (4.33)$$

4-42

$$2Na^+C_5H_5^- + NiCl_2 \xrightarrow{THF} NiCp_2 + 2NaCl \qquad (4.34)$$

Direct reactions of cyclopentadiene with metal halides in the presence of an amine acting as an acceptor of HCl are also useful.

$$2C_5H_6 + 2Et_2NH + FeCl_2 \xrightarrow{THF} FeCp_2 + 2Et_2\overset{+}{N}H_2Cl^- \qquad (4.35)$$

$$2C_5H_6 + 2Et_2NH + TiCl_4 \xrightarrow{THF} CP_2TiCl_2 + 2Et_2\overset{+}{N}H_2Cl^- \qquad (4.36)$$

Cyclopentadienylthallium, which can be readily prepared and stored, also provides a useful means for the preparation of various cyclopentadienyl complexes.

$$C_5H_6 + TlOH \xrightarrow{H_2O} TlC_5H_5 \downarrow + H_2O \qquad (4.37)$$

4-43

$$FeCl_2 + 2TlC_5H_5 \longrightarrow FeCp_2 + 2TlCl \qquad (4.38)$$

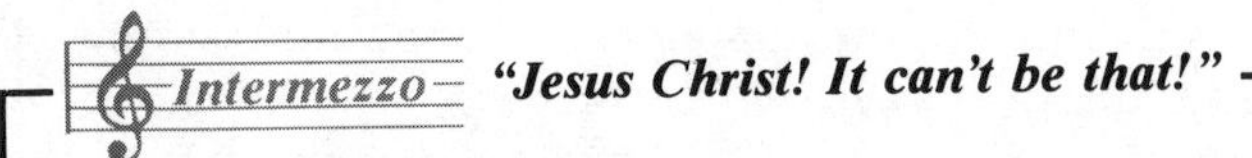

"Jesus Christ! It can't be that!"

So said Geoffrey Wilkinson to himself when, in January 1952, as an assistant professor at Harvard University, he read of a paper by Kealy and Pauson in a newly arrived issue of *Nature*. The paper[25a] reported an unsuccessful attempt to prepare fulvalene by oxidation of the cyclopentadienyl Grignard reagent:

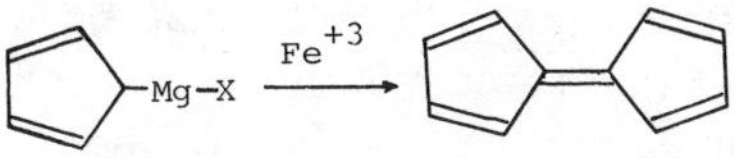

They reported that, instead, the reaction gave an extremely stable orange compound, for which they proposed the following resonance structures:

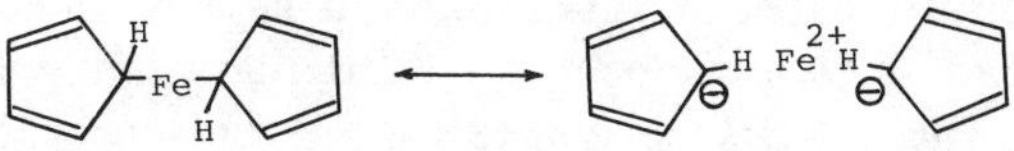

Wilkinson, who had some experience with organotransition metal compounds and knew of their instability, instantly realized the problem of formulating the structure of the new compound with metal–carbon σ bonds. While he was thinking of a way to prepare similar compounds, M. Rosenblum, a Ph.D. student in R. B. Woodward's group, came to his laboratory asking if Wilkinson had any ruthenium. Wilkinson immediately realized that Woodward was thinking along similar lines. It is a chemist's logic to try the synthesis of a compound similar to the newly prepared one. If one can synthesize the iron compound, why not try ruthenium, the element below iron in the periodic table? Wilkinson and Woodward sorted out their research projects at lunch at the Harvard Faculty Club on the following Monday.

Thus, the project started and within a very short period not only the iron but also the ruthenium and cobalt complexes were synthesized and their structures determined by the most modern analytical means available at the time, including IR and NMR spectroscopy. Equivalence of all the protons in the cyclopentadienyl ring bound to the iron atom was confirmed, leading to the proposal of the famous *sandwich* structure.

At the same time, on the other side of the Atlantic Ernst Otto Fischer was working rapidly on the same problem and reached the same conclusion as Wilkinson. He proposed a somewhat more rigorous but less appealing name, a double-cone structure, which was not used by others. Woodward, with M. C. Whiting, coined the name ferro*cene* for the new compound by analogy with ben*zene* because of the aromaticity of the complex.

Wilkinson reminisces that as a newly appointed assistant professor he had to work frantically during that period to prepare lectures and to keep his research going. He worked at least 12 hours a day and seldom returned to his lodgings before 2:00 A.M. The first joint paper with Woodward was written within a few weeks of starting work.

Thus began the hectic academic race of Wilkinson and Fischer, which eventually brought them jointly the Nobel prize in Chemistry in 1973.

Table 4.2 summarizes the metallocenes that have been prepared by this method. It may be noted that some metallocenes are diamagnetic, whereas the others show paramagnetism, corresponding to the presence of unpaired electrons, for example, 5 for $MnCp_2$. Some metallocenes are not completely characterized because of their insolubility, probably due to polymer formation. Furthermore, some complexes of composition MCp_2 may not have the simple sandwich structure of ferrocene but may be more complicated. For example, "titanocene" was formulated as $TiCp_2$ but was later revealed to be a hydride-bridged dimer having two connected cyclopentadienyl units spanning two titanium atoms. There is also evidence for the presence of another type of "titanocene," where the cyclopentadienyl ligand is acting as an η^1-ligand as well as an η^5-ligand.[27a]

4-44 **4-45**

Table 4.2. Some Bis(cyclopentadienyl)metal Complexes[a]

$TiCp_2$ [b]	VCp_2	$CrCp_2$	$MnCp_2$	$FeCp_2$	$CoCp_2$	$NiCp_2$
	[3]	[2]	[5]		[1]	[2]
$ZrCp_2$ [c]	$NbCp_2$ [d]	$MoCp_2$ [e]		$RuCp_2$	$RhCp_2$ [f]	$PdCp_2$ [f]
					[paramag]	
	$TaCp_2$ [d]	WCp_2 [f]		$OsCp_2$	$IrCp_2$ [f]	
					[paramag]	

[a] Numbers in square brackets indicate the numbers of unpaired electrons.
[b] For structures see **4-44** and **4-45**.
[c] Structure similar to **4-44**.
[d] Dimer with μ-H's and $\mu(\eta^1,\eta^5\text{-}C_5H_5)$.
[e] Not well characterized.
[f] Exists only as a reactive intermediate.

In addition to the metallocenes that have sandwich structures, a variety of complexes with one cyclopentadienyl ligand are known. The reactions of cyclopentadienyl sodium (**4-42**) with transition metal halides under CO pressure yield $CpM(CO)_n$-type complexes.[27b] The reaction of an alkali metal cyclopentadienyl (**4-42**) with metal carbonyl halides gives complexes $[CpM(CO)_x]_y$, where M = Pt, Rh, Ir, Ru, Os, or Mn; a typical example of this reaction is the synthesis of $CpRh(CO)_2$ (**4-46**).

$$[Rh(CO)_2Cl]_2 + 2Na^+C_5H_5^- \longrightarrow 2CpRh(CO)_2$$

4-42 **4-46**

In addition to the unsubstituted cyclopentadienyl ligand, cyclopentadienyl ligands substituted with methyl and other alkyl groups find increasing use, the most common being pentamethylcyclopentadienyl (C_5Me_5).[28] This ligand has several advantages. The introduction of five methyl groups onto the cyclopentadienyl ring increases the solubility of its complexes in organic solvents, and the electron density on the metal is also enhanced. The complication of hydrogen abstraction from the cyclopentadienyl ring, producing complexes such as **4-44** and **4-45**, is avoided. It should be added, however, that in some cases the methyl group in the C_5Me_5 ring is also susceptible to hydrogen abstraction.

Pentamethylcyclopentadiene can be used directly for preparing certain types of pentamethylcyclopentadienyl complexes.

MeM(CO)5 + [pentamethylcyclopentadiene] ⟶ [(C5Me5)M(CO)3] (4.39)

M = Mn, Re

4-47

Another route to pentamethylcyclopentadienyl complexes involves the reaction of transition metal compounds with Dewar benzene.[28]

MeC≡CMe ⟶ [hexamethyl Dewar benzene] ⟶ [Me5C5-C(Me)(H)X]

$\xrightarrow{RhCl_3 \cdot xH_2O}$ $[Rh(C_5Me_5)Cl_2]_2$ (4.40)

The most usual way, however, employs the preformed pentamethylcyclopentadienyl of alkali metals in combination with transition metal halides.[29]

$$2\,[C_5Me_5]^-\,Na^+ \xrightarrow{MCl_4} (\eta^5\text{-}C_5Me_5)_2MCl_2 \qquad (4.41)$$

M = Ti, Zr, Hf

4-48

f. η^6-Arene Metal Complexes

Benzene and other arenes form the most widely studied section of η^6-ligand transition metal chemistry.[30-33] The bis(η^6-arene) complexes can be prepared (i) by reduction of transition metal halides in the presence of aromatic hydrocarbons, the original method used by Fischer and co-workers in their first synthesis of bis(arene)chromium, (ii) by metal vapor synthesis, (iii) by displacement of other ligands by aromatic hydrocarbons from low-valent transition metal complexes, and (iv) by cyclotrimerization of acetylenes on transition metal complexes.

(1) *Reduction of a Metal Salt in the Presence of a Ligand*

Fischer and Hafner first prepared bis(η^6-benzene)chromium (**4-49**) by reducing chromium trichloride with aluminum powder in benzene in the presence of aluminum chloride. The reaction proceeds through the formation of a cationic bis(benzene)chromium complex that is then reduced with aqueous sodium dithionite to give the neutral complex **4-49**.

$$3CrCl_3 + 2Al + AlCl_3 + 6C_6H_6 \longrightarrow 3[(\eta^6\text{-}C_6H_6)_2Cr]^+[AlCl_4]^- \qquad (4.42)$$

$$[(\eta^6\text{-}C_6H_6)_2Cr]^+[AlCl_4]^- + S_2O_4^{2-} + 4OH^- \qquad (4.43)$$

$$\longrightarrow (\eta^6\text{-}C_6H_6)_2Cr + 2H_2O + 2SO_3^{2-} + AlCl_4^-$$

4-49

This method is of general use, and vanadium, molybdenum, and ruthenium analogs can be similarly prepared.

$$VCl_4 + 6C_6H_6 + Al \xrightarrow{AlCl_3} [(\eta^6\text{-}C_6H_6)_2V][AlCl_4]$$

$$3MoCl_5 + 6C_6H_6 + 4Al \xrightarrow{AlCl_3} 2[(\eta^6\text{-}C_6H_6)_2Mo][AlCl_4]$$

$$3RuCl_3 + 6C_6H_6 + Al \xrightarrow{AlCl_3} 3[(\eta^6\text{-}C_6H_6)_2Ru][AlCl_4]_2$$

(2) Cocondensation of Metal Atoms with Arenes

The metal–vapor synthesis was first used for synthesis of *bis(η^6-benzene) chromium.*[35] This synthetic method has proved to be particularly valuable for preparation of bis(η^6-arene) complexes that cannot withstand the Lewis acid used in the reductive method. A variety of bis(η^6-arene) complexes with functional groups has been synthesized successfully by the metal–vapor method. Other bis(arene)-type complexes such as $(C_6H_6)_2Ti$, $(C_6H_6)_2Nb$, and $(C_6H_6)_2W$, which are unavailable by other methods, have been prepared by the metal–vapor synthesis.

(3) Displacement of Other Ligands by Arenes

The direct reaction of an aromatic hydrocarbon with a metal carbonyl often gives an η^6-arene transition metal complex.

$$C_6H_5OMe + Cr(CO)_6 \longrightarrow (\eta^6\text{-}C_6H_5OMe)Cr(CO)_3 \quad (4.44)$$

One of the η^5-cyclopentadienyl ligands of ferrocene is readily replaced by benzene in the presence of $AlCl_3$ to give a cation that can be isolated as a stable salt with anions such as PF_6, BF_4, and BPh_4.

$$Cp_2Fe + C_6H_6 \xrightarrow{AlCl_3} [(\eta^5\text{-}C_5H_5)Fe(\eta^6\text{-}C_6H_6)]^+[AlCl_4]^- \quad (4.45)$$

4-50

(4) Cyclotrimerization of Acetylenes

The cyclotrimerization of disubstituted acetylenes in the presence of a transition metal can give η^6-arene complexes. For example, diphenylmanganese prepared *in situ* from manganese dichloride and phenylmagnesium bromide can be converted into an η^6-hexamethylbenzene complex (**4-51**) on treatment with but-2-yne.

$$MnCl_2 + 2PhMgBr \xrightarrow{MeC\equiv CMe} [(\eta^6\text{-}C_6Me_6)_2Mn]^+ \quad (4.46)$$

4-51

In general, however, the cyclic oligomerization of acetylenes is rarely used as a preparative route for specific η^6-arene transition metal complexes since the routes already described give better yields of products.

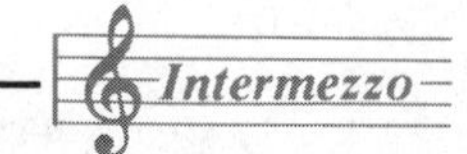

When Was the First Sandwich Complex Prepared?

Ferrocene was prepared and its sandwich-type structure established independently by the Wilkinson and Fischer groups in 1952. Fischer also prepared bis(η^6-benzene)chromium and confirmed its structure in 1955. Prior to these surprising discoveries the first sandwich complexes had been prepared without being recognized as such by Hein as long ago as 1919.[36] He obtained a dark brown solid by the reaction of $CrCl_3$ with C_6H_5MgBr and regarded it as a mixture of "polyphenylchromium compounds" in which the chromium atom is bonded to two to four phenyl groups, namely, $(Ph)_nCr^{0,1+}$ ($n = 2$, 3, or 4). Their true identity was not clarified until 35 years later. Hein's complex attracted the attention of H. H. Zeiss and M. Tsutsui, who after repeating the preparation over 50 times concluded that it was a mixture of η^6-arene complexes with the following structures.[37]

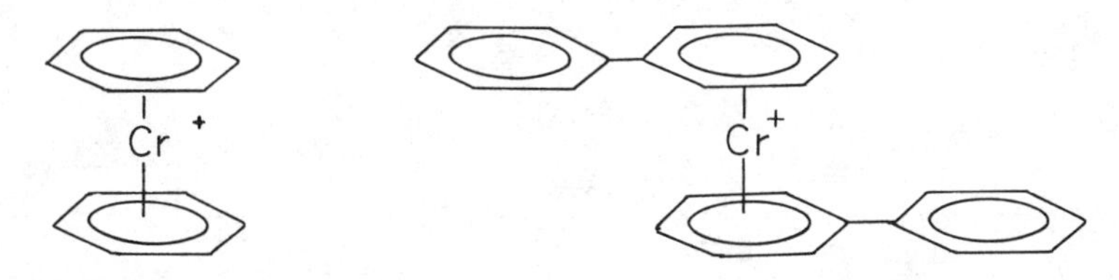

The paper, which was submitted to the *Journal of the American Chemical Society* in 1954, was at first rejected because of the lack of supporting evidence for the novel structures. The paper was accepted only after the publication of Fischer and Hafner's paper, which established the sandwich structure unequivocally by means of X-ray crystallography.[34]

g. η^7- and η^8-Bonded Polyene Complexes

The η^7-cycloheptatrienyl ligand is planar, and (η^7-cycloheptatrienyl)transition metal complexes[38,39] are formed readily by hydride abstraction from the corresponding η^6-cycloheptatriene complexes. The first η^7-cycloheptatrienyl complex was synthesized by hydride abstraction from tricarbonyl(η^6-cycloheptatriene)molybdenum with triphenylmethylium tetrafluoroborate.

$$C_7H_8 + M(CO)_6 \xrightarrow{-3CO} (\eta^6\text{-}C_7H_8)M(CO)_3 \xrightarrow{-H^-} [(\eta^7\text{-}C_7H_7)M(CO)_3]^+ \quad (4.47)$$

M = Cr, Mo, W

4-52

Triethyloxonium tetrafluoroborate can also be used as the hydride ion acceptor. In some cases the direct reaction of cycloheptatriene with a transition metal derivative gives an η^7-cycloheptatrienyl compound by a process involving a spontaneous expulsion of a hydride ion.

$V(CO)_6$ + R-cycloheptatriene $\xrightarrow{-H^-}$ $(\eta^7\text{-}C_7H_6R)V(CO)_3$ (4.48)

4-53

Complexes with η^8-cyclooctatetraenyl ligands can be prepared by the reaction of an alkali metal salt of cyclooctatetraene with transition metal halides. In this way, C_8H_8 complexes of Ti, V, Cr, Mo, Co, and Ni have been prepared.[40] X-ray analysis of the titanium complex $Ti_2(C_8H_8)_3$ **(4-54)** confirmed that this compound contains two planar eight-membered rings.

4-54

4.2. TRANSITION METAL COMPLEXES WITH METAL–CARBON σ BONDS

Transition metal complexes with metal–carbon σ bonds[41-46] were generally considered unstable before the 1950s, and only a few simple transition metal alkyls were known.[41] An increasing number of transition metal alkyls and aryls has been prepared since that time using supporting or stabilizing ligands such as cyclopentadienyl, CO, tertiary phosphines, and organic nitrogen bases like bipyridine, and complexes with the metal–carbon σ bond are now available for most transition metals.

Binary transition metal alkyls and aryls without supporting ligands are called homoleptic.[43] These are usually less stable than transition metal alkyls that also have supporting ligands. For example, $Ti(CH_3)_4$ decomposes thermally at −50 °C,[47] whereas its bipyridine adduct $Ti(CH_3)_4(bipy)$ is stable to 0 °C.[48] Some homoleptic transition metal alkyls containing special groups such as $CH_2C_6H_5$, CH_2SiMe_3, CH_2CMe_3, CH_2PMe_3, and 1-norbornyl are thermally more stable because in these cases low-energy thermolysis pathways involving β-hydrogen elimination are blocked. We deal with the thermolysis pathways in Section 6.5.

a. Transition Metal Alkyls

The following methods have been used in synthesis of transition metal alkyls and aryls.

1. Alkyl transfer reactions.
2. Treatment of anionic transition metal complexes with alkyl halides.
3. Oxidative addition.
4. Insertion reactions.
5. Elimination reactions.
6. Attack on coordinated ligand.

Of these methods, alkyl transfer reactions have been most widely employed.

(1) Synthesis by Alkyl Transfer Reactions

A variety of alkyl compounds of main-group elements serve as alkylating reagents. Most common are alkyllithium and Grignard reagents, but many other alkylmetal compounds have been used, notably those of aluminum, zinc, mercury, and tin. These are allowed to react with transition metal halides, acetylacetonates, or carboxylates. Since many transition metal alkyls are thermally unstable and oxygen and water sensitive, care must be taken with the reaction conditions, the combination of the alkylating reagents and the transition metal compound, as well as the supporting ligands. The choice of molar ratios and solvents is often crucial as well.

Generally speaking, main-group metal alkyls in which the alkyl groups are strongly carbanionic are the best alkylating reagents. For example, reactions of $CrCl_3$ with alkyl Grignard reagents give trialkylchromium compounds, whereas the alkylaluminum compounds give monoalkylchromium halides. A similar difference in alkylating powers has been observed in the alkylation of $TiCl_4$.

$$CrCl_3(thf)_3 + 3MeMgBr \longrightarrow CrMe_3(thf)_3 \quad (4.49)$$

$$CrCl_3(thf)_3 + AlR_3 \longrightarrow RCrCl_2(thf)_3 \quad (4.50)$$

$$TiCl_4 + 4MeLi \longrightarrow TiMe_4 \quad (4.51)$$

$$TiCl_4 + AlMe_3 \longrightarrow MeTiCl_3 \quad (4.52)$$

These are exchange reactions of the alkyl groups in the main group alkyls ($M'R_n$) with the halide ligands in the transition metal halides MX_m. The reactions probably proceed through bridged intermediates as shown below.

$$\text{M-X} + \text{R-M}' \rightleftarrows \text{M}\begin{smallmatrix}\text{X}\\ \\ \text{R}\end{smallmatrix}\text{M}' \rightleftarrows \text{M}\begin{smallmatrix}\text{X}\\ \\ \text{R}\end{smallmatrix}\text{M}' \longrightarrow \text{M-R} + \text{M}'\text{X} \quad (4.53)$$

Use of an excess of alkyllithium can sometimes lead to further alkylation of the neutral transition metal alkyls to produce lithium salts of ate complexes of transition metal alkyls. Examples include most transition metal complexes of group 4 to 10 metals such as $Li[TiMe_5]\cdot 2(dioxane)$, $[Li(dioxane)]_3[CrMe_6]$, $Li_2[PtMe_6]$.

Although alkyllithiums and Grignard reagents are most frequently used, it is sometimes difficult to remove the lithium or magnesium halides produced from the desired transition metal alkyls. One encounters difficulties in purification when the transition metal alkyls are water sensitive, so that the lithium or magnesium halides cannot be removed by treatment with water. In such cases reaction of alkylaluminum and zinc compounds with transition metal acetylacetonates in the presence of neutral ligands provides a useful alternative. The aluminum or zinc compounds formed by the ligand exchange reactions remain in solution, whereas the transition metal alkyls may deposit at low temperatures and can be readily isolated.

Some typical examples of synthesis of transition metal alkyls are given below.

$$TiCl_4 + 4PhCH_2MgCl \longrightarrow Ti(CH_2Ph)_4 \quad (4.54)$$

$$NbCl_5 + ZnMe_2 \longrightarrow NbMe_2Cl_3 \quad (4.55)$$

$$WCl_6 + 6AlMe_3 \longrightarrow WMe_6 \quad (4.56)$$

$$Ni(acac)_2 + AlR_2OEt + bipy \longrightarrow (bipy)NiR_2 \quad (4.57)$$

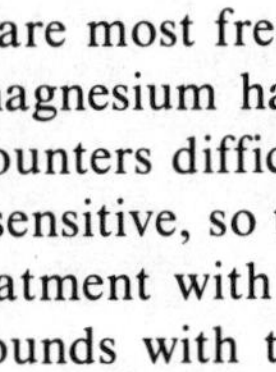

R = Me, Et, Pr^n, Bu^i

4-55

$$Fe(acac)_3 + AlR_2OEt + bipy \longrightarrow cis\text{-}FeR_2(bipy)_2 \quad (4.58)$$

R = Me, Et

4-56

$$NiCl_2 + MeLi + dppe \longrightarrow (Ph_2PCH_2CH_2PPh_2)NiMe_2 \quad (4.59)$$

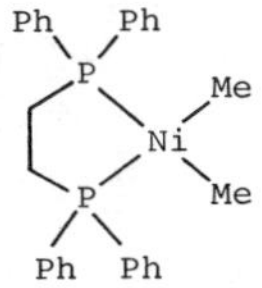

4-57

$$Ni(acac)_2 + Me_2AlOEt + PEt_3 \longrightarrow trans\text{-}NiMe_2(PEt_3)_2 \quad (4.60)$$

4-58

$$Pd(acac)_2 + R_2AlOEt + PR'_3 \longrightarrow trans\text{-}PdR_2(PR'_3)_2 \quad (4.61)$$

4-59

R = Me, Et, Pr $\quad PR'_3 = PPhMe_2$, etc.

$$PdCl_2 + RLi + PR'_3 \longrightarrow \underset{\textbf{4-60}}{cis\text{-}PdR_2(PR'_3)_2} \qquad (4.62)$$
$$R = Me, Et$$

Reactions (4.61) and (4.62) provide an interesting example of the stereochemistry of the alkyl complex determined by the nature of the alkylating agent. Details of the stereochemical consideration will be discussed in Section 6.1. Note also that employment of Grignard reagents in place of alkyllithium gives a monoalkylpalladium halide.

$$PdBr_2(PEt_3)_2 + MeMgBr \longrightarrow trans\text{-}MePdBr(PEt_3)_2 \qquad (4.63)$$

In contrast to the palladium dialkyls, which can take both trans and cis configurations, platinum dialkyls favor the *cis*-dialkyl configuration.

$$PtCl_2 + \underset{(RMgX)}{RLi} + PR'_3 \longrightarrow \underset{\textbf{4-61}}{cis\text{-}PtR_2(PR'_3)_2} \qquad (4.64)$$

Treatment of $Co(acac)_3$ with LiMe affords a trimethylcobalt complex (**4-62**).

$$Co(acac)_3 + MeLi \longrightarrow \underset{\textbf{4-62}}{mer\text{-}CoMe_3(PMe_3)_3^{\dagger}} \qquad (4.65)$$

As shown above, tertiary phosphines and other nitrogen bases such as 2,2′-bipyridine serve as useful supporting ligands. The choice of ligand still remains largely empirical. For some transition metal alkyls bipyridine acts as a good supporting ligand, whereas for others it is ineffective. The precise reason for the stabilizing effect of the supporting ligand is not clear. For example, tertiary phosphines are excellent stabilizing ligands for copper alkyls, which themselves are thermally very unstable and often decompose violently in the absence of the ligand. The tertiary phosphine-stabilized copper alkyls prepared in the following manner are much more stable and can be handled in the absence of air without danger of explosion.[49]

† For octahedral complexes ML_3X_3 having two types of ligands L and X there are two possible isomers, namely *mer*-ML_3X_3 and *fac*-ML_3X_3:

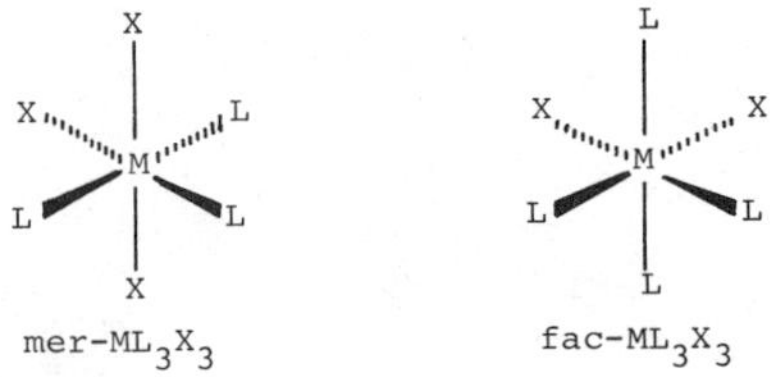

mer-ML_3X_3 fac-ML_3X_3

A meridional isomer (abbreviated *mer*) contains three identical ligands occupying the sites located on the meridian of the complex, and a facial isomer (*fac*) has the same three ligands occupying the sites on the face of the octahedron.

$$Cu(acac)_2 + R_2AlOEt + PR'_3 \longrightarrow \underset{\textbf{4-63}}{CuR(PR_3)_n} \quad n = 1\text{-}3 \tag{4.66}$$

On the other hand, 2,2′-bipyridine, which serves as an excellent stabilizing ligand for metal alkyls such as **4-55** and **4-56**, is totally ineffective for copper alkyls. Lithium dialkylcuprates prepared by the reaction of a copper halide with two equivalents of alkyllithium are more stable than neutral copper alkyls.

$$CuX + 2RLi \longrightarrow \underset{\textbf{4-64}}{[R_2Cu]Li} \tag{4.67}$$

In contrast to alkyllithium, which often gives transition metal ate complexes, trialkylaluminum, when employed as an alkylating agent, serves as a Lewis acid and tends to accept another alkyl group, forming a tetraalkylaluminate ion.[50]

$$Co(acac)_3 + AlR_3 + bipy \longrightarrow [Co(bipy)_2R_2]^+ AlR_4^- \tag{4.68}$$

4-65

Complex **4-65** probably forms by a reverse alkyl migration to AlR_3 from a trialkylcobalt intermediate formed in the reaction of $Co(acac)_3$, AlR_3, and bipy.

Reactions of transition metal acetylacetonates with alkylaluminum compounds probably proceed by stepwise displacement of the acetylacetonato groups by the alkyl groups of the aluminum alkyls. In fact, transition metal alkyls containing unexchanged acetylacetonato ligands can sometimes be isolated as intermediates under appropriate conditions.

$$Ni(acac)_2 + AlEt_2(OEt) + PPh_3 \longrightarrow Ni(acac)(CH_2CH_3)(PPh_3) \tag{4.69}$$

4-66

$$Co(acac)_3 + AlR_2(OEt) + L \longrightarrow Co(acac)R_2L_2 \tag{4.70}$$

L = tertiary phosphines

4-67

Although tertiary phosphines are generally excellent stabilizing ligands for transition metal alkyls, in certain cases they accelerate the thermolysis of transition metal alkyls,[51] so that addition of an excess of tertiary phosphines should be avoided.

(2) Treatment of Anionic Transition Metal Complexes with Alkyl Halides

This method is particularly useful for synthesis of transition metal alkyls containing carbonyl ligands. Electron-rich anionic transition metal complexes can be prepared either by reduction of neutral transition metal complexes or by treatment of neutral complexes with bases such as OH^- and NH_2^-.

$$Co_2(CO)_8 + 2Na/Hg \xrightarrow{THF} 2Na[Co(CO)_4]$$

$$Fe(CO)_5 + 3NaOH(aq) \longrightarrow Na[FeH(CO)_4](aq) + Na_2CO_3$$

$$Mo(CO)_6 + C_5H_5Na \longrightarrow Na[CpMo(CO)_3]$$

Reactions of these anionic transition metal complexes with alkyl halides give transition metal alkyls.

$$Na[CpMo(CO)_3] + RX \xrightarrow{-NaX} CpMo(CO)_3R$$

$$Na_2[Fe(CO)_4] + RX \xrightarrow{-NaX} \underset{\textbf{4-68}}{Na[RFe(CO)_4]}$$

Complex **4-68** is very electron rich, and its reactions with electrophiles provide various organic compounds.[52]

$$\underset{\textbf{4-68}}{Na[RFe(CO)_4]} \xrightarrow{L} Na[RCOFe(CO)_3L] \begin{cases} \xrightarrow{H^+} RCHO \\ \xrightarrow{R'X} RCOR' \\ \xrightarrow{H_2O,\ O_2} RCOOH \end{cases}$$

(3) Synthesis by Oxidative Addition Reactions

Reactions of low-valent transition metal complexes with alkyl halides lead to formal oxidation of the transition metal complexes, giving transition metal alkyls. The process is called oxidative addition and will be discussed in detail in Section 6.2. Some examples are given below.

$$RhCl(PPh_3)_3 + CH_3I \longrightarrow RhCl(I)(CH_3)(PPh_3)_2 + PPh_3$$

$$IrCl(CO)(PPh_3)_2 + RX \longrightarrow IrCl(X)(R)(PPh_3)_2$$

Oxidative addition reactions occur most readily for third-row elements (e.g., $Ir^I > Rh^I > Co^I$), and complexes in lower oxidation states are more reactive toward oxidative addition. Alkyl iodides are more reactive than bromides or chlorides. For d^{10} platinum(0) complexes oxidative addition of methyl iodide takes place in two steps, giving first a d^8 Pt(II) and then a d^6 Pt(IV) complex.

$$Pt(PPh_3)_3 + CH_3I \xrightarrow{-PPh_3} CH_3PtI(PPh_3)_2$$

$$\xrightarrow{CH_3I} (CH_3)_2PtI_2(PPh_3)_2$$

Oxidative additions involving C–H, C–O, and C–C bond cleavage can also give transition metal complexes with a metal–carbon σ bond. Details will be discussed in Section 6.2.

The metal atom vaporization technique can also be employed for the synthesis of certain metal alkyls. Alkyl halides oxidatively add to metal atoms to give alkyl metal halides.

$$Pd + CF_3I \longrightarrow \{PdI(CF_3)\}_n$$

$$Pd + RX + 2PR'_3 \longrightarrow PdR(X)(PR'_3)_2$$

These reactions require sophisticated equipment, and attempts have been made to generate reactive metals with chemical properties similar to those of the metal atoms prepared by vacuum vaporization technique by simpler methods. One method involves reduction of nickel, palladium, or platinum halides with potassium to give reactive slurries of the metals, which can be used for the preparation of $MR(X)L_2$ complexes.[53] Ultrasonic irradiation of a system containing zinc dust, alkyl halides, and transition metal halides is also applicable for preparation of certain transition metal alkyls.[54]

$$CF_3I + Zn + MX_n \xrightarrow[\text{irradiation}]{\text{ultrasonic}} CF_3MX$$

The ultrasonic irradiation is considered to promote the formation of active zinc metal, which reacts with the alkyl halide to produce alkylzinc halide *in situ.* The alkylzinc halides thus generated serve to alkylate the transition metal complexes, such as palladium and titanium. Electrochemical oxidation of a nickel or palladium anode in the presence of alkyl halides and stabilizing ligands has also been used for preparation of $MR(X)L_2$ complexes.[55]

(4) Synthesis Involving Insertion Reactions

A transition metal alkyl is available by insertion of an olefin into a metal–hydride bond. This is the reverse of the β-hydrogen elimination reaction of a transition metal alkyl having β-hydrogen atoms. The best-known example is insertion of ethylene under pressure into a Pt–H bond.[56]

$$\text{trans-}PtHBr(PEt_3)_2 + C_2H_4 \rightleftarrows \text{trans-}PtEt(Br)(PEt_3)_2 \tag{4.71}$$

Reaction (4.71) is reversible, and heating of the ethylplatinum complex gives the platinum hydride.

Olefin insertion into the Zr–H bond in $Cp_2ZrH(Cl)$ provides a useful method in organic synthesis.[57]

$$Cp_2Zr(Cl)(H) + CH_2{=}CHR \longrightarrow Cp_2Zr(Cl)(CH_2CH_2R) \qquad (4.72)$$

Insertion of olefins bearing electronegative substituents gives stable compounds that are not susceptible to the reverse β-hydrogen elimination.[58,59]

$$\text{trans-}RhH(CO)(PPh_3)_3 + C_2F_4 \longrightarrow \text{trans-}Rh(CF_2CF_2H)(CO)(PPh_3)_2 + PPh_3$$

$$Cp_2MoH_2 + NCCH{=}CHCN \longrightarrow Cp_2Mo(H)(CH(CN)CH_2CN)$$

Acetylenes also insert into M–H bonds to give metal vinyl complexes.[59]

$$Cp_2MoH_2 + MeO_2CC{\equiv}CCO_2Me \longrightarrow Cp_2Mo(H)[C(CO_2Me){=}C(H)CO_2Me]$$

Insertion of olefins and acetylenes into metal–halogen bonds is also known.[60]

$$AgF + CF_3CF{=}CF_2 \longrightarrow (CF_3)_2CFAg$$

$$AgF + CF_3C{\equiv}CCF_3 \longrightarrow CF_3CF{=}C(CF_3)Ag$$

Acetylene insertions into metal–halogen bonds may occur consecutively.[61]

$$PdCl_2L_2 \xrightarrow{RC\equiv CR} -Pd-C(R){=}C(R)-Cl \longrightarrow -Pd-C(R){=}C(R)-C(R){=}C(R)-Cl \longrightarrow \text{further products}$$

Insertion of methylene, produced from diazomethane, into a metal–hydride bond gives a transition methyl.

$$HMn(CO)_5 + CH_2N_2 \longrightarrow CH_3Mn(CO)_5 + N_2$$

Insertion into metal–chlorine bonds gives a route to chloromethyl derivatives.[62]

$$\text{trans-}IrCl(CO)(PPh_3)_2 + CH_2N_2 \longrightarrow \text{trans-}Ir(CH_2Cl)(CO)(PPh_3)_2$$

Ring opening of cyclic ethers involving insertion into a metal–hydride bond provides a special route to the preparation of a transition metal hydroxyethyl.[42]

$$HCo(CO)_4 + CH_2{-}CH_2\,(\text{-O-}) \longrightarrow HOCH_2CH_2Co(CO)_4$$

For preparation of simple alkyls having β-hydrogens, however, insertion into a metal–hydride bond has rather limited application since the process is reversible with liberation of β-elimination products.

(5) *Synthesis Involving Elimination Reactions*

Although seldom used in synthesis, some transition metal alkyls can be prepared by elimination of CO, CO_2, SO_2, and N_2 from a suitable precursor.

$$CH_3COMn(CO)_5 \xrightarrow{\Delta} CH_3Mn(CO)_5 + CO$$

$$p\text{-}CH_3C_6H_4SO_2IrCl_2(CO)(PPh_3)_2 \xrightarrow[-SO_2]{\Delta} p\text{-}CH_3C_6H_4IrCl_2(CO)(PPh_3)_2$$

$$C_6H_5N_2PtCl(PEt_3)_2 \xrightarrow[Al_2O_3]{-N_2} C_6H_5PtCl(PEt_3)_2$$

When thermal reactions do not proceed, irradiation of a complex by light sometimes gives a metal alkyl by an elimination process.

$$CH_3COFeCp(CO)_2 \xrightarrow{h\nu} CH_3FeCp(CO)_2 + CO$$

(6) *Synthesis by Attack on Coordinated Ligands*

External, nucleophilic attack on coordinated ligands such as alkenes, alkynes, CO, and carbenes transforms metal π complexes into metal complexes with metal–carbon σ bonds. This method is of considerable utility in organic synthesis. Some typical examples are given below, and detailed accounts will be given in Section 6.4.

$$[CpPd(C_2H_4)(PPh_3)]^+ + MeO^- \longrightarrow CpPd(PPh_3)CH_2CH_2OMe$$

$$(CH_2{=}CHCH_2CH_2NMe_2)PdCl_2 + MeOH \longrightarrow [(MeOCH_2CHCH_2CH_2NMe_2)Pd(\mu\text{-}Cl)]_2$$

$$[CpRe(CO)_2(NO)]^+ \xrightarrow{H^-} CpRe(CO)(NO)(CHO) \xrightarrow{H^-}$$

$$CpRe(CO)(NO)(CH_2OH) \xrightarrow{H^-} CpRe(CO)(NO)(CH_3)$$

$$(OC)_5Cr{=}C(OMe)(Me) + PMe_3 \longrightarrow (OC)_5Cr^- {-}C(OMe)(Me)(\overset{+}{P}Me_3)$$

$$[L_nMC{\equiv}O]^+ + 2HNR_2 \rightleftharpoons [L_nM{-}C({=}O){-}NR_2] + H_2\overset{+}{N}R_2$$

b. Transition Metal Aryls

Synthetic methods for the preparation of transition metal aryls are not essentially different from those for transition metal alkyls. Simple transition metal aryls with or without supporting ligands can be prepared by reactions of transition metal halides or acetylacetonates with aryllithium, arylmagnesium, or arylaluminum compounds.

$$CrCl_3(thf)_3 + PhMgBr \longrightarrow fac\text{-}CrPh_3(thf)_3$$

$$Mn(acac)_3 + PCy_3 + AlPh_3 \cdot Et_2O \longrightarrow Ph_2MnPCy_3$$

$$Ni(acac)_2 + PR_3 + AlPh_3 \cdot Et_2O \longrightarrow PhNi(acac)PR_3$$

$$PR_3 = PPh_3, PCy_3$$

Since $PhMXL_n$ complexes are generally more stable than their alkyl analogs, oxidative addition of aryl halides provides a convenient means for preparation of transition metal aryls.

$$Pd(PPh_3)_4 + PhI \longrightarrow I\text{—}Pd(PPh_3)_2\text{—}Ph + 2PPh_3$$

Oxidative addition involving a C–H bond cleavage at the ortho position of a phenyl ring in aryl phosphines gives aryl complexes as described in the following section.

c. Metallacycles

Metallacycles[63,64] may be regarded as special types of metal alkyls with cyclic structures. They can be prepared by the methods described in the preceding sections.

$$NiCl_2(PR_3)_2 + Li(CH_2)_4Li \longrightarrow (R_3P)_2Ni\langle CH_2\text{-}CH_2\text{-}CH_2\text{-}CH_2\rangle$$

$$Ni(cod)_2 + bipy + Br(CH_2)_4Br \longrightarrow (bipy)Ni\langle CH_2\text{-}CH_2\text{-}CH_2\text{-}CH_2\rangle$$

$$PtCl_2(cod) + BrMg(CH_2)_4MgBr \xrightarrow{PPh_3} (PPh_3)_2Pt\langle CH_2\text{-}CH_2\text{-}CH_2\text{-}CH_2\rangle$$

In addition to these straightforward synthetic methods, metallacycles can be made in a variety of ways involving oxidative additions with C–H and C–C bond cleavage. Oxidative addition involving C–C bond cleavage takes place with strained carbocyclic compounds.

$$[PtCl_2(C_2H_4)]_2 + \text{(phenylcyclopropane)} \xrightarrow[S\,(\text{solvent})]{-C_2H_4} S_2Cl_2Pt(CH_2CH(Ph)CH_2)$$

In the reaction of cubane with a rhodium carbonyl complex, the rhodacycle formed by oxidative addition involving C–C bond cleavage further undergoes subsequent CO insertion, which is followed by reductive elimination to give a cyclic ketone.[64]

$$[Rh_2(CO)_4Cl_2] + \text{cubane} \longrightarrow [\text{Rh(Cl)(CO) rhodacycle}] \longrightarrow Cl{-}Rh(CO){-}C(=O) \text{ acyl complex} \xrightarrow{PPh_3} \text{cyclic ketone } (C=O)$$

A representative oxidative addition process involving C–H bond cleavage to give a metallacycle is γ elimination, which produces a metallacyclobutane.

$$L_nM(R')CH_2CR_2CH_2H \xrightarrow{-R'H} L_nM(CH_2)_2CR_2$$

A metallacyclopentane can also be formed by C–C bond formation between two olefins coordinated to a metal. This is a particular mode of oxidative addition.

$$L_nM(C_2H_4)_2 \longrightarrow L_nM(CH_2)_4 \tag{4.73}$$

By this route catalytic formation of cyclobutane or butene-1 from ethylene is possible.

$$(Ph_3P)_nNi(CH_2{=}CH_2)_2 \rightleftharpoons (Ph_3P)_nNi(CH_2CH_2CH_2CH_2) \begin{cases} \xrightarrow{n\,=\,2} CH_2{-}CH_2{-}CH_2{-}CH_2 \text{ (cyclobutane)} \\ \xrightarrow{n\,=\,1} CH_3CH_2CH{=}CH_2 \end{cases}$$

For the metallacycle to be formed, it is necessary either that vacant sites be present or that such sites be readily created by ligand dissociation so that two olefins can be coordinated to the metal. When the coordinated ligands are reluctant to dissociate, the required dissociation may be achieved photochemically. Olefins with electronegative substituents tend to form a metallacycle with the substituents at the α positions.[65]

$$Fe(CO)_5 + CH_2{=}CH(CO_2Me) \xrightarrow{h\nu} (OC)_4Fe[CH(CO_2Me)CH_2CH_2CH(CO_2Me)] \qquad (4.74)$$

Metallacyclobutanes are important since they are believed to be intermediates in olefin metathesis reactions catalyzed by transition metal complexes (cf. Section 7.5).

Metallacyclobutanes may be formed by reversible cyclization processes between carbene complexes and olefins.

$$M{=}CHR + CH_2{=}CHR' \rightleftharpoons M({=}CHR)(CH_2{=}CHR') \rightleftharpoons M{-}CHR'{-}CH_2{-}CHR \rightleftharpoons CHR{=}CH_2 + M{=}CHR' \qquad (4.75)$$

When a metallacyclobutane decomposes, an olefin can be liberated by β-hydrogen elimination followed by reductive elimination. The following example, in which a stable metallacyclopentane is formed from the reaction of a carbene complex with ethylene, probably proceeds through such a process.[66]

$$CpCl_2Ta{=}CHCMe_3 + CH_2{=}CH_2 \longrightarrow CpCl_2Ta[CH_2CH_2CH(CMe_3)]$$

$$\xrightarrow{-CH_2{=}CHCH_2CMe_3} [CpCl_2Ta] \xrightarrow{2C_2H_4} CpCl_2Ta[CH_2CH_2CH_2CH_2]$$

Unsaturated metallacycles are formed by interactions of acetylenes with low-valent transition metal complexes. In a typical case an acetylene cobalt complex cyclizes with a second molecule of acetylene to a cobaltacyclopentadiene.[67]

$$Cp(PPh_3)Co(RC{\equiv}CR') \xrightarrow{RC{\equiv}CR'} Cp_2Co[CR'{=}CR{-}CR{=}CR'] \qquad (4.76)$$

4-69 **4-70**

In the cobaltacyclopentadiene formation the bulkier substituent R′ was found to occupy the α position to cobalt. Information regarding the regioselectivity in the formation of metallacyclopentadiene complexes is important in connection with regioselectivity in the products of catalytic cyclic oligomerizations of acetylenes.

There are numerous examples of synthesis of metallacycles containing phosphorus, nitrogen, and oxygen atoms. The best-known examples are internal metallation reactions involving the C–H bond cleavage at the ortho position of an aryl group bonded to a basic ligand, a process called *ortho-metallation.*[68]

$$CH_3Rh(PPh_3)_3 \xrightarrow[-CH_4]{\Delta} (PPh_3)_2Rh\text{–}C_6H_4\text{–}PPh_2$$

4-71

$$Cp_2Ni + PhN{=}NPh \longrightarrow [CpNi\text{–}C_6H_4\text{–}N{=}NPh] + C_5H_6$$

$$C_6H_5NHCOR + Pd(OAc)_2 \longrightarrow [Pd(\text{–}C_6H_4NHC(R){=}O)(OAc)]_2$$

The internal metallation or cyclometallation occurs also with ligands without aryl groups.

$$Br_2(Bu^tPr^n_2P)Pt[P(Bu^t)(Pr^n)CH_2CH_2CH_3] \xrightarrow{-HBr} Br(Bu^tPr^n_2P)Pt[CH_2CH_2CH_2P(Bu^t)(Pr^n)]$$

$$Fe(PMe_3)_4 \longrightarrow (Me_3P)_3Fe(H)(CH_2PMe_2)$$

Metallacycle formation is also observed in reactions of low-valent transition metal complexes with certain substituted olefins.

$$Ni(cod)_2 + CH_2{=}C(Me)CONH_2 \xrightarrow[-2cod]{PCy_3} [Cy_3PNi\cdots CH_2{=}C(Me)CONH_2] \longrightarrow Cy_3PNi\text{–}CH_2CH(Me)C({=}O)NH\text{–}$$

4-72

μ-Methylene complexes having a metal–metal bond may be regarded as special types of metallacycles. They are most frequently made from diazo compounds.[69] When mononuclear complexes are used as starting materials, metal–metal bond formation accompanies methylene bridging.

$$CpMn(CO)_2(thf) \xrightarrow[-N_2]{CH_2N_2} Cp(CO)_2Mn(\mu\text{-}CH_2)Mn(CO)_2Cp$$

4-73

$$Cp(CO)_2Co \xrightarrow[-N_2]{N_2CHCO_2Et} Cp(CO)Co(\mu\text{-}C(H)CO_2Et)Co(CO)Cp$$

Although most of reported μ-methylene complexes contain a metal–metal bond or an additional bridging ligand, a μ-methylene complex with no metal–metal bond has recently been reported.[70]

$$Na[CpRu(CO)_2] + \tfrac{1}{2}CH_2Cl_2 \longrightarrow \tfrac{1}{2}Cp(CO)_2Ru\text{-}CH_2\text{-}Ru(CO)_2Cp$$

This complex is much more reactive than analogous alkyl or metal–metal bonded μ-methylene complexes; for example, it readily inserts CO, affording a complex containing a Ru–CO–CH_2–Ru fragment.

A carbon atom can connect three transition metal atoms forming σ metal–carbon bonds to give a μ-alkylidyne complex. The best-known example is **4-74**, prepared by interaction of $Co_2(CO)_8$ with $CHCl_3$.[71]

$$Co_2(CO)_8 + CHCl_3 \longrightarrow (\mu_3\text{-CH})[Co(CO)_3]_3$$

4-74

Although the carbon atom in the μ-alkylidyne complex (**4-74**) may be formally regarded as sp^3 hybridized, there appears to be significant delocalization of electron density in the region of the apical carbon.[72]

d. Ylide Complexes

Ylides, which can be represented by resonance forms between the zwitterionic (A) and neutral olefinic (B) forms shown below are able to form organotransition metal complexes with metal–carbon bonds and metal–carbon multiple bonds.[73,74]

$$R_2\ddot{C}^{-}\text{—}E^{+}R_n \;(A) \longleftrightarrow R_2C{=}ER_n \;(B)$$

E = P, S, N etc.

Examples of ylide complexes with metal–carbon σ bonds are given below.

$$MeAuPMe_3 + CH_2PMe_3 \xrightarrow{-PMe_3} MeAuCH_2PMe_3$$

$$Ni(CO)_4 + CHRPR_3 \xrightarrow{-CO} R_3PCHRNi(CO)_3$$

$$NiMe_2(PMe_3)_2 + CH_2PMe_3 \xrightarrow{-PMe_3} Me_3P{-}Ni(Me)_2{-}CH_2PMe_3$$

Ylide complexes with metal–carbon σ bonds are usually more stable than simple metal alkyls. Bisylide complexes are also known.

$$2Me_3PCH_2 + Me_2Pt(cod) \xrightarrow{-cod} cis\text{-}Me_2Pt(CH_2PMe_3)_2$$

Ylide complexes are also available indirectly, by nucleophilic attack on coordinated carbene ligands.

$$Ph_2C{=}W(CO)_5 + PMe_3 \longrightarrow Me_3P{-}C(Ph)_2{-}W(CO)_5$$

Double ylides having two carbanionic donor sites can be generated on treatment of ylides with base.

$$MeR_2P{=}CH_2 \xrightarrow[-H\ base^+]{+base} R_2P(=CH_2)(\ddot{C}H_2^-) \longleftrightarrow R_2P(\ddot{C}H_2^-)(=CH_2) \longleftrightarrow R_2P^+(\ddot{C}H_2^-)(\ddot{C}H_2^-) \qquad (4.77)$$

These double ylides, having the resonance structures shown above, can serve either as bridging or as chelating groups.

$$1/2\,[Rh(cod)Cl]_2 \xrightarrow[-Bu^t_2Me_2PCl]{2Bu^t_2MePCH_2} (cod)Rh(CH_2)_2PBu^t_2$$

$$2/n\,[Me_3PMCl]_n + 4Me_3PCH_2 \xrightarrow[-2Me_4PCl]{-2PMe_3} Me_2P(CH_2{-}M{-}CH_2)_2PMe_2$$

M = Cu, Ag, Au

4-75

Complex **4-75** has a relatively short metal–metal distance, indicating a weak transannular metal–metal bond.

In addition to the examples shown above quite a variety of ylide complexes have been synthesized, although this rich chemistry has found rather limited synthetic or catalytic application to date.

4.3. TRANSITION METAL COMPLEXES WITH METAL-CARBON MULTIPLE BONDS

Carbenes and carbynes are unstable, short-lived chemical species that can be stabilized by forming complexes with transition metals.[75] Since the first planned synthesis and characterization of a stable transition metal–carbene complex by Fischer and Maasböl in 1964,[76] this field of organometallic chemistry has expanded rapidly. The synthesis, structures, and chemistry of transition metal complexes with metal-to-carbon multiple bonds have been studied very intensively, partly because of the uniqueness of the bonding and the structural diversity, but also because such complexes are involved in catalytic processes (e.g., olefin metathesis). They are also potential intermediates and precursors for the synthesis of organic and organometallic compounds.

a. Metal–Carbene Complexes

There are two types of metal–carbene complexes: one involving heteroatoms and the other without heteroatoms. Compounds in the first category are called Fischer carbene complexes and those in the second category are called Schrock complexes.

(1) Fischer Carbene Complexes

Carbene complexes containing oxygen can be prepared most conveniently by treating metal carbonyls with nucleophilic reagents following the original synthetic method of Fischer and Maasböl. The reaction involves external attack of a carbanionic nucleophile such as R^- in LiR (R = Mc, Ph, ctc.) at the carbon atom of a coordinated carbonyl to give an anionic acyl complex (**4-76**) that is converted into a carbene complex by electrophilic attack at the carbonyl oxygen atom.

$$Cr(CO)_6 + LiR \longrightarrow Li^+\left[(OC)_5Cr{-}C(=O)R\right]^- \;(\mathbf{4\text{-}76}) \xrightarrow{H^+} \left[(OC)_5Cr{=}C(OH)R\right] \;(\mathbf{4\text{-}77})$$

$$\mathbf{4\text{-}76} \xrightarrow{Me_3O^+BF_4^-} (OC)_5Cr{=}C(OCH_3)R \;(\mathbf{4\text{-}78}); \quad \mathbf{4\text{-}77} \xrightarrow{CH_2N_2} \mathbf{4\text{-}78} \tag{4.78}$$

The hydroxycarbene complex **4-77** available by acidification of **4-76** is unstable but can be converted into a methoxycarbene complex (**4-78**) on treatment with diazomethane. The methoxycarbene complex (**4-78**) can be obtained directly by treatment of **4-76** with trimethyloxonium tetrafluoroborate or methyl fluorosulfate. Fischer metal–carbene complexes are stabilized by resonance between the following hybrid structures.

$$M{=}C(OR')R \longleftrightarrow \bar{M}{-}\overset{+}{C}(OR')R \longleftrightarrow \bar{M}{-}C(R){=}\overset{+}{O}{-}R'$$

Thus, the carbon atom attached to the metal is generally susceptible to nucleophilic attack. Attack by nitrogen, sulfur, and carbon nucleophiles provides a convenient means of converting alkoxycarbene complexes into other carbene complexes with or without heteroatoms.

$$(OC)_5Cr{=}C(OMe)Me + H_2NEt \xrightarrow{-MeOH} (OC)_5Cr{=}C(NHEt)Me$$

$$(OC)_5W{=}C(OMe)Ph + PhLi \xrightarrow[-78\ ^\circ C]{} \left[(OC)_5\bar{W}{-}C(OMe)(Ph){-}Ph\right] Li^+$$

$$\xrightarrow[-78\ ^\circ C]{HCl} (OC)_5W{=}C(Ph)Ph$$

4-79

Since the hydrogen atoms attached to the carbon atom adjacent to the carbene carbon are acidic, treatment of the carbene complex with an alkyllithium followed by reaction with electrophiles gives functionalized metal–carbene derivatives.

$$(OC)_5Cr{=}C(OMe)CH_3 \xrightarrow[2)\ BrCH_2CO_2Me]{1)\ BuLi} (OC)_5Cr{=}C(OMe)CH_2CH_2CO_2Me$$

Since an isocyanide is isoelectronic with CO, an isocyanide complex is expected to show an analogous behavior to metal carbonyls. Indeed, isocyanide complexes react with nucleophiles to give aminocarbene complexes. Alcohols, thiols, and amines are known to attack the C–N bonds in isocyanide complexes.

$$cis\text{-}(RNC)PtCl_2(PEt_3) + R'OH \longrightarrow cis\text{-}Cl_2(PEt_3)Pt{=}C(OR')NHR$$

$R = Me, Ph;\quad R' = Me, Et, Pr^i$

Electron-rich olefins react with some transition metal complexes with cleavage of the double bond to give metal–carbene complexes.

$$Mo(CO)_6 + \left(\text{RN}\overset{}{\frown}\text{NR}\right)C{=}C\left(\text{RN}\frown\text{NR}\right) \longrightarrow (OC)_5Mo{=}C\left(\text{RN}\frown\text{NR}\right)$$

Other reactive molecules, such as diazo compounds, that are capable of liberating carbenes can also be used to synthesize carbene complexes.

$$(MeCp)Mn(CO)_2(thf) + N_2{=}CR_2 \longrightarrow (MeCp)(OC)_2Mn{=}CR_2$$

(2) *Schrock Carbene Complexes (Alkylidene Complexes)*

The first carbene complex without a heteroatom was prepared by Schrock in 1974.[77] Schrock calls complexes of this type alkylidene complexes. A number of group 5 and 6 metal alkylidene complexes have been prepared by α-elimination reactions.

$$M(CH_2CMe_3)_3Cl_2 \xrightarrow[-LiCl]{LiCH_2CMe_3} M(CH_2CMe_3)_4Cl$$

$$\xrightarrow{-CMe_4} \left[(Me_3CCH_2)_2(Cl)M{=}C(H)CMe_3\right] \xrightarrow[-LiCl]{LiCH_2CMe_3} (Me_3CCH_2)_3M{=}C(H)CMe_3$$

$$M = Nb,\ Ta$$

Steric congestion in the metal–neopentyl precursor is believed to be an important factor that helps to induce α-hydrogen abstraction.

Hydrogen abstraction may also be induced by presence of other ligands such as cyclopentadienyl and trimethylphosphine.

$$Ta(CH_2CMe_3)_2X_3 \xrightarrow[-CMe_4]{CpTl} CpX_2Ta{=}C(H)CMe_3 \xrightarrow[-TlX]{CpTl} Cp_2XTa{=}C(H)CMe_3$$

$$Ta(CH_2CMe_3)_2X_3 \xrightarrow[-CMe_4]{PMe_3} (Me_3P)_2X_3Ta{=}C(H)CMe_3$$

A metal–methylene complex, the simplest representative of the alkylidene complexes, can be prepared by the following route.

$$TaMe_3Cl_2 \xrightarrow{CpTl} CpTaMe_3Cl \xrightarrow{CpTl} Cp_2TaMe_3 \xrightarrow{Ph_3C^+BF_4^-}$$

$$[Cp_2TaMe_2]^+BF_4^- \xrightarrow{Ph_3PCH_2} Cp_2Ta(Me){=}CH_2$$

3-27 (cf. Fig. 3.21)

Bis(alkylidene) complexes have also been synthesized by addition of PMe_3 to solutions of mono(alkylidene) complexes.

$$(Me_3CCH_2)_3Ta{=}C\begin{matrix}H\\CMe_3\end{matrix} \xrightarrow[-CMe_4]{+2PMe_3} Me_3CCH_2\text{-}\overset{\displaystyle PMe_3}{\underset{\displaystyle PMe_3}{Ta}}\begin{matrix}{=}CHCMe_3\\{=}CHCMe_3\end{matrix}$$

These alkylidene complexes are formally in high oxidation states and show different reactivities from those of Fischer carbene complexes. The carbon atom bound to the metal is *nucleophilic* rather than electrophilic, with a few exceptions. Thus, alkylidene complexes may be compared with ylides:

$$L_nM{=}CH_2 \longleftrightarrow L_n\overset{+}{M}{-}\overset{-}{C}H_2 \quad \text{vs.} \quad R_3P{=}CH_2 \longleftrightarrow R_3\overset{+}{P}{-}\overset{-}{C}H_2$$

Some alkylidene complexes show reactivities analogous to those observed in Wittig reactions of ylides, consistent with the above resonance formulation. This will be discussed in Section 6.4.

b. Metal-Carbyne Complexes

Carbyne complexes containing a terminally bonded C-alkyl or C-aryl ligands were first synthesized in 1973 by treating Fischer carbene complexes with electrophiles.

$$(OC)_5M{=}C\begin{matrix}OR'\\R\end{matrix} \xrightarrow[-BX_2OR']{BX_3} \text{trans-}X(CO)_4M{\equiv}C{-}R + CO$$

4-80

These carbyne complexes were shown both spectroscopically and by X-ray structural analyses to have a metal–carbon triple bond. In the above reaction the Lewis acid is believed to attack the oxygen atom bound to the carbene carbon, thus converting the alkoxy group into a good leaving group and leaving the carbyne attached to the metal.

On the other hand, in reactions of Schrock alkylidene complexes to produce alkylidyne complexes, hydrogen atoms attached to the carbene carbons are abstracted by addition of bases.

$$Cp\overset{\displaystyle CHCMe_3}{\overset{\|}{\underset{\displaystyle Cl}{Ta}}}\text{-}CH_2CMe_3 \xrightarrow[-CMe_4]{2PMe_3} Cp\overset{\displaystyle CCMe_3}{\overset{|||}{\underset{\displaystyle Cl}{Ta}}}\text{-}PMe_3 \ (PMe_3)$$

By this method a complex containing an alkyl, an alkylidene, and an alkylidyne group in the same molecule has been synthesized. In this complex the tungsten atom is formally regarded as being in oxidation state 6.

$$WCl_6 + 6LiCH_2CMe_3 \longrightarrow [(Me_3CCH_2)_3W{\equiv}CCMe_3]_2$$

$$\xrightarrow{2PMe_3} Me_3CCH_2\text{-}\overset{\displaystyle PMe_3}{\underset{\displaystyle PMe_3}{W}}\begin{matrix}{\equiv}CCMe_3\\{=}CHCMe_3\end{matrix}$$

4-81

Schrock's mechanism for the formation of the alkylidyne complexes from the alkylidene complexes is as follows. According to a neutron diffraction study of the tantalum alkylidene complex $Me_3CCH{=}TaX_3(PMe_3)$, the Ta–C_α–C_β angle is 161°, which is much greater than that expected for an sp^2 carbon, whereas the Ta–C_α–H_α angle is 85° and the C_α–H_α bond length is 131 pm, the latter being longer than the expected value for the C–H distance to an sp^2-hybridized carbon atom. These results are taken to suggest that the H_α atom is pushed toward the tantalum atom so that it partially bridges the Ta and C_α atoms. This effect, which is due to the influence of the sterically very demanding tertiary butyl group, causes the H_α atom to be readily lost, giving the alkylidyne complex. A similar steric effect may also be operative in the formation of alkylidene complexes from alkyl complexes. The order of effectiveness of substituents attached to the carbene carbon in inducing α-hydrogen elimination is thought to be $Me_3C > Ph >> H$.

Some alkylidyne complexes show a tendency to dimerize with the alkylidyne carbons bridging two metal atoms.

$$NbCl_5 + Me_3SiCH_2Li \longrightarrow (Me_2SiCH_2)_2Nb(\mu\text{-}CSiMe_3)_2Nb(CH_2SiMe_3)_2$$

4.4. TRANSITION METAL HYDRIDES

There are many transition metal hydrides of nonstoichiometric composition, but here we are concerned with molecular transition metal hydrides having other supporting ligands.[78] Like transition metal alkyls, these hydride complexes are stable when they contain supporting ligands such as CO, CN, cyclopentadienyl, tertiary phosphine, and nitrogen bases. The first known complexes with M–H bonds were the hydridocarbonyls $H_2Fe(CO)_4$ and $HCo(CO)_4$ prepared by Hieber in the 1930s.

Preparative methods for transition metal hydrides include (1) direct reaction of dihydrogen with transition metal complexes, (2) reactions of transition metal complexes with other hydride sources, (3) hydrogen abstraction from organic entities, and (4) reactions of hydrogen compounds with transition metal complexes.

(1) Direct Reactions of Transition Metal Complexes with Molecular Hydrogen

Molecular hydrogen can oxidatively add to coordinatively unsaturated complexes of low-valent transition metals under relatively mild conditions (Section 6.2), whereas reactions with complexes of metals in higher oxidation states require higher pressures, reducing agents, or bases. The Ir(I) complex

$IrCl(CO)(PPh_3)_2$, known as Vaska's complex, reacts reversibly with H_2 at atmospheric pressure to give an Ir(III) dihydride complex.

$$\underset{\textbf{4-82}}{IrCl(CO)(PPh_3)_2} \underset{}{\overset{H_2,\ 1\ atm}{\rightleftharpoons}} IrH_2(CO)Cl(PPh_3)_2 \tag{4.79}$$

$$\underset{\textbf{4-83}}{RhCl(PPh_3)_3} \overset{H_2,\ 1\ atm}{\rightleftharpoons} RhH_2Cl(PPh_3)_3 \tag{4.80}$$

Complex **4-83** is called Wilkinson's complex, and it acts as a catalyst for olefin hydrogenation. Reaction (4.80) is important as an elementary step in these catalytic processes.

In contrast, the direct reaction of H_2 with $PtCl_2(PEt_3)_2$ requires more forcing conditions.

$$\text{cis-}PtCl_2(PEt_3)_2 \xrightarrow[EtOH,\ 90°C]{H_2,\ 50\ atm} \text{trans-}PtHCl(PEt_3)_2$$

In the presence of bases, similar reactions proceed under milder conditions.

$$RuCl_2(PPh_3)_3 + H_2 \xrightarrow[\text{toluene reflux}]{NEt_3} RuHCl(PPh_3)_3 + Et_3NHCl$$

Transition metal alkyls can also react directly with molecular hydrogen.

$$Cp_2ZrMe_2 \xrightarrow[25°C]{H_2,\ 30\ atm} Cp_2ZrH_2 + 2CH_4$$

Reactions of H_2 with some transition metal complexes involve cleavage of a metal–metal bond.

$$Co_2(CO)_8 \xrightarrow[25°C]{H_2,\ 30\ atm} HCo(CO)_4 \tag{4.81}$$

This reaction is important in relation to hydroformylation of olefins using $Co_2(CO)_8$ as a catalyst precursor.

(2) *Synthesis Using Boron or Aluminum Hydrides*

Reduction using commercially available metal hydrides such as $NaBH_4$ and $LiAlH_4$ provides a convenient laboratory method for the synthesis of transition metal hydride complexes.

$$RuCl_2(PPh_3)_3 + NaBH_4 \xrightarrow{PPh_3} RuH_2(PPh_3)_4$$

$$[Rh(en)_2Cl_2]^+ + NaBH_4 \xrightarrow[H_2O]{} [RhHCl(en)_2]^+$$

$$en = H_2NCH_2CH_2NH_2$$

$$IrHCl_2(PPh_3)_3 + LiAlH_4 \xrightarrow[THF]{} IrH_3(PPh_3)_3$$

In some cases, however, complexes containing BH_4 or AlH_4 entities are obtained instead of simple hydrides.

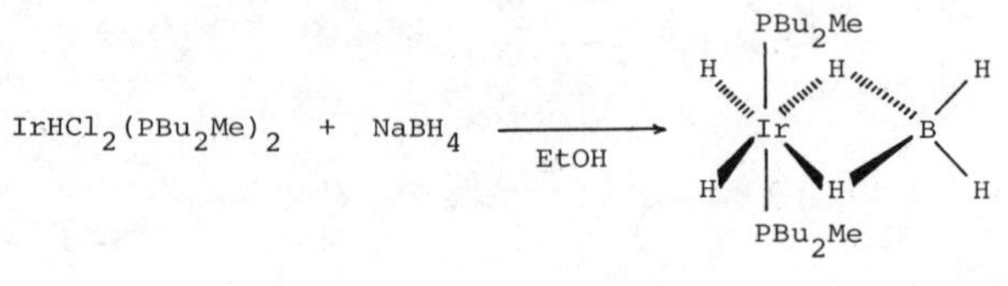

$$IrHCl_2(PBu_2Me)_2 + NaBH_4 \xrightarrow{EtOH}$$

PBu$_2$Me, H, H, H, Ir, B, H, H, H, PBu$_2$Me

(3) Synthesis by Hydrogen Abstraction from C–H Bonds

When a C–H bond in a transition metal complex is in the vicinity of the metal, hydrogen abstraction can give a transition metal hydride complex. The most well known process is β-hydrogen elimination, the reverse of the process of olefin insertion into an M–H bond, as shown in Eq. (4.71).

$$PtEtCl(PEt_3)_2 \underset{pressure}{\overset{\Delta}{\rightleftharpoons}} PtHCl(PEt_3)_2 + C_2H_4$$

$$CpFeEt(CO)_2(PPh_3) \xrightarrow{\Delta} CpFeH(CO)_2(PPh_3) + C_2H_4$$

If such β-hydrogen elimination takes place during the process of alkylation of transition metal complexes, transition metal hydrides can be produced.

$$RuCl_3 + PPh_3 + AlEt_3 \longrightarrow RuH_2(PPh_3)_4$$

$$Mo(acac)_3 + dppe + AlEt_3 \longrightarrow MoH(acac)(dppe)_2$$

Transition metal hydrides can be also formed by internal metallation.

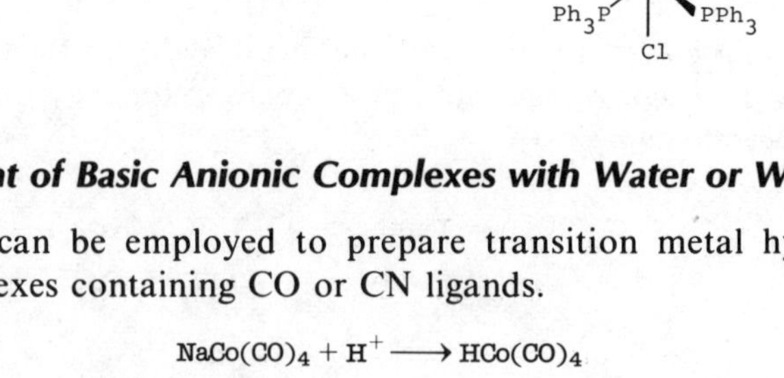

$IrCl(PPh_3)_3$ —heat→ PPh$_2$, H, Ir, Ph$_3$P, Cl, PPh$_3$

$IrCl(PPh_3)_3$ —hν→ PPh$_2$, H, Ir, Ph$_3$P, PPh$_3$, Cl

(4) Treatment of Basic Anionic Complexes with Water or Weak Acids

This method can be employed to prepare transition metal hydrides from anionic complexes containing CO or CN ligands.

$$NaCo(CO)_4 + H^+ \longrightarrow HCo(CO)_4$$

$$Na_2[Fe(CO)_4] \xrightarrow{H^+} Na[FeH(CO)_4] \xrightarrow{H^+} FeH_2(CO)_4$$

(5) *Addition of Hydrogen Compounds*

Oxidative addition of HX to neutral, low-valent transition metal complexes also affords transition metal hydrides.

$$Ni(PPh_3)_3 + HX \longrightarrow [NiH(PPh_3)_3]^+X^-$$

$$trans\text{-}IrCl(CO)(PPh_3)_2 + HSiCl_3 \longrightarrow IrHCl(CO)(SiCl_3)(PPh_3)_2$$

The latter process is believed to be one of the elementary steps in the catalytic hydrosilylation of olefins (cf. Section 8.1).

Many polynuclear hydride complexes are being synthesized as part of a study of cluster complexes (see Section 9.3). Preparative methods for these polynuclear hydrides are quite diverse.

4.5. METAL CARBONYLS AND RELATED COMPLEXES

Since Mond first discovered the formation of *$Ni(CO)_4$* from metallic nickel and carbon monoxide in 1890, a variety of metal carbonyls with and without other ligands has been prepared. Their bonding properties and structural aspects have been described in Section 3.3. Table 4.3 lists the known neutral, binary metal carbonyls.

a. Mononuclear Metal Carbonyls

Mononuclear metal carbonyls are all hydrophobic liquids or volatile solids, soluble to varying degrees in nonpolar solvents. Nickel and iron carbonyls can

Table 4.3. The Known Neutral, Binary Metal Carbonyls[a,b]

3	4	5	6	7	8	9	10	11
	Ti[c]	$V(CO)_6$	$Cr(CO)_6$	$Mn(CO)_{10}$	$Fe(CO)_5$		$Ni(CO)_4$	Cu[c]
					$Fe_2(CO)_9$	$Co_2(CO)_8$		
					$Fe_3(CO)_{12}$	$Co_4(CO)_{12}$		
	Zr[c]	Nb[c]	$Mo(CO)_6$	$Tc_2(CO)_{10}$	$Ru(CO)_5$	$Rh_2(CO)_8$	Pd[c]	Ag[c]
						$Rh_4(CO)_{12}$		
					$Ru_3(CO)_{12}$	$Rh_6(CO)_{16}$		
	Hf[c]	Ta[c]	$W(CO)_6$	$Re_2(CO)_{10}$	$Os(CO)_5$	$Ir_2(CO)_8$	Pt[c]	Au[c]
					$Os_3(CO)_{12}$	$Ir_4(CO)_{12}$		

[a] Source: K. F. Purcell and J. C. Kotz, *An Introduction to Inorganic Chemistry,* Holt-Saunders, Philadelphia, 1980, p. 509.
[b] In addition to the carbonyls tabulated here some unstable binary carbonyls prepared by low-temperature matrix isolation technique are known; G. A. Ozin and A. V. Voet, *Acc. Chem. Res.,* **6**, 313 (1973).
[c] Form only anionic carbonyl complexes or give carbonyl complexes only when the metal contains other ligands.

be prepared by direct interaction of carbon monoxide with the finely powdered metals.

$$Ni + CO \xrightarrow{\text{room temperature}} Ni(CO)_4$$

$$Fe + CO \xrightarrow[\text{pressure}]{250°C} Fe(CO)_5$$

Fertile Field

The discovery of nickel carbonyl in 1890 by Mond was accidental, like so many of the other discoveries listed in Table 1.1. Mond was making a detailed study of valve corrosion in the nickel vessels used in the Solvay process for the production of soda ash (Na_2CO_3) from NaCl, CO_2, and ammonia, a process invented in 1866. He found that the trouble was caused by the carbon monoxide present as an impurity in the carbon dioxide. Mond and co-workers not only confirmed that the reaction of metallic nickel with carbon monoxide produced $Ni(CO)_4$, a volatile colorless liquid, m.p. −25 °C, b.p. 43 °C, but also developed a process of dry refining of nickel ore utilizing the ready thermal decomposition of $Ni(CO)_4$ to metallic nickel. The discovery of $Fe(CO)_5$ followed that of $Ni(CO)_4$.

The properties of this strange class of compounds aroused the interest of other people. Walter Hieber, then at the University of Heidelberg, began to work on them, obtaining $Fe(CO)_5$ from A. Mittasch of B.A.S.F. Hieber has recalled Mittasch's emphatic warning of the danger of using these highly toxic substances and the comment that in this field one could only expect a great deal of trouble and results of little scientific value. [W. Hieber, *Adv. Organometal. Chem.,* **8,** 1 (1970).]

The metal carbonyls, which were regarded by many as curiosities at that time, had versatile and intriguing properties. The most notable applications were developed by Reppe in 1939, who successfully synthesized acrylic acid from acetylene using nickel carbonyl. The oxo process, olefin hydroformylation catalyzed by cobalt carbonyl, was also developed around the same time by Roelen.

Studies on metal carbonyls have not only produced a rich chemistry of enormous scientific value, contrary to Mittasch's prediction, but have also proved to be very useful.

Most other metal carbonyls are prepared by reducing metal halides or oxides in the presence of carbon monoxide.

$$CrCl_3 + RMgX + CO \longrightarrow Cr(CO)_6$$

$$VCl_3 + Na + CO \longrightarrow V(CO)_6$$

b. Polynuclear Metal Carbonyls

Binuclear metal carbonyls can be prepared by methods similar to those used for the preparation of mononuclear carbonyls.

$$CoCO_3 + H_2 + CO \xrightarrow[120\text{-}150^\circ C]{250\text{-}300\ atm} Co_2(CO)_8 + CO_2 + H_2O$$

$$Mn(acac)_3 + CO + AlEt_3 \longrightarrow Mn_2(CO)_{10}$$

In the latter reaction, organomanganese complexes are probably produced initially and are then decomposed by the excess of CO to give manganese carbonyl. In the following case CO itself acts as the reducing agent.

$$Re_2O_7 + CO \longrightarrow Re_2(CO)_{10} + CO_2$$

Formation of polynuclear carbonyls proceeds by initial dissociation of a carbonyl ligand in mononuclear carbonyls; this is followed by metal–metal bond formation or bridge formation by the CO ligands. For example, irradiation of $Fe(CO)_5$ with light causes liberation of a CO ligand and promotes metal–metal bond formation to give $Fe_2(CO)_9$.

$$2Fe(CO)_5 \xrightarrow{h\nu} Fe_2(CO)_9 + CO$$

Polynuclear metal carbonyls can be prepared by reaction of anionic metal carbonyls with metal carbonyl halides.

$$[Mn(CO)_x]^- + XM(CO)_y \xrightarrow{zCO} MnM(CO)_{x+y+z} + X^-$$

This approach is suitable for the preparation of heteronuclear metal carbonyls.

$$[Mn(CO)_5]^- + BrRe(CO)_5 \longrightarrow (OC)_5MnRe(CO)_5 + Br^-$$

The relative stabilities of polynuclear metal carbonyls with and without bridging carbonyl ligands are determined by the atomic radii of the metals. The stability of metal carbonyls tends to increase on descending the periodic table. For example, as can be seen from Figure 3.24, $Fe_3(CO)_{12}$ in the solid state has two bridging CO ligands, whereas the ruthenium and osmium analogs $Ru_3(CO)_{12}$ and $Os_3(CO)_{12}$ have no bridging CO ligand. A similar effect of the atomic radii can be seen when carbonyls of metals in the same row are compared. That $Co_2(CO)_8$ has an isomer with bridging CO ligands whereas $Mn_2(CO)_{10}$ has not may be explained by the fact that cobalt has a smaller atomic radius than manganese.

c. Anionic Metal Carbonyls

Anionic metal carbonyls can be prepared by various means, such as treatment of neutral metal carbonyls with alkali and reactions of metal–metal bonded carbonyl with alkali metals.

$$Fe(CO)_5 + 4OH^- \longrightarrow Fe(CO)_4^{2-} + CO_3^{2-} + 2H_2O$$

$$Mn_2(CO)_{10} + Na \longrightarrow Na^+[Mn(CO)_5]^-$$

$$Fe_3(CO)_{12} + Na \longrightarrow Na_2[Fe(CO)_4]^{2-}$$

$$Co_2(CO)_8 + Na/Hg \longrightarrow Na^+[Co(CO)_4]^-$$

Lewis bases often cause disproportionation,

$$3\ Mn_2(CO)_{10} + 12\ py \longrightarrow 2\ [Mn(py)_6]^{2+} + 4\ [Mn(CO)_5]^- + 10\ CO$$

although this method leads to the waste of one-third of the original metal carbonyl.

These anionic metal carbonyls (metal carbonylates) are useful starting materials for synthesizing other organotransition metal complexes. Typical reactions involve nucleophilic attack of the anion on a positive center.

$$Mn(CO)_5^- + RX \longrightarrow RMn(CO)_5 + X^-$$

$$Co(CO)_4^- + RCOX \longrightarrow RCOCo(CO)_4 + X^-$$

$$Na_2Fe(CO)_4 + RX \longrightarrow Na^+[RFe(CO)_4]^- \xrightarrow{CO} Na[RCOFe(CO)_4]^-$$

Some of these reactions have already been described in Section 4.2 in connection with the synthesis of transition metal alkyls. Some other examples are

$$Mn(CO)_5^- + ClCH_2CH{=}CH_2 \longrightarrow (OC)_5Mn{-}CH_2CH{=}CH_2 + Cl^-$$

$$CpW(CO)_3^- + ClSiMe_3 \longrightarrow CpW(CO)_3SiMe_3 + Cl^-$$

In addition to these mononuclear anionic metal carbonyls, a number of polynuclear anionic metal carbonyls have been prepared by methods similar to those employed for synthesizing mononuclear complexes.

$$Fe_2(CO)_9 + 4OH^- \longrightarrow [Fe_2(CO)_8]^{2-} + CO_3^{2-} + 2H_2O$$

$$Ni(CO)_4 + Na \xrightarrow{\text{liquid } NH_3} Na_2[Ni_5(CO)_{12}]$$

d. Hydridocarbonyls (Carbonyl Hydrides)

Hydridocarbonyl complexes can be prepared by (1) reactions of neutral metal carbonyls with molecular hydrogen, (2) acidification of anionic metal carbonyls, or (3) reactions of transition metal halides with hydridic reducing agents such as $NaBH_4$.

$$Mn_2(CO)_{10} + H_2 \longrightarrow HMn(CO)_5$$

See the other reactions described in Section 4.4(1).

$$Na[Co(CO)_4] + H^+(aq) \longrightarrow CoH(CO)_4 + Na^+$$

$$Fe(CO)_4I_2 + NaBH_4 \xrightarrow[THF]{} FeH_2(CO)_4$$

Many of the hydridocarbonyls are thermally unstable. For example, $FeH_2(CO)_4$ in the liquid state begins to decompose above about $-10\ °C$, giving H_2 and red $Fe_2H_2(CO)_8$, and $CoH(CO)_4$ decomposes above its melting point (−26 °C), giving H_2 and $Co_2(CO)_8$. They are relatively more stable in the gas phase, however, particularly when diluted with carbon monoxide. The hydrogen ligands in the carbonyl hydrides have dual character. Because of the presence of strongly electron-withdrawing CO ligands, carbonyl hydrides show acidic properties in water, in which they are slightly soluble.

$$HMn(CO)_5 = H^+ + [MnCO_5]^- \quad pK \sim 7$$

$$H_2Fe(CO)_4 = H^+ + [HFe(CO)_4]^- \quad pK_1 \sim 4$$

$$[HFe(CO)_4]^- = H^+ + [Fe(CO)_4]^{2-} \quad pK_2 \sim 13$$

$$HCo(CO)_4 = H^+ + [Co(CO)_4]^- \quad \text{strong acid}$$

Hydridotetracarbonylcobalt is as strong an acid as nitric acid.

In addition to behaving as acids, these hydridocarbonyls also show hydridic properties. They reduce olefins, and they all exhibit very-high-field NMR chemical shifts due to the strong shielding effect of the metal attached to the hydride ligand. The hydride ligand in the nonbridging hydrido metal carbonyls occupies a distinct coordination site, and the M–H distance is approximately the sum of the covalent radii (i.e., 160–170 pm).

e. Other Metal Carbonyls

Replacement of some of the CO ligands in binary metal carbonyls by other ligands leads to substituted metal carbonyl complexes.

$$Ni(CO)_4 + PR_3 \longrightarrow Ni(CO)_3(PR_3) + CO$$

$$Cr(CO)_6 + C_6H_6 \longrightarrow Cr(C_6H_6)(CO)_3 + 3\,CO$$

Photochemical activation is sometimes employed to aid in breaking M–CO bonds, but many reactions are carried out thermally.

Metal carbonyl halides are available either by direct reaction of metal halides with carbon monoxide, usually at higher pressure, or by halogen cleavage of dinuclear metal carbonyls.

$$RuI_3 + 2CO \xrightarrow{220\ °C} [Ru(CO)_2I_2]_n + I_2$$

$$Mn_2(CO)_{10} + Br_2 \longrightarrow 2\,Mn(CO)_5Br$$

The Discovery of Vaska's Complex

Among the many metal carbonyl complexes so far reported, Vaska's complex stands out, both for the variety of its reactions and for its historical significance in the development of the basic concepts of organotransition metal chemistry. The complex was discovered by Lauri Vaska when he was studying the reaction of $IrCl_3$ with triphenylphosphine in alcohol.[79]

$$IrCl_3 + PPh_3 + ROH \longrightarrow trans\text{-}IrCl(CO)(PPh_3)_2$$

4-82

Although the formation of other hydride complexes such as $IrHCl_2(PPh_3)_3$ and $IrH_2Cl(PPh_3)_3$ had been reported, the formation of complex **4-82**, which has a carbonyl ligand, had been overlooked. Vaska and co-workers have firmly established by ^{14}C labeling that the CO ligand in the complex comes from the alcohol.

The precise route by which Vaska's complex is formed has not been clarified, but in terms of the elementary reactions to be discussed in Chapter 6, a plausible sequence is as follows.

$$IrCl_3 + C_2H_5OH + L \xrightarrow{-HCl} CH_3CH(H)\text{-}O\text{-}IrCl_2L_n \xrightarrow[\beta\text{-elimination}]{-CH_3CHO} HIrCl_2L_n \xrightarrow{-HCl} IrClL_n \xrightarrow[\text{oxidative addition}]{+CH_3CHO}$$

$$CH_3\text{-}C(=O)\text{-}Ir(Cl)(H)\text{-}L_n \xrightarrow{\text{decarbonylation}} CH_3\text{-}Ir(Cl)(L)(L)(H)(CO) \xrightarrow[\text{reductive elimination}]{-CH_4} IrCl(CO)L_2 \ \textbf{4-82}$$

The low-valent Ir(I) complex was soon shown to have an extensive chemistry, as summarized in Figure 6.20. It reacts, often reversibly, with addenda such as H_2, RX, CO, acetylene, and O_2 to give either octahedral or penta-coordinated square pyramidal and trigonal bipyramidal complexes. The advent of Vaska's complex established the concepts of oxidative addition and reductive elimination in organotransition metal chemistry, concepts that have had a far-reaching impact on related fields such as catalytic chemistry and organic chemistry.

f. Isocyanide Complexes

Isocyanides (isonitriles) R—N≡C are structurally similar to carbon monoxide and are known to form complexes analogous to metal carbonyls.[80] Isonitriles can serve either as terminal ligands or as bridging ligands. Some of the coordination modes are shown below.

$M \leftarrow C{\equiv}NR$ terminal

$(M')(M)C{=}\ddot{N}\text{-}R$, $(M')(M)C{=}N\text{-}R$, $(M\text{—}M)C{\equiv}N\text{-}R$, $(M\text{—}M)(C{\equiv}N\text{-}R)\rightarrow M$ bridging

Isocyanides generally appear to act as stronger donors than CO. In complexes such as $[Ag(CNR)_4]^+$, $[Fe(CNR)_6]^{2+}$, and $[Mn(CNR)_6]^{2+}$, π bonding is considered to be of relatively little importance. The corresponding carbonyl analogs are not known. When bound to low-valent metals, however, isocyanide ligands can participate in extensive back donation. Examples of such complexes are $Cr(CNR)_4$ and $Ni(CNR)_4$. In $Ni(CNPh)_4$ the decrease in the CN bond stretching frequency on coordination from that in the free isocyanide is greater than the similar decrease in the CO bond stretching frequency in metal carbonyls. Low-valent metal–isocyanide complexes can be prepared as follows.

$$MoCl_3 + RNC + Mg \longrightarrow Mo(CNR)_6 + MgCl_2$$

$$Ni(cod)_2 + 4RNC \longrightarrow Ni(CNR)_4 + 2COD$$

$$Cp_2Ni \xrightarrow{RNC} [CpNi(CNR)]_2 \xrightarrow{RNC} Ni(CNR)_4$$

Like metal carbonyls, isocyanide complexes tend to form clusters (cf. Chapter 9).

g. Thiocarbonyl Complexes

Unlike carbon monoxide, the carbon monosulfide molecule CS exists only as an unstable species. But a number of thiocarbonyl complexes[81] that are analogous to carbonyl complexes are known, although pure binary thiocarbonyl complex $M(CS)_n$ has not been prepared. The bonding of CS with a metal is similar to but somewhat stronger than that of CO. Thiocarbonyl complexes can be generally prepared using CS_2 or $CSCl_2$. For example, $RhCl(PPh_3)_3$ reacts with CS_2 as shown below.

$$RhCl(PPh_3)_3 + CS_2 \longrightarrow RhCl(\eta^1\text{-}CS_2)(\eta^2\text{-}CS_2)(PPh_3)_2 \xrightarrow{MeOH} trans\text{-}RhCl(CS)(PPh_3)_2$$

Thiocarbonyl complexes can be prepared by reactions of anionic metal carbonyls with thiophosgen.

Complexes having terminal CS ligands show strong infrared CS stretching bands in the range of 1160–1400 cm^{-1}. The frequency varies depending on the charge and the oxidation state of the metal. Bridging CS ligands absorb in the region 1100–1160 cm^{-1}, whereas the noncoordinated CS molecule trapped in a matrix at -190 °C shows a stretching frequency at 1274 cm^{-1}.

The sulfur atom in the CS ligand shows a greater reactivity toward electrophiles than the oxygen in metal carbonyls, whereas the carbon atom in the CS complexes is more reactive to nucleophiles than that in metal carbonyls.

$$W(dppe)_2(CO)(CS) + Et_3O^+SO_3F^- \xrightarrow[-Et_2O]{} [W(dppe)_2(CO)(CSEt)]^+SO_3F^-$$

$$[CpFe(CO)_2(CS)]^+ + MeO^- \longrightarrow CpFe(CO)_2(C(=S)OMe)$$

An example of a preparation of a complex with bridging CS ligands is shown below.[82]

$$CpRu(CO)_2(CS)^+ + NaH \longrightarrow Cp(OC)Ru(\mu\text{-}CS)_2Ru(CO)Cp$$

h. Nitrosyl Complexes

Although nitrosyl complexes[83,84] have no direct metal-to-carbon bond and thus are not included in the organometallic compounds, there are many organometallic complexes having NO ligands, and it is useful to know something about the properties of nitrosyl complexes. The NO molecule has an unpaired electron in a π^* orbital. The loss of the unpaired electron gives the nitrosyl cation NO^+, which is isoelectronic with carbon monoxide. Thus, the NO ligand may be regarded as a three-electron donor, and a variety of complexes containing NO ligands are known, although few complexes having only NO ligands are well characterized.

Nitrosyl complexes can be prepared by replacement of CO with nitric oxide.

$$Fe(CO)_5 + 2NO \longrightarrow Fe(CO)_2(NO)_2 + 3CO$$

$$Co_2(CO)_8 + 2NO \longrightarrow 2Co(CO)_3(NO) + 2CO$$

In a chemical sense the NO ligand appears to be more strongly bonded to metal than CO, since substitution reactions on mixed carbonyl nitrosyl complexes usually result in displacement of CO in preference to NO. Examples of syntheses of nitrosyl complexes containing cyclopentadienyl ligands are given below.

$$Cp_2Ni + NO \longrightarrow CpNiNO + (C_5H_5)$$

$$CpMo(CO)_3H + NO \longrightarrow CpMo(CO)_2NO$$

$$CrCl_3 + CpNa + NO \longrightarrow CpCr(NO)_2Cl$$

$$\frac{1}{2}Pt_2(CO)_2Cl_4 + CpNa + NO \longrightarrow CpPtNO$$

In addition to complexes having linear MNO groups, some complexes that have bent MNO groups are known. The former complexes show stretching frequencies in the range 1800–1900 cm^{-1}, whereas the latter absorb at much lower frequencies, in the range 1525–1690 cm^{-1}.

The NO ligand also acts as a bridging group. An example of preparation of a complex with bridging and terminal NO ligands is

$$\begin{matrix} NO \\ | \\ CpCr{-}Cl \\ | \\ NO \end{matrix} \xrightarrow{NaBH_4} \text{Cp(ON)Cr}(\mu\text{-NO})_2\text{Cr(NO)Cp}$$

4.6. DINITROGEN COMPLEXES

It may seem somewhat awkward to include the complexes containing N_2 as well as those of NO in the category of organotransition metal complexes. However, it is convenient because N_2 is isoelectronic with CO, and many organometallic complexes containing the dinitrogen ligand are now known.[85]

The first N_2-containing complex was prepared in 1965 by reaction of hydrazine with $RuCl_3$.[86]

$$RuCl_3 + H_2NNH_2 \xrightarrow{N_2} \underset{\textbf{4-84}}{[Ru(NH_3)_5N_2]Cl_2}$$

The second example was also prepared by a chemical reaction of a nitrogen-containing compound.[87]

$$IrCl(CO)(PPh_3)_2 + RCON_3 \longrightarrow trans\text{-}IrCl(N_2)(PPh_3)_2$$

Later a cobalt–dinitrogen complex was prepared directly from N_2 in the following reaction.[88]

$$Co(acac)_3 + AlR_3 + PPh_3 \xrightarrow{N_2} \underset{\textbf{4-85}}{CoH(N_2)(PPh_3)_3}$$

Serendipity

As discussed in Chapter 1, many remarkable discoveries in the development of organometallic chemistry were made adventitiously. Examples include Zeise's complex, diethylzinc, ferrocene, and the Ziegler catalyst. Dinitrogen complexes provide yet another example to add to the long list.

Industrial nitrogen fixation for producing ammonia from N_2 and H_2 is carried out under very drastic conditions of temperature and pressure. Thus, the first demonstration of nitrogen fixation under atmospheric pressure at room temperature by Vol'pin and Shur in 1963 was quite surprising.[89] They found during their study of Ziegler catalysts that nitrogen gas employed as an "inert gas" to exclude air was introduced into the reaction product since on hydrolysis ammonia was formed. Because of the industrial importance of the Ziegler catalyst, many people had been working with similar systems under nitrogen, but no one had recognized previously that N_2 reacted with the system.

The first report of the N_2 complex **4-84** was disbelieved by the referees. It took much correspondence between the authors, the editor, and the referees before the paper was eventually accepted for publication. The second N_2-containing iridium complex was also not deliberately prepared. The cobalt complex **4-85** prepared in Japan[88a] was also made serendipitously. We were trying to prepare an alkylcobalt complex by treating cobalt acetylacetonate with alkylaluminum compounds in the presence of a tertiary phosphine under nitrogen using the Schlenk technique to be described in Chapter 5. I had learned this technique at the Max Planck Institute for Coal Research, which was then under the directorship of Professor Ziegler. Although argon was always used there as an inert gas, we used nitrogen in our laboratory simply because we could not afford to use argon all the time. Being poor can bring good luck, albeit very rarely.

We were lucky to be using a Toepler pump (see Section 5.1) to analyze the gas evolved on thermolysis of a compound we obtained. This complex had a sharp IR band at 2088 cm^{-1}. Thinking that the band might be due to a cobalt hydride possibly formed by β-hydrogen elimination of an intermediate ethylcobalt complex, an undergraduate student, S. Kitazume, collected the noncondensable gas at liquid nitrogen temperature by means of the Toepler pump and put it in for mass spectroscopic analysis. My first reaction on seeing the mass spectrum was that he must have made a mistake and let air in because there was a large m/e peak at 28 and a very small peak at 2 due to the H_2 we were looking for. A second look, however, showed that there was no peak above the background arising from O_2. This excluded the possibility of air leakage. Prompt elemental analysis did indicate that the compound contained nitrogen, which could only have come from the "inert" gas atmosphere.

Later I learned that this compound was called the "Pearl Harbor complex" in the United States and in Germany. Although it is not a very pleasant name for us Japanese, we can understand the surprise and chagrin of those people who were also very close to identifying the complex. It may not be necessary to add that Admiral Yamamoto, who attacked Pearl Harbor, is not my relative. Yamamoto is a very common name in Japan.

Following these findings there were many reports on the preparation of N_2 complexes. Figure 4.1 shows structures of representative N_2 complexes.

Most dinitrogen complexes contain N_2 coordinated to transition metal complexes linearly (end-on-manner). Like carbon monoxide in metal carbonyls, the dinitrogen can act as a weak electron donor and at the same time can accept electrons by back bonding, utilizing π^* orbitals. It can also serve as a bridging ligand connecting two metal atoms. Because dinitrogen is

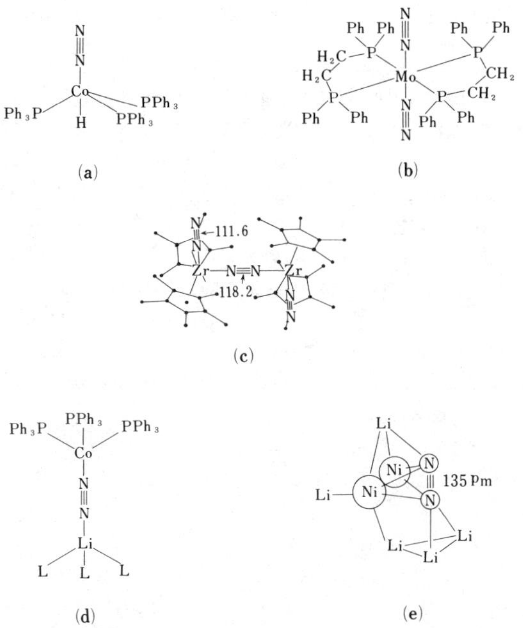

FIGURE 4.1. Structures of various dinitrogen complexes: (*a*) Structure of **4-85**; (*b*) structure of **4-86**; (*c*) structure of **4-87**; (*d*) structure of **4-88**, L = THF or Et_2O; (*e*) central part of a Ni–N_2 complex $[\{(PhLi)_6Ni_2(N_2)(Et_2O)_2\}_2]$ showing how N_2 is bound to two Ni atoms and neighboring Li atoms.

analogous to acetylene in having two orthogonal π orbitals, it can also bind to two metal atoms in a side-on manner as shown in Figure 4.1*e*.[90] Enhanced back donation from the metals to the ligated N_2 leads to elongation of the N–N distance, reflecting activation of the ligated dinitrogen.

Most dinitrogen complexes can be synthesized by reducing transition metal complexes under nitrogen or by displacing other labile ligands with N_2.

$$MoCl_3(thf)_3 + Na/Hg + dppe \xrightarrow{N_2} trans\text{-}Mo(N_2)_2(dppe)_2$$

4-86

$$Ru(NH_3)_5(H_2O)^{2+} + N_2 \rightleftarrows [Ru(NH_3)_5N_2]^{2+}$$

$$Cp^*_2ZrCl_2 + Na/Hg \xrightarrow[-NaCl]{N_2} [Cp^*_2Zr(N_2)]_2N_2$$

4-87

$Cp^* = C_5Me_5$ (pentamethylcyclopentadienyl)

The zirconium complex possesses both terminal and bridging N_2 ligands.

In limited examples N_2-containing complexes can be prepared by chemical transformation of other nitrogen-containing ligands.[91]

$$CpMn(CO)_2(N_2H_4) + H_2O_2/Cu^{2+} \longrightarrow CpMn(CO)_2N_2$$

Whereas the dinitrogen ligated to Co(I) as in **4-85** is not reactive toward protons, dinitrogen coordinated to early transition metals such as zirconium and molybdenum reacts with protonic acids to give ammonia and/or hydrazine. Reactions of N_2 complexes such as **4-86** or its tungsten analog with protons are believed[92] to proceed stepwise, as depicted in Figure 4.2.

The end-on coordinated dinitrogen ligand undergoes proton attack at the terminal nitrogen to give a diazenido complex (A). Attack by a second proton at the terminal dinitrogen leads to a hydrazido(2-) complex (B). These types of complexes have been isolated in several cases. The third proton may attack either at the terminal nitrogen to give ammonia or at the metal to give complex C, which may undergo the further proton attack at the metal-bonded nitrogen to liberate hydrazine.

The molybdenum and tungsten complexes of dinitrogen can also react with other hydrogen sources such as alcohols and metal hydrides to afford ammonia and hydrazine.[93]

The N_2-containing zirconium complex **4-87** also reacts with protons to liberate hydrazine.[94] Other early transition metal complexes are also known to react with dinitrogen to give ammonia or hydrazine upon hydrolysis of the reaction products. On the other hand, very few dinitrogen complexes of the later transition metals give ammonia or hydrazine on protonation. For example, treatment of $CoH(N_2)(PPh_3)_3$ (**4-85**) with acid simply causes release of N_2 without converting it into ammonia or hydrazine. However, complex **4-88** synthesized by reduction of **4-85** undergoes proton attack to give ammonia and hydrazine[95]

$$\underset{\textbf{4-85}}{CoH(N_2)(PPh_3)_3} \xrightarrow[THF]{BuLi} \underset{\textbf{4-88}}{(Ph_3P)_3CoN_2Li(thf)_3}$$

In the more electron rich Co(−I) complex **4-88** enhanced back donation from cobalt to N_2 seems to result in increased reactivity toward proton at the

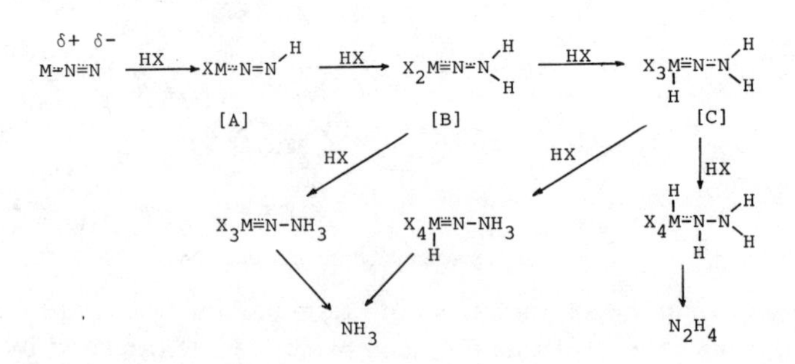

FIGURE 4.2. Reaction pathways in the protonation of coordinated N_2 ligand.

terminal nitrogen. This may be explained by an increased contribution from the resonance structure II shown below.

$$\underset{\mathbf{I}}{M{\leftarrow}N{\equiv}N} \longleftrightarrow \underset{\mathbf{II}}{M{=}\overset{+}{N}{=}\overset{-}{N}:} \tag{4.82}$$

The enhanced reactivity toward electrophiles of the terminal nitrogen atom is also evidenced by the reactions of the N_2-molybdenum and tungsten complexes with alkyl and acyl halides to give organic nitrogen complexes.

$$N_2{-}M{-}N_2 + RCOCl + HCl \xrightarrow{-N_2} Cl_2M{\leftarrow}N(H){=}NCOR$$

$$N_2{-}M{-}N_2 + RBr \xrightarrow{-N_2} Br{-}M{-}N{=}N{-}R \xrightarrow{H^+} [Br{-}M{=}N{-}N(H)R]^+$$

In contrast with the terminal nitrogen atom, which is rendered more electron rich as expressed in Eq. (4.82), the nitrogen atom bound to the metal is susceptible to nucleophilic attack. Reaction of methyllithium at the coordinated metal-bound nitrogen makes the terminal nitrogen more electron rich so that it can undergo electrophilic attack.[96]

$$CpMn(CO)_2(N_2) \xrightarrow[\text{THF} \ -30\ °C]{\text{MeLi}} CpMn(CO)_2(N(Me){=}N^-) \xrightarrow[\text{THF} \ -30\ °C]{Me_3O^+BF_4^-} \underset{\mathbf{4\text{-}89}}{CpMn(CO)_2(N(Me){=}N{-}Me)}$$

Treatment of complex **4-89** with N_2 under pressure liberates dimethyldiazene and restores the initial N_2 complex. This reaction suggests a potential catalytic method of converting the nitrogen molecule directly into organic nitrogen compounds.

Nitrogen Fixation. Nitrogen fixation is performed mainly by biological systems and by industrial processes that convert N_2 and H_2 into NH_3 under high pressures and temperatures. The reaction

$$\tfrac{1}{2}N_2 + \tfrac{3}{2}H_2 \rightleftarrows NH_3$$

is exothermic by 45.9 kJ mol^{-1}, but the high activation energy requires the process to be operated at high temperatures and pressures. The present industrial process is essentially the same as that developed by Haber and Bosch at the beginning of this century, although it is technologically more sophisticated. On the other hand, the biological processes proceed under much less drastic conditions; they reduce and cleave the N–N triple bond, which has the very high bond energy of 942 kJ mol^{-1}. The fixation is carried out by an enzyme called nitrogenase, which is contained in anaerobic bacteria such as *Clostridium pasteurianum*. The exact nature of the active site in nitrogenase has not been clarified. The metalloenzyme contains molybdenum and iron, and it is likely that the molybdenum serves as the nitrogen binding site into which electrons flow from the neighboring iron-containing entities. Recent EXAFS (extended X-ray absorption fine structure) studies suggest the presence of a cubelike cluster involving molybdenum and iron bridged by sulfur. An interesting aspect of nitrogenase activity in relation to organometallic chemistry is that it can reduce substrates other than N_2, such as acetylene, RNC, and RCN, as well as N_3^- and N_2O. The reduction of acetylene to ethylene has been used for monitoring nitrogenase activity.

Figure 4.3 is a schematic diagram of the nitrogen fixation process involving the presumed molybdenum active site. In this scheme the dinitrogen binds the reduced form of molybdenum in the N_2-fixing active site and is activated to react with protons. Whether the dinitrogen bonds to the metal in the end-on or side-on manner[97] is yet to be clarified. Upon successive protonations at the ligand dinitrogen, ammonia may be formed, whereas the electron flow from molybdenum to nitrogen converts the molybdenum into high oxidation states. The oxidized molybdenum atom receives electrons from electron transfer sites such as ferredoxin and reverts to the reduced state to bind again with N_2, and the catalytic cycle proceeds to fix nitrogen. However, the exact process occurring in the enzyme, which has a molecular weight of several hundred thousand, is difficult to establish. The information gained from the chemistry of low-molecular-weight model complexes has helped to clarify the enzymatic nitrogen fixation processes.

4.7. CO_2 AND CS_2 COMPLEXES

The insertion of carbon dioxide into the metal–carbon bonds of organometallic compounds of main-group elements such as Grignard reagents has been

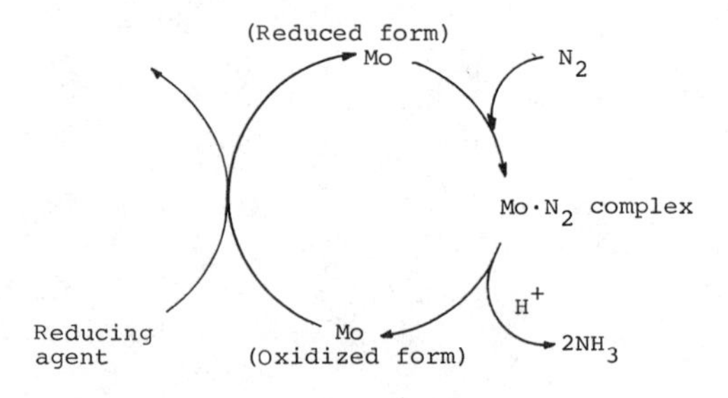

FIGURE 4.3. Schematic diagram of biological dinitrogen fixation assuming molybdenum as the N_2 activating center.

long known. It gives a carboxylate that can be liberated as carboxylic acid on hydrolysis. The process has been utilized in organic synthesis. In contrast, reactions of CO_2 with transition metal complexes[98] had not been explored before the dinitrogen complexes were discovered. The dinitrogen complex **4-85,** for example, reacts with CO_2, liberating N_2 to form a formate complex that can be regarded as having been formed by CO_2 insertion into the Co–H bond.

$$\underset{\textbf{4-85}}{CoH(N_2)(PPh_3)_3} + CO_2 \longrightarrow Co(O_2CH)(PPh_3)_3 + N_2$$

A similar insertion of CO_2 into a Ru–H bond is also known.

$$RuH_2(N_2)(PPh_3)_3 + CO_2 \longrightarrow RuH(O_2CH)(PPh_3)_3$$

Insertion of CO_2 into a Cu–C bond gives carboxylates.

$$NCCH_2Cu(PBu_3)_x + CO_2 \longrightarrow NCCH_2CO_2Cu(PBu_3)_x$$

$$RCu(PR'_3)_2 + CO_2 \longrightarrow RCO_2Cu(PR'_3)_2$$

These reactions are thought to be preceded by coordination of the CO_2 molecule to the transition metal involved. There are two types of CO_2 complexes known to date. In one type the CO_2 ligand is bound to the transition metal through one of the two carbonyl groups in a side-on manner.[99]

$$Ni(PCy_3)_3 + CO_2 \xrightarrow{-PCy_3} (Cy_3P)_2Ni(\eta^2\text{-}CO_2)$$

$$(\eta^5\text{-}C_5H_4Me)_2Ta(CH_2SiMe_3)Cl + CO_2 \xrightarrow[-NaCl]{Na} (\eta^5\text{-}C_5H_4Me)_2Ta(\eta^2\text{-}CO_2)(CH_2SiMe_3)$$

$$Fe(PMe_3)_4 + CO_2 \rightleftharpoons (Me_3P)_4Fe(\eta^2\text{-}CO_2)$$

In the other type the electron-deficient center carbon atom in CO_2 is bonded to an electron-rich transition metal complex, Co(pr-salen)K.[98h,100]

$$[Co(pr\text{-}salen)]^- K^+ + CO_2 \rightleftharpoons \underset{\textbf{4-90}}{Co(pr\text{-}salen)(CO_2)K}$$

Co(pr-salen) **4-90**

In complex **4-90** one oxygen atom in the coordinated CO_2 molecule interacts with the alkali metal cation. The reaction of Co(pr-salen)K with CO_2 is reversible.

In addition to these examples, an intriguing complex in which two molecules of CO_2 are joined together and are attached to a transition metal has been synthesized and characterized.[101]

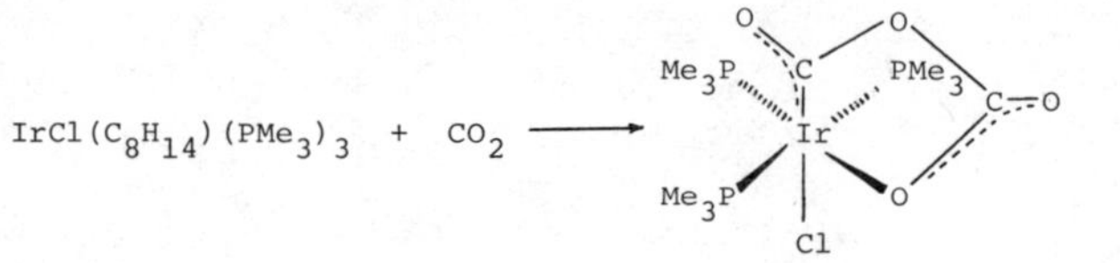

$$IrCl(C_8H_{14})(PMe_3)_3 + CO_2 \longrightarrow$$

4-91

A complex such as **4-91** may have been formed in two steps, and an entity such as the "dimeric CO_2" ligand tends to disproportionate giving the carbonyl and the carbonate complexes.

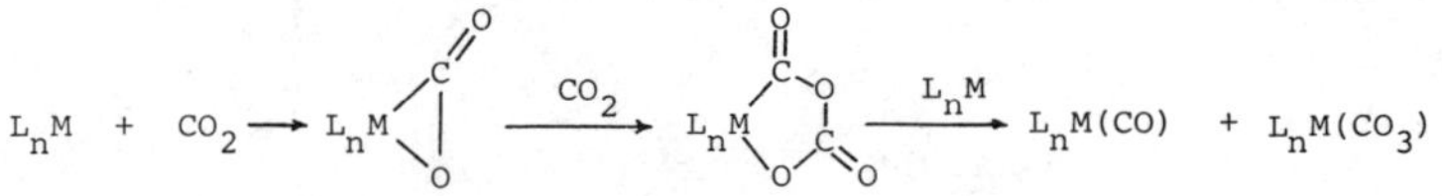

$$L_nM + CO_2 \longrightarrow L_nM(CO_2) \xrightarrow{CO_2} \xrightarrow{L_nM} L_nM(CO) + L_nM(CO_3)$$

However, carbonate or hydrogen carbonate complexes can also be formed by interaction of a small amount of adventitious water in the reaction of CO_2 with transition metal complexes. Therefore, care must be taken in the characterization of CO_2 complexes without unequivocal X-ray structural analysis.

CO_2 can be inserted into a metal-allylic bond to give a carboxylate.[102]

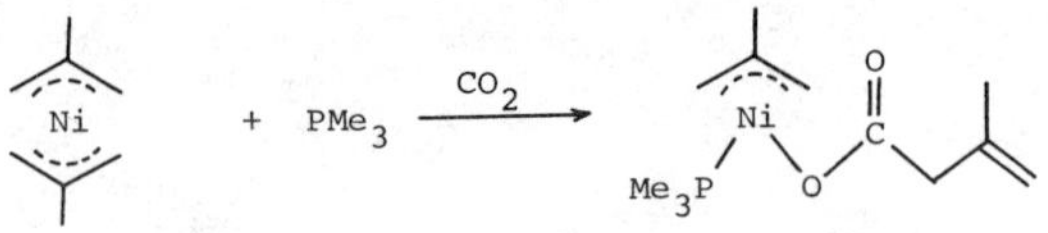

Ni + PMe_3 $\xrightarrow{CO_2}$ Ni, Me_3P

Such a CO_2 insertion into a metal–carbon bond may be involved in catalytic conversion of butadiene and CO_2 into lactones and carboxylic acids.[103]

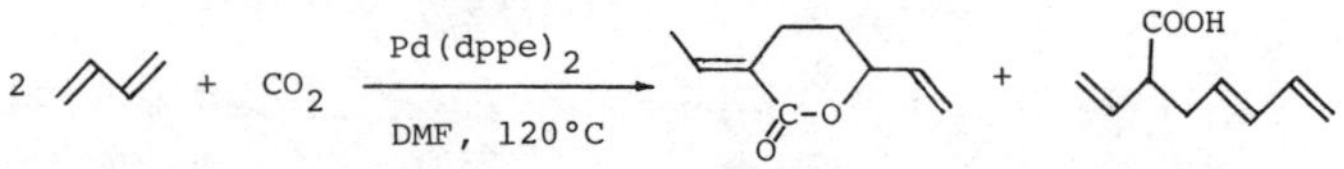

2 + CO_2 $\xrightarrow[\text{DMF, 120°C}]{\text{Pd(dppe)}_2}$ + COOH

Acetylene dimerization also may be coupled with CO_2 insertion.[104]

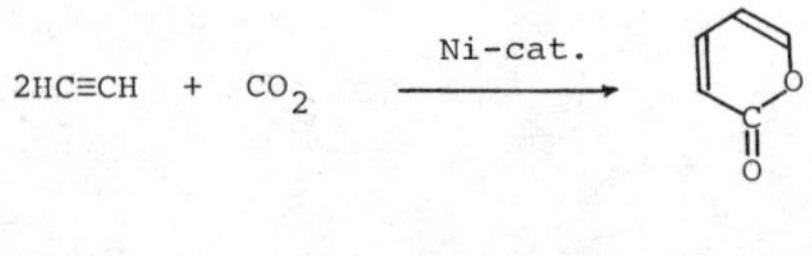

$$2HC{\equiv}CH + CO_2 \xrightarrow{\text{Ni-cat.}}$$

Although thermodynamically favorable catalytic reactions involving CO_2 as a starting material are limited, in special cases, such as those above, CO_2 can act as a component of catalytic synthetic reactions.

Reactions CS_2 and COS, which are analogs of CO_2, are attracting increasing attention. Examples of CS_2 complexes are given below.[105,106]

$$Pt(PPh_3)_2(PhC{\equiv}CPh) + CS_2 \longrightarrow (Ph_3P)_2Pt(\eta^2\text{-}S{=}C{-}S) + PhC{\equiv}CPh$$

$$Cp_2Nb(\eta^3\text{-}C_3H_5) + CS_2 \longrightarrow Cp_2Nb(CH_2CH{=}CH_2)(\eta^2\text{-}S{-}C{=}S)$$

The η^2-CS_2 complexes thus formed are sometimes susceptible to electrophilic attack at the sulfur.

$$(Ph_3P)_2(OC)_2Ru(\eta^2\text{-}CS_2) + MeI \longrightarrow [(Ph_3P)_2(OC)_2Ru(\eta^2\text{-}C(SMe)S)]^+ I^-$$

In some cases carbene complexes are formed:

$$(Ph_3P)_2Pt(\eta^2\text{-}CS_2) + 2MeI \longrightarrow [(Ph_3P)_2IPt{=}C(SMe)_2]^+ I^-$$

4.8. DIOXYGEN COMPLEXES

Complexes containing coordinated O_2 are important in two respects.[107] One is related to biological function and the other to the oxidation of organic compounds promoted by transition metal compounds.

The dioxygen molecule can bind a transition metal complex in two ways: end-on coordination and side-on coordination. There has been a controversy concerning the mode of O_2 coordination in hemoglobin and related biological oxygen carriers. Complexes exhibiting both types of O_2 coordination are now known, but model studies using iron porphyrin complexes support the end-on angular binding in biological systems as originally proposed by Pauling. The dioxygen molecule undergoes two reduction steps on interaction with a transition metal.

$$O_2 \underset{}{\overset{+e}{\rightleftharpoons}} \underset{\text{superoxo type}}{O_2^-} \underset{}{\overset{+e}{\rightleftharpoons}} \underset{\text{peroxo type}}{O_2^{2-}}$$

Coordination of O_2 is called oxygenation. An oxygen carrier such as hemoglobin must bind O_2 reversibly if it is to work. The dioxygen bound to iron in hemoglobin is carried by blood and then given up at the site where it is to be used.

The first example of oxygenation was discovered in a complex formed by coordination of a Schiff's base of acetylacetone or its analogs to planar, divalent cobalt.

$$\text{Co(acacen)} + \text{DMF} + O_2 \underset{}{\overset{K}{\rightleftharpoons}} \text{Co(acacen)(dmf)}O_2 \qquad K = 21 \; (-10\ ^\circ\text{C})$$

In these O_2 complexes electron transfer takes place from Co(II) to the bound O_2 to give a Co(III)–superoxide complex in which O_2 is bent. Interaction of this complex with another Co(II) complex gives a dinuclear μ-peroxo cobalt complex.

$$\text{B–Co} + O_2 \rightleftharpoons \text{B–Co–O–O} \xrightarrow{\text{B–Co}} \text{B–Co–O–O–Co–B}$$

Some transition metal complexes bind O_2 reversibly only at low temperatures or under special conditions. Iron porphyrins are believed to lose their O_2 binding ability because of dimerization and subsequent inversible oxidation to iron(III) μ-oxo complexes that can no longer bind O_2.

$$\text{L–Fe}^{II}\text{–O}_2 + \text{Fe–L} \longrightarrow \text{L–Fe–O–O–Fe–L} \longrightarrow \text{L–Fe–O–Fe–L}$$

In the hemoglobin environment the protein surrounding the active site prevents dimerization and provides a hydrophobic site for approach of the O_2 molecule. Studies using model iron–porphyrin systems designed to mimic part of the functions of the O_2 binding site have provided important informations regarding the O_2 binding mode. Figure 4.4 shows some of the models that have been designed to prevent dimerization of the iron porphyrins while still leaving room for O_2 binding.

These studies showed the dioxygen molecule to be bound end-on to the iron active center. Figure 4.4*d* reproduces schematically how O_2 is bound to a model iron porphyrin system that has four "picket fence" substituent groups on one side of the porphyrin plane to prevent dimerization. With another ligand such as imidazole coordinated below the porphyrin ring, the dioxygen is bound to iron in a bent manner as revealed by X-ray crystallography.

Side-on bonded O_2 complexes are also known for transition metal complexes. The first example of this type to be recognized was the complex $IrCl(O_2)(CO)(PPh_3)_2$, which is formed by the reversible reaction of O_2 with Vaska's complex.

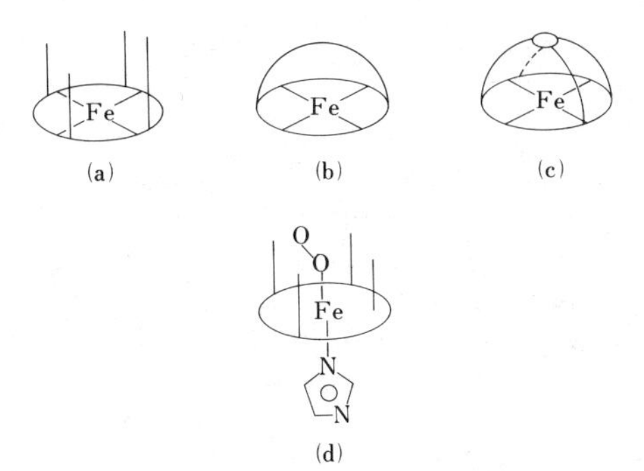

FIGURE 4.4. Representative examples of synthetic oxygen carriers. (*a*) Picket-fence; (*b*) strapped; (*c*) capped; (*d*) mode of O_2 coordination to iron atom in the picket-fence model.

$$Ph_3P{-}Ir(Cl)(CO){-}PPh_3 + O_2 \rightleftharpoons Ir(Cl)(CO)(PPh_3)_2(O_2)$$

This dioxygen adduct may be considered as a distorted octahedral Ir(III) complex if O_2 is regarded as a bidentate, dianionic ligand, although it is possible to consider the adduct as a trigonal bipyramid Ir(I) if O_2 is regarded as a monodentate, neutral ligand. The duality is similar to that discussed for olefin and acetylene complexes [Sections 4.1*a,b*].

In addition to direct oxygenation, transition metal dioxygen complexes or peroxo complexes may be also prepared via other routes.[108] One approach involves treatment of metal oxides with hydrogen peroxide in the presence of a ligand such as hexamethylphosphoric triamide (HNPA).[109]

$$MoO_3 + 2H_2O_2 + HMPA \longrightarrow MoO(O_2)_2(hmpa)(H_2O) + H_2O$$

4-92

Complex **4-92** has a pentagonal, bipyramidal structure with HMPA occupying the equatorial position in the same plane as the two peroxo ligands.[109] The complex is quite effective for selective epoxidation of olefins.

The superoxide ion (O_2^-) can be also used for the preparation of μ-peroxo complexes having nonconjugated diolefins as ligands.[110]

$$(cod)Rh(\mu\text{-}Cl)_2Rh(cod) + 2KO_2 \xrightarrow{-O_2,\ -2KCl} (cod)Rh(\mu\text{-}O_2)Rh(cod)$$

4-93

Interesting reactions associated with organometallic chemistry and oxidation reactions promoted by transition metal complexes are formation of metallacycle complexes on treatment of peroxo complexes with ketones and olefins.[112]

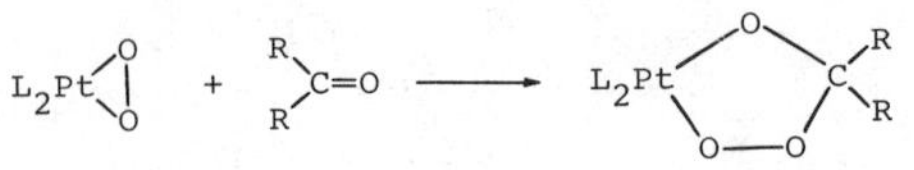

4-94

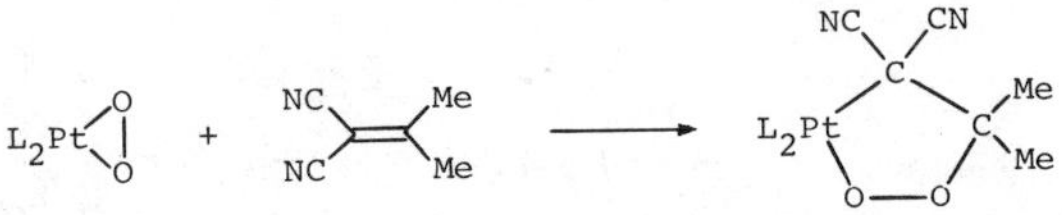

4-95

These reactions may be relevant to the oxidation of organic compounds promoted by transition metal complexes in nonradical reaction mechanisms, as will be discussed in Chapter 7.

REFERENCES

1. For references concerning the π complexes, see the general references in Chapter 1, this volume.
2. G. B. Kauffmann, ed., *Classics in Coordination Chemistry,* Part 2, Dover Publishing, New York, 1976.
3. R. Cramer, *Inorg. Chem.,* **4,** 445 (1965).
4. (a) E. O. Fischer and H. Werner, *Metal π-Complexes,* Vol. I, Elsevier, Amsterdam, 1966; (b) M. Herberhold, *Metal π-Complexes,* Vol. II, Elsevier, Amsterdam, 1972 (part 1), 1974 (part 2).
5. B. Bogdanovic, M. Kröner, and G. Wilke, *Justus Liebigs Ann. Chem.,* **699,** 1 (1966).
6. (a) E. Koerner von Gustorf and F.-W. Grevels, *Fortschr. Chem. Forsch.,* **13,** 366 (1969); (b) M. Wrighton, *Chem. Rev.,* **74,** 401 (1974); (c) G. L. Geoffroy and M. S. Wrighton, *Organometallic Photochemistry,* Academic Press, New York, 1979.
7. D. J. Cole-Hamilton and G. Wilkinson, *Nouv. J. Chim.,* **1,** 141 (1977); S. Komiya and A. Yamamoto, *Bull. Chem. Soc. Jpn.,* **49,** 2553 (1976).
8. J. R. Sweet and W. A. G. Graham, *J. Am. Chem. Soc.,* **105,** 305 (1983).
9. (a) F. G. A. Stone, *Acc. Chem. Res.,* **14,** 318 (1981); (b) J. D. Ittel, *Inorg. Synth.,* **17,** 117 (1977).
10. (a) J. R. Blackborrow and D. Young, *Metal Vapor Synthesis in Organometallic Chemistry,* Springer-Verlag, Berlin, 1979; (b) P. L. Timms, *Adv. Inorg. Chem. Radiochem.,* **14,** 121 (1972); (c) P. L. Timms and T. W. Turney, *Adv. Organometal. Chem.,* **15,** 53 (1977); (d) M. L. H. Green, *J. Organometal. Chem.,* **200,** 119 (1980); (e) K. Klabunde, *Acc. Chem. Res.,* **8,** 393 (1975); (f) M. J. McGlinchey, in *The Chemistry of the Metal–Carbon Bond,* F. R. Hartley and S. Patai, eds., Wiley-Interscience, New York, 1982, p. 534.
11. (a) G. A. Ozin, *Acc. Chem. Res.,* **10,** 2 (1977); (b) M. Moskovits and G. A. Ozin, *Cryochemistry,* Wiley-Interscience, New York, 1976.
12. G. S. Lewandos, in *The Chemistry of the Metal–Carbon Bond,* F. R. Hartley and S. Patei, eds., Wiley-Interscience, New York, 1982, p. 287.

13. (a) G. Marr and B. W. Rochett, in *The Chemistry of the Metal–Carbon Bond,* F. R. Hartley and S. Patai, Eds., Wiley-Interscience, New York, 1982; (b) M. A. Bennett, *Adv. Organometal. Chem.,* **4,** 353 (1966).

14. H. Yasuda, Y. Kajihara, K. Mashima, K. Nagasuna, K. Lee, and A. Nakamura, *Organometallics,* **1,** 388 (1982); *Chem. Lett.,* 519 (1981).

15. (a) U. Dorf, K. Engel, and G. Erker, *Organometallics,* **2,** 462 (1983) and references therein; (b) S. S. Wreford and J. F. Whitney, *Inorg. Chem.,* **20,** 3918 (1981) and references therein.

16. (a) F. L. Bowden and A. B. P. Lever, *Organometal. Chem. Rev.,* **3,** 227 (1968); (b) A. Efraty, *Chem. Rev.,* **77,** 691 (1977).

17. (a) P. Powell, in *The Chemistry of the Metal–Carbon Bond,* F. R. Hartley and S. Patai, eds., Wiley-Interscience, New York, 1982, p. 325; (b) M. I. Lobach, B. D. Babitskii, and V. A. Kormer, *Russ. Chem. Rev.,* **36,** 1158 (1967); (c) J. Powell, in *MPI International Review of Science,* Series 1, Vol. 6, Butterworths, London, 1972, p. 273.

18. G. Wilke, B. Bogdanovic, P. Hardt, P. Heimbach, W. Keim, M. Kröner, W. Oberkirch, K. Tanaka, E. Steinrücke, D. Walter, and H. Zimmermann, *Angew. Chem. Int. Ed. Engl.,* **5,** 151 (1966).

19. R. F. Heck, *Organotransition Metal Chemistry,* Academic Press, New York, 1974.

20. M. Brookhart, T. H. Whitesides, and J. M. Crockett, *Inorg. Chem.,* **15,** 1550 (1976), and references therein.

21. (a) J. Ishizu, T. Yamamoto, and A. Yamamoto, *Chem. Lett.,* 1091 (1976); (b) T. Yamamoto, J. Ishizu, and A. Yamamoto, *J. Am. Chem. Soc.,* **103,** 6863 (1981).

22. T. Yamamoto, O. Saito, and A. Yamamoto, *J. Am. Chem. Soc.,* **103,** 5600 (1981).

23. H. Werner and H. Kraus, *Chem. Ber.,* **113,** 1072 (1980).

24. D. Walter and G. Wilke, *Angew. Chem.,* **78,** 941 (1966).

25. (a) T. J. Kealy and P. J. Pauson, *Nature,* **168,** 1039 (1951); (b) S. A. Miller, J. A. Tebboth, and J. F. Tremaine, *J. Chem. Soc.,* 632 (1952).

26. (a) J. M. Birmingham, *Adv. Organometal. Chem.,* **2,** 365 (1964); (b) M. Rosenblum, *Chemistry of the Iron Group Metallocenes,* Wiley-Interscience, New York, 1965; (c) G. Marr and B. W. Rockett, in *The Chemistry of the Metal–Carbon Bond,* F. R. Hartley and S. Patai, eds., Wiley-Interscience, New York, 1982, p. 443.

27. (a) G. Pez, *Adv. Organometal. Chem.,* **19,** 1 (1981); (b) R. B. King, *Adv. Organometal. Chem.,* **2,** 157 (1964).

28. P. M. Maitlis, *Acc. Chem. Res.,* **11,** 301 (1978).

29. T. C. McKenzie, R. D. Sanner, and J. E. Bercaw, *J. Organometal. Chem.,* **102,** 457 (1975).

30. G. Marr and R. B. Rockett, in *The Chemistry of the Metal–Carbon Bond,* F. R. Hartley and S. Patai, eds., Wiley-Interscience, New York, 1982, p. 463.

31. M. A. Bennett, in *Rodd's Chemistry of Carbon Compounds,* Vol. 3, S. Coffey, ed., Elsevier, Amsterdam, 1974, p. 357.

32. R. P. A. Sneeden, *Organochromium Compounds,* Academic Press, New York, 1975.

33. W. A. Silverthorn, *Adv. Organometal. Chem.,* **13,** 47 (1975).

34. E. O. Fischer and W. Hafner, *Z. Naturforsch.,* **10B,** 665 (1955); E. O. Fischer and W. Hafner, *Z. Anorg. Chem.,* **286,** 246 (1956); H. H. Zeiss and M. Tsutsui, *J. Am. Chem. Soc.,* **79,** 3062 (1957).

35. P. L. Timms, *Chem. Commun.,* 1033 (1969).

36. F. Hein, *Ber.,* **52,** 195 (1919).

37. H. H. Zeiss and M. Tsutsui, *J. Am. Chem. Soc.,* **79,** 3062 (1957).

38. M. A. Bennett, *Adv. Organometal. Chem.,* **4,** 353 (1966).

39. T. A. Stephenson, in *International Review of Science, Inorganic Chemistry,* H. J. Mays, eds., Series 1, Vol. 6, p. 401 (1972) and Series 2, Vol. 6, p. 287 (1975), Butterworths, London.

40. H. Breil and G. Wilke, *Angew. Chem. Int. Ed. Engl.*, **5,** 898 (1966).
41. F. A. Cotton, *Chem. Rev.*, **55,** 551 (1955).
42. G. W. Parshall and J. J. Mrowca, *Adv. Organometal. Chem.*, **7,** 157 (1968).
43. P. J. Davidson, M. F. Lappert, and R. Pearce, *Chem. Rev.*, **76,** 219 (1976).
44. R. R. Schrock and G. W. Parshall, *Chem. Rev.*, **76,** 243 (1976).
45. R. J. Puddephatt, in *The Chemistry of the Metal–Carbon Bond,* F. R. Hartley and S. Patai, eds., Wiley-Interscience, New York, 1982, p. 245.
46. R. Taube, H. Drevs, and D. Steinborn, *Z. Chem.*, **18,** 425 (1978).
47. (a) K. Clauss and C. Beermann, *Angew. Chem.*, **71,** 627 (1959); (b) H. J. Berthold and G. Groh, *Z. Anorg. Allg. Chem.*, **319,** 239 (1963); *Angew. Chem.*, **78,** 495 (1966).
48. K.-H. Thiele and J. Müller, *Z. Chem.*, **4,** 273 (1964).
49. A. Miyashita and A. Yamamoto, *Bull. Chem. Soc. Jpn.*, **50,** 1102 (1977).
50. S. Komiya, M. Bundo, T. Yamamoto, and A. Yamamoto, *J. Organometal. Chem.*, **174,** 343 (1979).
51. (a) P. S. Braterman, R. J. Cross, and G. B. Young, *J. Chem. Soc., Dalton Trans.*, 1310 (1976); 1982 (1977); (b) S. Komiya, Y. Abe, T. Yamamoto, and A. Yamamoto, *Organometallics.*, **2,** 1466 (1983).
52. J. P. Collman, *Acc. Chem. Res.*, **8,** 342 (1975).
53. R. D. Rieke, *Acc. Chem. Res.*, **10,** 301 (1977).
54. T. Kitazume and N. Ishikawa, *Yuki Gosei Kyokaishi (Jpn.),* **41,** 432 (1983); J. Am. Chem. Soc., **107,** 5186 (1985).
55. J. H. Habeeb and D. G. Tuck, *J. Organometal. Chem.*, **139,** C17 (1977).
56. J. Chatt, R. S. Coffey, A. Gouph, and D. T. Thompson, *J. Chem. Soc.*, **A,** 190 (1968).
57. J. Schwarz, *Pure App. Chem.*, **52,** 733 (1980); J. Schwarz and J. A. Labinger, *Angew. Chem.*, **88,** 402 (1976).
58. H. C. Clark, *J. Organometal. Chem.*, **200,** 63 (1980).
59. A. Nakamura and S. Otsuka, *J. Am. Chem. Soc.*, **95,** 7262 (1973).
60. (a) W. T. Miller, Jr. and R. J. Burnard, *J. Am. Chem. Soc.*, **90,** 7367 (1968); (b) W. T. Miller, R. H. Snider, and R. J. Hummel, *J. Am. Chem. Soc.*, **91,** 6532 (1969).
61. P. M. Maitlis, *J. Organometal. Chem.*, **200,** 161 (1980).
62. F. D. Mango and I. Dvoretzky, *J. Am. Chem. Soc.*, **88,** 1654 (1966).
63. (a) R. H. Grubbs, *Prog. Inorg. Chem.*, **24,** 1 (1978); (b) R. J. Puddephatt, *Coord. Chem. Rev.*, **33,** 149 (1980).
64. (a) J. Halpern, in *Organic Syntheses via Metal Carbonyls,* Vol. II, I. Wender and P. Pino, eds., Wiley, New York, 1977, p. 705; (b) K. C. Bishop III, *Chem. Rev.*, **76,** 461 (1976); (c) L. Cassar, P. E. Eaton, and J. Halpern, *J. Am. Chem. Soc.*, **92,** 3515 (1970).
65. F. W. Grevels, D. Schulz, and E. Koerner von Gustorf, *Angew. Chem.*, **86,** 585 (1974); F. W. Grevels, U. Feldhoff, J. Leitlich, and C. Krüger, *J. Organometal. Chem.*, **118,** 79 (1976).
66. S. J. McLain, C. D. Wood, and R. R. Schrock, *J. Am. Chem. Soc.*, **101,** 4558 (1979).
67. (a) Y. Wakatsuki, O. Nomura, K. Kitaura, K. Morokuma, and H. Yamazaki, *J. Am. Chem. Soc.*, **105,** 1907 (1983); (b) A. Stockis and R. Hoffmann, *J. Am. Chem. Soc.*, **102,** 2952 (1980); (c) S. Otsuka and A. Nakamura, *Adv. Organometal. Chem.*, **14,** 245 (1976).
68. (a) G. W. Parshall, *Acc. Chem. Res.*, **3,** 139 (1970); **8,** 113 (1975); (b) M. I. Bruce, *Angew. Chem. Int. Ed. Engl.*, **16,** 73 (1977); (c) I. Omae, *Coord. Chem. Rev.*, **32,** 235 (1980).
69. (a) W. A. Herrmann, *Angew. Chem. Int. Ed. Engl.*, **17,** 800 (1978); (b) W. A. Herrmann, *Adv. Organometal. Chem.*, **20,** 159 (1982).
70. Y. C. Lin, J. C. Calabrese, and S. S. Wreford, *J. Am. Chem. Soc.*, **105,** 1679 (1983).
71. D. Seyferth, *Adv. Organometal. Chem.*, **14,** 97 (1976).

72. D. C. Miller and T. B. Brill, *Inorg. Chem.*, **17,** 240 (1978).

73. L. Weber, in *The Chemistry of the Metal–Carbon Bond,* F. R. Hartley and S. Patai, eds., Wiley-Interscience, New York, 1982.

74. H. Schmidbaur, *Acc. Chem. Res.*, **8,** 62 (1975); *Pure Appl. Chem.*, **50,** 19 (1978); H. Schmidbaur and J. R. Mandl, *Angew. Chem. Int. Ed. Eng.*, **16,** 646 (1977).

75. (a) H. Fischer, in *The Chemistry of the Metal–Carbon Bond,* F. R. Hartley and S. Patai, eds., Wiley-Interscience, New York, 1982, pp. 181, 233; (b) E. O. Fischer, *Adv. Organometal. Chem.*, **14,** 1 (1976); (c) R. R. Schrock, *Acc. Chem. Res.*, **12,** 98 (1979); (d) D. J. Cardin, *Chem. Soc. Rev.*, **2,** 99 (1973); (e) D. J. Cardin, B. Cetinkaya, and M. F. Lappert, *Chem. Rev.*, **72,** 545 (1972); (f) F. A. Cotton and C. M. Lukehart, *Prog. Inorg. Chem.*, **16,** 243 (1972); (g) C. P. Casey, *Transition Metal Organometallics in Organic Synthesis,* Vol. 1, H. Alper, ed., Academic Press, New York, 1976, p. 189; (h) F. J. Brown, *Prog. Inorg. Chem.*, **27,** 1 (1980); (i) E. O. Fischer and U. Schubert, *J. Organometal. Chem.*, **100,** 59 (1975); (j) E. O. Fischer, U. Schubert, and H. Fischer, *Pure Appl. Chem.*, **50,** 857 (1978); (k) K. H. Dötz, H. Fischer, P. Hofmann, F. R. Kreissl, U. Schubert, and K. Weiss, *Transition Metal Carbene Complexes,* Verlag Chemie, Weinheim, 1983.

76. E. O. Fischer and A. Massböl, *Angew. Chem. Int. Ed. Engl.*, **3,** 580 (1964).

77. R. R. Schrock, *J. Am. Chem. Soc.*, **96,** 6796 (1974).

78. (a) M. L. H. Green and D. L. Jones, *Adv. Inorg. Chem. Radiochem.*, **7,** 115 (1965); (b) G. L. Geoffroy and J. R. Lehman, *Adv. Inorg. Chem. Radiochem.*, **20,** 189 (1977); (c) E. L. Muetterties, ed., *Transition Metal Hydrides,* Dekker, New York 1971; (d) R. Bau, ed., *ACS Advances in Chemistry Series,* No. 167, 1978; (e) H. D. Kaesz and R. B. Saillant, *Chem. Rev.*, **72,** 231 (1972); (f) J. P. McCue, *Coord. Chem. Rev.*, **10,** 265 (1973); (g) D. M. Roundhill, *Adv. Organometal. Chem.*, **13,** 273 (1975); (h) D. Giusto, *Inorg. Chim. Acta Rev.*, 91 (1972); R. Bau and T. F. Koetzle, *Pure Appl. Chem.*, **50,** 55 (1978); A. P. Ginsberg, *Transition Metal Chem.*, **1,** 111 (1965).

79. L. Vaska and J. W. Luzio, *J. Am. Chem. Soc.*, **83,** 2784 (1961).

80. (a) P. M. Treichel, *Adv. Organometal. Chem.*, **11,** 21 (1973); (b) Y. Yamamoto, *Coord. Chem. Rev.*, **32,** 193 (1980).

81. (a) I. M. Butler, *Acc. Chem. Res.*, **10,** 359 (1977); (b) P. V. Yaneff, *Coord. Chem. Rev.*, **23,** 183 (1977).

82. T. A. Wnuk and R. J. Angelici, *Inorg. Chem.*, **16,** 1173 (1977).

83. (a) W. P. Griffith, *Adv. Organometal. Chem.*, **7,** 211 (1968); (b) F. Bottomley, *Acc. Chem. Res.*, **11,** 158 (1978); (c) R. Eisenberg and C. D. Meyer, *Acc. Chem. Res.*, **8,** 26 (1975); (d) J. H. Enemark and R. D. Feltham, *Coord. Chem. Rev.*, **13,** 339 (1974); (e) N. G. Connelly, *Inorg. Chim. Acta. Rev.*, **6,** 48 (1972); (f) K. G. Caulton, *Coord. Chem. Rev.*, **14,** 317 (1975).

84. J. McCleverty, *Chem. Rev.*, **79,** 53 (1979).

85. (a) W. Newton, J. R. Postgate, and C. Rodriguez-Barruecco, eds., *Recent Developments in Nitrogen Fixation,* Academic Press, New York, 1977; (b) R. W. Hardy, F. Bottomley, and R. C. Burns, eds., *A Treatise on Dinitrogen Fixation,* Wiley, New York, 1979; (c) J. Chatt, J. R. Dilworth, and R. L. Richards, *Chem. Rev.*, **78,** 589 (1978); (d) E. I. Stiefel, *Prog. Inorg. Chem.*, **22,** 1 (1977).

86. A. D. Allen and C. V. Senoff, *Chem. Commun.*, 621 (1965).

87. J. P. Collman and J. W. Kang, *J. Am. Chem. Soc.*, **88,** 3459 (1966).

88. (a) A. Yamamoto, S. Kitazume, L. S. Pu, and S. Ikeda, *Chem. Commun.*, 79 (1967); (b) A. Misono, Y. Uchida, and T. Saito, *Bull. Chem. Soc. Jpn.*, **40,** 700 (1967); (c) A. Sacco and M. Rossi, *Chem. Commun.*, 316 (1967).

89. M. E. Vol'pin and V. B. Shur, *Akad. Nauk SSSR,* **156,** 1102 (1964).

90. C. Krüger and Y.-H. Tsay, *Angew. Chem. Int. Ed. Engl.*, **12,** 998 (1973).

91. D. Sellmann, *Angew. Chem.*, **83,** 1017 (1971).

92. T. Takahashi, Y. Mizobe, M. Sato, Y. Uchida, and M. Hidai, *J. Am. Chem. Soc.*, **102,** 7461 (1980).

93. H. Nishihara, T. Mori, T. Saito, and Y. Sasaki, *Chem. Lett.*, 667 (1980); M. Hidai, T. Takahashi, I. Yokotake and Y. Uchida, *Chem. Lett.*, 645 (1980).

94. R. D. Sanner, J. M. Manriquez, R. E. Marsh, and J. E. Bercaw, *J. Am. Chem. Soc.*, **98,** 8351 (1976).

95. A. Yamamoto, K. Miura, T. Ito, H. L. Chen, K. Iri, F. Ozawa, K. Miki, T. Sei, N. Tanaka, and N. Kasai, *Organometallics,* **2,** 1429 (1983).

96. D. Sellmann and W. Weiss, *J. Organometal. Chem.*, **160,** 183 (1978); *Angew. Chem. Int. Ed. Engl.*, **17,** 269 (1978).

97. G. N. Schrauzer, *Angew. Chem.*, **87,** 579 (1975).

98. (a) R. P. A. Sneeden, in *Comprehensive Organometallic Chemistry,* G. Wilkinson, F. G. A. Stone, and E. W. Abel, eds., Pergamon Press, Oxford, 1982, Chapter 50; (b) T. Ito and A. Yamamoto, in *Organic and Bio-Inorganic Chemistry of Carbon Dioxide,* S. Inoue and N. Yamazaki, eds., Kodansha, Tokyo, 1980; (c) R. P. A. Sneeden, *J. Mol. Cat.*, **17,** 349 (1982); (d) R. Eisenberg and D. E. Hendricksen, *Adv. Catal.*, **28,** 79 (1979); (e) M. E. Vol'pin and I. S. Kolomnikov, *Organometal. React.*, **5,** 313 (1975); (f) M. E. Vol'pin and I. S. Kolomnikov, *Pure Appl. Chem.*, **33,** 567 (1973); (g) B. Denise and R. P. A. Sneeden, *Chemtech,* 108 (1982); (h) C. Floriani, *Pure Appl. Chem.*, **55,** 1 (1983); (i) C. Floriani, *Pure Appl. Chem.*, **54,** 59 (1982); (j) D. J. Darensbourg and R. A. Kudaroski, *Adv. Organometal. Chem.*, **22,** 129 (1983).

99. (a) M. Aresta and C. F. Nobile, *J. Chem. Soc. Dalton Trans.*, 708 (1977); (b) G. S. Bristow, P. B. Hitchcock, and M. F. Lappert, *J. Chem. Soc. Chem. Commun.*, 1145 (1981); (c) H. H. Karsch, *Chem. Ber.*, **110,** 2213 (1977).

100. (a) M. Pasquali, C. Floriani, A. Chiesi-Villa, and C. Guastini, *J. Am. Chem. Soc.*, **101,** 4740 (1979); (b) S. Gambarotta, F. Arena, C. Floriani, and P. F. Zanazzi, *J. Am. Chem. Soc.*, **104,** 5082 (1982).

101. T. Herskovitz and L. J. Guggenberger, *J. Am. Chem. Soc.*, **98,** 1615 (1976).

102. P. W. Jolly, S. Stobbe, G. Wilke, R. Goddard, C. Krüger, J. C. Sekutowski, and Y.-H. Tsay, *Angew. Chem. Int. Ed. Engl.*, **17,** 124 (1978).

103. Y. Inoue, Y. Sasaki, and H. Hashimoto, *Bull. Chem. Soc. Jpn.*, **51,** 2375 (1978).

104. H. Hoberg and V. G. Burkhart, *Angew. Chem.*, **94,** 75 (1982).

105. P. V. Yaneff, *Coord. Chem. Rev.*, **23,** 183 (1977).

106. I. S. Butler, *Acc. Chem. Res.*, **10,** 359 (1977).

107. (a) B. D. Jones, D. A. Summeville, and F. Basolo, *Chem. Rev.*, **79,** 139 (1979); (b) L. Vaska, *Acc. Chem. Res.*, **9,** 175 (1976); (c) A. B. P. Lever and H. B. Gray, *Acc. Chem. Res.*, **11,** 348 (1978); (d) G. A. Ozin, *Acc. Chem. Res.*, **10,** 21 (1977); (e) F. Basolo, B. M. Hoffman, and J. A. Ibers, *Acc. Chem. Res.*, **8,** 384 (1975); (f) R. S. Drago and B. B. Cordon, *Acc. Chem. Res.*, **13,** 353 (1980); (g) T. Yoshida, K. Tatsumi, and S. Otsuka, *Pure Appl. Chem.*, **52,** 713 (1980); (h) J. A. Valentine, *Chem. Rev.*, **73,** 236 (1973); (i) J. P. Collman, *Acc. Chem. Res.*, **10,** 265 (1977); (j) G. McLendon and A. E. Martell, *Coord.* Chem. Rev., **19,** 1 (1976); (k) G. Henrici-Olivé and S. Olivé, *Angew. Chem.*, **86,** 1 (1974); (l) T. G. Traylor, *Acc. Chem. Res.*, **14,** 102 (1981); (m) A. E. Martell, *Acc. Chem. Res.*, **15,** 155 (1982); (n) T. Smith and J. R. Pilbrow, *Coord. Chem. Rev.*, **39,** 295 (1981).

108. (a) H. Mimoun, *Pure Appl. Chem.*, **53,** 2389 (1981); (b) H. Mimoun, *Angew. Chem. Int. Ed. Engl.*, **21,** 734 (1982); (c) Y. Moro-oka and H. Suzuki, *Yuki Gosei Kagaku Kyokaishi* (*J. Synth. Org. Chem. Jpn.*), **41,** 316 (1983).

109. H. Mimoun, I. Seree de Roch, and L. Sajus, *Bull. Soc. Chim. Fr.*, 1489 (1969).

110. H. Suzuki, K. Mizutani, Y. Moro-oka, and T. Ikawa, *J. Am. Chem. Soc.*, **101,** 748 (1979).

111. (a) R. Ugo, F. Conti, S. Cenini, R. Mason, and G. B. Robertson, *Chem. Commun.*, 1498 (1968); (b) R. Ugo, G. M. Zanderighi, A. Fusi, and D. Carreri, *J. Am. Chem. Soc.*, **102,** 3745 (1980).

112. R. A. Sheldon and J. A. van Doorn, *J. Organometal. Chem.*, **94,** 115 (1975).

CHAPTER FIVE

Experimental Techniques in Organometallic Chemistry

Many organometallic compounds are air sensitive and tend to decompose if not handled properly, so one might prefer to avoid dealing with them. Yet they have such a diverse, intriguing, and challenging chemistry that to avoid them could mean a great loss of opportunity. Although handling air-sensitive compounds may at first appear cumbersome, under proper precautions and with appropriate manipulation tools and techniques, they may be handled almost as easily as ordinary compounds. Taming a shrew can be enjoyable.

The first part of this chapter provides an outline of manipulation techniques for air-sensitive compounds. Since we now have excellent reference books specifically dealing with the techniques, we confine ourselves to giving an idea of essential techniques for handling air-sensitive compounds and leave the detail of the techniques to these references.[1-7] However, we shall describe in more detail some techniques not covered by these references that should be useful to those interested in handling air-sensitive compounds. Organometallic chemists tend to favor tried and trusted techniques. Thus, this is a somewhat personal account of the techniques I acquired at the Max Planck Institut für Kohlenforschung at Mülheim, where I was trained as an organometallic chemist, together with some modifications resulting from our group's experiences at the Tokyo Institute of Technology.

The second part of this chapter is concerned with various physicochemical methods used in studying organometallic compounds, with emphasis on NMR spectroscopy. The reason for this is that NMR spectroscopy is an eminently suitable technique for the study of organometallic compounds, particularly of their dynamic properties in solution.

5.1. MANIPULATION OF AIR-SENSITIVE COMPOUNDS

For carrying out experiments with exclusion of air the following techniques are employed: (1) vacuum line technique, (2) Schlenk technique, and (3) glove-box technique. Depending on the objectives and air sensitivity of the compounds handled, one technique or a combination of techniques are used. From our own experience, techniques using the vacuum line and Schlenk glassware are sufficient and convenient for most purposes of handling air-sensitive compounds. Glove-box techniques have some merits in handling air-sensitive solids, but well-designed, reliable glove boxes are expensive and most manipulations using glove boxes can be carried out using a vacuum line and Schlenk techniques. Furthermore, manipulations in the glove box are scarcely different from those practiced in organic synthesis. Therefore, we are concerned here only with combinations of vacuum line and Schlenk techniques. Those interested in glove-box techniques are referred to references 1 and 2.

Techniques Using Schlenk Glassware

The basic and simplest Schlenk tube is shown in Figure 5.1*a*. With the Schlenk tube one can transfer a solid or liquid in an atmosphere of an inert

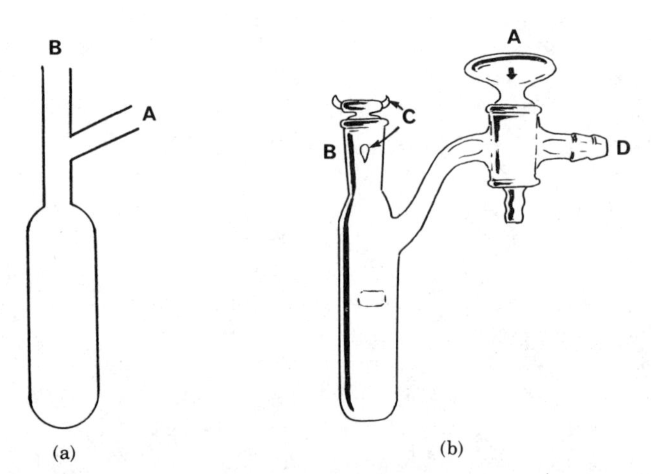

FIGURE 5.1. Handling of air-sensitive compounds. (*a*) Prototype of a Schlenk tube. (*b*) Modified type of Schlenk tube with a tailed tap A, standard taper B, and hooks (C) for holding a stopper with rubber bands. (*b'*) Transfer of liquid; A, stopcock; B, standard taper $\frac{25}{42}$ or $\frac{15}{35}$; C, N_2 inlet; D, hypodermic syringe; E, locking device. (*c*) Filtration with a glass tube bridge A equipped with a glass frit C. The tube without C can be used for liquid transfer; D, bored rubber stopper. (*d*) Cylindrical filtration apparatus with a jacket A for keeping the sample below or above room temperature. (*e*) Filtration with filter paper. (*f*) Filtration using a catheter. (*g*) Transfer of a solid sample (i). (*h*) Transfer of a solid sample (ii). (*i*) Transfer of a small amount of solid sample; A, polyethylene film. (*j*) Filtration in vacuum; C, glass frit.

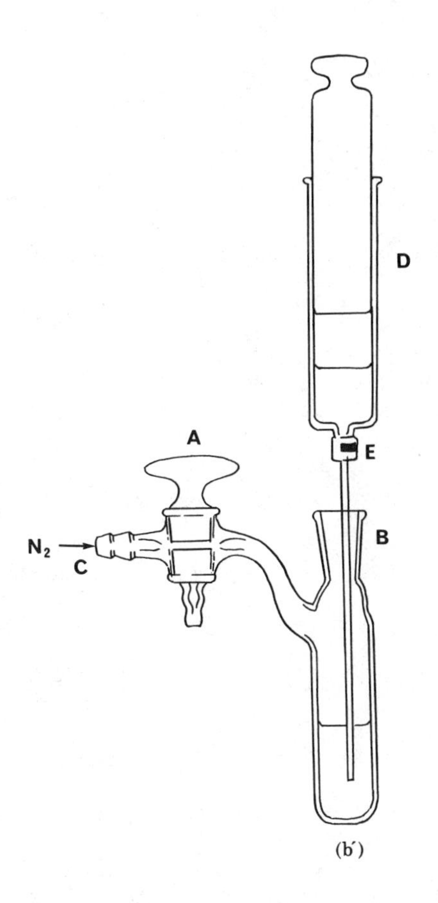

(b′)

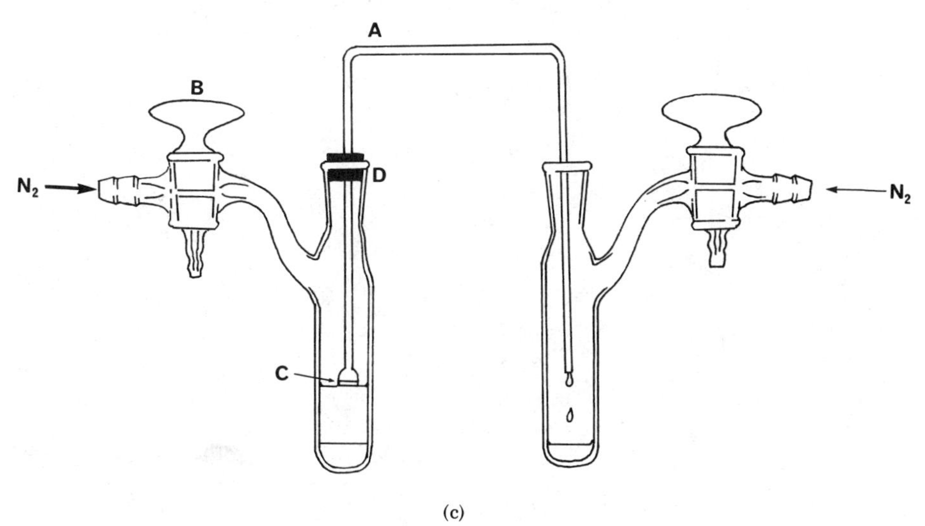

(c)

Figure 5.1. (*Continued*)

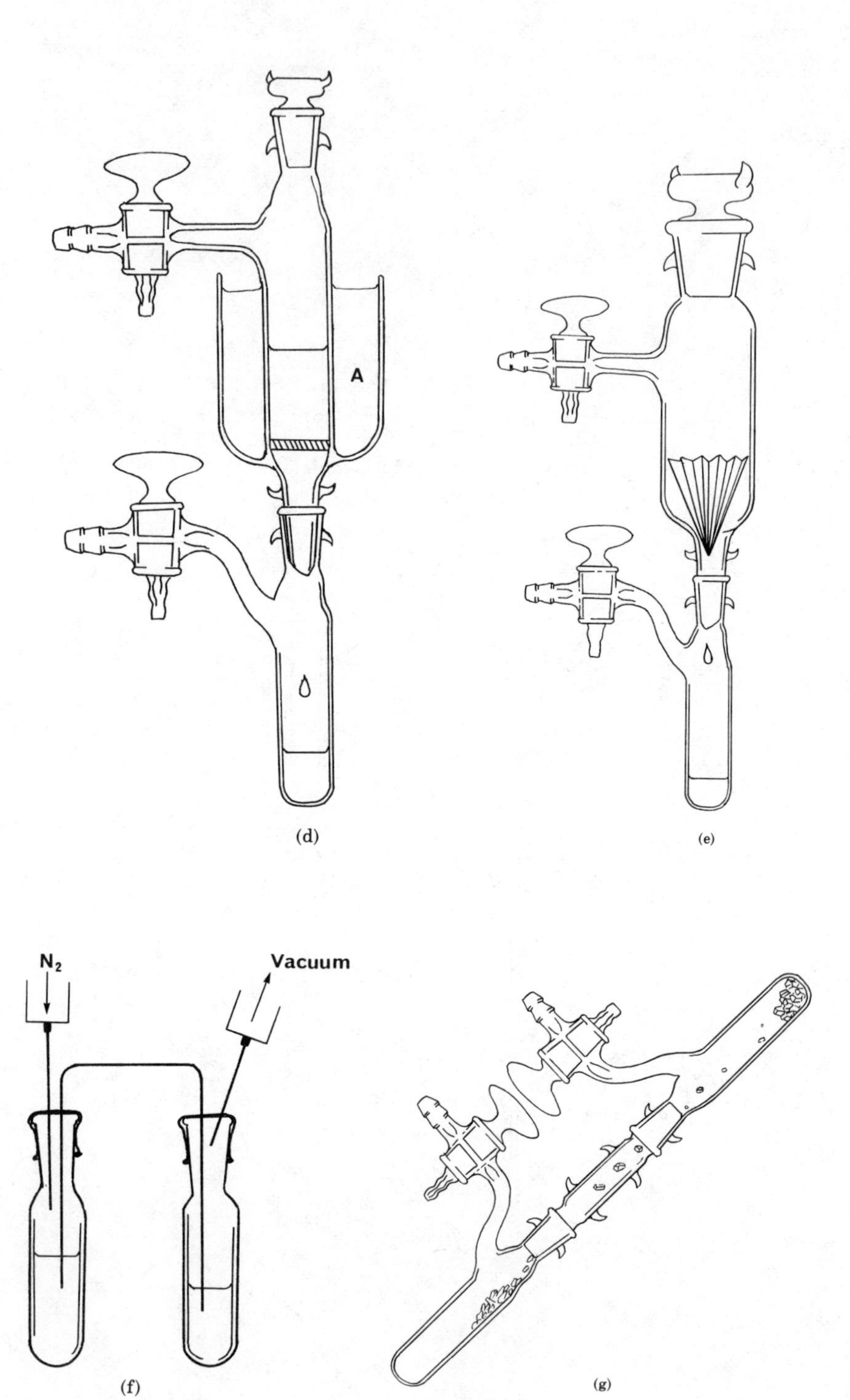

Figure 5.1. (*Continued*)

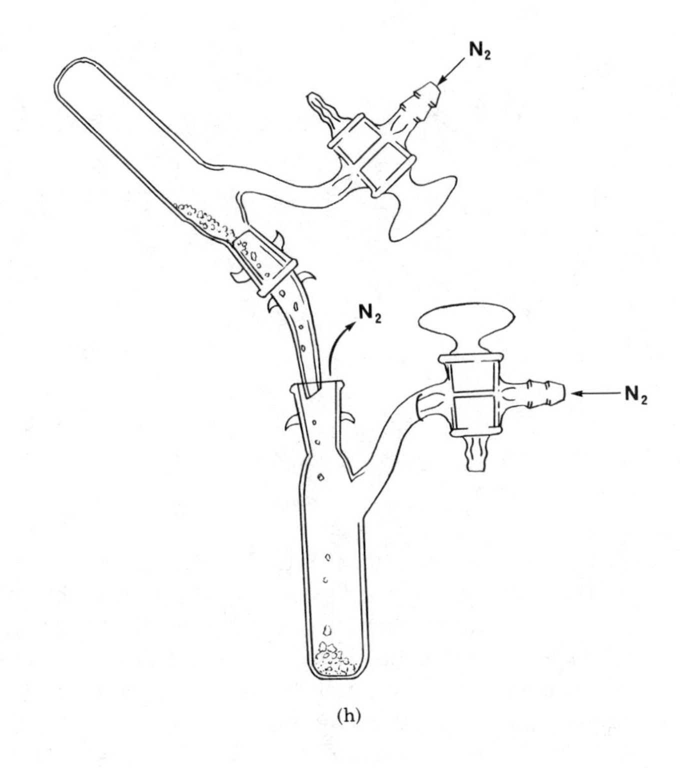

(h)

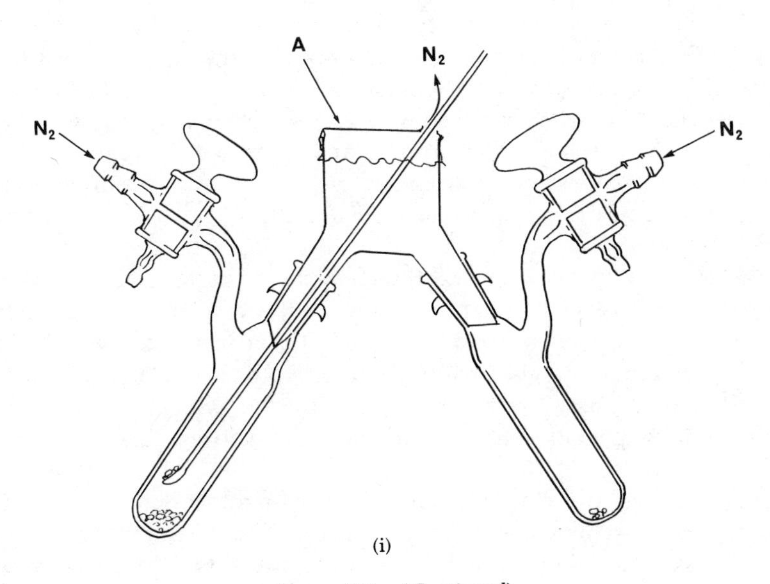

(i)

Figure 5.1. (*Continued*)

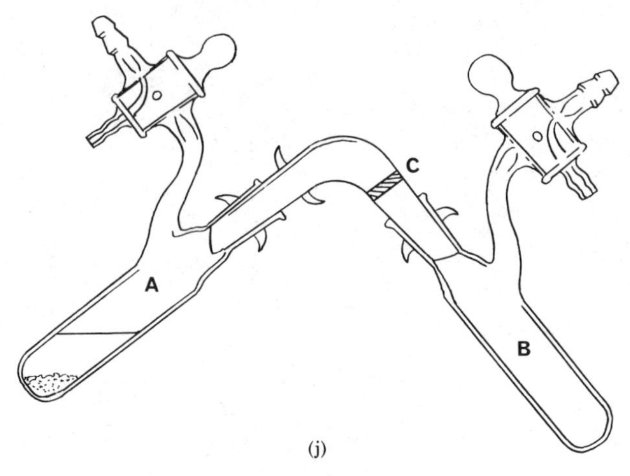

(j)

Figure 5.1. (*Continued*)

gas, such as nitrogen and argon, that is introduced from the side inlet A and escapes from the outlet B. A modified version of this tube is shown in Figure 5.1*b*. The inlet is equipped with a stopcock (A) and the main opening has a standard taper (B), which has small glass ears (C) to hold the stopper or other component with rubber bands or small metal springs. Manipulation of air-sensitive compounds is carried out as follows.

The Schlenk tube is stoppered and evacuated by pumping through D. By introducing the inert gas through A the tube is filled with the inert gas. The usual operation is to repeat the evacuation and the filling cycle a few times. Figure 5.1*b* shows the Schlenk tube fitted with a two-way stopcock that can be used both for evacuation and for filling with inert gas. The tap is turned through 90° to let gas pass through the tail part and then is turned through another 90° to allow gas into the flask. Although this type of tailed tap is quite convenient, it can be replaced by a simpler, less expensive two-way stopcock by providing vacuum-tight tubing and a three-way stopcock, which is connected to a vacuum line and an inert gas source. Transfer of liquid can be performed by using a hypodermic syringe with a long needle as shown in Figure 5.1*b*′.

Transfer of a liquid sample and filtration can be accomplished as shown in Figure 5.1*c*. A U-shaped glass tube (A) is used as a syphon for transferring a liquid sample by applying a moderate pressure of inert gas through the side arm via stopcock B. If a glass-frit (C) is fused onto the transfer tube, a sample can be filtered and the filtrate purged into a second Schlenk tube. After filtration the remaining solid sample is washed with a solvent in which the sample is not too soluble and filtration is repeated. By putting a stopper on and pumping through the side arm or by attaching the Schlenk tube to the vacuum line, the sample can be vacuum dried. This filtration method is convenient for small amounts of samples, since the weight of the dried sample can be measured immediately after vacuum drying and no transfer to another flask is required for storage. However, because of the small filtration area of the glass frit C, the filtration of a large amount of a solid sample can be slow; in that case usage of a fritted cylindrical funnel as shown in Figure 5.1*d* is recommended. Figure 5.1*d* shows a special type of cylindrical fritted funnel with a special jacket (A) to keep the sample below or above room temperature. To remove a small amount of solid by filtration, a folded filter paper can be used, as in Figure 5.1*e*.

A liquid sample can also be transferred by using a stainless steel tube (catheter tubing) and serum bottle caps as shown in Figure 5.1*f*.

The ends of the tube are inserted through the rubber septa, and a pressure differential is used to transfer the liquid. This method is preferred by some researchers since the catheter is more flexible than the glass tube (A) used in Figure 5.1*c* and a simple, one-hole glass bottle can be used instead of the more expensive Schlenk tube. By using cappable centrifuge tubes, precipitates can also be separated by centrifugation and washed by decantation using solvent added through a syringe.

Transfer of a solid sample can be carried out as shown in Figures 5.1*g,h*. By connecting two Schlenk tubes via a glass joint as in Figure 5.1*g* or by using an adaptor as in Figure 5.1*h*, solid samples can be transferred under inert gas. Transfer of a smaller amount of solid sample can be performed as shown in Figure 5.1*i* by using a two-legged trouserlike adapter[8] and a long spatula.

Air-sensitive compounds can also be manipulated in vacuum instead of under an inert gas. Figure 5.1*j* shows a method of filtration in vacuum. After setting up the apparatus, which consists of two Schlenk tubes (A and B) joined by a fritted elbow (C), the solid-containing liquid in Schlenk tube A is frozen by using dry ice or liquid nitrogen and the whole system is evacuated. After a few cycles of thawing, freezing, and pumping, the system can be completely made free of the inert gas. The liquid sample in A can then be transferred to B by tipping the connected Schlenk tubes. To wash or extract the precipitate, the solid is shaken from the frit and the solvent is distilled back to A by cooling the leg and by allowing the other to warm. This method is convenient for repeated recrystallization or extraction in vacuum.

For experiments in which leakage or contamination by grease must be completely avoided, an apparatus using vacuum seal-offs and break-seals is used. For details of sealed-tube techniques, the reader is referred to reference 1. A stopcock fitted with a Teflon barrel, which does not require grease, is also useful. Although special devices or precautions may be required in some special cases, the apparatus and associated manipulations shown in Figure 5.1 or modifications thereof should be sufficient for most synthetic purposes. These operations are suitable for handling very air-sensitive compounds. If the compounds involved are not so air sensitive, some of the above manipulations can be modified or omitted. The method in Figure 5.1*f*, which employs serum caps and a one-opening flask, is less expensive than the others.

Salute to Pioneers in Organometallic Chemistry

Handling very air-sensitive compounds requires great patience because of the many precautions that have to be taken. A split second of carelessness may instantly destroy an important sample that has been synthesized with painstaking effort. It is, however, through these failures, disconcerting as they may be, that an organometallic chemist is trained to perfection.

If we think of the difficulties and risks that Frankland and his coworkers encountered in handling inflammable alkylzinc compounds over a century ago, one can only wonder at how they worked. How did they cope with the dangerous compounds that spontaneously catch fire on contact with air and react with water explosively using such rudimentary apparatus?

Frankland describes in his report how he distilled dimethylzinc in an apparatus filled with dry hydrogen (!) in Bunsen's laboratory. When he tried the action of water on the residue, a greenish-blue flame several feet long shot out of the tube—and diffused an abominable odor throughout the laboratory.

We also wonder how Zeise reached the correct formulation of his platinum–ethylene complex when none of the modern experimental techniques, such as X-ray crystallography, gas chromatography, IR, NMR, and mass spectroscopy, was available.

We owe them homage.

5.2. CHARACTERIZATION OF ORGANOMETALLIC COMPOUNDS

The methods used to characterize organometallic compounds are not vastly different from those for ordinary organic compounds. The basic problem is how to handle and weigh air-sensitive compounds into physicochemical apparatus. Operations may be devised by combining handling principles described in the previous section.

a. Chemical Analysis

Elemental analysis is usually the first step in the complete characterization of a newly synthesized compound.[9,10] Although instrumental methods such as NMR spectroscopy may provide important information regarding the organic entities present in organometallic compounds, neglect of elemental analysis may lead to erroneous conclusions. Depending on the nature of the compound, microscopic and macroscopic methods are employed for elemental analysis. In microscopic elemental analysis using a conventional combustion apparatus, the most important problem is how to weigh the air-sensitive compound and introduce it into the combustion apparatus. A platinum boat (P) to hold the sample is first tared together with a weighing bottle S (Fig. 5.2*a*). Both are put into a container that has an airtight top and a greaseless, transparent, ground glass taper. The weighing bottle S with the boat is put into a Schlenk tube, and after displacing the air with the inert gas, the Schlenk tube is connected through a trouser-type joint, as shown in Figure 5.1*i*, with the other Schlenk tube containing the air-sensitive compound to be analyzed. In this setup the required amount of the sample is transferred to the boat in S, which is then topped snugly. The sample contained in S is now ready to be weighed outside the Schlenk tube. After weighing, S is put in a cylindrical glass container, shown in Figure 5.2*a* in an atmosphere of nitrogen and the boat is taken from S to C. By removing the top (A), the container is connected with the elemental analysis apparatus (M), and the boat (P) can be slid into the combustion apparatus by pushing it with a rod inserted from B.

If the sample is thermally unstable, it cannot be analyzed as described above and a macroscopic method must be used. The sample transferred to a precooled Schlenk tube is quickly weighed and subjected to hydrolysis or thermolysis, and the metal content is determined by the usual analytical methods.[9] If the sample evolves gases on hydrolysis or thermolysis, the amounts of gases evolved on decomposition can be measured by means of a gas burette, manometer, or Toepler pump. For analysis of gases the vacuum

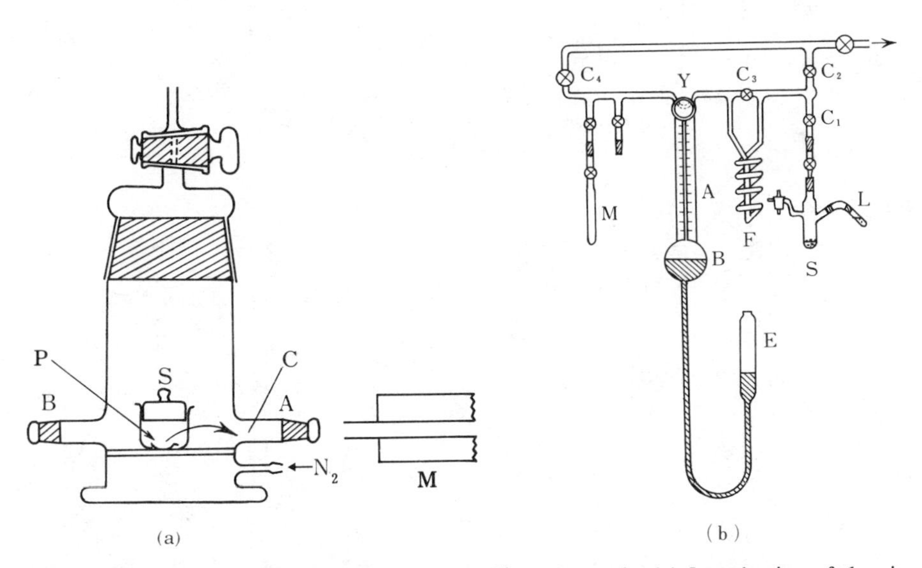

FIGURE 5.2. Analytical methods of organometallic compounds. (*a*) Introduction of the air-sensitive compound into the combustion apparatus. (*b*) Conceptual sketch of a Toepler pump. (*c*) How to prepare a KBr disc containing an air-sensitive compound; A, polyethylene film. (*d*) T-shaped joint with a two-way stopcock. (*e*) Apparatus for molecular weight determination of air-sensitive compounds by a cryoscopic method. (*f*) How to prepare an NMR sample of an air-sensitive compound.

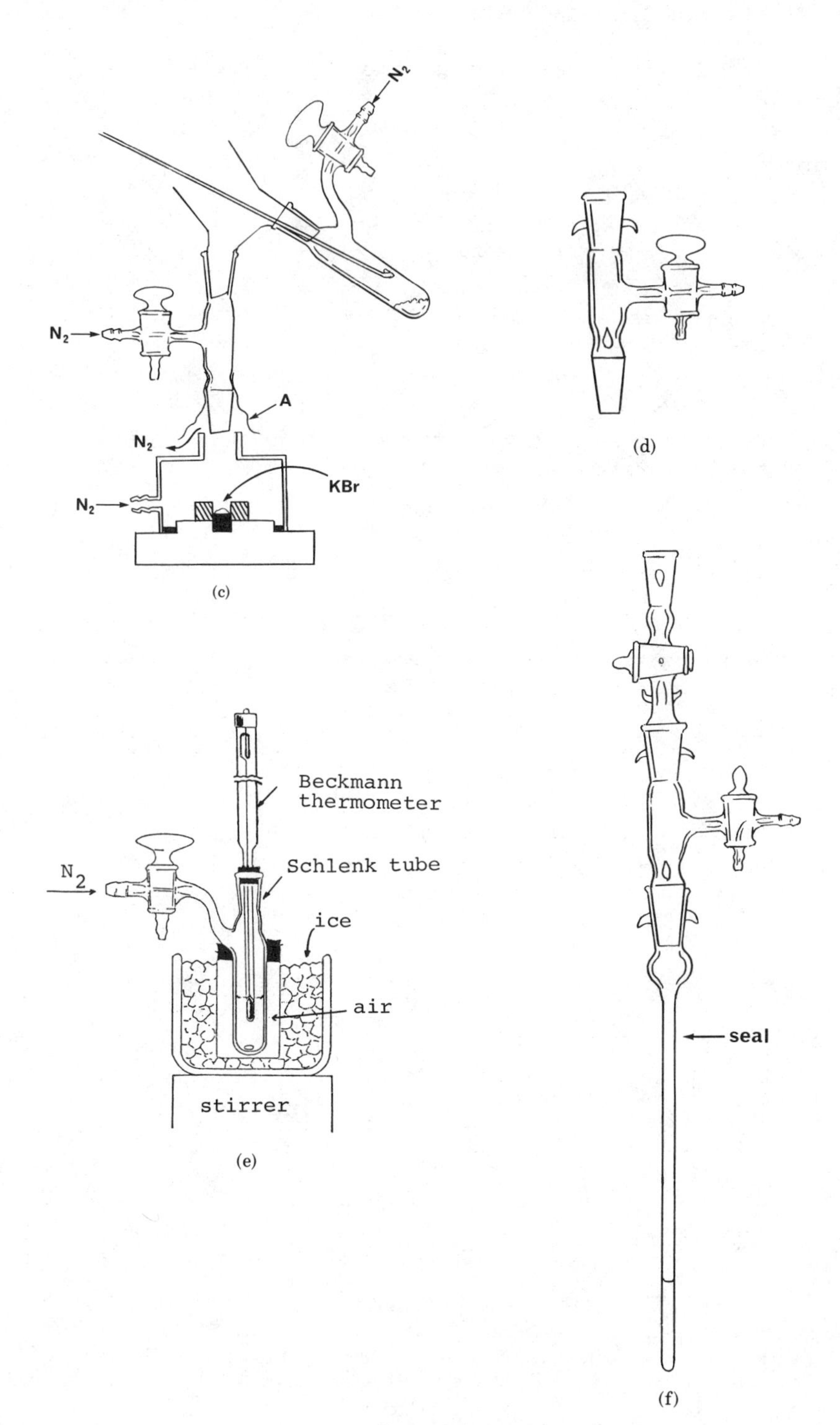

Figure 5.2. (*Continued*)

line technique is essential. In the vacuum line a gas condensable at liquid nitrogen temperature can be freely transferred from one site to another by use of a cold trap, and the amount of gas can be measured precisely by means of a mercury manometer. The gas is then identified by gas chromatography and mass and IR spectroscopy. Gases such as N_2 and H_2, which are noncondensable at liquid nitrogen temperature, can be transferred in the vacuum line and their volumes determined with the aid of a Toepler pump.

There are two types of Toepler pumps, one automatic, the other manual. Figure 5.2*b* is a sketch of the less expensive, manual type. The mercury-filled manometer in the center of the vacuum line is the Toepler pump, which consists of a small-bored gas burette A, a two-way, three-exit stopcock Y, and mercury reservoirs B and E. By slowly raising or lowering the reservoir E, which is connected to the atmosphere, the mercury level can be raised to A or lowered below B; thus pumping motion can be achieved.

For illustration, let us see how a metal alkyl complex can be analyzed. The metal–alkyl bond may be hydrolyzed by concentrated sulfuric acid, and the gases evolved may be determined as follows. The solid sample is placed in an atmosphere of inert gas in a Schlenk tube S having a side arm L containing sulfuric acid. The Schlenk tube S is connected with a vacuum line, and the whole system is evacuated by opening the stopcocks C_1, C_2, C_3, and C_4. When the required degree of vacuum is reached, all the stopcocks are closed, and the sulfuric acid in L is poured onto the sample contained in S by rotating the elbow. Appropriate cooling during hydrolysis by sulfuric acid may sometimes be required to avoid thermal decomposition. When hydrolysis is over, the Schlenk tube is cooled by a liquid nitrogen trap. By opening C_1 noncondensable gases such as H_2 and N_2 are allowed to diffuse into the vacuum line, which is connected with a helical tube (F) cooled with a liquid nitrogen trap. After turning the stopcock Y so as to connect the Toepler pump and the sample-containing side, the mercury level is lowered to introduce the noncondensable gases into A.

The stopcock Y is then closed, the mercury level is raised very slowly, and the volume of the noncondensable gases under atmospheric pressure is read off by means of the scale on A. After lowering the mercury level again and turning the stopcock Y to connect the manometer with the left side of the vacuum line, the gas sample in the manometer is pushed up into the left side by raising the mercury level to stopcock Y. By repeating this cycle, the total volume of the noncondensable gas sample is measured. At the same time it is transferred from the right to the left side of the vacuum line, where it can be collected in tube M for analysis by gas chromatography and mass spectroscopy. By replacing M with a gas cell for IR analysis, the gas sample can also be subjected to IR spectroscopy. When the noncondensable gases have been analyzed, the cold traps S and F are replaced by dry ice baths, and the measurement and transfer operations are repeated. By this means gases such as low-boiling hydrocarbons, which are condensable at liquid nitrogen temperature but noncondensable at dry ice temperature, can be determined.

The other ligands liberated on hydrolysis of the sample in S may be extracted with a suitable solvent and analyzed by gas chromatography. The metal component is analyzed chemically as an aqueous solution of appropriate concentration.

A slightly modified procedure can be used to determine the thermolysis products of air-sensitive organometallic compounds. The sample may be put into the Schlenk tube S without the side arm L and may be subjected to thermolysis by putting S in a hot bath. The rest of the procedure is as described above.

b. Infrared and Raman Spectroscopy

The IR spectra[11-17] of solid samples of air-sensitive compounds can be measured in KBr disks or Nujol mulls. The KBr method is handier, but care should be taken when preparing KBr disks of compounds that are susceptible to hydrolysis, since it is difficult to remove traces of water from KBr. The solid sample mixed with finely ground KBr powder can be transferred onto a piece of the polished steel drum placed in a die for preparing the KBr disk, as shown in Figure 5.2*c*, using a long spatula. The other part of the steel drum is placed on top of the KBr sample and pressure is applied to make the disk. Once the disk has been prepared, an air-sensitive sample in the KBr pellet usually shows little sign of decomposition on its exposure to air.

Liquid samples can be introduced into an IR liquid cell using a syringe. In this operation it is convenient to use a T-shaped joint with a tailed two-way stopcock (Fig. 5.2*d*).

Infrared absorption bands characteristic of organometallic compounds are observed in the following regions: $\nu(C—H)_s$, 2750–2950 cm^{-1}; $\nu(C—H)_{asym}$, 2810–3050 cm^{-1}; $\nu(M—H)$, 1800–2000 cm^{-1}; $\nu(C\equiv O)$ of metal carbonyls, 1850–2125 cm^{-1} (terminal) and 1750–1850 cm^{-1} (bridging); $\delta(CH_3)_{sym}$, 1100–1350 cm^{-1}; $\delta(CH_3)_{asym}$, 1300–1475 cm^{-1}; $\rho(CH_3)$, 650–975 cm^{-1}; $\nu(M-C)$ 400–600 cm^{-1}. The intensity and number of these bands depend on the number of chemical bonds, the nature of the metal complexes, and the symmetry of the molecule. Depending on the symmetry, Raman spectroscopy may provide complementary information, although, unfortunately, many organometallic compounds are decomposed by the intense laser radiation.

In the IR spectra of olefin–metal complexes the C=C stretching vibration of the olefin is shifted to low frequency on coordination. A convenient measure of the modification of the C=C bond order is the summed percentage lowering (SPL) of bands due to the coupled modes $\nu(C=C)$, $\delta(CH_2)_{scis}$, and $\delta(CH)_{bend}$.[18] For example, the SPL values of $[Ag(olefin)]^+BF_4^-$ are 4–4.5%, whereas those of $Pt(olefin)Cl_2$ are 10–14%. The SPL value of 1,5-cyclooctadiene complexes was found to increase in the order Pd(II) $<$ Pt(II) $<$ Rh(I).[19,20] Metal–olefin stretching frequencies are observed in the range 385–500 cm^{-1} for Pt–C_2H_4 and 270–290 cm^{-1} for Ag–C_2H_4 complexes.[18]

The IR spectra of volatile compounds can be measured in gas cells, and the IR spectra of gases evolved on hydrolysis or thermolysis of organometallic compounds provide a convenient means for their characterization. This is particularly useful when deuterium-containing olefins are evolved on thermolysis, since mass spectrometric analysis does not provide unequivocal information about the isomers of deuterated olefin.

c. Molecular Weight

In principle, any suitably modified standard method of molecular weight determination may be applied to air-sensitive compounds. If the organometallic compound is soluble in a suitable solvent, the cryoscopic method will be the first choice because it is easy to design an inexpensive molecular weight determination apparatus properly modified to exclude air. Figure 5.2*e* illustrates a simple apparatus for the cryoscopic method. The sample solution can be prepared and transferred to the Schlenk tube as described in Section 5.1. In a stream of inert gas a Beckmann thermometer can be placed on top of the Schlenk tube so that the thermometer bulb is immersed in the solution. The solution is stirred magnetically from below through an air bath and an ice bath. The freezing point measurement can be performed as in the usual cryoscopic method. The ebullioscopic method with a proper design to exclude air also is useful and allows a wider selection of solvents as far as the solubility is concerned. However, this method is not suitable for thermally unstable compounds, and foaming of the solution sometimes hinders measurement. Several makes of apparatus are available commercially for molecular weight determination by vapor pressure osmometry, although it is difficult to completely exclude air from them. In the isothermal distillation method air-sensitive samples in appropriate solutions may be introduced into the apparatus under rigorous exclusion of air, and the molecular weights may be determined after the apparatus has been sealed off. Since the system needs days or weeks to reach equilibrium, this method is not suitable for thermally unstable compounds. A detailed description of the method is given in reference 1.

d. Mass Spectroscopy

Progress in mass spectroscopy[21-26] has enabled high-molecular-weight organometallic compounds to be studied. In addition to the common method of electron impact ionization, other methods such as chemical ionization, desorption methods, and other ionization methods are gaining increasing importance. If the parent ion peak can be identified and confirmed by recognition of the polyisotopic pattern of the metal in the organometallic complex, the molecular weight of the complex can be determined. The fragmentation pattern also provides important information about the composition of the complex.

Mass spectrometric analysis of volatile material evolved on thermolysis and/or hydrolysis of organometallic compounds provides useful information about the organic ligands present in the complexes and is a useful complement to direct elemental analysis.

e. X-Ray Diffraction and Neutron Diffraction Analysis

Since most organometallic compounds contain heavy metal atoms, the heavy-atom method of phase determination in X-ray crystallography can be used.[27,28] It is fortunate that the rapid progress in the synthesis of organotransition metal compounds coincided with advances in X-ray crystallography and especially with the very rapid improvement in computing facilities. It is truly marvellous that one can recognize the structures of complex molecules as if one can see them. Today it would be unthinkable to study large molecules such as metal clusters without the aid of X-ray structural determination. Although the precise manipulation of air-sensitive compounds for X-ray analysis is not described here, a combination of the methods mentioned in the previous sections together with special devices allows a single crystal to be picked up from a Schlenk tube under inert gas atmosphere, put into a capillary, and sealed for examination by X-ray analysis.

Small atoms such as hydrogen attached to a heavy metal atom are difficult to locate precisely by X-ray diffraction. Neutron diffraction is more suitable for this purpose, but unfortunately neutron beam sources of sufficient flux are not readily available, and the crystals need to be larger than those used in X-ray analysis.

f. Visible and Ultraviolet Spectroscopy (Electron Absorption Spectroscopy)

In Section 2.3 we dealt with the electronic absorption spectra[29,30] of transition metal complexes having d electrons. The magnitude of the ligand field splitting is estimated from the energy required for the d–d transition. However, the energy levels of the d orbitals cannot be estimated from these data. On the other hand, when combined with information from charge transfer bands, the energy levels of d orbitals can be estimated indirectly. In some cases one can also obtain information on the reactivities of organometallic compounds from electronic absorption spectra, as shown by the following example.

Alkylnickel complexes $NiR_2(bipy)$ (R = Me, Et) contain the bipyridine ligand, which has an extensively conjugated π system. The complexes are dark green owing to $d \rightarrow \pi^*$ charge transfer absorption. The electronic spectrum of free bipyridine has two absorption bands (A, B) in the UV region arising from electronic excitation from the π orbital to the π^* orbitals (Fig. 5.3*a*). In NiR_2-(bipy) these two absorption bands are shifted to the low-frequency visible region and are observed at 410 and 650 nm; these bands account for the deep green color. These bands correspond to $d \rightarrow \pi^*$ electronic excitations A′ and

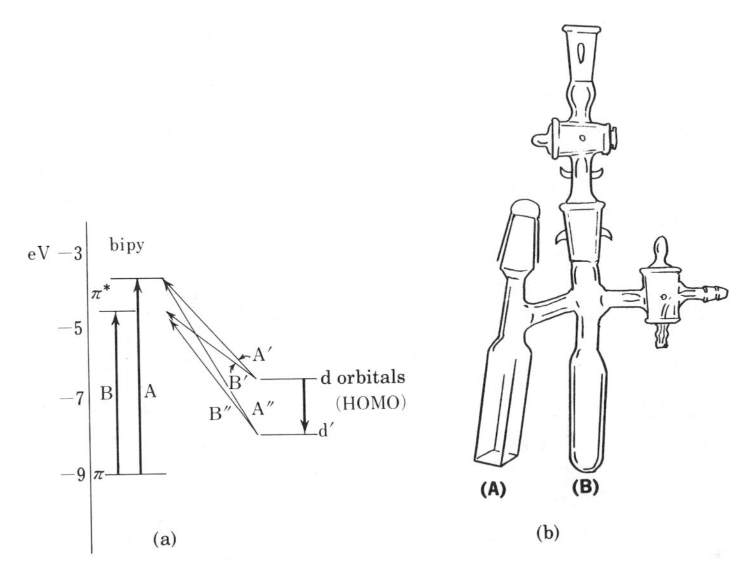

FIGURE 5.3. (*a*) Energy diagram showing the $\pi \rightarrow \pi^*$ and $d \rightarrow \pi^*$ excitations of 2,2′-bipyridine and the nickel–bipyridine complexes (for explanation see text). (*b*) Optical cell (A) for measurement of UV and visible spectra of a solution of an air-sensitive compound.

B′ in Figure 5.3*a*. Therefore, if the energy levels of the π^* orbitals of the bipyridine can be estimated, the energy levels of the d orbitals (HOMO) also can be estimated indirectly.[31]

On interaction of these complexes with acrylonitrile at low temperature (−78 °C), unstable orange solids containing $R_2Ni(bipy)$ and acrylonitrile in a 1 : 1 ratio are produced.

$$R_2Ni(bipy) + CH_2{=}CH(CN) \underset{}{\overset{K_1}{\rightleftharpoons}} R_2Ni(bipy)(H_2C{=}CHCN) \tag{5.1}$$

5-1 (green) **5-2** (orange)

In complex **5-2** acrylonitrile is believed to be coordinated to nickel through the double bond. Extensive back bonding from the d orbitals of nickel to the π^* orbital of acrylonitrile in **5-2** causes the energy levels of the d orbitals to decrease. Consequently, the energy required for electronic excitation from the d orbitals of nickel to the π^* orbitals of bipyridine (A″, B″) is increased, thus shifting the charge transfer bands toward the UV region. This accounts for the orange color of the complex.

A second example of the application of the electronic absorption spectroscopy is the kinetic study of reaction (5.2) with olefins monitored by the absorption spectrum of reactants and products.

$$R_2Ni(bipy) + CH_2{=}CHX \rightleftharpoons \left[R_2Ni(bipy)(H_2C{=}CHX) \right] \quad (5.2)$$

$$\xrightarrow{-\ R\text{–}R} Ni(CH_2{=}CHX)_n(bipy)$$

$n = 1, 2$ (orange)

5-3

The reactions with various olefins proceed with clear isosbestic points at room temperature and are of first order in the concentration of **5-1** and of olefin. Equilibrium (5.1) at room temperature is believed to lie well to the left, and the reductive elimination of R–R from complex **5-2** is rate determining.

The olefin–Ni(0) complexes (**5-3**) are orange in the solid state, but in solution they turn green. On addition of olefin to the solution the orange color is restored. The color change is due to the following equilibrium.

$$\underset{\text{(green)}}{Ni(bipy)(solvent)} + olefin \overset{K_2}{\rightleftharpoons} \underset{\textbf{5-3}\ \text{(orange)}}{Ni(bipy)(olefin)} + solvent \quad (5.3)$$

As a third example of the application of electronic absorption spectroscopy, the spectral change due to Eq. (5.3) can be observed with varying concentration of the added olefin, and the formation constant K_2 of the olefin π complex (**5-3**) can be obtained. Furthermore, by observing the change of K_2 at various temperatures, the enthalpy and entropy changes of Eq. (5.3) can be derived.

These are a few illustrations of applications of electronic absorption spectroscopy to the chemistry of organometallic compounds. The electronic absorption spectra of the air-sensitive compounds can be measured in a cell of the type shown in Figure 5.3*b*. The sample is placed in B, and the cell is evacuated. The solvent for the UV measurement is introduced by trap-to-trap distillation from a vacuum line that contains a solvent reservoir. When the required amount of solvent has been collected in B, the cell is detached from the vacuum line and the sample is dissolved in the solvent. A part of the solution may be transferred into a compartment (A) consisting of an optical cell by tipping the system. If the solution is too concentrated for the spectroscopic measurement, part of the solution may be returned to B by tipping the cell, and solvent alone can be condensed into A from B by cooling A by a dry ice trap. Liquid nitrogen should not be used in this operation, since it would crack the quartz optical cell. Olefins for the reactions can be added in a stream of nitrogen from the top of the cell by means of a microsyringe.

g. Photoelectron Spectroscopy

By means of electronic absorption spectroscopy, one can estimate the energy difference betwen the HOMO and LUMO energy levels, but the absolute values of these energy levels have to be estimated from other data, such as ionization potentials and electron affinities. Photoelectron spectroscopy[32,33] provides direct information regarding the molecular orbital levels of organometallic complexes. Applications are still limited, but more development is certainly expected.

In photoelectron spectroscopy, photons of high energy such as X-rays and vacuum UV rays impinge on samples and cause the inner or valence electrons to ionize. The energy of the ionizing photons is known from their frequency ($E = h\nu$) and the kinetic energy (E_k) of the photoionized electrons may be measured. The difference between $E_{h\nu}$ and E_k is the amount of energy (the binding energy) that must be provided to overcome the electron binding by the nucleus. As an illustration of this technique, Figure 5.4 shows the photoelectron spectrum of ferrocene.[34]

Assignment of each peak in the spectrum is made by comparison with the calculated results of molecular orbital energy levels. In the photoelectron spectrum of ferrocene the peaks of the lowest energies A and A' in the range of 6.88–7.23 eV have been assigned to the a'_{1g} and e_{2g} orbitals, respectively (cf. MO diagrams in Fig. 3.20). The assignment of peaks at higher energies is still a matter of some disagreement arising from the different approaches in the

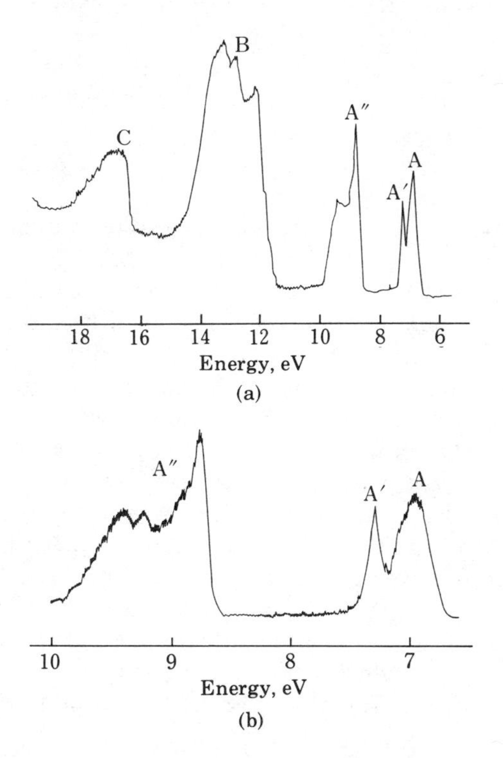

FIGURE 5.4. Photoelectron spectrum of ferrocene (*a*) Whole spectrum. (*b*) High-resolution spectrum of the portions A, A′, and A″ in (*a*). (From Evans et al.[34])

approximation of the MO methods employed for the calculation. As these are refined, better correspondence with the experimental results should be reached. It is not yet possible to correlate MO and photoelectron spectroscopy so as to make detailed arguments about the effect of ligands on the energy levels.

h. Mössbauer Spectroscopy

Mössbauer spectroscopy[35] also provides us with certain information about the electronic environment at the nucleus of an atom. However, because of all the transition elements iron is almost the only one that is conveniently studied by the Mössbauer technique, we shall not treat the topic here. The reader is referred to reference 35.

i. Dipole Moments

Dipole moments should provide important information about charge distributions in transition metal complexes. A classic example is the distinction between *cis*- and *trans*-$Pt(NH_3)_2Cl_2$; the former has a dipole moment, whereas the dipole moment of the latter is zero. However, because commercial instruments are not readily available, dipole moment studies of organotransition metal complexes are quite limited. Nevertheless, Chatt and his co-workers[36] have used the method to obtain useful information on the charge distribution in various transition metal complexes, including some organotransition metal complexes.

Measurements of dipole moments of square planar complexes *cis*-$MCl_2(PR_3)_2$ and of octahedral complexes *mer*-$MCl_3(PR_3)_3$ reveal that the group dipole moment of P–M–Cl falls in the range of $6.8 \pm 0.9D$ regardless of the metals and tertiary phosphines. The results indicate the presence of a positive charge $+0.3e$ on P and of a negative charge $-0.3e$ on Cl. Similar measurements on transition metal complexes having other ligands showed that no ligand has a charge greater than $0.3e$. Pyridines were found to possess as much positive charge as tertiary phosphines. The charges of CH_3, CO, H, N_2, NO^+, and PF_3 were close to zero. On the other hand, O^{2-}, N^{3-}, and $RCOO^-$ have about the same magnitude of negative charges as halogens.

These results are in agreement with the trend derived empirically from the reactivities of transition metal complexes. For example, if we compare *cis*-$PtMe_2(PEt_3)_2$ with *cis*-$PtI_2(PEt_3)_2$, negative charge should flow from the phosphine ligands through platinum to the iodide atoms for the latter complex, whereas in the methylplatinum complex the negative charge should be accumulated on platinum. Therefore, the methylplatinum complex should be more electron rich than the platinum diiodide complex, even though the metal atom in both cases is in the +2 oxidation state. This is reflected in reactivities of these Pt(II) complexes; *cis*-$PtMe_2(PEt_3)_2$ oxidatively adds methyl iodide, whereas *cis*-$PtI_2(PEt_3)_2$ does not. The results are also consistent with the information derived from X-ray photoelectron spectroscopy.

j. Magnetic Susceptibility and ESR Spectroscopy

The significance of magnetic susceptibility[37] regarding the electron configurations of transition metal complexes was dealt with in Section 2.1. The magnetic susceptibility of a compound is determined by measuring the force exerted on it when it is placed in a nonhomogeneous magnetic field. The more paramagnetic the compound, the more strongly will it be drawn into the stronger part of the field. Gouy's method is the simplest of the methods measuring the magnetic susceptibility. In this method the sample to be examined is hung by a fine nonferrous chain from a balance in the field of an electromagnet, and the weight change of the sample on application of the magnetic field is measured. If the sample is diamagnetic, it experiences a small weight loss, whereas a paramagnetic compound becomes heavier. The magnetic susceptibility can be calculated from the weight change by comparing it with that of a standard sample. When magnetic susceptibility of an air-sensitive compound is measured, it may be transferred to a capsule with an airtight top as shown in Figure 5.1*i*, and the weight change of the sample can be measured by hanging the capsule in an electromagnet. Various methods have been devised to increase sensitivity, such as measuring the horizontal displacement of a sample on application of the magnetic field instead of measuring the weight change. To calculate the magnetic moment, it is necessary to correct the magnetic susceptibility of a paramagnetic substance for the diamagnetic contribution. The diamagnetic contribution, being smaller than the paramagnetism, can be neglected in complexes of small molecular weight. But when the complex has large ligands, the diamagnetic contribution cannot be ignored. From the corrected molar magnetic susceptibility, χ_m^{corr}, an effective magnetic moment, μ_{eff}, can be calculated. The number of unpaired electrons in the compound can then be derived from Eq. (2.1).

NMR spectroscopy can also be used to measure the magnetic susceptibility.[38]

Electron spin resonance (ESR)[39] spectroscopy provides a powerful means of detecting the presence of unpaired electrons and of examining their environment.[40,41] As a result of their spin, unpaired electrons behave as tiny magnets and line up parallel or antiparallel to an applied magnetic field. These two alignments will have slightly different energies, and transitions from one to the other can be detected by applying the energy of transition in the form of radio frequency electromagnetic radiation. In transition metal complexes the unpaired electron interacts with the ligand nuclei and gives rise to hyperfine splitting of the resonance, which gives clues about the environment of the unpaired electrons in the compounds examined. The information provided by the ESR technique is usually much less quantitative than that

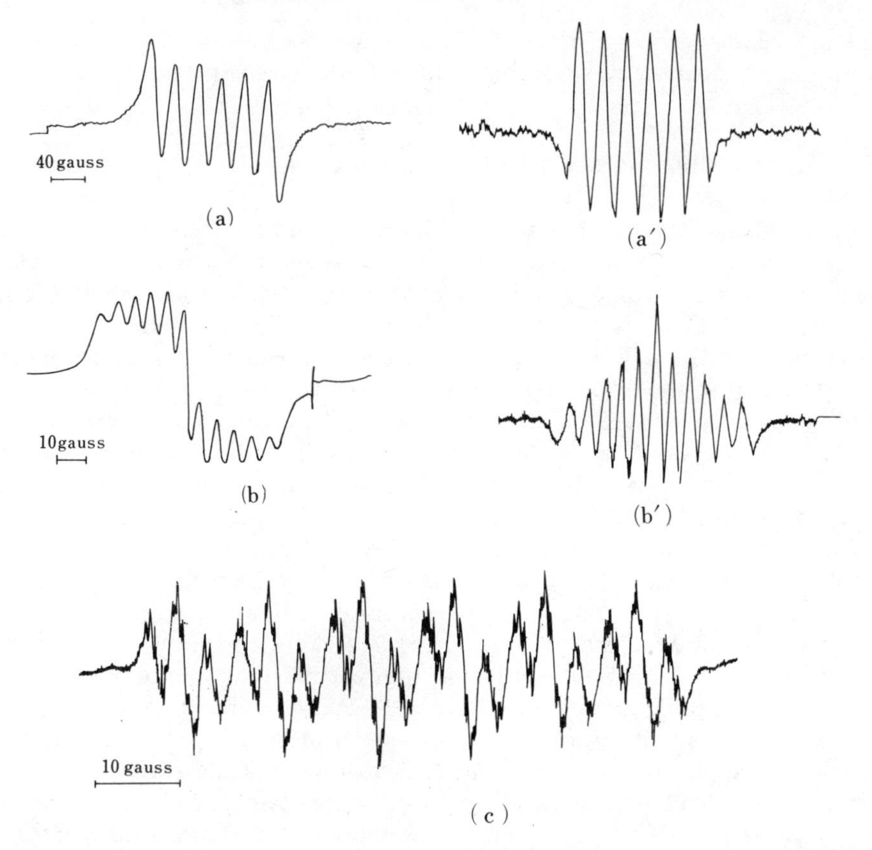

FIGURE 5.5. ESR spectra of Ti(III)–Al complexes. (*a*) ESR spectrum of complex **5-4**, the first derivative. (*a′*) The second derivative of (*a*). (*b*) ESR spectrum of reaction product of **5-4** with $AlCl_{3'}$, the first derivative. (*b′*) The second derivative of (*b*). (*c*) ESR spectrum of complex **5-5**. (From Henrici-Olivé and Olivé,[42] Hüthig and Wepf Verlag, Basel, publishers.)

available from the NMR method, but it can be applied effectively to paramagnetic species, as illustrated in the following example.[42]

Ziegler catalysts are usually composed of complicated mixtures of various titanium and aluminum components, and identification of the chemical species in these catalysts is not an easy task (cf. Section 7.2). However, the Ti(III) species produced by reaction of a Ti(IV) compound with an alkylaluminum has an unpaired electron that serves as a convenient probe. On mixing Cp_2TiCl_2 with $AlEtCl_2$, the ESR spectrum observed undergoes a series of changes, eventually giving the spectrum shown in Figures 5.5*a,a′*. The sextet hyperfine structure† arises from interaction of the unpaired electron on Ti(III)

† When an unpaired electron interacts with a neighboring nucleus of nuclear spin moment *I*, the otherwise single resonance line of the unpaired electron is split to give the hyperfine structure of $2I + 1$ lines.

with aluminum having a nuclear spin moment I of $\frac{5}{2}$. The result suggests the presence of the binuclear, chloride bridged complex **5-4**, which has also been identified by the X-ray structural analysis of a sample prepared separately.

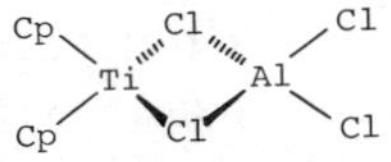

Cp Cl Cl
Ti Al
Cp Cl Cl

5-4

Complex **5-4** is further transformed on reaction with $AlCl_3$ in the presence of a trace amount of water to a species giving rise to the spectrum shown in Figures 5.5*b*,*b′*. The spectral change suggests the formation of a species that has a Ti(III) atom flanked by two aluminum atoms.

On the other hand, the reaction of Cp_2TiCl_2 with sodium and $AlCl_3$ in tetrahydrofuran gives a dinuclear, hydride bridged species (**5-5**).

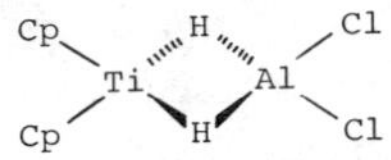

Cp H Cl
Ti Al
Cp H Cl

5-5

The ESR spectrum of **5-5** measured with dilution in THF[42c] is shown in Figure 5.5*c*. The spectrum is composed of a sextet that is further split into triplets of 1 : 2 : 1 intensity ratio. The sextet arises from interaction of the unpaired electron in Ti(III) with the aluminum ($I = \frac{5}{2}$), and the splitting into triplets is due to interaction with the two bridging hydrides ($I = \frac{1}{2}$). Each triplet is seen to be further split into multiplets due to the interaction with 10 hydrogen atoms of the cyclopentadienyl ligands to give the 11 lines ($2nI + 1 = 11$).

This case shows that although paramagnetic compounds are not amenable to NMR studies, their ESR spectra occasionally provide useful information. However, because of the very high sensitivity of the ESR method, one should be careful to check that the ESR spectrum obtained comes from the main chemical species and not from a minor component present as an impurity. A sample may be introduced into an ESR tube in the same way as into an NMR tube; this will be described later.

5.3. NMR SPECTROSCOPY

Since most organometallic compounds are diamagnetic and many of them exhibit dynamic behavior on the NMR time scale, NMR spectroscopy provides the single most useful tool for the characterization of diamagnetic compounds and for studying their behavior in solution. This justifies a separate treatment of NMR spectroscopy. We leave the treatment of principles and the general discussion of NMR spectroscopy to well-established texts and confine ourselves to matters closely related to organometallic compounds.[43,44]

Chemical shifts, peak intensities, spin–spin coupling constants, and peak shapes provide the most important clues to characterize the compounds and to study their behavior in solutions. For example, in the ^{1}H NMR spectra of some ethyl metal compounds, the methyl and the methylene groups show triplet and quartet resonances, respectively, in a relative intensity of 3 : 2 at different field strengths (chemical shifts), reflecting the difference in the environment of the respective protons. An important feature of NMR spectroscopy relevant to organometallic compounds is the variation of NMR spectra with temperature, which yields important information concerning dynamic behavior in solution. Also, in recent years the study of nuclei other than the proton has gained increasing importance. The ^{13}C and ^{31}P NMR spectra are now measured routinely in many laboratories concerned with organotransition metal compounds, and studies using other nuclei such as ^{2}D, ^{15}N, ^{14}N, ^{103}Rh, and ^{195}Pt are increasing.

Air-sensitive samples can be loaded into NMR tubes using a T-shaped joint (Fig. 5.2*d*) in an atmosphere of inert gas, as shown in Figure 5.2*f*. The sample may be dissolved in an appropriate solvent in a Schlenk tube and transferred into the NMR tube by means of a syringe, together with a reference such as tetramethylsilane (TMS). The NMR tube may be sealed off for complete security to exclude the air. For less air-sensitive compounds NMR tubes stoppered with airtight Teflon plugs may be used.

a. Studies of Metal Alkyls

Metal alkyls of main elements have been the subject of ^{1}H NMR studies.[45] In metal alkyls protons attached to a carbon directly bound to a metal are affected by the metal to a greater extent than other protons further from the metal. For example, the CH_2 protons attached to the α carbon (the one bound directly to the metal in metal ethyls) experience a shielding effect due to the electropositive metal, and the resonance peaks are shifted to higher magnetic field, whereas the CH_3 protons attached to the β carbon are affected to a lesser degree. The situation is opposite to that encountered in the spectra of ethyl alcohol or ethyl halides, which have a more electronegative group or atom attached to the ethyl group. In this case the CH_2 group is deshielded relative to the CH_3 group, and this gives rise to the familiar quartet (CH_2)–triplet (CH_3) spectral pattern.

Figure 5.6 shows ^{1}H chemical shifts of the CH_3 and CH_2 groups [referred to TMS ($\delta = 0$) with the lower shift positive] plotted against the chemical shift difference between the CH_3 and CH_2 groups. It can be seen that the CH_2 signal of lithium ethyl, which has a very electropositive metal attached to the CH_2 group, appears at a much higher field than that of the CH_3 group. As the electronegativity of the element attached to the ethyl group increases, the difference Δ between the chemical shifts of the CH_3 and CH_2 protons decreases and converges to zero, at which point a singlet resonance would appear. Further increase in the electronegativity of the element attached to the ethyl

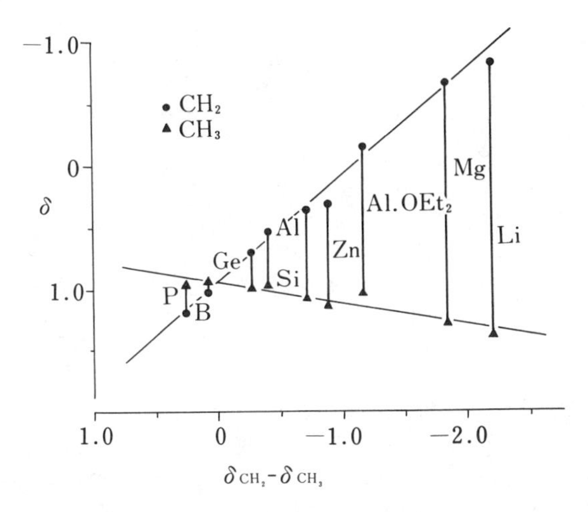

FIGURE 5.6. Proton chemical shifts of the CH_3 and CH_2 groups of ethyl groups in various ethyl–metal compounds plotted against the chemical shift difference between the CH_3 and CH_2 groups.

group causes the CH_2 group to be shifted further to the lower field. The following relationship between the electronegativity χ of the element attached to the ethyl group and the chemical shift difference Δ has been derived.[46]

$$\chi = 9.62\Delta + 2.07 \tag{5.4}$$

There are some limited applications of this relationship to organotransition metal complexes. When transition metal ethyl complexes having different ligands are compared, the chemical shift difference Δ between the CH_2 and CH_3 groups will reflect the effect of the ligand attached to the metal. For example, the $\Delta(\delta_{CH_2} - \delta_{CH_3})$ values of $(CH_3CH_2)NiCl(bipy)$, $(CH_3CH_2)Ni(acac)PPh_3$, and $(CH_3CH_2)_2Ni(bipy)$ are 0.50, 0, and −0.30, respectively. The electronegativity values calculated from Eq. (5.4) are 2.4, 2.1, and 1.9, respectively. The electronegativity value of 1.9 in $(CH_3CH_2)_2Ni(bipy)$ is close to 1.8, the electronegativity (Allred–Rochow value) of Ni(0). Although the metal atom in both EtNiCl(bipy) and $Et_2Ni(bipy)$ is formally in the +2 oxidation state, the results suggest that the nickel atom in $Et_2Ni(bipy)$ is more electron rich than that in EtNiCl(bipy), and the former may in fact be closer to the Ni(0) oxidation state.

As discussed previously, addition of acrylonitrile to the electron-rich complex $Et_2Ni(bipy)$ at low temperature gives the acrylonitrile-coordinated π complex $Et_2Ni(bipy)(CH_2{=}CHCN)$ (**5-2**) as shown in Eq. (5.1). In this π complex extensive back donation takes place from the electron-rich nickel atom to acrylonitrile, causing further deshielding of the protons at the CH_2 groups in the ethyl groups. The electronegativity of the nickel atom in **5-2** calculated from Eq. (5.4) yields a value of 2.6, which is as large as that of carbon (2.5). The value indicates that the electrons in the ethyl groups are considerably withdrawn toward nickel. The information obtained here is consistent with the experimental fact that addition of acrylonitrile, which interacts strongly with nickel, enhances the rate of the concerted Ni–Et bond cleavage and coupling reaction to give butane. This is consistent with the view that two negatively charged ethyl groups are in an unfavorable situation to couple with each other, but the circumstances become more favorable for coupling if electron density flows from the ethyl groups to the nickel entity. NMR spectroscopy can also be used to estimate the equilibrium constants and thermodynamic parameters that govern the equilibrium of Eq. (5.1).[47]

When the metal atom in an organotransition metal complex has a nuclear spin, spin–spin couplings are observed between the metal and the ^{1}H, ^{13}C, or ^{31}P nuclei of the organic groups attached to the metal. The magnitude of the coupling constant J provides information about the metal–ligand bond.† For example, since platinum contains ^{195}Pt with $I = \frac{1}{2}$ in a natural abundance of 33.8%, the ^{1}H NMR is split on coupling with the ^{195}Pt nucleus to give a triplet with satellites in a 1 : 4 : 1 intensity ratio. The magnitude of $^{2}J(^{195}Pt\text{–}H)$, the coupling constant between the platinum and the protons attached to the carbon directly bonded to platinum, is quite large. The coupling constant diminishes as the number of bonds intervening between the metal and the proton increases, as shown in the following example.[48]

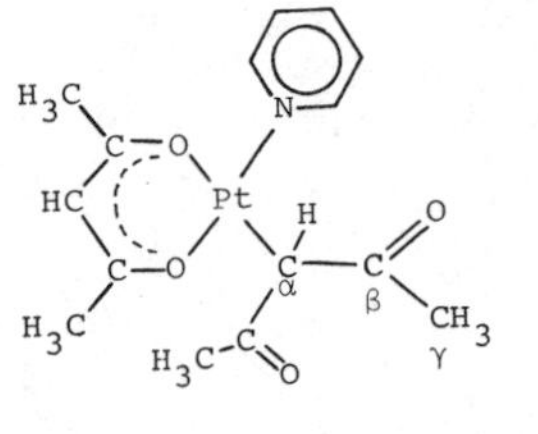

$|^{2}J(Pt\text{–}H_{\alpha})| = 129$ Hz
$|^{4}J(Pt\text{–}H_{\gamma})| = 9$ Hz
$|^{1}J(Pt\text{–}^{13}C_{\alpha})| = 619$ Hz
$|^{2}J(Pt\text{–}^{13}C_{\beta})| = 65$ Hz

1-4

The same trend is seen in $J(Pt\text{–}^{13}C)$ values.

Since there are many important tertiary phosphine-containing transition metal complexes, ^{31}P NMR spectroscopy provides an important method for studying the structures and behavior of phosphine-containing complexes.[49] Information concerning the interaction of the phosphine ligand and a metal is available from one-bond coupling constant $^{1}J(M\text{–}^{31}P)$. Figure 5.7 shows the relationship between $^{1}J(Pt\text{–}P)$ and d(Pt–P), the platinum and phosphorus bond length determined for various tertiary phosphine–platinum complexes.[50]

† J denotes the spin–spin coupling constant in hertz. The left superscript on J represents the number of bonds intervening between the two interacting nuclei.

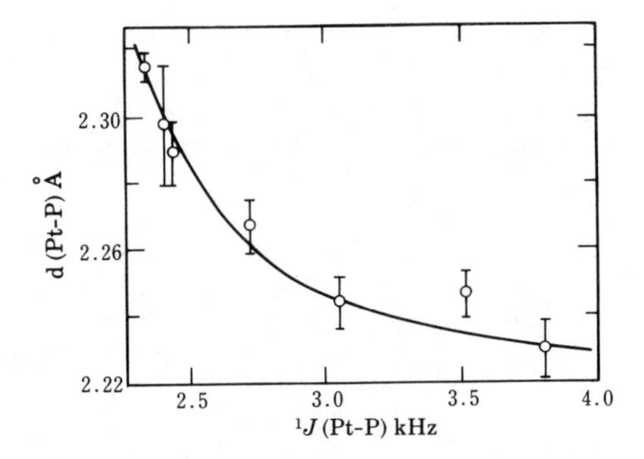

FIGURE 5.7. Relationship between ^1J(Pt–P) values and Pt–P bond distances (d) in platinum complexes containing various phosphine ligands. (From Mather et al.[50])

It is seen that those complexes having larger ^1J(Pt–P) values have smaller Pt–P separations and presumably stronger Pt–P bonds. Therefore, one can use the ^1J(Pt–P) value as an indirect measure of the Pt–P bond strength, at least for series of related tertiary phosphine complexes.

Spin–spin coupling between magnetic nuclei reflects interaction between valence electrons. It has been pointed out that the most important factors in determining the coupling constant are the magnitudes of the valence state s orbitals of the interacting nuclei.[51-60] For example, an sp^2 orbital has more s character than an sp^3 orbital. The decrease in the s character in one nucleus is transmitted to the other nucleus, resulting in a decrease of the coupling constant. It is usually not justified to infer greater bond strength from a larger coupling constant. However, the relationship shown in Figure 5.7 allows us to use ^1J(Pt–P) as an indirect measure of bond strength and trans influence, as discussed in the following section.

b. Trans Influence Studied by NMR

The trans effect is the labilizing effect of a ligand (the trans-directing ligand T) exerted on a ligand trans to it (A) in ligand displacement reactions, mainly of square planar complexes. A detailed treatment of the trans effect will be given in Section 6.1, but it suffices here to mention that it is a *kinetic* influence that enhances the rate of ligand displacement.[57] On the other hand, the trans influence is the relative *static* influence of a ligand T in the square planar complex shown below, which weakens the bond M–A trans to T in the equilibrium state of the complex.

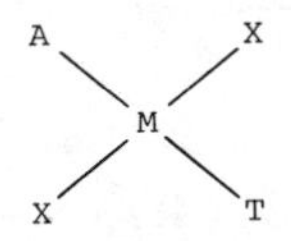

The most direct means of observing the trans influence is to study by X-ray or neutron diffraction analysis the effect on a given metal–ligand bond of placing different ligands trans to it. However, since the effects are small, precise measurement of the bond lengths is required. Bond lengths in *trans*- and *cis*-$PtCl_2(PEt_3)_2$ are shown below.[50,54]

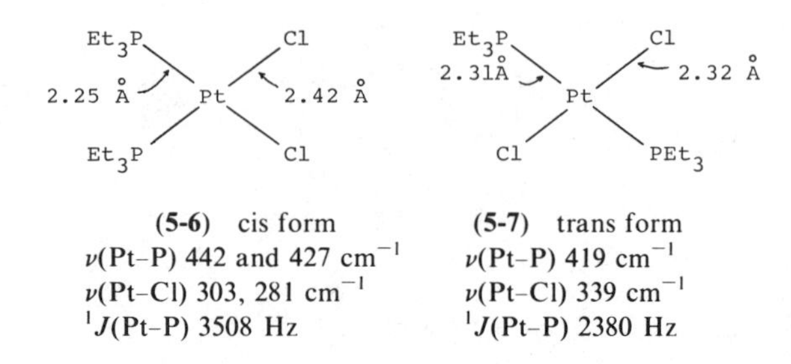

(5-6) cis form
ν(Pt–P) 442 and 427 cm^{-1}
ν(Pt–Cl) 303, 281 cm^{-1}
^{1}J(Pt–P) 3508 Hz

(5-7) trans form
ν(Pt–P) 419 cm^{-1}
ν(Pt–Cl) 339 cm^{-1}
^{1}J(Pt–P) 2380 Hz

It can be seen that the Pt–P bond length for triethylphosphine trans to triethylphosphine is greater than that for triethylphosphine trans to chloride. Correspondingly, the Cl ligand trans to PEt_3 has a longer Pt–Cl bond than that trans to Cl. There is also evidence for a cis influence, which has been suggested to arise from steric effect, but it is usually much smaller than the trans influence and can be neglected in most cases.

The trans influence in the above complexes is also reflected in their IR spectra. The metal–ligand stretching vibration of the ligand trans to PEt_3 is observed at a lower frequency than those of the same ligand trans to Cl.

The ^{1}J(Pt–P) value for PEt_3 trans to PEt_3 (2380 Hz) in the ^{31}P NMR spectrum of *trans*-$PtCl_2(PEt_3)_2$ is much smaller than the ^{1}J(Pt–P) value (3508 Hz) for the corresponding cis complex having PEt_3 trans to the Cl ligand.

These results indicate that the PEt_3 ligand has a greater trans influence than the Cl ligand.

Let us now examine the trans influence in organoplatinum complexes using the ^{1}J(Pt–P) value as a measure. The ^{1}J(Pt–P) value of PEt_3 trans to the CH_3 group in the following complex is 1719 Hz, whereas the ^{1}J(Pt–P) value of PEt_3 trans to Cl is 4179 Hz.

Et3P CH3
Pt
Et3P Cl

The big difference in the ^{1}J(Pt–P) values occurs because the trans influence of the methyl group is greater than that of Cl, and this in turn reflects the decrease in the s character of the orbital used by platinum in forming the bond with the PEt_3 ligand trans to CH_3.

The trans influence is also observed in ^{2}J(Pt–H) in ^{1}H NMR spectra. The ^{2}J(Pt–H) value for the methyl group bound to platinum in the position trans to X in complex **5-8** and the ^1J(Pt–P) value of P trans to X in complex **5-9** shows the same dependence on ligand X.

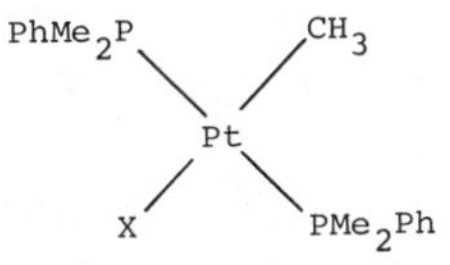

5-8 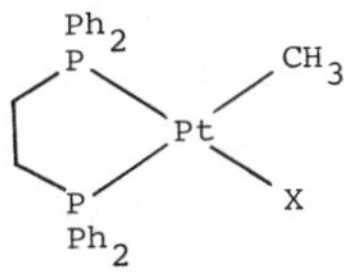 **5-9**

The trans influence observed in these complexes is in the following order: $CH_3 > P(OPh)_3 \sim PPh_3 > CN > CO > AsPh_3 \gtrsim NO_2 \sim SPh > SbPh_3 > SCN > C_5H_5N > NCO \sim I > O_2CCH_3 > Cl > CH_3CN > ONO_2$.

A similar relation is observed between ^1J(Pt–C) in complex **5-8** and ^1J(Pt–P) in complex **5-9**. The order of trans influence of ligands bonded directly through the Pt–C bonds is as follows: C_6H_9(cyclohexenyl) $\sim C_2H_5 > C_6H_5 > CH_2Ph \sim CH_3 > CF_3 > CH_2COCH_3 > CH_2CN > C{\equiv}CPh \sim CH_2NO_2 > \eta^1\text{-}C_5H_5 > CN \sim NCO > CH(COCH_3)_2 > CH(CN)_2 > CO > C(COCH_3)_3 > C(CN){=}C(CN)_2$.

Two explanations for the trans influence have been advanced. One is the effect of a π-bonding ligand, which tends to weaken the bond trans to it. With this view, two π-bonding ligands vying for the same d orbital of the metal atom tend to labilize each other. Such an influence might account for the trans influence of tertiary phosphines or related π-bonding ligands, but obviously it cannot be invoked for hydrido and alkyl ligands, which are good trans directors. The second explanation of the trans influence is illustrated in Figure 5.8. With this view, the two ligands L and X, which are situated in trans positions to each other, compete for a larger share of the metal σ orbitals. The ligand that competes more effectively can better utilize the σ orbital to weaken the bonding of the ligand trans to it. A similar reasoning can be applied to account for the decrease in the ^1J(Pt–P) value for the tertiary phosphine ligand trans to X. When the σ-bonding character of the Pt–X bond is strong, it will cause a decrease in the s character in the Pt–L bond, resulting in a decrease in the ^1J(Pt–P) coupling constant.

As mentioned when discussing the trans influence in $Pt(PEt_3)_2Cl_2$ complexes, the metal–ligand stretching vibrations observed in the IR spectra can also be used as a measure of the trans influence. Both the ν(Pt–H) band in the hydridoplatinum complexes and the ν(Pt–C) band in alkylplatinum complexes are influenced by the ligands trans to them. For example, the ν(Pt–H) band in

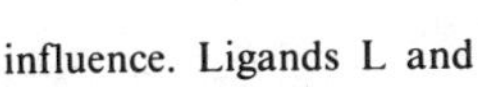

FIGURE 5.8. Cause of the trans influence. Ligands L and X attached to Pt compete for the electron density for forming σ bonds. Increase in the share for L causes decrease in the share for X.

trans-$PtH(X)(PEt_3)_2$ is observed in the region of 2050–2250 cm^{-1}, the $\nu(Pt-CH_3)$ band in *trans*-$PtMe(X)(PEt_3)_2$ is observed in the region of 520–565 cm^{-1}, and both bands vary in parallel manner as X is changed (Fig. 5.9).[58] The order of the trans influence of X is in agreement with the order derived from the coupling constants of the methylplatinum complexes.

Since the trans influence measured by $^1J(M-P)$ reflects the changes in the s-orbital bond order between M and P, it is expected that the trans influence in complexes of square planar Pt(II) and octahedral Pt(IV) may be different. In fact, the $^1J(Pt-P)$ in *trans*-$PtCl_2(PBu_3)_2$ is found to be 2395 Hz, whereas $^1J(Pt-P)$ in octahedral *trans*-$PtCl_4(PBu_3)_2$ is 1474 Hz. The difference has been accounted for in terms of the different s characters. These should be in the ratio of $\frac{1}{4}$: $\frac{1}{6}$ since in terms of valence bond theory[59] the Pt(II) complex is formally dsp^2 hybridized, whereas the Pt(IV) complex is d^2sp^3 hybridized. However, this s-character argument does not hold for Pt(0) complexes (assumed to be sp^3 hybridized) and Pt(II) complexes (dsp^2 hybridized), and the validity of the s-character argument associated with the valence bond theory is open to some doubt. Nevertheless, although a theory to explain the trans influence has not yet been established unequivocally, $^1J(M-P)$ values still serve as useful measures of the trans influence.

c. NMR of π Complexes

Coordination of olefins or conjugated hydrocarbons to a transition metal affects the ^{13}C and ^1H NMR chemical shifts of the coordinated ligands.[61,62] For example, the ^1H NMR chemical shift of coordinated ethylene in Zeise's salt, $K[Pt(C_2H_4)Cl_3]$, is shifted to δ 4.7 from the value of δ 6.0 for free ethylene. The shielding effect of the transition metal is more pronounced in olefin

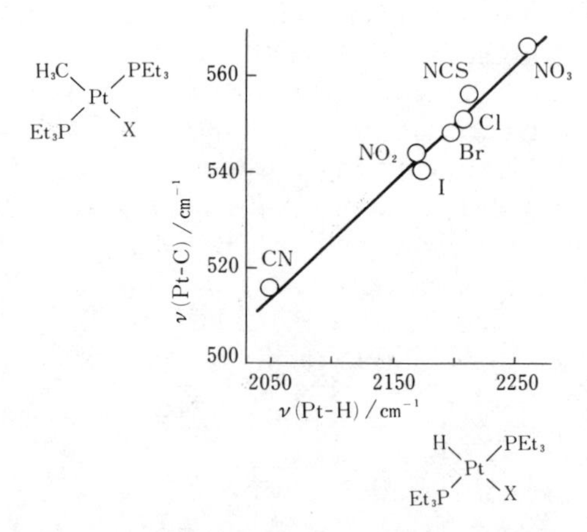

FIGURE 5.9. Trans influence observed in IR spectra. Influence of the change of X on ν(Pt-C) and ν(Pt-H) bands in *trans*-$PtMeX(PEt_3)_2$ and *trans*$PtH(X)(PEt_3)_2$, respectively. (From Allen and Pidcock.[58])

complexes of low-valent transition metals. The ^{1}H NMR chemical shifts of the inequivalent protons of ethylene in $CpRh(C_2H_4)_2$ (cf. **5-12**) are observed at δ 2.75 and 1.0. On the other hand, in an ethylene–Ag^+ complex the protons of the coordinated ethylene are slightly deshielded to δ 6.1 relative to those in free ethylene. The ^{13}C NMR spectra of olefin complexes also exhibit a similar trend.

The ^{1}H and ^{13}C nuclei in π complexes of substituted olefins are also shielded relative to those in the uncomplexed ligands. In the ^{13}C NMR spectra of Ni(olefin)$(PR_3)_2$ complexes the vinylic carbons of the coordinated olefins are more shielded in those complexes having the more basic phosphines, possibly reflecting the greater back donation from Ni(0) to the coordinated olefins in such complexes.[63] A similar trend is observed in the proton NMR spectra. In platinum–olefin complexes ^1J(Pt–C) values can be utilized as measures to probe the nature of the olefin–metal bonding. It has been found that the ^1J(Pt–C) values of η^2-carbons are much smaller than ^1J(Pt–C) values of σ-bonded carbons.[64]

A typical ^{1}H NMR spectrum of an η^3-allyltransition metal complex that is not undergoing molecular rearrangement[65] is shown in Figure 5.10.

Three types of protons, H_a, H_b, and H_c, in an η^3-allyl complex such as $[Pd(\eta^3\text{-}C_3H_5)Cl]_2$ (**5-10**) are observed in an intensity ratio of 1 : 2 : 2. The H_a proton at the lowest field is split into a multiplet by coupling with the H_b and H_c protons, whereas the H_b and H_c protons give rise to two doublets split by coupling with the H_a proton. Since the H_b–H_c coupling is small, H_b and H_c are observed as simple doublets. This spectral pattern changes on warming due to the dynamic behavior in solution, as discussed in the following section.

d. NMR of Fluxional Molecules

When one observes dynamic processes, the available information is restricted by the means used for observation. For example, if a photograph is to be

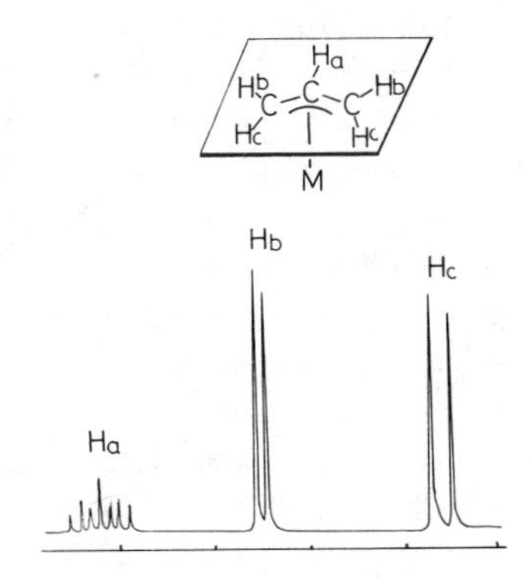

FIGURE 5.10. Low-temperature ^{1}H NMR spectrum of an η^3-allyl complex (**5-10**) indicating absence of exchange between H_b and H_c.

taken of a body moving at high speed, a camera with a high-speed shutter is required, whereas to take a motion picture of an extremely slow-moving object, the film speed in the camera should be slow enough that the motion of the object can be observed. Crudely speaking, a somewhat similar situation exists in the study of the dynamic behavior of atoms and molecules by physical methods. Diffraction methods, having the time scale 10^{-18}–10^{-20} sec, are suitable for taking "pictures" of atoms and molecules in stopped motion, but they do not provide information about molecular movement. The time scale of NMR 10^{-1}–10^{-9} sec is in the right range to detect the motion of a molecule in solution.

When a molecule rearranges from one configuration to another stereochemically equivalent configuration, the molecule is called *fluxional.*[66] For example, an η^1-cyclopentadienyl ligand attached to a metal by a metal–carbon σ bond can undergo intramolecular rearrangements as shown below; the complex of such a ligand is called fluxional. If this process occurs rapidly on the NMR time scale, the ^{1}H and ^{13}C nuclei will appear to be equivalent.

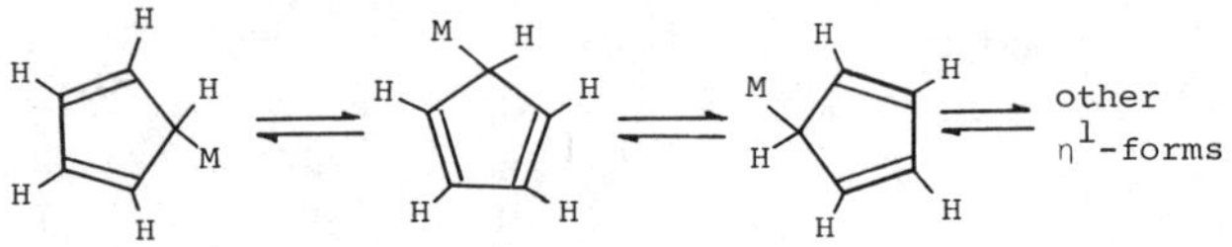

other η^1-forms (5.5)

When the molecular rearrangement occurs between stereochemically nonequivalent configurations, as shown below, the complex is called *stereochemically nonrigid.*[67]

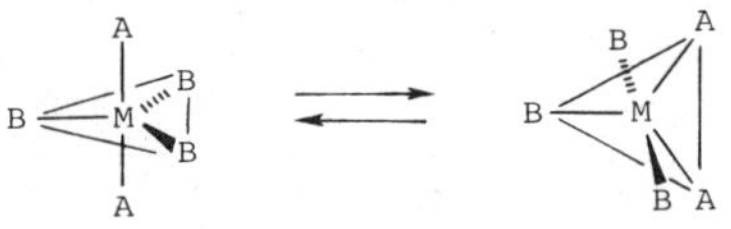

(5.6)

However, because of the difficulty in strictly differentiating between the two terms, they are often used as synonyms.

When a solution containing a fluxional molecule is cooled, the rate of the fluxional process decreases and may become slow on the NMR time scale below a certain temperature. If one can "freeze" the fluxional molecule in this way, nuclei that are now in a different stereochemical environment in the frozen molecule give rise to NMR signals at different magnetic fields. If complete freezing can be achieved, each NMR signal will be observed as a sharp peak or as a multiplet peak due to spin–spin coupling, and the half-width of that peak will not become smaller on further cooling of the solution. For example, the ^{1}H NMR spectrum of the η^1-C_5H_5 ligand shown in Eq. (5.5) shows an A_2B_2X pattern with a 2 : 2 : 1 relative intensity at very low temperatures. On raising the temperature the half-width of each peak increases and the multiplets collapse to a single, average line at a certain temperature, called the coalescence temperature. On further raising the temperature a sharp peak

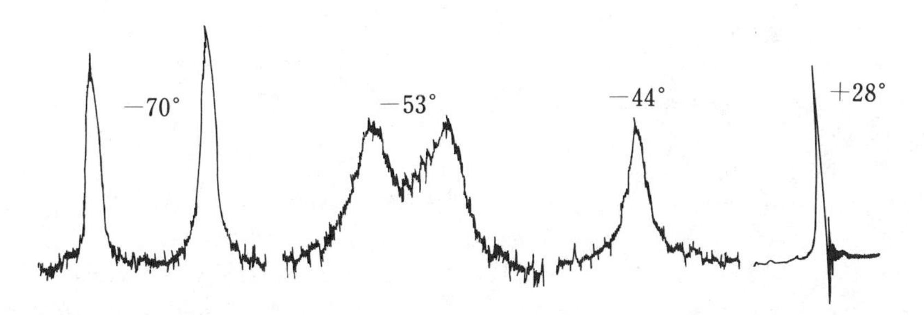

FIGURE 5.11. Variation of ^1H NMR of $Cp_2Fe(CO)_4$ with temperature [Reproduced from F. A. Cotton and G. Wilkinson, *Advanced Inorganic Chemistry*, Wiley, New York, 1980, by permission].

develops in a position corresponding to a weighted average of the peaks observed in the frozen state.

In the ^1H NMR spectrum of Cp_2Hg the ring protons on the cyclopentadienyl groups are observed as a sharp singlet at δ 5 at room temperature, owing to the rapid molecular rearrangement of Eq. (5.5). Wilkinson, who first discovered the phenomenon, advanced the idea of the σ ring moving around by a 1,2-shift, later termed *ring whizzing*.[68] This name was later replaced by the terms fluxionality and stereochemical nonrigidity.

The rearrangement reactions are rate processes with activation energies. When these activation energies are in the range of 25–100 kJ mol^{-1}, the rates of the rearrangements can be brought into the range of 10^2–10^5 sec^{-1} at temperatures between −150 and +150 °C, a convenient range for the variable-temperature NMR technique. Figure 5.11 shows the variation of the ^1H NMR spectrum of $Cp_2Fe_2(CO)_4$ with temperature.[43b]

The $Cp_2Fe_2(CO)_4$ molecule exists in solution as a mixture of cis and trans isomers with bridging CO groups as shown below.

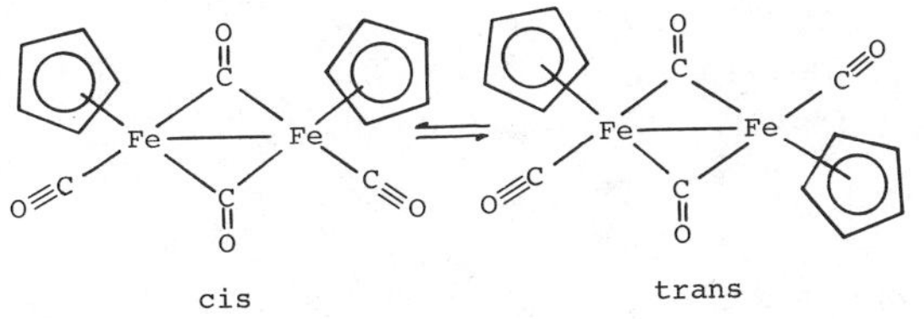

cis ⇌ trans (5.7)

At −70 °C the ^1H NMR resonances for the Cp ring protons are observed at different positions. On raising the temperature the two peaks broaden and finally coalesce to give a sharp singlet peak at +28 °C at a position intermediate between the peak positions at −70 °C. The ^{13}C NMR spectra of the CO groups show that interchange of the bridging and terminal CO groups accompanies the cis–trans interconversion. To account for these observations, it has been proposed that the CO bridges open in a concerted way to give a nonbridged intermediate $Cp(OC)_2Fe–Fe(CO)_2Cp$. Subsequent rotation about the Fe–Fe bond followed by reclosing of the bridges produces either a cis or trans isomer and is accompanied by scrambling of the bridging and terminal CO groups.

Complexes of cyclic polyenes are often fluxional.[69] In the ^{13}C NMR spectrum of the cyclooctatetraene complex $Fe(C_8H_8)(CO)_3$, two singlets arising from the C_8H_8 and the carbonyl ligands are observed at room temperature. At −134 °C the signals due to the C_8H_8 ligand now appear as four peaks, two of which are observed at rather low magnetic fields. The low-temperature spectrum suggests that the $Fe(CO)_3$ moiety is coordinated to four-carbon atoms through two conjugate double bonds, that is, to a 1,3-diene (**5-11**), in agreement with the structure found by X-ray single-crystal analysis.

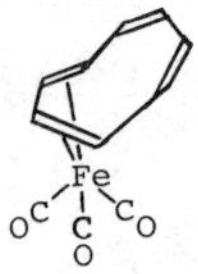

5-11

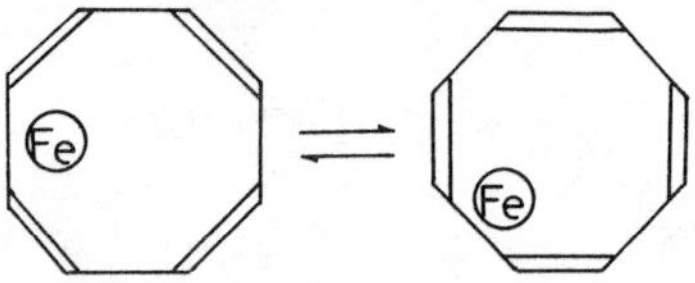

(5.8)

Detailed analysis of the NMR spectra at various temperatures indicates the occurrence of 1,2-shift, as shown in Eq. (5.8).[70] It has also been shown that the intramolecular rearrangement of the η^4-C_8H_8 ring and the scrambling of the three CO ligands are taking place in the same temperature range with an activation energy of 35 kJ mol^{-1}.

The ^1H NMR spectrum of the η^3-allyl complex shown in Figure 5.10 is that of a rigid molecule. The ^1H NMR of $Zr(C_3H_5)_4$ at −66 °C in $CFCl_3$ shows an analogous AM_2X_2 spectral pattern.[70a] On raising the temperature the two doublets broaden and eventually collapse into one doublet, whereas the multiplet assigned to the methine H_a proton is converted into a quintet at −20 °C, as shown in Figure 5.12. These changes indicate that the H_b and H_c protons are exchanging rapidly, giving four equivalent protons at the higher temperature. The interconversions probably proceed through η^1-allyl intermediates, as shown by the mechanism included in Figure 5.12.

$CpRh(C_2H_4)_2$ (**5-12**) provides another example of a fluxional molecule.[71]

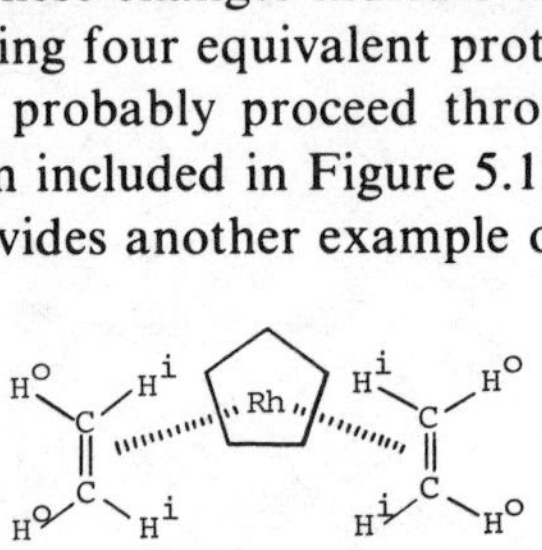

5-12

Two ethylene ligands are coordinated to the rhodium atom perpendicular to the plane connecting the center of the cyclopentadienyl ligand, the Rh atom, and the center of the C–C double bonds of the two coordinated ethylene ligands. This arrangement makes the outer protons H^o and inner protons H^i (four each) nonequivalent. On warming, the ethylene ligands begin

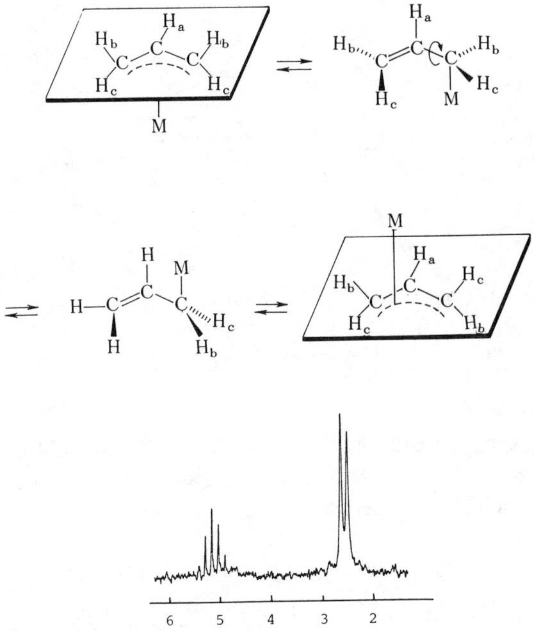

FIGURE 5.12. ^1H NMR spectrum of a fluxional allyl complex $Zr(C_3H_5)_4$ (bottom) and a possible mechanism for exchange of H_b and H_c protons (top).

to rotate around the axes connecting the Rh and the two ethylene molecules, like two propellers on a double-engined plane, thus making the H^o's and H^i's equivalent. Since ^2J(Rh–H) coupling is retained, the ethylene ligands must remain bound to rhodium, thus excluding an alternative mechanism, namely dissociation and recoordination of ethylene, which would also make H^o and H^i equivalent. Such dissociation does in fact take place at higher temperatures. An activation energy of 62.8 kJ mol^{-1} has been estimated for rotation of the ethylene ligands in $CpRh(C_2H_4)_2$.

Variable-temperature NMR spectra coupled with computer-simulated line shape analysis provide much information concerning dynamic processes. Assuming that each absorption has a Lorenzian line shape and putting the appropriate chemical shifts, coupling constants, half-widths, and life times (T) for the various configurations into a computer, simulated spectra are drawn. By comparison with the experimental spectra, the validity of the assumptions and the data provided may be examined. An example of such a study is shown below.

Figure 5.13*a* shows the actual ^1H NMR spectra of $CpCu(PEt_3)$ in SO_2 measured at various temperatures.[72] At -70 °C an A_2B_2X spectrum with a 2 : 2 : 1 intensity ratio is observed. Clearly the cyclopentadienyl ring is η^1 bonded with copper at this temperature. On warming, molecular rearrangement is accelerated so that the peaks broaden and eventually collapse to a single sharp peak

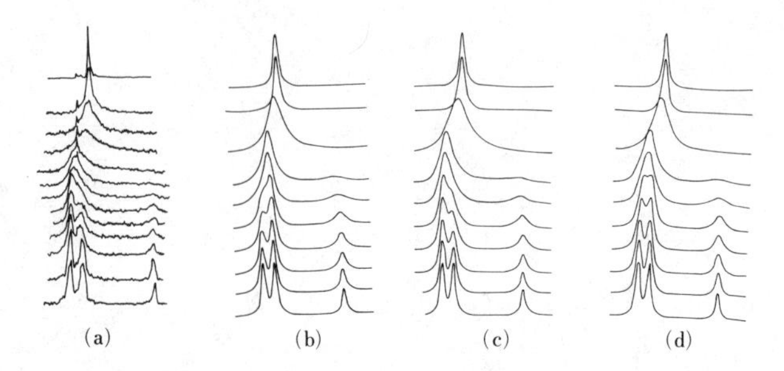

FIGURE 5.13. Variable-temperature ^{1}H NMR spectra of $CpCu(PEt_3)$ (*a*) and computer-simulated spectra (b–d). (Reprinted from G. M. Whitesides and J. S. Flemming, *J. Am. Chem. Soc.*, **89**, 2855. Copyright 1967 American Chemical Society.)

at −1 °C. Two alternative molecular rearrangement mechanisms can be envisaged to account for these spectral changes. One is a 1,2-shift, as shown in Eq. (5.5), the other is a 1,3-shift, as shown in Eq. (5.9).

(5.9)

The results of computer simulations are shown in (*b*)–(*d*) of Figure 5.13 beside the experimental spectra in (*a*). The simulated spectra in (*c*), assuming the 1,3-shift mechanisms, are quite similar to that in (*a*). Although the possibility of a 1,2-shift mechanism, which give the simulated spectra in (*b*), cannot be excluded completely because of the assumptions used in the computation, it may be safely concluded that the molecular rearrangement mechanism is not the random-shift process that gives the simulated spectra in (*d*).

Another example of a fluxional molecule is the *trihapto*-benzylmolybdenum complex **5-13**.[73] Complex **5-13** can be prepared by photolysis of the σ-benzylmolybdenum complex $CpMo(CO)_3(\sigma\text{-}CH_2C_6H_5)$ with liberation of 1 mol of CO. The complex is in an η^3-bonded form below −30 °C, where the ^{1}H NMR spectrum shows the presence of nonequivalent ortho phenyl protons H_a and H_b (the meta protons also are nonequivalent), whereas above 60 °C these protons become equivalent. The molecular rearrangement probably proceeds via a short-lived σ-benzyl intermediate.

5-13

The β elimination of transition metal alkyls and its reserve process, olefin insertion into an M–H bond, are important elementary processes in relation to catalytic processes involving olefins. An ethylene hydrido complex (**5-14**) can be obtained by protonation of $Mo(C_2H_4)_2(dppe)_2$ with CF_3COOH. The fluxional behavior of complex **5-14** was demonstrated by the observation of variable-temperature NMR spectra in CH_2Cl_2 solution.[74] The process is proposed to proceed by the ethylene insertion to give the ethylmolybdenum complex **5-15** and deinsertion processes of Eq. (5.10).

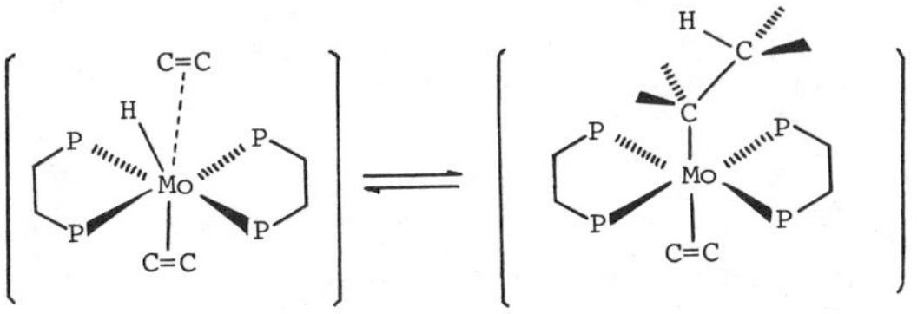

(5.10)

5-14 **5-15**

The dynamic behavior of five-coordinate complexes occupies a particularly prominent part of the study of fluxional compounds. Because the energy difference between the trigonal bipyramid (tbp) and square pyramid (spy) structures is small, the transformation from one structure to the other may take place with a small activation energy. Whether a five-coordinate complex favors the tbp or spy structure is a delicate problem, and factors such as the nature of the ligand and number of d electrons[75] should be considered. For ML_5 complexes having identical ligands, conversion from one structure to the other takes place particularly readily.

For example, $Fe(CO)_5$ has the tbp structure in the solid state as confirmed by X-ray diffraction, a method with a short time scale. Even at −50 °C, ^{13}C NMR spectroscopy of $Fe(CO)_5$ in solution does not separately distinguish the two apical and three equatorial carbons, which should be in different environments in the frozen state. This is because the apical and equatorial ligands are rapidly interchanging their coordination sites. The interchange must be intramolecular and cannot involve ligand dissociation or bimolecular exchange processes because ^{57}Fe–^{13}C coupling is retained.

Two mechanisms for the tbp–spy interconversion are considered here. One is the Berry mechanism[68b,76] (Fig. 5.14) and the other is the "turnstile" rotation mechanism[78] (Fig. 5.15). In Berry's pseudorotation mechanism the initial tbp complex (*a*) transforms into another form of tbp complex (*c*) with the apical and equatorial vertices interchanged. This can be achieved not with actual rotation but by relatively small and simple angle deformation motions.

In the Berry pseudorotation mechanism, as shown in Figure 5.14, the angle L_1ML_2 closes from 180°, and simultaneously the angle L_4ML_5 opens from 120° without moving the pivot L_3, giving the square pyramid in (*b*), where L_1, L_4, L_2, and L_5 form the basal set. This configuration is the transition state for tbp–tbp pseudorotation. Continuation of the movements of the L_4, L_5 set

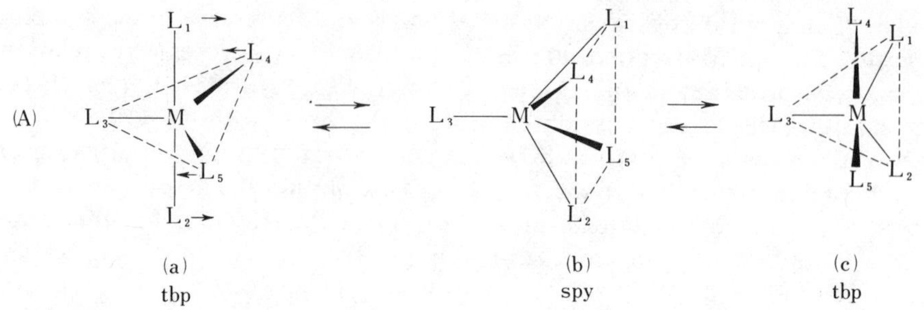

FIGURE 5.14. The Berry pseudorotation mechanisms for interconversion of tbp complexes.

toward L_3 and of the L_1, L_2 set away from L_3 brings the complex into another tbp configuration (Fig. 5.14*c*) having L_4 and L_5 at apical positions and L_1, L_2, and L_3 at equatorial positions of the tbp configuration. The resultant tbp configuration looks as if it were derived by rotation of the initial tbp in (*a*) by 90° about the L_3–M axis to give (*c*), but in fact the process can be achieved by small changes of the bond angles with L_3 as the pivot. Therefore, it is understandable that the interconversion can take place with low activation energies. Of course, starting from (*c*), the process can be repeated with L_1 or L_2 as the pivot so as to bring L_3 also to a new apical position.

In the alternative turnstile rotation three of the five ligands in a tbp complex rotate by 60° without moving the other two ligands (one apical and the other equatorial),† as shown in Figure 5.15. In this rotation ligands 1 and 2, which were at equatorial and apical positions, respectively, are transformed into ligands at apical and equatorial positions in the new tbp without moving their positions.[78]

On the basis of NMR spectroscopy one cannot differentiate whether the Berry mechanism or the turnstile mechanism is operative in these rearrangement reactions, and it can be shown that they are permutationally indis-

† According to those who originally proposed the mechanism, the turnstile mechanism involves the turning of the three ligands 3, 4, and 5 by 120° and simultaneous rotation of ligands 1 and 2 by 180° in the same direction. Such a process is equivalent in the net result with the process described in this text.

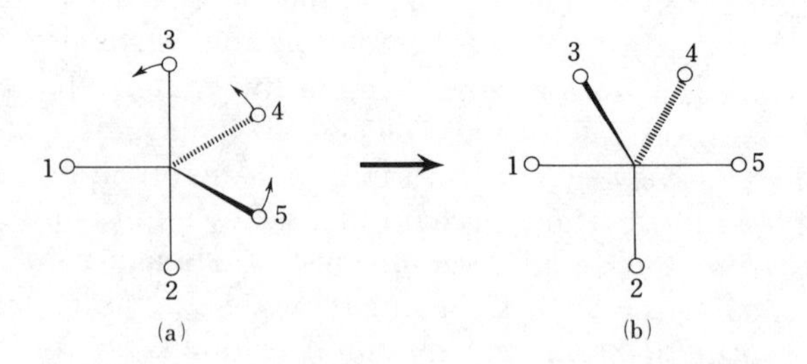

FIGURE 5.15. Ligand interchange by turnstile rotation.

tinguishable. Theoretically, however, the Berry mechanism seems to be the lower activation pathway.

Intramolecular rearrangements without metal–ligand bond dissociation are rather rare among octahedral complexes. The dynamic behavior of *cis*-$MH_2(PR_3)_4$ complexes provides an example. The preservation of $^{31}P-^{1}H$ coupling during the rearrangement suggests that the process is nondissociative.[43b]

Fluxional complexes are abundant among seven- or eight-coordinate complexes and metal clusters. Fluxionality is the rule rather than the exception for these complexes. Detailed treatment of the topic, however, is beyond the scope of this text.

Although NMR studies of organometallic complexes provide us with a wealth of important information, it is sometimes desirable to measure the NMR spectra in the solid state and to compare them with the solution spectra and with the structures established by X-ray analysis. With the recently developed combined techniques of cross-polarization and magic angle spinning, high-resolution measurements of ^{31}P NMR spectra of various phosphine-containing complexes have been carried out.[79] This technique enables heterogeneous catalyst systems to be studied, and further development in this field is expected.

REFERENCES

1. D. F. Shriver, *The Manipulation of Air-Sensitive Compounds,* McGraw-Hill, New York, 1969.
2. R. B. King, *Organometallic Syntheses,* Vol. 1, *Transition-Metal Compounds,* Academic Press, New York, 1965; J. J. Eisch, *Organometallic Syntheses,* Vol. 2, *Nontransition-Metal Compounds,* Academic Press, New York, 1981.
3. H. C. Brown, *Organic Syntheses via Boranes,* Wiley-Interscience, New York, 1975.
4. C. J. Barton, in *Techniques of Inorganic Chemistry,* Vol. III, H. B. Jonassen and A. Weissberger, eds., Wiley-Interscience, New York, 1963.
5. S. Herzog, J. Dehnert, and K. Luhder, in *Techniques in Inorganic Chemistry,* Vol. VII, H. B. Jonassen and A. Weissberger, eds., Wiley-Interscience, New York, 1968.
6. Houben-Weyl's *Methoden der Organischen Chemie,* Vol. XIII, *Organometallic Compounds,* Georg Thieme Verlag, Stuttgart, 1970–78.
7. W. Schlenk and A. Thal, *Chem. Ber,* **46,** 2843 (1913).
8. K. Ziegler, H. G. Gellert, H. Martin, K. Naged, and J. Schneider, *Justus Liebigs Ann. Chem.,* **589,** 119 (1954).
9. O. Schwarzkopf and F. Schwarzkopf, *Characterization of Organometallic Compounds,* Part I, Wiley, New York, 1969.
10. T. R. Crompton, in *The Chemistry of the Metal–Carbon Bond,* Vol. 1, F. R. Hartley and S. Patai, eds., Wiley-Interscience, New York, 1982.
11. M. J. Taylor, in *The Chemistry of the Metal–Carbon Bond,* Vol. 1, F. R. Hartley and S. Patai, eds., Wiley-Interscience, New York, 1982, p. 775.
12. E. Maslowsky, Jr., *Vibrational Spectra of Organometallic Compounds,* Wiley-Interscience, New York, 1977.

13. K. Nakamoto, *Infrared and Raman Spectra of Inorganic and Coordination Compounds,* 3rd ed., Wiley-Interscience, New York, 1978.
14. K. Nakamoto, in *Characterization of Organometallic Compounds,* Chemical Analysis Monograph, Vol. 26, Part 1, M. Tsutsui, ed., Wiley-Interscience, New York, 1969.
15. A. J. Downs, in *Spectroscopic Methods in Organometallic Chemistry,* W. O. George, ed., Butterworths, London, 1970.
16. G. Davidson, *Organometal. Chem. Rev.,* **A8,** 303 (1972).
17. H. P. Fritz, *Adv. Organometal. Chem.,* **1,** 239 (1964).
18. D. B. Powell, J. G. Scott, and N. Sheppard, *Spectrochim. Acta,* **28A,** 327 (1972); D. B. Powell and T. J. Leedham, *Spectrochim. Acta.,* **28A,** 337 (1972).
19. M. A. Meester, D. J. Stufkens, and K. Vrieze, *Inorg. Chim. Acta,* **21,** 251 (1977).
20. D. W. Wertz and M. A. Moseley, *Inorg. Chem.,* **19,** 705 (1980).
21. T. R. Spalding, in *The Chemistry of the Metal–Carbon Bond,* F. R. Hartley and S. Patai, eds., Wiley-Interscience, New York, 1982.
22. M. R. Litzow and T. R. Spalding, *Mass Spectrometry of Inorganic and Organometallic Compounds,* Elsevier, Amsterdam, 1973.
23. T. R. Spalding, in *Spectroscopic Methods in Organometallic Chemistry,* W. D. George, ed., Butterworths, London, 1970.
24. R. W. Kiser, in *Characterization of Organometallic Compounds,* Part I, M. Tsutsui, ed., Wiley, New York, 1964.
25. D. B. Charalambous, ed., *Mass Spectrometry of Metal-Containing Compounds,* Butterworths, London, 1975.
26. M. I. Bruce, *Adv. Organometal. Chem.,* **6,** 273 (1968).
27. A. D. Redhouse, in *The Chemistry of the Metal–Carbon Bond,* F. R. Hartley and S. Patai, eds., Wiley-Interscience, New York, 1982.
28. N. C. Baenziger, in *Characterization of Organometallic Compounds,* Part I, M. Tsutsui, ed., Wiley, New York, 1969.
29. B. G. Ramsey, *Electronic Transitions in Organometallics,* Academic Press, New York, 1969.
30. T. R. Crompton, in *The Chemistry of the Metal–Carbon Bond,* F. R. Hartley and S. Patai, eds., Wiley-Interscience, New York, 1982.
31. T. Yamamoto, A. Yamamoto, and S. Ikeda, *J. Am. Chem. Soc.,* **93,** 3350, 3360 (1971).
32. R. F. Fenske, *Prog. Inorg. Chem.,* **21,** 179 (1976).
33. A. H. Cowley, *Prog. Inorg. Chem.,* **26,** 45 (1979).
34. S. Evans, M. L. H. Green, B. Jewitt, A. F. Orchard, and C. F. Pygall, *J. Chem. Soc. Faraday II,* **68,** 1847 (1972).
35. T. C. Gibb, in *Spectroscopic Methods in Organometallic Chemistry,* Vol. IV, W. O. George, ed., Butterworths, London, 1970.
36. J. Chatt and G. J. Leigh, *Angew. Chem. Int. Ed. Engl.,* **17,** 400 (1978).
37. L. N. Mulay and J. T. Dehn, in *Characterization of Organometallic Compounds,* Part II, M. Tsutsui, ed., Wiley, New York, 1971.
38. D. F. Evans, *J. Chem. Soc.,* 2003 (1959).
39. Landolt-Börnstein, *Magnetic Properties of Coordination and Organometallic Transition Metal Compounds,* Springer-Verlag, Weinheim E. König, ed., Vol. II/2 (1966), E. König and G. König, eds., Vol. II/8 (1976).
40. F. J. Smentowski, in *Characterization of Organometallic Compounds,* Part II, M. Tsutsui, ed., Wiley, New York, 1971.
41. P. B. Ayscough, in *Spectroscopic Methods in Organometallic Chemistry,* W. O. George, ed., Butterworths, London, 1970.

42. (a) G. Henrici-Olivé and S. Olivé, *Angew. Chem.*, **79,** 764 (1967); (b) *Makromol. Chem.*, **121,** 70 (1969); (c) *J. Organometal. Chem.*, **19,** 309 (1969).

43. (a) J. A. Davies, in *The Chemistry of the Metal–Carbon Bond,* F. R. Hartley and S. Patai, eds., Wiley-Interscience, New York, 1982; (b) L. M. Jackman and F. A. Cotton, eds., *Dynamic Nuclear Magnetic Resonance Spectroscopy,* Academic Press, New York, 1975; (c) A. Steigel and H. W. Spiess, *Dynamic NMR Spectroscopy,* Springer-Verlag, Weinheim, 1978; (d) P. S. Pregosin and R. Kunz, *^{31}P and ^{13}C NMR of Transition Metal Phosphine Complexes,* Springer-Verlag, Weinheim, 1979; (e) J. J. Kaplan and G. Fraenkel, *NMR of Chemically Exchanging Systems,* Academic Press, New York, 1980; (f) E. L. Muetterties, ed., *Transition Metal Hydrides,* Marcel Dekker, New York, 1971.

44. (a) M. L. Maddox, S. Stafford, and H. D. Kaesz, *Adv. Organometal. Chem.*, **3,** 1 (1965); (b) J. M. Faller, *Adv. Organometal. Chem.*, **16,** 211 (1977); (c) K. Vrieze and P. W. N. M. van Leeuwen, *Prog. Inorg. Chem.*, **14,** 1 (1971); (d) B. E. Mann, *Adv. Organometal. Chem.*, **12,** 135 (1974); (e) F. A. Cotton, *Acc. Chem. Res.*, **1,** 257 (1968); (f) C. W. Spangler, *Chem. Rev.*, **76,** 187 (1976); (g) E. L. Muetterties and W. D. Phillips, *Adv. Inorg. Chem. Radiochem.*, **4,** 231 (1962); (h) M. H. Chisholm and S. Godleski, *Prog. Inorg. Chem.*, **20,** 299 (1976); (i) E. L. Muetterties, *MTP Int. Rev. Sci. Series 1,* Vol. 9, Butterworths, London, 1972; (j) R. G. Kidd, in *Characterization of Organometallic Compounds,* Part II, M. Tsutsui, ed., Wiley, New York, 1971; (k) W. McFarlane, in *Spectroscopic Methods in Organometallic Chemistry,* W. O. George, ed., Butterworths, London, 1970; (l) B. E. Mann, in *Comprehensive Organometallic Chemistry,* Vol. 3, Pergamon Press, Oxford, 1982.

45. N. S. Ham and T. Mole, *Prog. Nucl. Mag. Res. Spect.*, **4,** 101 (1969).

46. P. T. Narashimhan and M. T. Rogers, *J. Am. Chem. Soc.*, **82,** 5983 (1960).

47. T. Yamamoto, Y. Nakamura, and A. Yamamoto, *Bull. Chem. Soc. Jpn.*, **49,** 191 (1976).

48. T. Ito, T. Kiriyama, Y. Nakamura, and A. Yamamoto, *Bull. Chem. Soc. Jpn.*, **49,** 3257 (1976).

49. J. D. Nixon and A. Pidcock, *Ann. Rev. NMR Spectroscopy,* **2,** 345 (1969).

50. G. G. Mather, A. Pidcock, and G. J. N. Rapsey, *J. Chem. Soc. Dalton Trans.*, 2095 (1973).

51. N. F. Ramsey, *Phys. Rev.*, **91,** 303 (1953).

52. A. Pidcock, R. E. Richards, and L. M. Venanzi, *J. Chem. Soc. A,* 1707 (1966).

53. T. G. Appleton, H. C. Clark, and L. E. Manzer, *Coord. Chem. Rev.*, **10,** 335 (1973).

54. A. Pidcock, in *Catalytic Aspects of Metal Phosphine Complexes,* E. C. Alyea and D. W. Meek, eds., (Advances in Chemistry Series, Vol. 196), American Chemical Society, Washington, DC, 1982.

55. T. G. Appleton, M. H. Chisholm, H. C. Clark, and L. E. Manzer, *Inorg. Chem.*, **11,** 1786 (1972).

56. T. G. Appleton and M. A. Bennett, *Inorg. Chem.*, **17,** 738 (1978).

57. C. H. Langford and H. B. Gray, *Ligand Substitution Processes,* Benjamin, New York, 1965.

58. F. H. Allen and A. Pidcock, *J. Chem. Soc. A,* 2700 (1968).

59. P. S. Pregosin and L. M. Venanzi, *Chem. Britain,* 276 (1978); 162 (1968).

60. P. S. Pregosin and R. W. Kunz, *NMR Basic Principles Prog.*, **16,** 1 (1979).

61. M. Herberhold, *Metal π-Complexes,* Vol. II, Elsevier, Amsterdam, 1974.

62. B. E. Mann, *Adv. Organometal. Chem.*, **12,** 135 (1974).

63. T. Yamamoto, I. Ishizu, S. Komiya, Y. Nakamura, and A. Yamamoto, *J. Organometal. Chem.*, **171,** 103 (1979).

64. B. E. Mann and B. F. Taylor, *^{13}C NMR Data for Organometallic Compounds,* Academic Press, London, 1981.

65. G. E. Coates, M. L. H. Green, P. Powell, and K. Wade, *Principles of Organometallic Chemistry,* Methuen, London, 1968.

66. F. A. Cotton, *Acc. Chem. Res.*, **1,** 257 (1968).
67. E. L. Muetterties, *Acc. Chem. Res.*, **3,** 266 (1970).
68. (a) G. Wilkinson, *J. Organometal. Chem.*, **100,** 273 (1975); (b) F. A. Cotton, *J. Organometal. Chem.*, **100,** 29 (1975).
69. G. Daganello, *Transition Metal Complexes of Cyclic Polyolefins,* Academic Press, New York, 1979.
70. F. A. Cotton and D. L. Hunter, *J. Am. Chem. Soc.*, **98,** 1413 (1976); (a) J. K. Becconsall, B. E. Job, and S. O'Brien, *J. Chem. Soc. (A),* 423 (1967).
71. R. Cramer, J. B. Kline, and J. D. Roberts, *J. Am. Chem. Soc.*, **91,** 2519 (1969).
72. G. M. Whitesides and J. S. Flemming, *J. Am. Chem. Soc.*, **89,** 2855 (1967).
73. R. B. King and A. Fronzaglia, *J. Am. Chem. Soc.*, **88,** 709 (1966).
74. J. W. Byrne, H. O. Blaser, and J. A. Osborn, *J. Am. Chem. Soc.*, **97,** 3871 (1975).
75. A. R. Rossi and R. Hoffmann, *Inorg. Chem.*, **14,** 365 (1975).
76. R. S. Berry, *J. Chem. Phys.*, **32,** 933 (1960).
77. J. R. Shapley and J. A. Osborn, *Acc. Chem. Res.*, **6,** 305 (1973).
78. R. Ugi, D. Marquading, K. Klusacek, P. Gillespie, and F. Ramirez, *Acc. Chem. Res.*, **4,** 288 (1971).
79. G. E. Maciel, D. J. O'Donnell, and R. Greaves, in *Catalytic Aspects of Metal Phosphine Complexes,* E. C. Alyea and D. W. Meek, eds. (Advances in Chemistry Series Vol. 196), American Chemical Society, Washington, DC, 1982.

CHAPTER 6

Fundamental Processes in Reactions of Organotransition Metal Complexes

Organotransition metal complexes exhibit a range of unique reactions that are not observed in ordinary, purely organic or inorganic compounds. An understanding of the characteristics and underlying principles of these reactions helps to clarify the mechanisms of various transition-metal-catalyzed reactions and further to design new synthetic reactions, catalytic as well as stoichiometric.

The reactions of organotransition metal complexes, including those of some transition metal complexes that are convertible to organotransition metal complexes *in situ*, may be broadly classified as follows.[1] Some of these reaction types are shown in Figure 6.1.†

1. Ligand coordination and dissociation (in this category are included ligand displacement reactions).
2. Oxidative addition and reductive elimination (oxidative C–C coupling and reductive C–C bond breaking also are included).

† Electron transfer reactions are regarded as one of the important elementary steps in reactions of transition metal complexes.[2] However, because they are usually rapid processes that precede the slower atom transfer processes, which are included in the present classification, we do not include the electron transfer process in the categories of fundamental reactions involving atom transfer reactions, although electron transfer may occur in any of the four categories of reactions discussed.

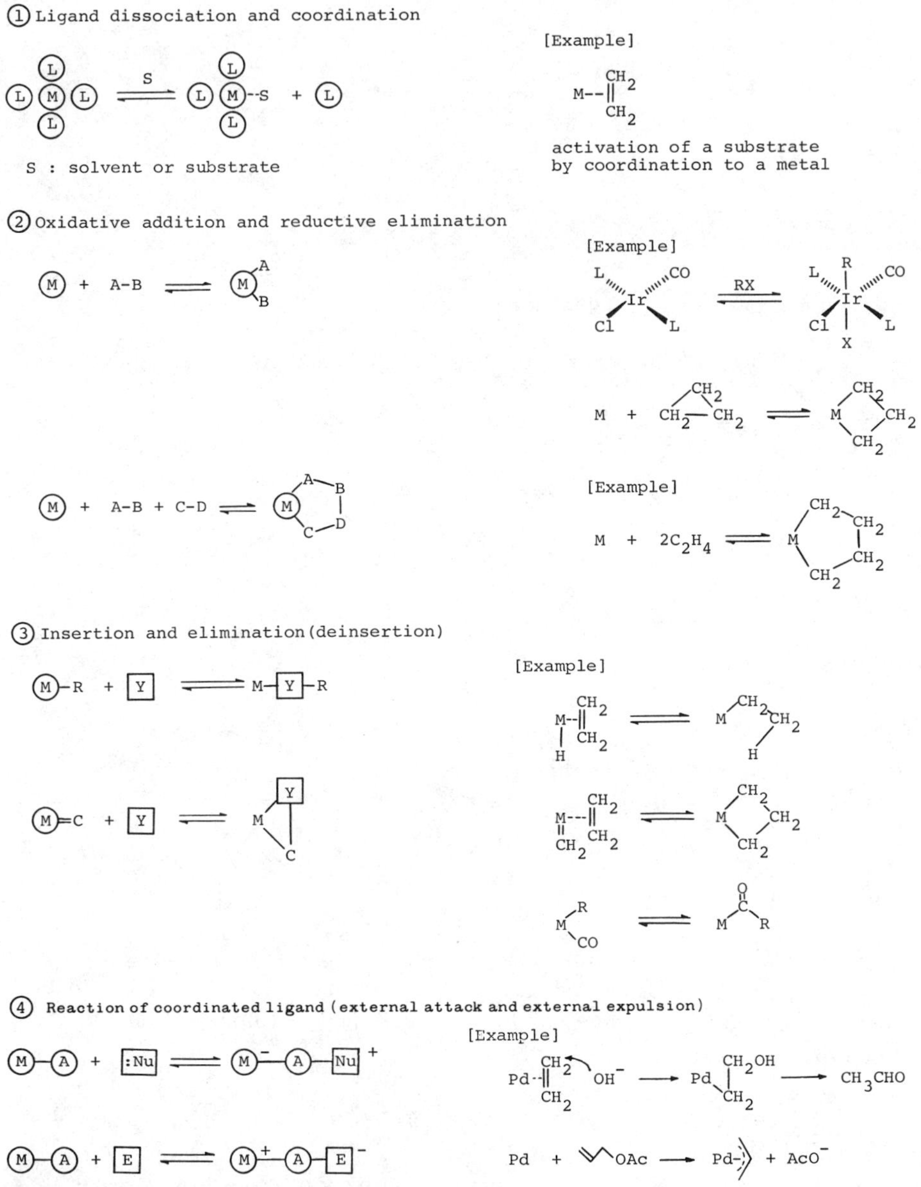

① Ligand dissociation and coordination

[Example]

$M\text{--}\|(CH_2)(CH_2)$

activation of a substrate by coordination to a metal

S : solvent or substrate

② Oxidative addition and reductive elimination

[Example]

RX

M + CH_2—CH_2—CH_2 ⇌ M(CH_2)$_3$

[Example]

M + $2C_2H_4$ ⇌ M(CH_2)$_4$

③ Insertion and elimination (deinsertion)

[Example]

④ Reaction of coordinated ligand (external attack and external expulsion)

[Example]

$Pd\text{--}\|(CH_2)(CH_2)$ + OH^- → $Pd(CH_2OH)(CH_2)$ → CH_3CHO

Pd + ⟋⟍OAc → Pd-allyl + AcO^-

FIGURE 6.1. Classification of fundamental reactions of organotransition metal complexes. Ligands are omitted in categories 2–4.

3. Insertion and elimination (deinsertion).
4. Reactions of coordinated ligands.

As in any classification, the above four categories have some borderline areas that are difficult to define uniquely. However, too much compartmentalization would make the classification itself meaningless. Thus, the above classification seems to be the minimum essential usable in discussions of organo-transition metal chemistry.

6.1. LIGAND COORDINATION AND DISSOCIATION

Many reactions of transition metal complexes in solution are accompanied by ligand coordination or dissociation. Except for intramolecular ligand rearrangement or electron transfer, which are not included here as fundamental reactions, coordination or dissociation of a ligand, including a solvent molecule, are necessary for a reaction to occur with any transition metal complex. Essential prerequisites for transition-metal-catalyzed reactions are the coordination of a substrate to the transition metal complex and its activation by the metal. Also, a product formed on a transition metal complex catalyst must leave the catalytic site as soon as possible in order to free the site for the incoming substrate. Thus, homogeneous reactions catalyzed by transition metal complexes will always involve ligand substitution or replacement reactions.

Some transition metal complexes are susceptible to substitution (with half-lives of less than 30 sec) and are called substitution *labile*, whereas other complexes are substitution inert (with half-lives greater than 30 sec).[3-5] Inorganic substitution reactions, like their organic counterparts, can be classified as *nucleophilic* substitution (S_N) reactions and electrophilic substitution (S_E) reactions. However, one must take account of the oxidation states of the transition metal complexes involved. Transition metal complexes in higher oxidation states may behave as Lewis acids, and these complexes usually react with nucleophiles. On the other hand, electron-rich transition metal complexes in lower oxidation states behave as Lewis bases and thus react with electrophiles.

Ligand substitutions involving transition metal complexes in higher oxidation states (S_N reactions) are classified as S_N1 and S_N2 reactions. An S_N2 reaction is one involving a bimolecular rate-determining step in which one nucleophilic reagent displaces another ligand that was originally attached to the central metal atom.

$$L' + M{-}L \longrightarrow [L' \text{ ---- } M \text{ ---- } L]^{\ddagger} \longrightarrow L'{-}M + L \qquad (6.1)$$

In this type of reaction, the coordination number is increased by one unit at the transition state.

On the other hand, in an S_N1 reaction the reaction proceeds by a two-step mechanism, the first step being a slow unimolecular process that generates a coordinatively unsaturated species.

$$ML_n \underset{k_{-1}}{\overset{k_1}{\rightleftarrows}} ML_{n-1} + L \tag{6.2}$$

This process is followed by a rapid coordination of a nucleophilic ligand.

$$ML_{n-1} + L' \xrightarrow[\text{fast}]{k_2} ML_{n-1}L' \tag{6.3}$$

It should be noted here that the rate law can be dependent on the concentration of L′ even if the ligand dissociation step expressed by Eq. (6.2) constitutes the rate-determining step. If one applies the well-known steady-state approximation for $[ML_{n-1}]$ in Eqs. (6.2) and (6.3), the reaction rate r can be expressed as

$$r = \frac{k_1 k_2 [ML_n][L']}{k_{-1}[L] + k_2[L']} \tag{6.4}$$

Thus, the rate depends on [L′] as well as on $[ML_n]$, but when $k_2[L']$ becomes very large compared to $k_{-1}[L]$, $k_2[L']$ may be canceled from the numerator and denominator, leaving

$$r_{k_2[L'] \to \infty} = k_1[ML_n]$$

and the rate becomes independent of [L′].

The classification into the S_N1 and S_N2 reactions is convenient, particularly in relation to organic reaction mechanisms, but one should note that S_N1 and S_N2 mechanisms are extreme cases, and many real cases are not so clear-cut. Furthermore, the S_N1 and S_N2 terminology can be misleading, particularly when electron-rich, nucleophilic transition metal complexes are involved. In contemporary discussions of ligand displacement kinetics the terms dissociative (*D*) and associative (*A*) mechanisms are employed. In a dissociative reaction the activation energy for ligand substitution is determined primarily by the energy required to break the bond to the leaving group, whereas in an associative reaction bond making to an entering group is the main factor determining the activation energy.[6]

The other type of process that is important in organometallic chemistry is ligand exchange via a bimolecular process[7] in which the exchanging ligands bridge both metal atoms, either as a transition state or as an intermediate [Eq. (6.5)]. The process is said to proceed by a four-center mechanism.

$$M\text{-}L + M'\text{-}L' \rightleftharpoons M\langle{}^{L}_{L'}\rangle M' \longrightarrow M\text{-}L' + M'\text{-}L \tag{6.5}$$

Alkyl transfer or transmetallation reactions are believed to proceed by such a mechanism. It should be added that in some cases ligand dissociation may also be involved in Eq. (6.5) where the other ligands are omitted for simplificaiton.

a. Ligand Displacement Reactions in Square Planar Complexes

In contrast to the numerous studies carried out on nonorganometallic complexes, examples of kinetic studies on ligand displacement reactions of organotransition metal complexes are still limited. Relatively well studied among them is ligand displacement in square planar complexes. Many d^8 transition metal complexes [Ni(II), Pd(II), Pt(II), Rh(I), Ir(I), and Au(III)] form four-coordinate square planar complexes, especially with strong-field ligands, and most of the ligand displacement reactions at these complexes proceed by an associative mechanism. This is as expected, because square planar complexes have coordination sites above and below the molecular plane. (We shall discuss the dissociative pathway later.)

The rate law for the reaction

```
  L                      L
  |                      |
T-M-X  +  Y  ------>   T-M-Y  +  X          (6.6)
  |                      |
  L                      L
```

can usually be expressed as two terms:

$$\text{Rate} = (k_s + k_Y[Y])[ML_2TX] \qquad (6.7)$$

where X and Y are leaving and entering ligands, respectively. The ligand displacement reaction is usually *stereospecific*, that is, the entering ligand Y replaces the leaving ligand X at the site trans to the ligand T, the original geometry being maintained. The two-term rate law [Eq. (6.7)] indicates the presence of two parallel reaction paths, one involving the participation of the solvent (k_s term) and the other of the entering ligand Y ($k_Y[Y]$ term). The reaction mechanism consistent with rate equation (6.7) and the stereospecificity is shown in Figure 6.2.

In this mechanism the entering ligand Y attacks the square planar complex from above (or below) the molecular plane to form first a square pyramidal complex, which then rearranges to a trigonal bipyramidal (tbp) complex. Displacement of X by Y in the tbp complex regenerates a square planar complex in which the original stereochemistry is retained. The k_s pathway also involves the formation of a trigonal bipyramidal transition state, except that the solvent is the entering group that is later displaced by Y.

Various factors affect the reactivity of square planar complexes. One of the important factors affecting the rate of ligand displacement is the *trans effect*. The trans effect is defined as the effect of a coordinated ligand [T in Eq. (6.6)] on the *rate of substitution* of ligands [X in Eq. (6.6)] trans to it. The trans effect was particularly well studied in Russia early in the development of coordination chemistry because of the large deposits of platinum ore there.[8]

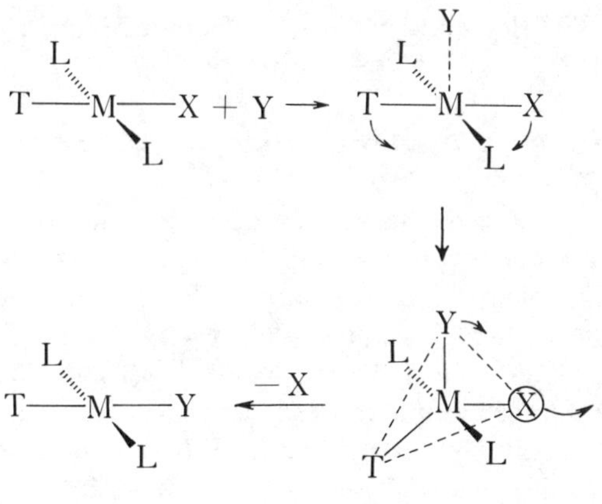

FIGURE 6.2. Ligand substitution reaction mechanism of a square planar complex involving square pyramidal and trigonal bipyramidal intermediates. The initial trans configuration is retained in the product.

By comparing a large number of ligand displacement reactions of Pt(II) complexes, the following order in the trans effect has been established: H_2O, OH, NH_3, py $<$ Cl, Br $<$ SCN, I, NO_2, C_6H_5 $<$ CH_3, $SC(NH_2)_2$ $<$ H, PR_3 $<$ C_2H_4, CN, CO.

The same order holds for other square planar complexes.

The trans effect series has proved very useful in rationalizing the known synthetic procedures for square planar platinum complexes. A classic example is the synthesis of the cis and trans isomers of $PtCl_2(NH_3)_2$. The cis isomer can be synthesized by treating $[PtCl_4]^{2-}$ ion with ammonia.

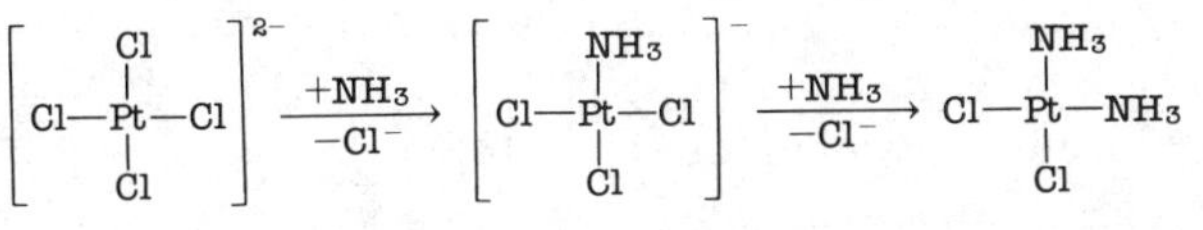

Displacement of the first Cl^- by NH_3 simply generates $[PtCl_3(NH_3)]^-$, but in the displacement of the second Cl^- the trans effect plays an important role. Since the trans-labilizing effect of Cl^- is greater than that of NH_3, displacement of the second Cl^- ligand by NH_3 occurs in the position trans to Cl^- to give *cis*-$PtCl_2(NH_3)_2$.

The trans isomer can be prepared by treating $[Pt(NH_3)_4]^{2+}$ with Cl^-.

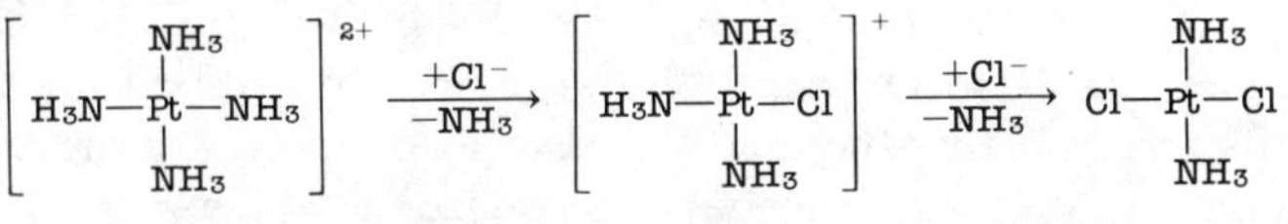

In this case, also because the trans-labilizing effect of Cl^- is greater than that of NH_3, the second Cl^- enters trans to the already present Cl^- ligand, producing *trans*-$PtCl_2(NH_3)_2$.

A recent development of relevance to stereochemistry of square planar complexes is the finding of the carcinostatic effect of *cis*-$PtCl_2(NH_3)_2$. The trans isomer proved ineffective. The specific interaction of the stereochemically stable cis isomer with DNA is believed to be connected with the anticancer action of the platinum complex.[9]

Among organotransition metal complexes, ligand displacement reactions in square planar Pt(II), Pd(II), and Ni(II) complexes have been studied in some detail.

$$\text{trans-M(PEt}_3)_2(\text{o-tolyl})\text{Cl} + \text{py} \longrightarrow \text{trans-[M(PEt}_3)_2(\text{o-tolyl})\text{py]}^+\text{Cl}^- \quad (6.8)$$

$$\text{M} = \text{Ni(II), Pd(II), Pt(II)}$$

The rates of displacement of Cl^- by pyridine in the above reactions were found to decrease in the following order:[10]

	Ni(II) > Pd(II) > Pt(II)
Relative rate:	$5 \times 10^6 : 10^5 : 1$

Comparison of *trans*-$Ni(PEt_3)_2$(*o*-tolyl)Cl and *trans*-$Ni(PEt_3)_2$(mesityl)Cl in reaction (6.8) revealed that the rate of displacement of Cl^- in the more sterically hindered mesityl nickel complex, which has two ortho methyl groups, is 1/6000 of the rate in the *o*-tolylnickel complex. The sharp decrease in the rate reflects steric hindrance to pyridine coordination in the rate-determining step, as shown in Figure 6.2.

Furthermore, in displacement of Cl^- ligand in *trans*-$Pt(PEt_3)_2RCl$ (R = phenyl, *o*-tolyl, mesityl) by other groups that are less nucleophilic than pyridine, such as NO_2^-, Br^-, and I^-, the rate law displays only a first-order term, rate = $k_1[Pt(PEt_3)_2RCl]$ for R = mesityl, k_1 being identical for all the entering groups.[10b] This has been taken as evidence that severe steric hindrance has blocked the associative pathway completely and that a dissociative mechanism is now operative. However, it has been argued that a solvent-assisted associative mechanism explains the experimental results.[12]

Five-coordinate intermediates are assumed to be involved in the cis–trans isomerization of many square planar Pt(II) and Pd(II) complexes, although, in contrast, analogous Ni(II) complexes can isomerize via tetrahedral intermediates. The more stringent constraint of square planar geometry on Pt(II) and Pd(II) complexes excludes a mechanism involving tetrahedral species as an intermediate in thermal isomerization. The cis–trans isomerization of $Pt(PR_3)_2X_2$ can be catalyzed by addition of PR_3 or X^-, and the mechanism has been the subject of considerable debate.[10b-14] For the cis–trans isomerization of $Pt(PR_3)_2X_2$ catalyzed by PR_3, the mechanism shown in Figure 6.3 involving five-coordinate intermediates has been proposed.

In this mechanism addition of PR_3 to *cis*-$Pt(PR_3)_2X_2$ gives a five-coordinate tbp intermediate (A). Displacement of X from the tbp intermediate

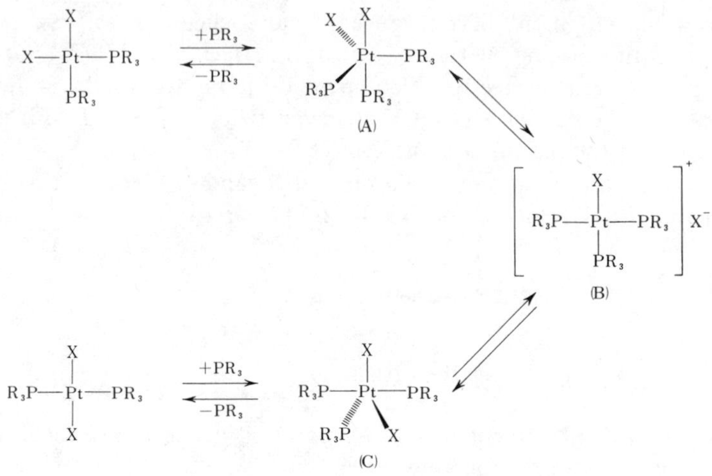

X—Pt(PR_3)(X)—PR_3 ⇌ (+PR_3 / −PR_3) (A) ⇌ $[R_3P—Pt(X)(PR_3)—PR_3]^+X^-$ (B) ⇌ (C) ⇌ (+PR_3 / −PR_3) $R_3P—Pt(X)(X)—PR_3$

FIGURE 6.3. Cis–trans isomerization mechanism of a square planar $Pd(PR_3)_2X_2$ complex promoted by PR_3.

by PR_3 may give an ionic, square planar intermediate (B). Further attack of X^- on the ionic intermediate (B) gives the tbp intermediate (C) from which the trans isomer may be formed by loss of PR_3.

In this isomerization mechanism the character of the putative ionic intermediate (B) may be considerably altered by various factors such as the nature of the ligands and solvents. In polar solvents stepwise displacement of X takes place and the ionic intermediate (B) can sometimes be characterized (Fig. 6.4*a*) as a stable entity.[11] In solvents of intermediate polarity an ion pair may be formed (Fig. 6.4*b*). In nonpolar solvents the isomerization is considered to proceed through a Berry pseudorotation mechanism (cf. Section 5.3), as shown in Figure 6.4*c*.

When two-term kinetics [Eq. (6.7)] is observed, another reaction pathway may operate, namely, coordination of a solvent molecule prior to the coordination of L. We shall not discuss the mechanism of the halide-ion-catalyzed reaction, since this is not directly relevant to organometallic reactions.

Although well-characterized examples are still limited, a dissociative mechanism for isomerization should also be considered for certain d^8 square planar complexes, particularly for more electron-rich, organometallic complexes. For cis–trans isomerization in the absence of added X^- or PR_3 the following dissociative pathway through T-shaped three-coordinate intermediates is certainly a possibility.

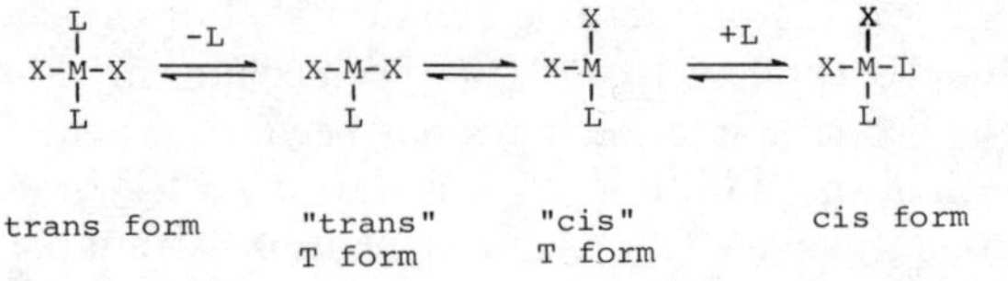

$$\text{X-M(L)(L)-X} \xrightleftharpoons{-L} \text{X-M(L)-X} \rightleftharpoons \text{X-M(X)(L)} \xrightleftharpoons{+L} \text{X-M(X)(L)-L} \tag{6.9}$$

trans form — "trans" T form — "cis" T form — cis form

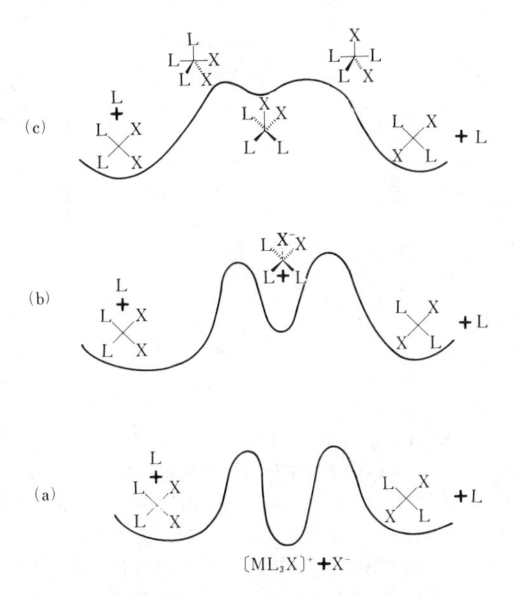

FIGURE 6.4. Cis–trans isomerization of a square planar complex promoted by addition of L. (*a*) In a highly polar solvent a stepwise displacement involving an ionic intermediate takes place. (*b*) In a solvent of medium polarity an ion pair is formed. (*c*) In a nonpolar solvent isomerization proceeds by pseudorotation mechanism. (From Anderson and Cross.[11a])

Supporting kinetic evidence for the intermediacy of a three-coordinate T-shaped species has been obtained for $R_2R'Au(PR_3'')$[15] and *cis*- and *trans*-$PdMe_2(PR_3)_2$ complexes.[16,17] According to MO theoretical calculations,[15,18] in this type of isomerization the energy barrier between the trans and cis T-shaped isomers may vary depending on the metal and the ligands. A windshield wiper movement of an alkyl group (polytopal rearrangement) from the trans T-shaped isomer via a Y-shaped isomer to the cis T-shaped isomer [Eq. (6.10)] has been found to be a low-energy process for $R_2R'Au(PR_3'')$[15] and a high energy process for $PdMe_2(PR_3)_2$.[18]

$$\text{X-}\underset{\underset{\text{L}}{|}}{\text{M}}\text{-X} \rightleftharpoons \text{X}\diagdown\underset{\underset{\text{L}}{|}}{\text{M}}\text{-X} \rightleftharpoons \overset{\overset{\text{X}}{|}}{\underset{\underset{\text{L}}{|}}{\text{M}}}\text{-X} \qquad (6.10)$$

For the trans–cis isomerization of $PdMe_2(PR_3)_2$, an intermolecular reaction pathway proceeding through a methyl-bridged intermediate has been proposed.[17a] The isomerization is effectively blocked by addition of PR_3 but is accelerated by addition of *cis*-$PdMe_2(PR_3)_2$. Employment of deuterium-labeled methyl groups established the occurrence of the intermolecular methyl group exchange. Those results are compatible with the following isomerization mechanism involving a methyl-bridged intermediate.

Me–Pd(L)(L)–Me (trans) $\underset{}{\overset{-L}{\rightleftharpoons}}$ [L–Pd(Me)(Me)] $\xrightarrow{\text{Me}'\text{–Pd(L)(Me}')\text{–L}}$ [L, Me–Pd(μ-Me)(μ-Me′)Pd(Me′)(L)(L)]

$\xrightarrow{\text{Me–Pd(Me}')\text{(L)–L}}$ [L–Pd(Me)–Me′] $\overset{L}{\rightleftharpoons}$ L–Pd(L)(Me)–Me′

A bridged dinuclear platinum complex consisting of square planar and trigonal bipyramidal entities with hydride bridges similar to the above assumed intermediate has been recently characterized.[17b]

Dissociative mechanisms in the reactions of square planar complexes were not recognized until recently, and the assumption of a T-shaped intermediate may seem strange for some readers. That T-shaped species have lower energies than Y-shaped C_{2v} or triangular D_{3h} forms is supported by MO theoretical considerations.[1b,15,18] In fact, the T-shaped structure has been established by X-ray crystallography for the complex $[Rh(PPh_3)_3]^+$.[19]

The cis–trans isomerization of $PdMe_2(PR_3)_2$ complexes has also been found to be catalyzed by added LiMe.[20] Trialkyl- and tetraalkylpalladate species have been detected in this process, and the isomerization is proposed to proceed by stepwise ligand replacement via a trialkylpalladate species, as shown in Figure 6.5.

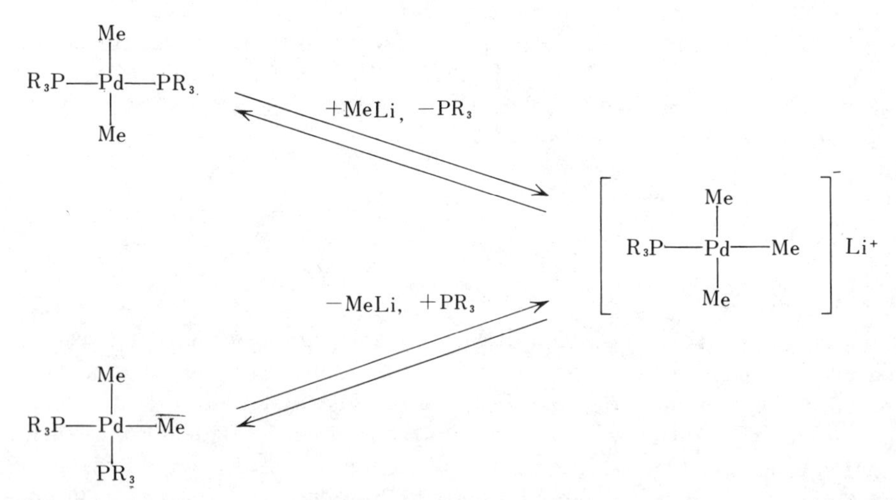

FIGURE 6.5. Trans–cis isomerization mechanism via a palladate intermediate promoted by addition of LiMe.

This mechanism is reminiscent of the cis–trans isomerization mechanism of $Pt(PR_3)_2X_2$ (Fig. 6.3) with the difference that the charge on the intermediate square planar ionic species is negative instead of positive.

For $M(PR_3)_2X_2$ complexes equilibria between the cis and trans isomers can be studied by spectroscopic methods and by measuring the dipole moment of the mixture.[21]

$$\begin{array}{c}\mathrm{X}\\ |\\ \mathrm{R_3P{-}M{-}PR_3}\\ |\\ \mathrm{X}\end{array} \rightleftharpoons \begin{array}{c}\mathrm{X}\\ |\\ \mathrm{R_3P{-}M{-}X}\\ |\\ \mathrm{PR_3}\end{array}$$

Such studies have led to the following conclusions[11a,21]: (1) the trans isomer is favored by a strong entropy effect ($\Delta S = 60$ J mol^{-1} K^{-1}) arising from the release of the solvent molecules accompanying the transformation from the polar cis isomer to the nonpolar trans isomer; (2) the enthalpy term favors the cis isomer ($\Delta H = -20$ kJ mol^{-1}) since the trans influence destabilizes the trans configuration where the PR_3 ligands of high trans influence are situated in mutually trans positions.

Since the trans effect is a kinetic phenomenon that depends on activation energies, both the ground state and the activated state for a reaction must be considered to explain it.[3]

At least in part ligands such as C_2H_4, CO, CN^-, NO, and PR_3, which are all capable of forming π bonds with a transition metal, are good trans directors because these ligands decrease the activation energy in the associative mechanism for ligand displacement reactions of square planar complexes by π-bond formation in the transition state. Figure 6.6 depicts how the activated complex of a trigonal bipyramidal configuration can be stabilized by the presence of a π-acceptor ligand L.[3] This receives electrons from the metal in the activated complex in which both the incoming ligand Y and the leaving ligand X are simultaneously bound to the metal.

This reasoning cannot be applied, however, to the trans effect of ligands unable to form π bonds with the metal. Typical trans-labilizing groups in this category are hydride and alkyl ligands. The reason that these ligands have a strong trans effect lies in the bond-weakening effect in the ground state. This ground-state effect is static in nature, and we have already discussed it in terms of *trans influence* in Section 5.3. Although for ligands such as hydride and methyl the trans influence may be the sole reason for the trans effect, it is difficult to decide for other π-bonding ligands such as olefins which of the two effects plays the dominant role in labilizing the trans ligand.

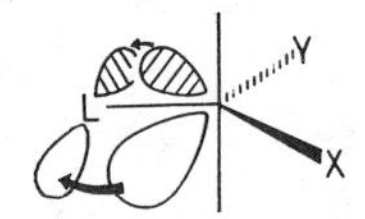

FIGURE 6.6. Trigonal bipyramidal activated complex involved in displacement of ligand X by ligand Y in a *trans*-MA_2LX complex.

Cis ligands also exert a certain labilizing effect, although it is less important than the trans effect.

b. Steric Influence of Ligands

Although the steric effects of ligands on the stability or reactivity of transition metal complexes are usually difficult to quantify, steric effects of tertiary phosphine ligands can be estimated by their cone angle. This parameter was proposed by Tolman on the basis of his investigation of the dissociation equilibria of tertiary phosphine- or tertiary phosphite-coordinated Ni(0) complexes.[23]

$$NiL_4 \underset{}{\overset{K_d}{\rightleftarrows}} NiL_3 + L \tag{6.11}$$

The dissociation constant K_d [Eq. (6.11)] depends on the steric bulk of the tertiary phosphine or phosphite ligand as expressed by the cone angle defined by a conical surface as measured in Figure 6.7. For a tertiary phosphine ligand PR_3 with the same R groups the cone angle θ can be measured by means of a space-filling model (such as the CPK model) as in Figure 6.7. The space-filling model is placed on a wooden block by means of a pin so that the distance between the center of the phosphorus and the center of the metal (Ni) is 2.28 Å (the sum of the covalent bond radii of P and the metal, 2.57 cm in the CPK model). By rotating the model of PR_3 around the M–P bond axis, the angle θ of the vertex of the cone in which the outer atom in the PR_3 ligand is enclosed is measured.

For ligands that have θ exceeding 180°, the cone angle can be estimated by means of trigonometry, as shown in Figure 6.8.

When the substituents on the tertiary phosphine ligand are different, the effective cone angle θ can be estimated by measuring $\theta/2$ for each respective R group, as in Figure 6.9, and by summing it according to the equation

$$\theta = \frac{2}{3} \sum_{i=1}^{3} \frac{\theta_i}{2} \tag{6.12}$$

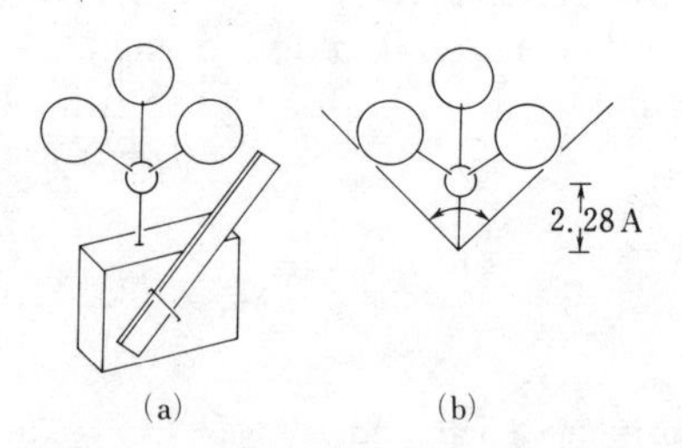

FIGURE 6.7. Measurement of cone angle.

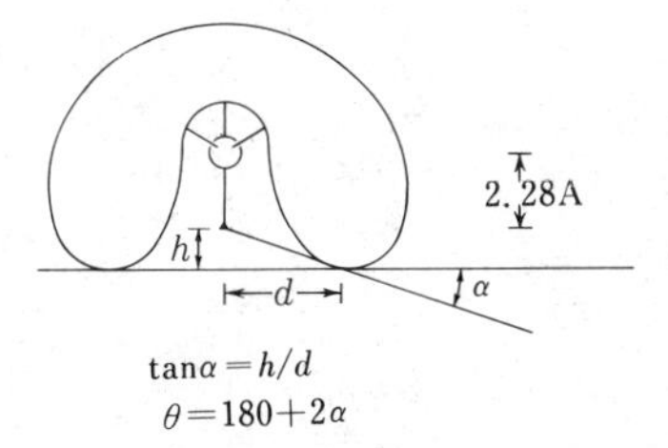

FIGURE 6.8. Measurement of cone angle when $\theta > 180°$.

Table 6.1 tabulates the cone angles and the dissociation constants K_d [Eq. (6.11)] for representative tertiary phosphines and phosphites.

As can be seen from the table, the increase in K_d reflects the steric bulk of the phosphine or phosphite ligands as expressed by the cone angle θ. When θ is difficult to measure by means of the steric model, it can be estimated indirectly by measuring the extent of substitution of the CO ligands in $Ni(CO)_4$ by PR_3. This can be accomplished by adding eight equivalents of PR_3 to $Ni(CO)_4$ in benzene and heating the solution in a sealed tube at 80 °C. The extent of CO substitution can be deduced by observing the infrared ν(CO) bands of the nickel carbonyl complexes in the solution. Since θ and the ease of CO substitution by L are linearly related, one can estimate θ. Table 6.1 includes the θ values derived in this manner.

Tricyclohexylphosphine (PCy_3) is a flexible ligand that has some spaces between the cyclohexyl groups. When more than two flexible ligands of this type are coordinated to a metal, the organic groups may pack by intermeshing with each other. Therefore, the value of θ for a flexible ligand such as PCy_3 derived from the symmetrical space-filling model may be bigger than the real value. In such a case the value derived by the indirect method of IR spectroscopy may be more appropriate.[24]

The steric bulkiness of ligands as represented by the cone angle influences both the stability and the reactivity of complexes. Figure 6.10 illustrates these effects on the decarbonylation rate of an acetylmolybdenum complex. The first-order rate constant k_1 for decarbonylation increases with an increase in θ,

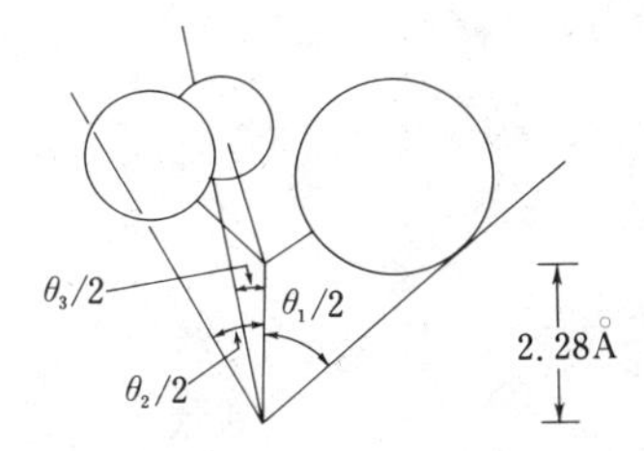

FIGURE 6.9. Measurement of cone angle of a tertiary phosphine ligand where all substituents are different.

Table 6.1. Cone Angle θ and Dissociation Constant K_d for Tertiary Phosphine and Phosphite Complexes of Nickel,

$$NiL_4 \underset{}{\overset{K_d}{\rightleftarrows}} NiL_3 + L \text{ (in benzene, at 25 °C)}$$

L	K_d, M	θ
$P(OEt)_3$	$<10^{-10}$ (70°)	109
PMe_3	$<10^{-9}$ (70°)	118
$P(O\text{-}p\text{-}C_6H_4Cl)_3$	2×10^{-10}	128
$P(O\text{-}p\text{-}C_6H_4CH_3)_3$	6×10^{-10}	128
$P(O\text{-}i\text{-}Pr)_3$	2.7×10^{-5}	130
PEt_3	1.2×10^{-5}	132
$P(O\text{-}o\text{-}C_6H_4CH_3)$	4.0×10^{-2}	141
$PMePh_2$	5.0×10^{-2}	136
PPh_3	Large	145
$P(i\text{-}Pr)_3$	Large	160
PBz_3	Large	165
PCy_3	Large	170
$P(t\text{-}Bu)_3$	Large	182
$P(o\text{-}C_6H_4CH_3)_3$	Large	194
$P(mesityl)_3$	Large	212

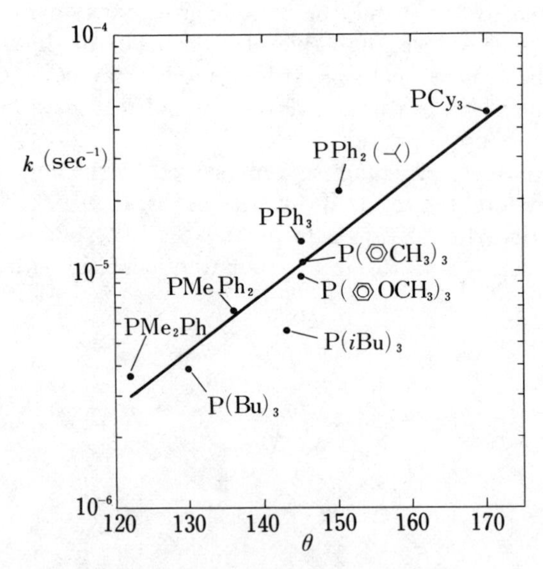

FIGURE 6.10. Ligand steric effect on decarbonylation:

$$CpMo(CO)_2L(COCH_3) \xrightarrow{k} CpMo(CO)_2L(CH_3) + CO$$

reflecting the ease of dissociation of L from the intermediate, coordinatively saturated acetyl complex, $CpMo(CO)_2L(COCH_3)$. The dissociated compound has a vacant site, which allows decarbonylation to proceed.[25]

A steric effect of the organic moiety bound to the $Co(CO)_4$ group on replacement of a CO ligand by triphenylphosphine has been reported[26]:

$$RCOCo(CO)_4 + PPh_3 \longrightarrow RCOCo(CO)_3(PPh_3) + CO$$

The reaction is first order and independent of the PPh_3 concentration, suggesting the operation of the following dissociative pathway.

$$RCOCo(CO)_4 \underset{k_{-1}}{\overset{k_1}{\rightleftharpoons}} RCOCo(CO)_3 + CO$$

$$RCOCo(CO)_3 + PPh_3 \xrightarrow{k_2} RCOCo(CO)_3(PPh_3)$$

Application of the steady-state approximation to the concentration of the $RCOCo(CO)_3$ intermediate leads to the rate expression:

$$\text{rate} = \frac{k_1 k_2[RCOCo(CO)_4][PPh_3]}{k_{-1}[CO] + k_2[PPh_3]}$$

which leads to a first-order rate equation when $k_{-1}[CO] << k_2[PPh_3]$.

The substituent attached to the acyl group has an electronic as well as a steric effect on the ligand displacement reaction. Particularly noteworthy was the increase in the rate constants in the order of $CH_3CO < (CH_3)_2CHCO << (CH_3)_3CCO$, the ratio being 1 : 2.1 : 86. The effect on the ligand displacement rate constant has been accounted for by a steric effect that will become more pronounced with each successive methyl group substitution.

c. Electronic Influence of Ligands

Various trivalent compounds of phosphorus, arsenic, antimony, and bismuth serve as ligands to transition metals. These ligands will donate electrons to the metal complexes, at the same time accepting electrons from metals by π back bonding through d orbitals. Among organic nitrogen ligands, aliphatic amines serve only as σ-donating ligands, but conjugated aromatic ligands such as 2,2′-bipyridine serve as π-bonding ligands.

The extent of donation from the lone pairs in these ligands to the metal and back donation from the metal to the ligands depends on the nature of the substituents on these group 15 elements. For $P(alkyl)_3$, π-acceptor ability is very low, but it becomes important when more electronegative groups are attached to the phosphorus atom. The net flow of electrons is estimated by observing the stretching frequency of other ligands, such as the CO attached to the transition metal. Attempts to estimate the electronic effects of tertiary phosphines and phosphites have been made by means of IR data of complexes such as $Ni(CO)_3L$.[27,28]

To investigate the electronic effect of substituents in tertiary phosphines, Tolman measured the $\nu(CO)$ bands of $Ni(CO)_3L$ complexes for 70 different ligands (L) in CH_2Cl_2 solutions. The $Ni(CO)_3L$ complex of the most basic ligand $P(t\text{-}Bu)_3$ had its $\nu(CO)$ band (A_1 symmetry) at 2056.1 cm^{-1} and it was found that a parameter χ reflecting the electronic effect could be expressed by the following equation.

$$\nu(CO)_{A_1} = 2056.1 + \sum_{i=1}^{3} \chi_i \ cm^{-1} \tag{6.13}$$

The values of the parameter χ for substituents in representative tertiary phosphines and phosphites are: *t*-Bu, 0.0; *n*-Bu, 1.4; Et, 1.8; Me, 2.6; Ph, 4.3; H, 8.3; OPh, 9.7; Cl, 14.8; F, 18.2; and CF_3, 19.6.

The frequency of the A_1 $\nu(CO)$ band in $Ni(CO)_3L$ [$L = P(OR)_3$] is greater by 5.2 cm^{-1} than that for $L = PR_3$ regardless of the R group, reflecting the smaller basicity of $P(OR)_3$ than of PR_3. On stepwise substitution of the phenyl groups in PPh_3 with other atoms or groups, the $\nu(CO)$ in $Ni(CO)_3(PPh_{3-n}X_n)$ is observed to change linearly, as shown in Figure 6.11.

Thus, it can be seen that the substitution parameter χ reflects the electronic influence of tertiary phosphines and phosphites with reasonable sensitivity.

For reactions of tertiary phosphine or phosphite complexes proceeding through dissociative pathways, the factors that control ligand dissociation are also important in determining the reactivity of these complexes. Clearly, further fundamental kinetic and thermodynamic studies of ligand dissociation

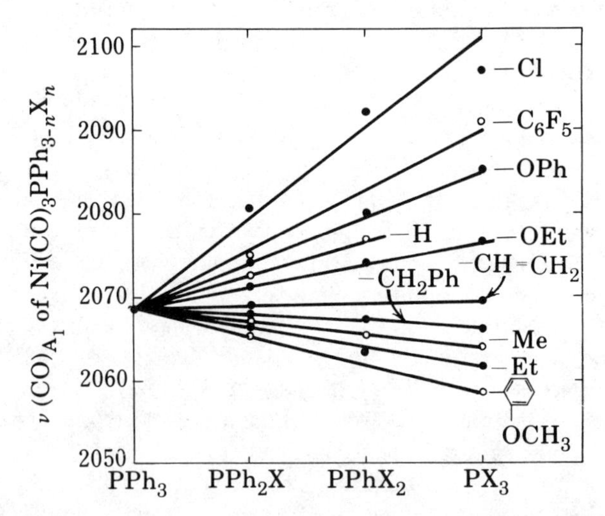

FIGURE 6.11. Electronic effect of substitution of phenyl groups in $Ni(CO)_2(PPh_{3-n}X_n)$ by substituent X on $\nu(CO)_{A_1}$. (Reprinted with permission from C. A. Tolman, *J. Am. Chem. Soc.* **92,** 2953. Copyright 1970 American Chemical Society.)

are required. A summary of some of the limited studies of these factors follows.

Coordinated benzonitrile in $PdCl_2(PhCN)_2$ is susceptible to substitution by various ligands; thus, this is a convenient complex with which to study their coordinating abilities. The equilibria of the reactions in Eq. (6.14) have been studied and the enthalpy changes ($-\Delta H$) summarized in Table 6.2 have been obtained.[29]

$$PdCl_2(PhCN)_2 + 2L \rightleftarrows PdCl_2L_2 + 2PhCN \tag{6.14}$$

The results indicate that the coordinating abilities of the ligands L decrease in the order $PPh_3 > P(OPh)_3 >$ py $>$ cod. It is also seen that the bidentate ditertiary phosphines such as dppe show greater coordinating ability than the monotertiary phosphine PPh_3.

Depending on the tertiary phosphine ligand, $M(PR_3)_4$ complexes (M = Ni, Pd, Pt) dissociate in two steps.

$$M(PR_3)_4 \overset{K_1}{\rightleftarrows} M(PR_3)_3 + PR_3 \tag{6.15}$$

$$M(PR_3)_3 \overset{K_2}{\rightleftarrows} M(PR_3)_2 + PR_3 \tag{6.16}$$

Usually $K_1 > K_2$ and the $M(PR_3)_2$ complex is present in solution only to a very small extent. However, because of its highly unsaturated nature, the two-coordinate species can be quite reactive, and its effect cannot always be neglected in reactions, even though its concentration is very low. In these dissociations both steric and electronic factors are involved. When a very bulky tertiary phosphine ligand is employed, K_2 can be very large. For example, the dissociation of $Pt(PEt_3)_4$ in THF gives mainly $Pt(PEt_3)_3$ ($K_1 = 0.5\ M$),

Table 6.2. Enthalpy Change in the reaction[a]

$PdCl_2(PhCN)_2 + 2L \rightleftarrows PdCl_2L_2 + 2PhCN$

L	$-\Delta H$ (kJ mol^{-1})
py	109 ± 3
tmeda[b]	114 ± 7
dpae[c]	147 ± 3
dppe[d]	211 ± 7
PPh_3	163
cod[e]	54

[a] From ref. 29.
[b] *N,N,N',N'*-Tetramethylethylenediamine.
[c] 1,2-Bis(diphenylarsino)ethane.
[d] 1,2-Bis(diphenylphosphino)ethane.
[e] 1,5-Cyclooctadiene.

whereas $Pt[P(i\text{–}Pr)_3]_3$ in the same solvent exists mainly as $Pt[P(i\text{–}Pr)_3]_2$ ($K_2 = 0.14\ M$).[30] Because of the high reactivity of the coordinatively highly unsaturated PtL_2 complex, water can oxidatively add to it, and the solution becomes so alkaline that it can be used to catalyze the hydration of nitriles.[30]

$$PtL_2 + H_2O \rightleftarrows PtH(OH)L_2 \overset{S}{\rightleftarrows} PtH(S)L_2 + OH^-$$

S = solvent

In certain types of ligand replacement reactions involving coordinatively saturated cyclopentadienyl complexes, some evidence suggesting a molecular rearrangement of the cyclopentadienyl ligand from η^5 to η^3 coordination has been provided. By slippage of the Cp ligand, a coordination site is created for the incoming ligand. The displacement rate is decreased on increase of the cone angle of the tertiary phosphine used for the reaction.[31a]

$$CpRh(CO)_2 \xrightarrow{+L} (\eta^3\text{-}C_5H_5)Rh(CO)_2L \xrightarrow{-CO} CpRh(CO)L$$

L = PR_3

Further rearrangement from the η^3 to η^1 bonding has been observed in reactions of other coordinatively saturated cyclopentadienyl transition metal complexes.[31b,c]

d. Coordination of Olefins to Transition Metal Complexes

Olefins form π complexes with transition metals in both high and low oxidation states. On complexation, the characters, and thus the reactivities, of the olefins are modified, so that olefins undergo many reactions that are mediated by transition metal complexes. Studies on transition metal–olefin bonds are therefore of particular significance for understanding these reactions.

The olefin–metal π bond is constituted of two synergic interactions involving olefin-to-metal dative bonding and metal-to-olefin back bonding, as discussed in Section 3.3. In terms of MO theory the strength of the metal–olefin bond is determined by the relative energies between the metal d orbitals and the π and π^* orbitals of the olefin, as shown in Figure 6.12. Depending on the relative energies of these orbitals, there are three conceivable situations: (a) the olefin-to-metal dative bond is of prime importance; (b) back donation from the filled metal d orbitals to the olefin π^* orbital is dominant; and (c) both dative and back bonding interactions contribute to bond formation.

Case (a) The Olefin-to-Metal Dative Bond Is Dominant. In olefin–metal complexes formed between metals of relatively high oxidation states and olefins having electron-donating substituents, olefin-to-metal electron donation plays a dominant role in determining the olefin–metal bond strength. The higher the energy of the olefin π orbital (HOMO) and the lower the energy of the unoccupied metal d orbital (LUMO), the stronger the HOMO–LUMO interaction, resulting in greater stabilization of the orbital of the newly formed metal–olefin σ bond (Fig. 6.12*a*). When the d orbitals are occupied, the upper s and p orbitals may be utilized, as in Ag(I)–olefin complexes.

The following equilibria for olefin–Ag^+ complexes have been studied.[32]

$$Ag^+ + \text{olefin} \underset{}{\overset{K_1}{\rightleftarrows}} Ag(\text{olefin})^+$$

$$K_1 = [Ag(\text{olefin})^+]/[Ag^+][\text{olefin}]$$

In complexes of Ag(I) with olefins, $CH_2{=}CHX$, having various substituents X, the stability constant K_1 decreases with an increase in the Hammett constant σ_m of X, as shown in Figure 6.13.

The value ρ derived from the slope in Figure 6.13 is -5.07. The correlation indicates that the more electron donating is the substituent, the more electron rich the olefin becomes, so that the olefin-to-metal dative bond is strengthened. A similar correlation has been found for Ag(I) complexes with styrenes that have various substituents in the benzene ring.[33] In this case a linear rela-

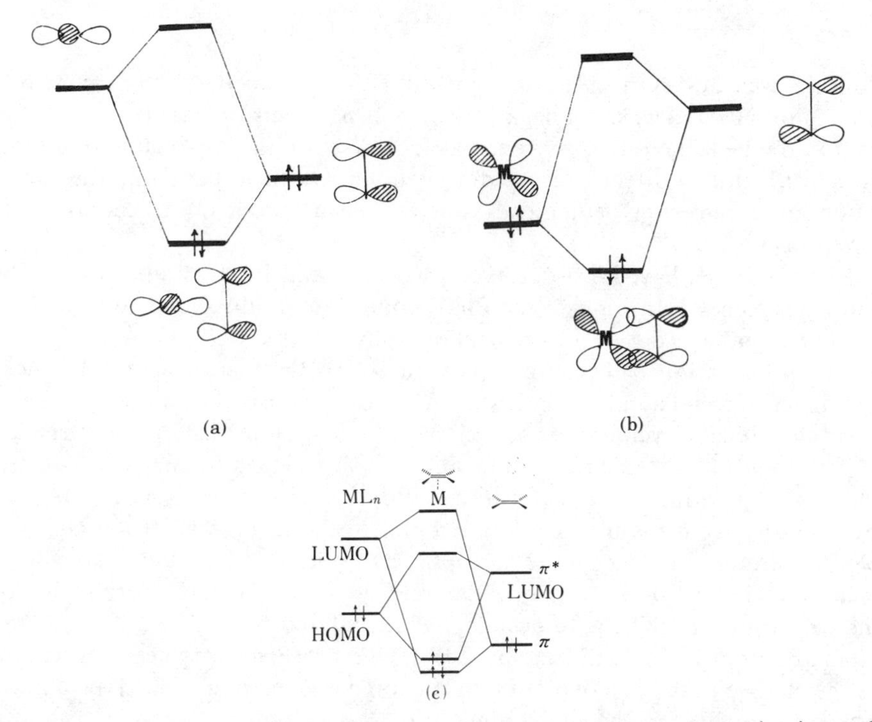

FIGURE 6.12. Energy level relationship in donation and back donation interaction in metal–olefin-to-complexes; (*a*) Olefin-to-metal dative bond; (*b*) metal–olefin back bonding; (*c*) donation and back donation.

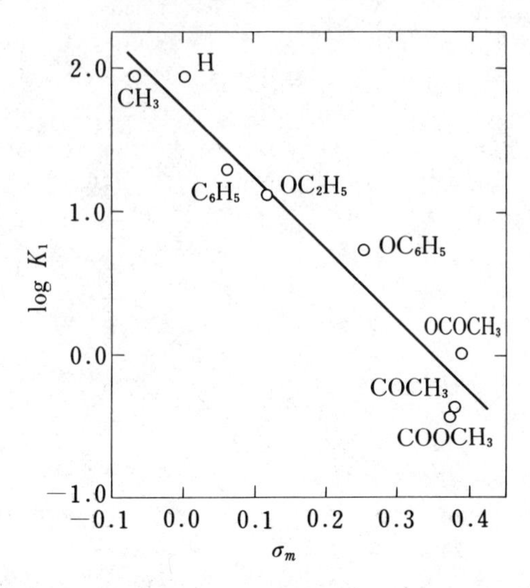

FIGURE 6.13. Relationship between σ_m value and stability constant K in complexes of Ag^+ and $CH_2{=}CHX$. (Reprinted with permission from T. Fueno et al., *J. Am. Chem. Soc.* **87,** 170. Copyright 1965 American Chemical Society.)

tionship was observed between log K and the Hammett constant σ with a slope ρ of -0.77. The small negative value of ρ reflects the fact that a substituent in the benzene ring of styrene has less effect on the Ag–olefin bond than does a substituent directly on the double bond. On the other hand, the correlation for styrene–Ag(I) complexes is better because steric influences are minimized in this case.

Similar trends have been observed for Pd(II) and Pt(II) complexes of substituted styrenes.[34] In some of the Pt(II) complexes double-bond coordination has been established by X-ray crystallography.[35]

In complexation of cyclic olefins with Ag(I), the thermochemically measured olefin–metal bond strength was found to increase in the following order[36]: *cis*-cyclooctene < cyclohexene < cycloheptene < cyclopentene < norbornene.

The trend indicates that complexation to the metal is associated with the relief of ring strain. In an olefin-to-metal dative bond there is a drift of electron density from the olefinic double bond to the metal, and the complexed olefin therefore becomes more susceptible to nucleophilic attack. As will be mentioned in Section 7.4, the key step in the Hoechst–Wacker process, that is, the oxidation of ethylene to acetaldehyde catalyzed by the Pd(II) ion, is the attack on coordinated ethylene by OH^-. This process is a typical illustration of the change in the reactivity of ethylene on complexation to Pd(II) from the usual nucleophilic to the electrophilic.

The olefin–metal dative bond may be regarded as a bond formed between a Lewis base (olefin) and a Lewis acid (metal). Thus, this bond is weakened by the presence of another, more electron-donating ligand attached to the metal complex, which makes the complex less acidic.

Case (b) Metal-to-Olefin Back Bonding Is Dominant. Back bonding between the metal and the olefin becomes stronger as the energy levels of the filled metal d orbitals (HOMO) and the π^* orbital of the olefin (LUMO) become closer, thus forming a more stable metal–olefin π-bonding orbital, as shown in Figure 6.12*b*. The olefin π^* orbital is usually higher in energy than the metal d orbital. Therefore, as the energy of the filled metal d orbital is raised, the olefin–metal π bond is strengthened. Olefins that have electron-withdrawing substituents have the lower π^* orbitals and are therefore capable of stronger back bonding with the metal.

Stability constants have been measured for some olefin complexes of group 8–10 transition metals, and these generally indicate the importance of the back bonding. The stability constants K_2 of rhodium–olefin complexes in the following equilibria are summarized in Table 6.3.

$$Rh(acac)(C_2H_4)_2 + olefin \underset{}{\overset{K_2}{\rightleftarrows}} Rh(acac)(C_2H_4)(olefin) + C_2H_4 \quad (6.17)$$

$$K_2 = [Rh(acac)(C_2H_4)(olefin)][C_2H_4]/[Rh(acac)(C_2H_4)_2][olefin]$$

As Table 6.3 shows, the π complexes formed between the more electron-withdrawing olefins and Rh(I) have higher K_2 values, reflecting the dominance of back bonding in determining the character of olefin–Rh(I) complexes.

Similar trends have been observed in olefin–Ni(0) and olefin–Pd(0) complexes. Stability constants of the following Ni(0) complexes have been measured and thermodynamic parameters estimated, as summarized in Table 6.4.

Table 6.3. Equilibrium Constants for the Reaction[a]

$$Rh(acac)(C_2H_4)_2 + olefin \overset{K_2}{\rightleftarrows} Rh(acac)(C_2H_4)(olefin) + C_2H_4$$

Olefin	K_2
i-C_4H_8	3.5×10^{-4}
Trans-C_4H_8	2.0×10^{-3}
Cis-C_4H_8	4.1×10^{-3}
$CH_2{=}CHOCH_3$	1.8×10^{-2}
$CH_2{=}CHCH_3$	7.8×10^{-2}
$CH_2{=}CHCl$	1.7×10^{-1}
$CH_2{=}CHF$	3.2×10^{-1}
Trans-$CHF{=}CHF$	1.2
Cis-$CHF{=}CHF$	1.6
$CH_2{=}CHCN$	>50
$CF_2{=}CF_2$	59
$CHF{=}CF_2$	88

[a]From ref. 37.

Table 6.4. Equilibrium Constants and Thermodynamic Parameters for the Reaction[a,b]

$$\mathrm{Ni(bipy)(solvent) + olefin \overset{K}{\rightleftharpoons} Ni(bipy)(olefin) + solvent}$$

Olefin	log K	$\Delta G°$ (kJ mol^{-1})	$\Delta H°$ (kJ mol^{-1})	$\Delta S°$ (J mol^{-1} deg^{-1})
OC—CH=CH—CO (bridged by —O—)	4.52	−25.9	−53.6	−92
CH_2=$CHCONH_2$	3.31	−19.0	−74.5	−185
CH_2=CHCN	2.82	−16.2	−66.9	−169
CH_2=$C(CH_3)CN$	2.37	−14.6	−59.0	−152
CH_2=CHCHO	3.62	−20.6	−78.5	−197
CH_2=$CHCOCH_3$	2.58	−14.8	−56.7	−142
CH_2=$CHCOOCH_3$	2.49	−14.3	−56.7	−144
CH_2=$C(CH_3)CO_2CH_3$	1.07	−6.1	−36.4	−101
CH_2=$CHOCOCH_3$	−0.68	3.8	—	—

[a]From ref. 38.

$$\mathrm{Ni(bipy)(solvent) + olefin \overset{K}{\rightleftharpoons} Ni(bipy)(olefin) + solvent} \tag{6.18}$$

A linear relationship has been observed between the measured log K values and Alfrey-Price's *e* values, which reflect the electron-withdrawing abilities of the substituents attached to the double bond. It was also observed that the lower the π^* orbital of the olefin, the greater the value of log K.[38]

Similar relationships have also been observed for tertiary phosphite Ni(0)–olefin complexes in the following equilibria.[39]

$$\mathrm{Ni(PR_3)_2 + olefin \overset{K}{\rightleftharpoons} Ni(PR_3)_2(olefin)} \tag{6.19}$$

$$\mathrm{R = O\text{-}o\text{-}tolyl\ group}$$

Figure 6.14 shows the linear relationship between log K and $\sum\sigma_p^+$, the sum of the Hammett σ_p^+ of the substituents attached to the olefin. The ρ value derived from the slope is +5.5. The relationship is opposite to that observed in Figure 6.13.

Figure 6.15 shows the relationship between log K and the energy levels of the olefin π^* orbitals. These results underscore the relationship shown in Figure 6.12 that the lower the LUMO of the olefin, the greater the interaction with the filled metal d orbitals and the higher the stability of the π complex.

The strengths of the Ni(0)–olefin π bonds have also been measured by a thermochemical method.[40] Figure 6.16 indicates the linear relationship between the $-\Delta H$ value for the following reaction and the Hammett constant of the substituent.

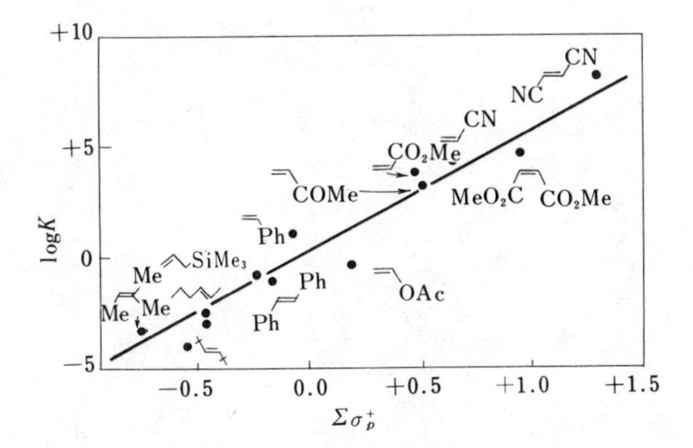

FIGURE 6.14. Relationship between the Hammett constant σ_p^+ of the substituent on olefin and the equilibrium constant K in Eq. (6.19) at 25 °C. (Reprinted with permission from C. A. Tolman, *J. Am. Chem. Soc.* **96,** 2780. Copyright 1974 American Chemical Society.)

$$\mathrm{Ni(PPh_3)_2(olefin)} + (\mathrm{NC})_2\mathrm{C{=}C}(\mathrm{CF_3})_2 \underset{}{\overset{-\Delta H}{\rightleftarrows}} \mathrm{Ni(PPh_3)_2((NC)_2C{=}C(CF_3)_2)} + \mathrm{olefin} \tag{6.20}$$

In π-complex formation an electron-withdrawing substituent contributes to the strengthening of the bond by enhancing back donation, whereas an electron-donating substituent has the opposite effect. Additivity of substituent contributions was observed and the following contributions to the stabilities of olefin complexes were noted (kJ mol^{-1}): Me(−13) < Ph(−3) <F(−2) < H(0) < CN(20) < CO_2R(22) < COMe(25) < CF_3(25) < NO_2(27).

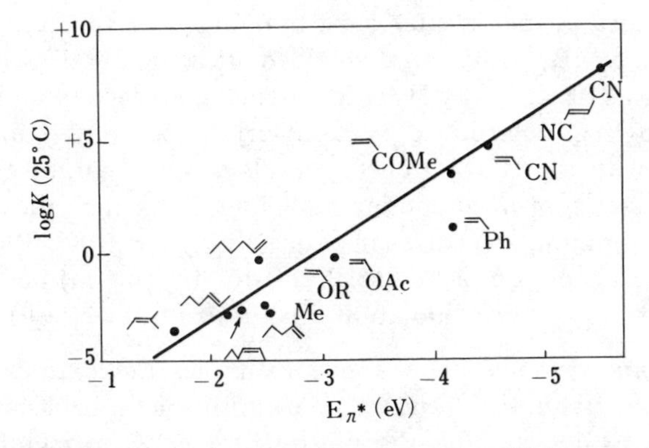

FIGURE 6.15. Relationship between LUMO of olefin and equilibrium constant in Eq. (6.19). (Reprinted with permission from C. A. Tolman, *J. Am. Chem. Soc.* **96,** 2780. Copyright 1974 American Chemical Society.)

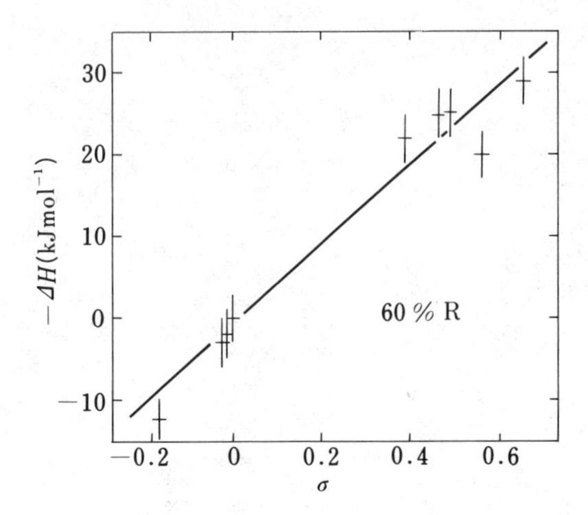

FIGURE 6.16. Relationship between σ value and $-\Delta H$ in the reaction

$$\mathrm{Ni(PPh_3)_2(olefin) + (NC)_2C{=}C(CF_3)_2 \rightleftharpoons Ni(PPh_3)_2((NC)_2C{=}C(CF_3)_2) + olefin}$$

(Reprinted with permission from S. D. Ittel, *Inorg. Chem.*, **16**, 2589. Copyright 1977 American Chemical Society.)

In related isonitrile-coordinated olefin complexes, Ni(t-BuNC)$_2$(olefin), a linear relationship is observed between the $-\Delta H$ value of Eq. (6.20) and ν_{NC}, one of the NC stretching bands of the coordinated isonitrile ligands. The ν_{NC} band shifts to a higher frequency as the back bonding is increased.[40,41]

The back bonding is enhanced as the energy of the metal HOMO increases (cf. Figure 6.12*b*). The proton chemical shift, $\nu_{C=C}$, and the C=C bond length of ethylene in some of its transition metal complexes have been used to assess the extent of back bonding (Table 6.5).[42]

The table indicates that greater back bonding causes the ethylene protons, as observed by NMR, to be more shielded. It also causes a decrease in the $\nu_{C=C}$ frequency as observed by IR spectroscopy and leads to an increase in the C=C bond length as determined by X-ray crystallography. Tolman correlates back bonding with the HOMO energy levels of metal atoms estimated from their ground-state ionization potentials[43] and finds that the propensity to engage in back bonding decreases in order Ni(0) $\gtrsim$ Fe(0) $\gtrsim$ Pt(0) $>$ Rh(I) $>$ Pt(II). For a series of complexes M{P(O-o-tolyl)$_3$}$_2$(olefin) he also infers the following order in back-bonding ability: Ni(0) $\gtrsim$ Pt(0) $>$ Pd(0).

(Case c) Intermediate. The above examples are extreme cases where either olefin-to-metal dative bonding or metal-to-olefin back bonding plays a dominant role in determining the nature of the olefin–metal bond. Intermediate cases are conceivable for π complexes where both dative and back-bonding contributions are considerable. Figure 6.12*c* illustrates the energetic

Table 6.5. NMR, IR, and Structural Data of Metal-Olefin Complexes and Ethylene Oxide

	Ground State IP of Atomic M^{n+} (eV)[a]	δ^b	$\nu_{C=C}$	$d_{C=C}$ (pm)
Ethylene oxide (CH_2—CH_2 bridged by O)	13.6	2.8	1487	147.2
$(C_2H_4)Ni\{P(O\text{-}p\text{-tolyl})_3\}_2$	7.6	1.9	1487	146 (2)
$(C_2H_4)Fe(CO)_4$	7.9	2.5	1510	146 (6)
$(C_2H_4)Pt(PPh_3)_2$	9.0	2.6	—	143
$(C_2H_4)Rh(acac)$	18.1	3.1	1524	141 (3)
$(C_2H_4)PtCl_3^-$	18.6	4.7	1526	135 (1)
$(C_2H_4)Ag^+$	21.5	5.7	1583	—
C_2H_4	—	5.3	1623	133.7 (d)

[a] Ionization potentials from C. E. Moore, "Atomic Energy Levels," National Bureau of Standards, Circular, No. 467, Vol. I, 1949.
[b] Chemical shifts referred to tetramethylsilane ($\delta = 0$).

correlation where both dative and back-bonding interactions are operative. If the olefin π orbital (together with the π^* orbital) is raised by introduction of an electron-donating substituent or the metal LUMO is lowered (together with the metal HOMO), the olefin-to-metal dative interaction plays the dominant role in determining the metal–olefin bond strength, as seen in Case (a). On the other hand, if the metal orbitals are raised by a drop in oxidation state or by coordination of electron-donating ligands relative to the olefin LUMO, which itself may be lowered by introducing electron-withdrawing substituent, then Case (b) is attained.

In a determination of the equilibrium constants for the reaction[44]

$$PtCl_3(n\text{-}C_{10}H_{21}CH{=}CH_2)^- + CH_2{=}CHC_6H_4X$$

$$\overset{K}{\rightleftarrows} PtCl_3(CH_2{=}CHC_6H_4X)^- + n\text{-}C_{10}H_{21}CH{=}CH_2 \qquad (6.21)$$

it was observed that the K value decreased with an increase in the Hammett σ constant for substituent X to a certain σ value; then the stability constant K again increased with σ, that is, the plot of K versus σ was U-shaped.

This result has been interpreted as showing the competing effects of the substituents. Strongly electron-donating substituents X may increase the electron density at the double bond of the styrene, causing the dative bond to predominate, whereas electron-withdrawing substituents may cause the back-bonding contribution to be dominant.

Although the existence of such intermediate cases is plausible, the magnitude of the stability constants spans only a limited range, so the conclusion should be regarded with caution. In fact, in a somewhat related study on the following equilibria, log K was observed to decrease linearly as the Hammett σ value of the para-substituent Y increased, the value of ρ being -1.32, indicating that the greater the donor ability of the olefin, the stronger the olefin–Pt(II) bond.[34b]

$$\left[\text{Me–C}\begin{smallmatrix}CH_2\\ CH_2\end{smallmatrix} \text{—Pt}\begin{smallmatrix}PPh_3\\ NCPh\end{smallmatrix} \right]^+ + CH_2{=}CH\text{–}C_6H_4\text{–}Y \overset{K}{\rightleftharpoons} \left[\text{Me–C}\begin{smallmatrix}CH_2\\ CH_2\end{smallmatrix} \text{—Pt}\begin{smallmatrix}PPh_3\\ (CH_2{=}CH\text{–}C_6H_4\text{–}Y)\end{smallmatrix} \right]^+ + PhCN$$

We must await further studies to confirm the existence of the intermediate case.

When the substituents attached to the double bond can also coordinate to a transition metal, such as the COOR, CHO, and CN groups, three binding modes can be envisaged: (1) coordination solely through the olefinic double bond, (2) coordination solely through the substituent group, and (3) coordination through both the olefinic double bond and the substituent. Depending on the nature of the substituted olefin and the transition metal entity, these three modes have been observed, for example, in acrylonitrile complexes.[45]

For low-valent metal–olefin complexes the back bonding interaction between the olefinic double bond and the metal d orbitals is very strong, and many examples of olefin complexes bound solely through the double bond are known. Figure 6.17 shows one such example.[46]

The nickel complex is planar, with the nickel atom being at the center of the triangle formed by two phosphorus atoms and the center of the C–C double bond of ethyl methacrylate. The ethoxycarbonyl group does not interact with nickel. The C–C double-bond length of 141.2 pm shows considerable lengthening of the double bond on complexation.

On coordination to a transition metal, hydrogen atoms or other substituents attached to the olefinic double bond bend away from the metal. In addition, the C–C double-bond length is increased to approach that of a C–C single bond. Thus, the configuration around the carbon approaches that

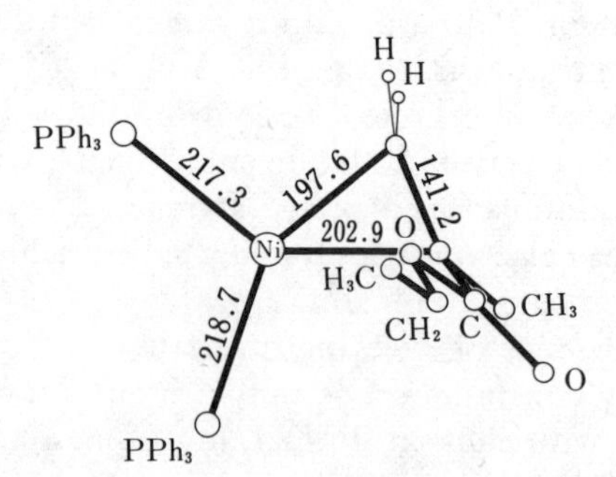

FIGURE 6.17. Molecular struture of ethyl methacrylate–nickel complex, $Ni(PPh_3)_2(CH_2{=}CMeCOOEt)$, unit in picometers. (From Komiya et al.[46])

expected for sp^3 hybridization. These changes in the nature of the bonding can be well accounted for in terms of MO theory. Various approaches utilizing semiempirical and ab initio MO methods[47-50,52] have attempted to explain both the manner of olefin coordination, that is, whether it adopts an upright or an in-plane mode, and the bond distances and bond angles in the π complexes. As shown in Figure 6.18, ethylene is coordinated to Pt(II) in Zeise's complex in an upright mode,[51a] whereas ethylene[51b] and tetracyanoethylene (TCNE)[51c] bind in-plane to Ni(0).

In these theoretical approaches that in-plane coordination is favored for Ni(0)–olefin complexes where the back bonding plays a dominant role has been reasonably explained. On the other hand, the upright binding mode in Zeise's complex seems to arise mainly from steric reasons. Figure 6.18*a* includes a comparison of the experimental values of bond lengths and bond angles (in parentheses) with the results calculated by ab initio method.[49a] The agreement is reasonable. The bending back of the hydrogen atoms of ethylene on complexation is explicable in terms of MO theory mainly as a means of lowering the energy of the π^* orbital, thus facilitating the back bonding.[50]

There are limited kinetic studies on the complexation of olefins with transition metal complexes. The rate of the following reaction is second order—first order in the concentration of Ir(I) complex and in that of the olefin.[53]

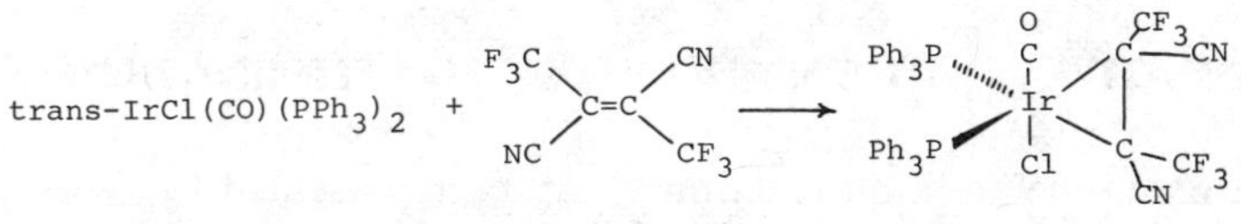

trans-$IrCl(CO)(PPh_3)_2$ + $(F_3C)(NC)C{=}C(CN)(CF_3)$ ⟶ $(Ph_3P)_2Ir(CO)(Cl)[C(CF_3)(CN)–C(CF_3)(CN)]$

In the reaction of the square planar complex $NiR_2(bipy)$ with olefins, a second-order rate law was also observed, and the result was interpreted in terms of initial coordination of the olefin and subsequent reductive elimination of the R–R to give olefin–Ni(0) complexes [cf Eq. (5.2)].

In the reactions of olefins with the coordinatively saturated octahedral complexes, *cis*-$FeR_2(bipy)_2$, a dissociative mechanism involving partial dissociation of the bipy ligands has been proposed to interpret the kinetics.[54]

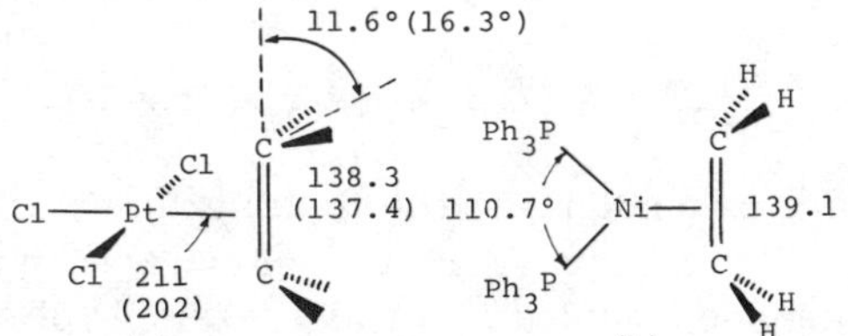

FIGURE 6.18. Comparison of results of MO calculations with experimental values in Pt(II)– and Ni(0)–ethylene complexes. Experimental values in parenthesis.

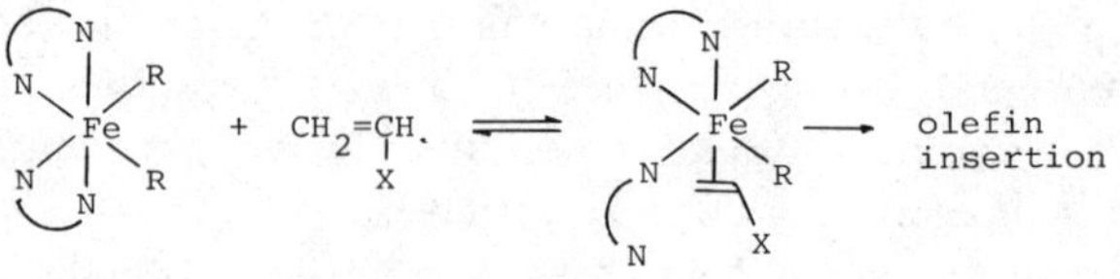

$$Fe(N_4)R_2 + CH_2{=}CHX \rightleftharpoons Fe(N_3)(\eta^2\text{-}CH_2{=}CHX)R_2 \longrightarrow \text{olefin insertion} \tag{6.22}$$

π-Complex formation with a transition metal is also possible for compounds having C–O double bonds and other unsaturated groups. Figure 6.19 shows an example[55a] of a formaldehyde molecule π-bonded to rhenium. The C–O double-bond length (137.4 pm) in this complex is larger than that in free formaldehyde, 121 pm. In some cases the C–O double bond is even further elongated to 158.4 pm, approaching that of a metallaoxirane ring[55d]

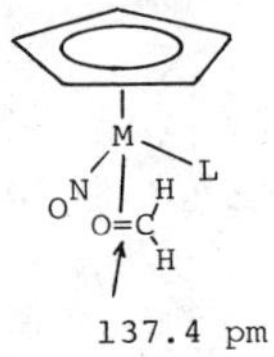

137.4 pm

FIGURE 6.19. Example of π complex of formaldehyde with a transition metal (Re).

6.2. OXIDATIVE ADDITION AND REDUCTIVE ELIMINATION

A typical oxidative addition reaction[56] may be represented by

$$L_nM + A\text{–}B \rightleftharpoons L_nM\begin{smallmatrix}A\\B\end{smallmatrix} \tag{6.23}$$

In this reaction the compound A–B adds to the transition metal complex with cleavage of the A–B bond. In addition to the above type of oxidative addition, where both the oxidation state and coordination number of the metal increase by 2, the following type of reactions, where the oxidation state and coordination number are increased by 1, are also included in the category of oxidative addition.

$$2L_nM + A\text{–}B \rightleftharpoons L_nM\text{–}A + L_nM\text{–}B \tag{6.24}$$

Oxidative addition is not restricted to transition metal complexes. The formation of Grignard reagents by reaction of magnesium metal with alkyl or aryl halides also may be regarded as a classic example of oxidative addition.

In the reverse process of oxidative addition, which is called reductive elimination, both oxidation number and coordination number are decreased. Oxidative addition and reductive elimination are very important elementary processes that are involved in many useful catalytic and stoichiometric reactions in organic synthesis. In oxidative addition a particular bond in a substrate is cleaved, often under mild conditions, with the formation of a reactive organotransition metal complex, which can undergo further reactions. In reductive elimination, on the other hand, a particular bond such as C–C, C–halogen, C–H, or C–O, is formed. By combining these bond-breaking and bond-making processes with other elementary processes, such as insertion and β elimination, a variety of useful synthethic methods have been devised. These will undoubtedly be further developed to replace many of the conventional, less specific synthetic processes.

Oxidative addition and reductive elimination processes have a relatively short history. They were recognized in studies of reactions of Vaska's complex, $IrCl(CO)(PPh_3)_2$, and Wilkinson's complex, $RhCl(PPh_3)_3$, but the importance of these ubiquitous reactions has gained rapid recognition since then.

a. Oxidative Addition: An Overview

Figure 6.20 illustrates various reactions of Vaska's complex, a typical, square planar, 16-electron d^8 metal complex. Addition of methyl iodide to Vaska's complex yields an octahedral, 18-electron, d^6 methyliridium(III) complex, process (i) in Figure 6.20.

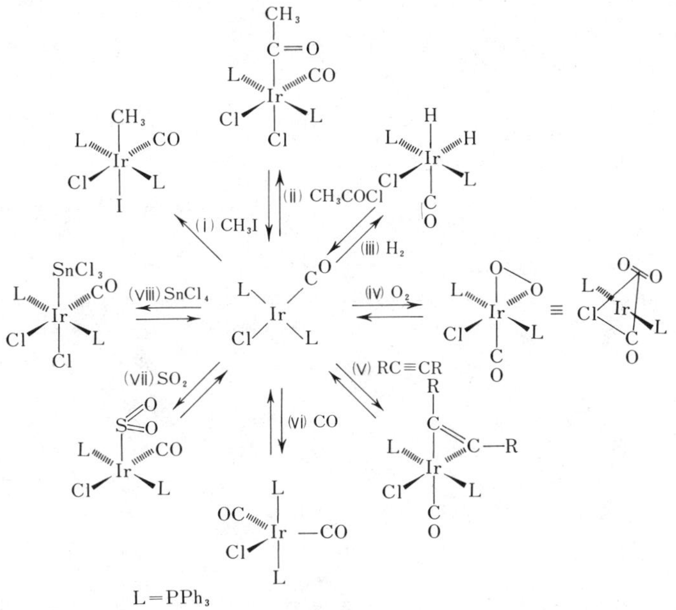

L=PPh_3

FIGURE 6.20. Various addition reactions of Vaska's complex.

The octahedral complex thus obtained has a trans configuration. Similar trans additions are observed in the oxidative addition of acetyl chloride (ii) to give an acetyliridium complex.

In the addition of H_2[process (iii)], a *cis*-dihydridoiridium(III) complex is formed. Thus, the starting iridium complex can be regarded as having been "oxidized by the action of molecular hydrogen." The apparent discrepancy with the common definition of reduction, in which H_2 acts as a reducing agent, arises because the hydride ligand is assigned a formal negative charge so the complex $IrH_2Cl(CO)L_2$ formed by addition of H_2 to the Ir(I) complex is regarded as containing iridium in the oxidation state of +3.

Whether processes (iv) and (v), involving the addition of O_2 and acetylene, should be regarded as oxidative additions depends on one's viewpoint. The O_2 adduct may be regarded either as a distorted octahedral Ir(III) complex with a single O–O bond or as a trigonal bipyramidal Ir(I) complex where O_2 is π-bonded through the double bond to Ir(I). In the latter sense formation of the O_2 adduct is simply an addition reaction and not "oxidative" addition. The reactions with acetylenes or olefins are treated similarly. To be consistent with the definition of oxidation states adopted for olefin complexes, it seems better to exclude the side-on additions of O_2, acetylenes, and olefins from the category of oxidative addition reactions.

In Figure 6.20 are also included simple additions of CO and SO_2 [processes (vi) and (vii)] to give a trigonal bipyramidal and square pyramidal Ir(I) complexes. In the reaction with $SnCl_4$[process(viii)] an Ir(III) complex with an Ir–Sn bond is formed. Similarly, complexes with Ir–Hg or Ir–Si bonds are formed on treatment of Vaska's complex with $HgCl_2$ or R_3SiH.

The following reactions are representative of processes represented by Eq. (6.24).[56c]

$$2[Co^{II}(CN)_5]^{3-} + A{-}B \longrightarrow [Co^{III}(CN)_5A]^{3-} + [Co^{III}(CN)_5B]^{3-}$$

A—B: H_2, H_2O, Br_2, H_2O_2, CHI_3, ICN, etc.

In many reactions of this type radical mechanisms are considered to be operative.

Intramolecular oxidative addition is also a well-known process. The best-known example is internal oxidative addition of a coordinated aromatic ligand, a process called orthometallation.[57] The following example shows the cleavage of a C–H bond in coordinated triphenylphosphine.

$$IrCl(PPh_3)_3 \xrightarrow{\Delta} Ph_3P{-}Ir(H)(PPh_3)_2Cl \quad (\text{orthometallated } C_6H_4) \qquad (6.25)$$

Similar reactions involving cleavage of a C–H bond also take place with triarylphosphites[58] and trialkylphosphines.[59]

$$IrHCl_2[P(OPh)_3]_3 \xrightarrow{\Delta} (PhO)_2P\text{—}Ir(Cl)(P(OPh)_3)\text{—}(C_6H_4O)P(OPh)_2 \quad (6.26)$$

Internal metallation has also been observed in complexes containing organic nitrogen bases.[60]

$$C_6H_5CH_2NR_2 + Li_2PdCl_4 \longrightarrow [Pd(\mu\text{-}Cl)(C_6H_4CH_2NR_2)]_2 \quad (6.27)$$

In reactions (6.25)–(6.27) coordination of phosphorus or nitrogen to the metal helps to bring the ortho carbon atoms into the vicinity of the transition metal and to enhance interaction between the C–H bond and the metal by the chelate effect.

b. Oxidative Addition of H_2

Oxidative addition of H_2 to low-valent transition metal complexes always proceeds in a cis fashion. In this reaction the strong H–H bond, which has a dissociation energy of 436 kJ mol^{-1}, is readily cleaved on interaction with a transition metal complex. The reaction takes place, often reversibly, under very mild conditions, such as at room temperature and atmospheric pressure. This fact suggests that a concerted process such as that shown in Figure 6.21 may be operative.

In this process H_2 is considered to approach the transition metal complex in a side-on manner. Although direct breaking of the H–H bond requires the large bond dissociation energy D_{H-H} (436 kJ mol^{-1}) as shown in Figure 6.21, the cleavage of the H–H bond can take place in a concerted manner provided

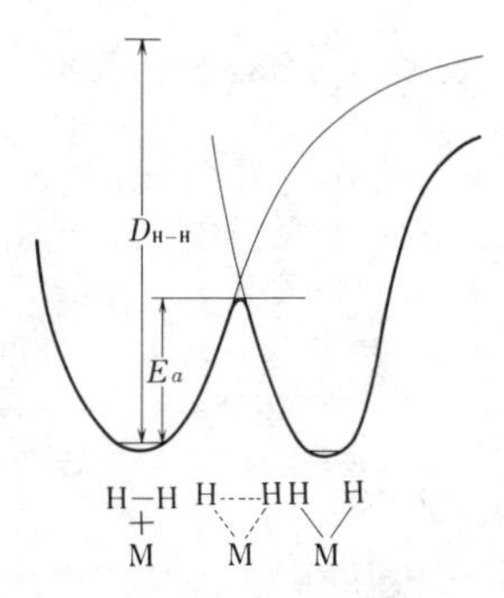

FIGURE 6.21. Dissociation of H_2 and its oxidative addition to a transition metal complex.

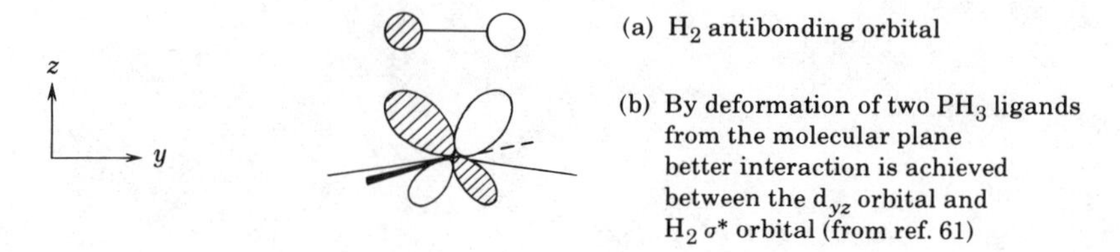

z
y
(a) H_2 antibonding orbital
(b) By deformation of two PH_3 ligands from the molecular plane better interaction is achieved between the d_{yz} orbital and H_2 σ^* orbital (from ref. 61)

FIGURE 6.22. Interaction of $RhCl(PH_3)_3$ with H_2 as computed by EHMO.

the thermodynamics for the formation of the metal dihydride complex is favorable. In other words, the activation energy for cleaving the H–H bond is reduced as a consequence of M–H bond formation by a concerted process. The mean bond energy of a transition metal hydride bond is estimated to be in the range of 200–300 kJ mol^{-1}. Therefore, twice the M–H bond dissociation energy is comparable with D_{H-H}. Taking into account the energy required for the configurational change from square planar to octahedral and the negative entropy of binding H_2, one can see that it is thermodynamically possible for the reaction to proceed reversibly.

According to extended Hückel MO calculations on the reaction of a model complex $RhCl(PH_3)_3$ with H_2, a π interaction of a metal d_{yz} orbital with a σ^* orbital of H_2 leads to effective cleavage of the H–H bond and concomitant formation of two Rh–H bonds. The interaction is further enhanced by bending the phosphine ligands away from planarity, as shown in Figure 6.22.[61] The postulated π interaction is reminiscent of the olefin–metal interaction.

A related study of oxidative addition of H_2 to a model complex $Pt(PH_3)_2$ has been made by the ab initio MO method. The transition state for the reaction, which was estimated by an MO method called the energy gradient approach, showed that the activation barrier for the reaction corresponds mainly to a deformation of the P–Pt–P angle, which allows the π interaction with the σ^* orbital of H_2 to develop.[62]

c. Oxidative Addition Involving Carbon–Halogen Bond Cleavage

Oxidative addition of alkyl and aryl halides (RX) has been extensively studied and has been utilized for organic syntheses with transition metal complexes. According to stereochemical studies, RX adds to transition metal complexes in a trans manner. In some cases stereochemical inversion of the configuration of asymmetric carbon atom in the alkyl group bonded to halide has been observed, but in other cases racemization occurs, suggesting the occurrence of radical processes. The nonradical process proceeding with stereochemical inversion is regarded as a nucleophilic attack of the alkyl group by the electron-rich transition metal.[63-65]

In kinetic studies on the oxidative addition of alkyl halides to Vaska's complex,

$$IrCl(CO)L_2 + RX \longrightarrow IrR(CO)XClL_2$$

a second-order rate law is observed and an electron-donating substituent in the para position of the triphenylphosphine ligand causes an increase in the reaction rate. The activation entropies for the reactions with alkyl halides show rather large negative values: for CH_3I, about -120 J mol^{-1} deg^{-1}, and for $PhCH_2Cl$, in the range of -200 to -250 J mol^{-1} deg^{-1}. These large negative entropies have been taken to indicate a polar asymmetric transition state.[65]

The following types of activated complexes have been proposed as transition states for the oxidative addition of alkyl halides.

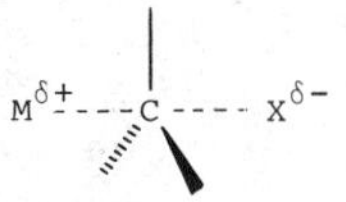

$M^{\delta+}$---C---- $X^{\delta-}$

(A)

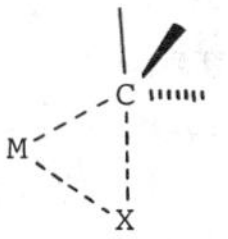

M---C---X

(B)

The activated complex (A) is reminiscent of that invoked in S_N2 substitution reactions in organic chemistry and requires no further comment. Transition state (B), invoked to explain the polar asymmetric transition state, involves some difficulty in accounting for the stereochemical inversion observed in the oxidative addition of benzyl halide derivatives to $Pd(PPh_3)_4$. Stille[64] has proposed a concerted attack, as shown in Figure 6.23. In this scheme palladium is assumed to approach the asymmetric carbon from the opposite side of the sterically demanding phenyl group so as to cause inversion through a trigonal pyramidal transition state, as depicted in the figure.

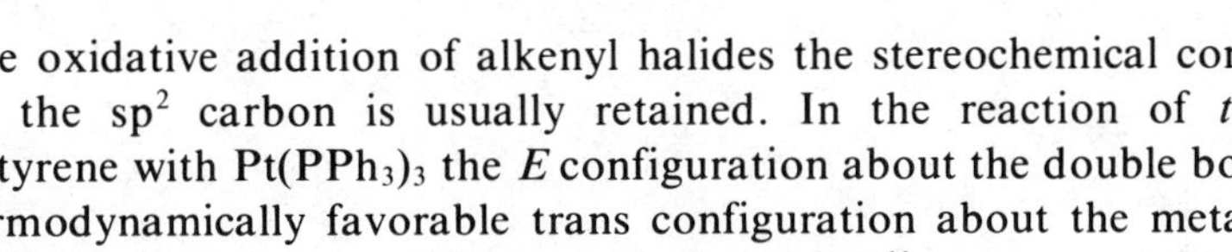

FIGURE 6.23. Postulated mechanism to account for the steric inversion of a benzyl halide derivative. In the transition state the palladium atom approaches the asymmetric carbon from the opposite side of the sterically demanding phenyl group.

In the oxidative addition of alkenyl halides the stereochemical configuration at the sp^2 carbon is usually retained. In the reaction of *trans*-β-bromostyrene with $Pt(PPh_3)_3$ the E configuration about the double bond and the thermodynamically favorable trans configuration about the metal atom were established by means of X-ray crystallography.[66]

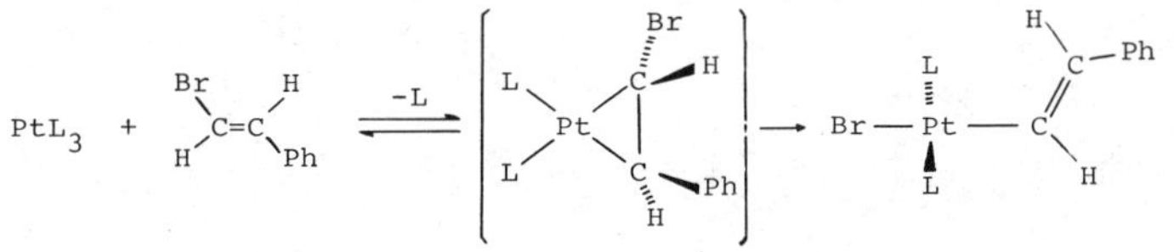

$$PtL_3 + (Br)(H)C{=}C(H)(Ph) \underset{}{\overset{-L}{\rightleftharpoons}} [L_2Pt(\eta^2\text{-}CH(Br){=}CH(Ph))] \longrightarrow Br{-}PtL_2{-}C(H){=}C(H)Ph \quad (6.28)$$

Similar retention of stereochemistry about the sp^2 carbon atom was observed in the oxidative addition of *cis*-β-bromostyrene. The oxidative addition was not inhibited by the presence of radical inhibitors.

In the oxidative additions of aryl halides to Pd(0) complexes, a nucleophilic substitution mechanism has been proposed.[63,67]

$$PdL_4 + p\text{-}XC_6H_4R \rightleftharpoons [L_2Pd(X)C_6H_4R]^{+} \xrightarrow{\text{rds}} p\text{-}(PdL_2X)C_6H_4R \qquad (6.29)$$

The reaction is first order in the concentrations of the palladium complex and aryl halide, and the reactivity decreases in the order: PhI > PhBr >> PhCl. The rate constants increase as the substituent R in Eq. (6.29) becomes more electron withdrawing and a Hammett linear free energy relationship is obeyed with $\rho = +2$. The results are consistent with nucleophilic attack of palladium on the aromatic carbon.

In some reactions of alkyl and aryl halides with low-valent transition metal complexes, however, considerable evidence has been provided in support of radical mechanisms.[68-70] The following mechanism illustrates a typical radical pathway.

A free radical is generated on interaction of an organic halide with a low-valent transition metal complex.

$$L_nM + RX \longrightarrow L_nMX + R\cdot$$
$$\text{(A)}$$

The radical produced then attacks the metal complex to form an organometallic species.

$$L_nM + R\cdot \longrightarrow L_nMR$$
$$\text{(B)}$$

Further reaction of B with RX gives the oxidative addition product (C) and liberates R· as the carrier of the chain reaction.

$$\underset{\text{(B)}}{L_nMR} + RX \longrightarrow \underset{\text{(C)}}{L_nM(R)(X)} + R\cdot$$

In the first step of the reaction of the transition metal with an organic halide a one-electron transfer from the metal to RX may take place.[68,69] If the R· radical rapidly diffuses away from the solvent cage, the chain reaction shown above may occur.

$$L_nM + RX \longrightarrow \underset{\text{(D)}}{[L_nM^{+}\cdot RX^{-}]} \longrightarrow \underset{\text{(E)}}{[L_nMX \cdots\cdots R\cdot]}$$

On the other hand, if the radical pair (E) in the solvent cage rearranges to give the oxidative addition product, no chain reaction is initiated. However, this case is difficult to distinguish from the nonradical mechanism.

In reaction products in which the initial stereochemistry at the carbon atom has been lost, it is likely that oxidative addition proceeds by a radical mechanism. Other criteria such as ESR measurement of the spin-trapped species, CIDNP (chemically induced dynamic nuclear polarization), and inhibition experiments using a radical trap are often employed to infer radical mechanisms. However, care should be taken in interpreting the results of these experiments, since radicals might be formed by perturbing the system, for example, by addition of a spin-trapping agent.

The reaction type may vary from one complex to another and depends on substrate, solvent, and impurities such as oxygen. One can state, however, that no free radical mechanism is involved where the stereochemical inversion or retention has been proved.

The mechanism of oxidative addition of methyl tosylate or $MeOSO_2F$ is similar to that of CH_3I.[71,72]

d. Oxidative Addition Involving C–H Bond Cleavage

If a specific C–H bond in an organic compound is cleaved on interaction with a transition metal complex to give an oxidative addition product, the organometallic complex thus formed will have reactive M–C and M–H bonds.

$$\mathrm{C{-}H + M} \rightleftharpoons \overset{\mathrm{C}\quad\mathrm{H}}{\diagdown\!\!\mathrm{M}\!\!\diagup} \tag{6.30}$$

This type of reaction may be taking place in the cracking of hydrocarbons catalyzed by noble metals such as platinum at high temperatures. If one could selectively induce C–H bond cleavage under mild conditions, there would be a tremendous range of applications. A reaction that causes C–H bond cleavage is sometimes referred to as *activation* of the C–H bond. We have already dealt with the internal metallation of a coordinated ligand [cf. Eqs. (6.25)–(6.27)]. This process is assisted through a chelating effect caused by coordination of the ligand containing heteroatoms such as nitrogen and phosphorus. To accomplish C–H bond cleavage of a nonactivated C–H bond by an intermolecular process is a more difficult task, but considerable progress has been made quite recently.

Carbon–hydrogen bond cleavage does occur in some aromatic compounds that do not contain a heteroatom because the substrate–metal interaction is facilitated by coordination through the aromatic ring. Naphthalene coordinated to iron, ruthenium, or osmium complexes containing 1,2-bis(dimethylphosphino)ethane (dmpe) ligands undergoes C–H bond cleavage to give a hydrido-naphthyl complex, which reacts further with various substituted benzenes giving hydrido-aryl complexes.[73]

$$(dmpe)_2M\text{-}(C_{10}H_8) \longrightarrow (dmpe)_2M(H)(C_{10}H_7) \xrightarrow[-C_{10}H_8]{C_6H_5X} (dmpe)_2M(H)(C_6H_4X) \qquad (6.31)$$

Carbon–hydrogen bond cleavage of aromatic compounds has been reported for other transition metal complexes as well.[74]

$$(\eta^6\text{-}C_6H_6)Os(PMe_3)(C_2H_4) \xrightarrow{PMe_3} [(Me_3P)_2Os(C_2H_4)(C_6H_6)] \xrightarrow{PMe_3} (Me_3P)_3Os(H)(C_6H_5)(C_2H_4) \xrightarrow[-C_2H_4]{PR_3} (Me_3P)_3Os(H)(C_6H_5)(PR_3) \qquad (6.32)$$

By combining C–H bond activation with oxidative addition of molecular hydrogen, certain metal complexes can catalyze the exchange of D_2 with aromatic C–H bonds.[75]

$$Cp_2TaH_3 \underset{}{\overset{-H_2}{\rightleftharpoons}} Cp_2TaH \overset{ArD}{\rightleftharpoons} Cp_2Ta(H)(D)(Ar) \rightleftharpoons Cp_2TaD + ArH \qquad (6.33)$$

An essential requirement for C–H bond activation is creation of an activation site in a transition metal by forming a coordinatively unsaturated, reactive species. The coordinatively unsaturated species may be produced by removing other ligands, either chemically or photochemically. Photoinduced elimination of H_2 from coordinatively saturated dihydride species has been observed to give the coordinatively unsaturated, reactive species that can undergo oxidative addition with C–H bonds.[76]

$$Cp_2WH_2 \xrightarrow[-H_2]{h\nu} [Cp_2W] \xrightarrow{C_6H_5CH_3} Cp_2W(H)(C_6H_4CH_3) + Cp_2W(CH_2Ph)(C_6H_4CH_3) \qquad (6.34)$$

This type of C–H activation is facilitated by making the transition metal complex more electron rich. Thus, the pentamethylcyclopentadienyl group C_5Me_5 (abbreviated Cp′) is a better ligand than the unsubstituted cyclopentadienyl group for causing the C–H bond activation. Employment of a basic, strongly coordinating ligand such as PMe_3 also enhances the reactivity.[77]

$$Cp'Ir(PMe_3)H_2 \xrightarrow[C_6H_6]{h\nu,\ -H_2} Cp'(PMe_3)Ir(H)(Ph)$$

$$Cp' = \eta^5\text{-}C_5Me_5$$

On the other hand, introduction of an electron-withdrawing substituent such as fluorine into hydrocarbons facilitates their oxidative addition by making the site of nucleophilic attack more electron deficient.[78]

$$PtL_2 + C_6F_5H \longrightarrow trans\text{-}PtH(C_6F_5)L_2$$
$$L = PCy_3$$

Very recently even "nonactivated" alkanes have been induced to undergo oxidative addition. The reactive species may be created by photoirradiation[77,79,80] or by chemical reactions.[81] In the chemical reactions hydrido ligands interact with olefins, and the hydrogenated product leaves the coordination sphere, thus creating coordinative unsaturation.

$$Cp'Ir(CO)_2 \xrightarrow[-CO]{h\nu} [Cp'Ir(CO)] \begin{cases} \xrightarrow{CMe_4} Cp'(CO)Ir(H)(CH_2CMe_3) \\ \xrightarrow{C_6H_{12}} Cp'(CO)Ir(H)(C_6H_{11}) \end{cases} \qquad (6.36)$$
$$Cp' = C_5Me_5$$

$$ReL_2H_7 + C_5H_{12} \text{ (n-pentane)} + CH_2{=}CHCMe_3 \xrightarrow{-C_2H_5CMe_3} L_2H_3Re(\eta^4\text{-pentadiene}) \xrightarrow{L'} \text{1,3-pentadiene} \qquad (6.37)$$
$$L = PAr_3 \qquad L' = P(OMe)_3$$

The previously observed H–D exchange reaction of alkanes with CH_3COOD–D_2O catalyzed by K_2PtCl_4 may well proceed by a mechanism involving the C–H bond cleavage and formation of hydrido-alkyl species.[82]

In comparison with the C–H activation of nonactivated alkanes, activation of an allylic C–H bond should be easier, and indeed this process is believed to be involved in some olefin isomerization processes catalyzed by transition metal complexes (cf. Section 8.1). Allylic C–H activation has been demonstrated in the following reactions of well-characterized complexes.[83]

$$(CH_2{=}CHCH_3)Pt(PR_3) \rightleftharpoons (\eta^3\text{-}CH_2CHCH_2)Pt(H)(PR_3) \qquad (6.38)$$
$$R = tert\text{-}Bu,\ Cy$$

The C–H bond attached to a double bond also is activated selectively. The C–H bonds in methacrylate[84] and vinyl acetate[85] are cleaved readily.

$$RuH_2L_4 + 2CH_2{=}C(Me)CO_2R \longrightarrow L_3RuH[CH{=}C(Me)C(OR){=}O] + Me_2CHCOOR \qquad (6.39)$$
$$L = PPh_3$$

$$MeCuL_2 + CH_3COOCH{=}CH_2 \xrightarrow{-CH_4} L_2Cu(CH{=}CH{-}O{-}C(CH_3){=}O) \quad (6.40)$$

$L = PPh_3$

The C–H bonds of ethylene can be cleaved by cluster complexes.[86]

$$M_3(CO)_{12} + C_2H_4 \longrightarrow (OC)_3M(\mu\text{-}H)(\mu_3\text{-}C{=}CH_2)M(CO)_3M(CO)_3 \quad (6.41)$$

$M = Ru, Os$

Such C–H bond activation may be facilitated by interaction of the olefin with more than two transition metal atoms in cluster complexes (cf. Section 9.3).

The C–H bonds in aldehydes are also cleaved on interaction with low-valent transition metal complexes to give hydrido-acyl complexes. The previously described iron dmpe complex reacts with benzaldehyde to give a hydrido-benzoyliron complex.[87]

$$Fe(dmpe)_2 + PhCHO \longrightarrow (dmpe)_2Fe(H)(COPh) \quad (6.42)$$

Oxidative addition of aliphatic aldehydes is usually followed by decarbonylation and other reactions such as reductive elimination of alkane and β-hydrogen elimination.[87c]

$$RhClL_3 \xrightarrow{RCHO} RhH(COR)ClL_3 \xrightarrow{-L} RhH(COR)ClL_2 \longrightarrow R{-}H + RhCl(CO)L_2 \quad (6.43)$$

$L = PMe_3$

The decarbonylation of aldehydes promoted by transition metal complexes has some synthetic utility, as will be described in Section 8.2.[88]

Oxidative addition of formaldehyde gives a hydrido-formyl complex, a species that may play an important role in the reduction of CO by H_2.[89]

$$[IrL_4]PF_6 + (CH_2O)_x \longrightarrow [L_4Ir(H){-}C(=O)H]PF_6 \quad (6.44)$$

Compounds having an active hydrogen atom, such as acetone, methyl acetate, methyl cyanoacetate, and acetonitrile, oxidatively add to electron-rich, coordinatively unsaturated complexes.[73c]

$$[Fe(dmpe)_2] + HZ \longrightarrow (dmpe)_2Fe(H)(Z) \qquad (6.45)$$

$HZ = CH_3COCH_3, CH_3CO_2CH_3, NCCH_2CO_2CH_3, CH_3CN$, etc.

e. Oxidative Addition Involving C–O Bond Cleavage

In comparison with oxidative additions involving C–halogen and C–H bond breakage, oxidative additions of carboxylic esters, acid anhydrides, ethers, and carbonates have received only limited attention until recently.[90] Carboxylic esters have two sites that could undergo C–O bond cleavage.

$$[M] + R\text{-}C(=O)\overset{A}{\downarrow}O\overset{B}{\downarrow}R' \xrightarrow{A} R\text{-}C(=O)\text{-}M\text{-}OR' ; \quad \xrightarrow{B} R\text{-}C(=O)\text{-}O\text{-}M\text{-}R' \qquad (6.46)$$

Allyl esters are cleaved at the allyl–O bond to give η^3-allyl–carboxylato complexes, which may undergo disproportionation as shown below.

$$Ni(cod)_2 + CH_3COOCH_2{=}CH_2 \longrightarrow [CH_3CO_2Ni(\eta^3\text{-}CH_2CHCH_2)] \xrightarrow{\text{disproportionation}} Ni(OAc)_2 + Ni(\eta^3\text{-}C_3H_5)_2 \qquad (6.47)$$

The intermediate allyl–carboxylato complex can be intercepted by adding tertiary phosphines.[91]

$$Ni(cod)_2 + CH_3CO_2C_3H_5 + L \longrightarrow (\eta^3\text{-}CH_2CHCH_2)Ni(OAc)(L) \qquad (6.48)$$

$L = PCy_3, PPh_3, PEtPh_2$

Oxidative addition of allyl formate is followed by decarboxylation and generation of propylene. Thus, allyl formate is catalytically degraded to propylene and CO_2 on interaction with a $Ni(cod)_2$–PPh_3 mixture.[91]

$$HCOOCH_2CH{=}CH_2 \xrightarrow{Ni(cod)_2-PPh_3} CH_3CH{=}CH_2 + CO_2 \qquad (6.49)$$

Pd(0) complexes also promote the cleavage of allyl carboxylates.[92]

$$PdL_2 + CH_3CO_2CH_2CH{=}CH_2 \longrightarrow (\eta^3\text{-}CH_2CHCH_2)Pd(OAc)(L) + L\text{-}Pd(\mu\text{-}C_3H_5)(\mu\text{-}O_2CCH_3)Pd\text{-}L \qquad (6.50)$$

The η^3-allyl ligand thus formed can be attacked by nucleophiles.

Reaction of the allylic acetate with Pd(0) complexes proceeds with inversion at carbon, and the nucleophilic attack on the η^3-allylic ligand also proceeds with inversion, resulting in overall stereochemical retention. Based on these processes, palladium-catalyzed organic syntheses have been developed.[90b,93] The following example illustrates introduction of a nucleophile into a steroid.

$$\text{steroid–CH(Me)(OAc)} \xrightarrow{[Pd^0]} \eta^3\text{-allyl–Pd(OAc)} \xrightarrow{^-CH(CO_2Me)_2} \text{steroid–CH(Me)CH(CO_2Me)_2} \quad (6.51)$$

Examples of acyl–O bond cleavage of carboxylic esters are still limited. The process is promoted by the interaction of esters with zero-valent transition metal complexes or with alkyl- or hydridotransition metal complexes.

Phenyl propionate undergoes C–O bond cleavage on interaction with a Ni(0) complex, according to the following stoichiometry.[94]

$$Ni(cod)_2 + C_2H_5CO_2Ph + 2PAr_3 \longrightarrow C_2H_4 + PhOH + Ni(CO)(PAr_3)_3 \quad (6.52)$$

The reaction rate is first order in the concentration of the Ni(0) complex and increases with increasing basicity of the tertiary phosphine ligand. In a reaction with parasubstituted phenyl propionates, the reaction rate increases as the substituent becomes more electron withdrawing. The reaction is believed to be initiated by nucleophilic attack of the nickel on the carbonyl carbon followed by the acyl–OPh bond cleavage. In contrast to the reaction of allyl carboxylate, where the allyl group acts as the leaving group, the phenoxy group serves as the leaving group in this type of reaction. The propionyl–phenoxo complex then undergoes decarbonylation and elimination reactions as shown below.

$$Ni(cod)_2 + C_2H_5CO_2Ph + nL \longrightarrow [L_nNi(OPh)(COC_2H_5)] \xrightarrow{-CO} [L_nNi(OPh)(C_2H_5)] \quad (6.53)$$

$$\xrightarrow{-C_2H_4} [L_nNi(OPh)(H)] \xrightarrow[-PhOH]{+CO} Ni(CO)L_3$$

In support of this scheme, $Ni(CH_3)(OPh)(bipy)$ has been isolated from the reaction of $Ni(cod)_2$ with phenyl acetate in the presence of bipyridine.

Aryl carboxylates undergo acyl–OAr bond cleavage on reaction with alkyl and hydridotransition metal complexes such as $RhH(PPh_3)_4$,[95] $RuH_2(PPh_3)_4$,[95] $CoH(N_2)(PPh_3)_3$,[96] and $CoCH_3(PPh_3)_3$. Two types of pathway are conceivable for these reactions. One is direct oxidative addition of the aryl carboxylate to the transition metal complex and the other is insertion of the carbonyl group of the aryl carboxylate into the M–H or M–R bond followed by OAr group abstraction by the transition metal.

$$L_nMR + R'-\underset{\underset{O}{\|}}{C}-OAr \longrightarrow L_nM\overset{\overset{R}{|}}{\underset{}{}}\text{(O–C(R)(R')–OAr)} \longrightarrow L_nMOAr + R\underset{\underset{O}{\|}}{C}R' \quad (6.54)$$

$R = CH_3, H$

Evidence supporting the second possibility was obtained in the reaction of CF_3COOPh with $CoH(N_2)(PPh_3)_3$, which gave the insertion product.[96]

$$CoH(N_2)(PPh_3)_3 + CF_3CO_2C_6H_5 \xrightarrow{-N_2} (Ph_3P)_3Co-O\underset{\underset{OPh}{|}}{C}HCF_3 \quad (6.55)$$

The related reaction of $CoH(N_2)(PPh_3)_3$ with CH_3CO_2Ph gave $Co(OPh)(PPh_3)_3$ owing to aryloxo group abstraction by cobalt, according to Eq. (6.54).

Cleavage of the C–O bond also occurs in reactions of acid anhydrides with low-valent transition metal complexes.[97]

$$Ni(cod)_2 + RCOOCOR + nL \longrightarrow (RCO)(L)Ni(L)(OCOR) \quad (6.56)$$

The reaction is followed by decarbonylation and other reactions depending on the substrate, ligand, and reaction conditions. Cyclic anhydrides also oxidatively add to Ni(0) to form metallacycles. The reaction is followed by decarbonylation and, depending on the ligand, may also be accompanied by ring contraction.[98]

$$Ni(cod)_2 + bipy + \text{glutaric anhydride } (O{=}C{-}CH_2{-}CH_2{-}CH_2{-}C({=}O){-}O) \longrightarrow (bipy)Ni\text{[C(=O)CH}_2\text{CH}_2\text{CH}_2\text{C(=O)O]}$$

$$\xrightarrow{-CO} (bipy)Ni\text{[CH}_2\text{CH}_2\text{CH}_2\text{C(=O)O]} \xrightarrow{dppe} (dppe)Ni\text{[CH(CH}_3\text{)CH}_2\text{C(=O)O]} \quad (6.57)$$

Carbonic esters also undergo C–O bond cleavage when they oxidatively add to Pd(0) complexes.[96] The reaction is followed by decarboxylation and reductive elimination. Tsuji utilized this reaction in organic synthesis.[99]

$$L_nPd + ROC(=O)R' \longrightarrow L_nPd(R)(OC(=O)R') \xrightarrow{-CO_2} L_nPd(R)(OR') \quad (6.58)$$

R = allyl group

In addition to allyl carboxylates, other allylic compounds such as allyl ethers and allyl alcohols are cleaved at the C–O bonds on interaction with low-valent transition metal complexes. The C–O bond in allyl phenyl ether is readily cleaved on interaction with $Ni(cod)_2$ and PPh_3 to give a π-allyl–nickel phenoxide.[91]

$$Ni(cod)_2 + CH_2{=}CHCH_2OPh + PPh_3 \longrightarrow (\pi\text{-}C_3H_5)Ni(PPh_3)(OPh) \quad (6.59)$$

The reaction of diallyl ether with $Ni(cod)_2$ and tertiary phosphines at or below room temperature gives Ni–allyl ether π complexes that undergo a C–O bond cleavage reaction at higher temperatures, liberating propylene.[91]

$$Ni(cod)_2 + nL + (CH_2{=}CHCH_2)_2O \longrightarrow Ni(\eta^2\text{-diallyl ether})L$$

$$L = PPh_3,\ PCy_3 \qquad \xrightarrow{\Delta} \left[(\pi\text{-allyl})Ni(OC_3H_5)L \right] \longrightarrow CH_3CH{=}CH_2 \quad (6.60)$$

The C–O bond cleavage in allylic alcohols takes a somewhat different course. Allyl alcohol undergoes a dismutation on treatment with $Ni(cod)_2$ and PPh_3 to give the products shown in reaction (6.61) in a 1 : 1 : 1 ratio.[91]

$$2CH_2{=}CHCH_2OH + Ni(cod)_2 + 2PPh_3 \longrightarrow$$
$$CH_2{=}CHCH_3 + Ni(CH_2{=}CHCHO)(PPh_3)_2 + H_2O \quad (6.61)$$

These results can be accounted for by assuming intermediates having $Ni{-}OCH_2CH{=}CH_2$ and $NiCH_2CH{=}CH_2$ entities, which may generate the acrylaldehyde complex and propylene on β-hydrogen transfer.

The C–O bonds in vinyl esters, β-lactones, and oxiranes can also be cleaved.[91,100,101] In the last case the reaction is followed by hydrogen transfer to give a hydrido-β-ketoalkyl complex.[102]

$$RCH\overset{O}{-}CH_2 + IrL_4Cl \xrightarrow{-L} \left[R\text{-}CH\text{-}CH_2 \atop O{-}IrL_3Cl \right] \longrightarrow RC(=O)CH_2IrL_3Cl(H) \quad (6.62)$$

R = H, Me, Ph $\qquad$ L = PMe_3, C_8H_{14}

f. Other Oxidative Additions

In addition to the reactions mentioned above, oxidative additions involving other bonds including C–C, C–N, C–P, O–H, C–S, S–H, N–O, N–H, O–S, and others are gradually attracting increasing attention.

Among the oxidative addition involving the C–C bond cleavage, reactions of strained carbocyclic compounds to give metallacycles have attracted particular interest. In Section 4.2 we dealt with the cleavage of cyclopropanes and cubane to give metallacycles. Other examples of oxidative additions of strained carbocycles are shown below.[103]

$$Pt(PPh_3)_2L + (NC)_2C\text{—}C(CN)_2\ (\text{bridged by } CRR') \longrightarrow (Ph_3P)_2Pt[C(CN)_2CRR'C(CN)_2] \quad (6.63)$$

$L = 2PPh_3,\ C_2H_4 \qquad R = H, Me, Et, Ph$

$$Pt(PPh_3)_3 + \text{benzocyclobutenedione} \longrightarrow (Ph_3P)_2Pt[\text{C(O)C(O)C}_6H_4] \quad (6.64)$$

An interesting reversible C–C bond cleavage has been reported recently.[104]

$$(CO)_2Ru\text{—}Ru(CO)_2 \text{ (fulvalene)} \underset{\Delta}{\overset{h\nu}{\rightleftharpoons}} \text{Ru(CO)}_2\text{ / Ru(CO)}_2 \text{ isomer} \quad (6.65)$$

Cleavage reactions involving C–CN bonds have been relatively well studied. Benzonitrile behaves somewhat similarly to chlorobenzene in its reactions with zero-valent group 8–10 metal complexes, and *trans*-cyanophenyl compounds have been obtained.[105]

$$Pt(PEt_3)_3 + PhCN \longrightarrow trans\text{-}Pt(PEt_3)_2Ph(CN) \quad (6.66)$$

$$Pt(PEt_3)_3 + F\text{-}C_6H_4\text{-}CN \longrightarrow Pt(FC_6H_4)(PEt_3)(Et_3P)(CN) \quad (6.67)$$

The C–CN bond in aliphatic nitriles is also cleaved by nickel,[106] platinum,[107] and cobalt[108] complexes.

$$Pt(PPh_3)_4 + CH_3C(CN)_3 \xrightarrow{-2PPh_3} (Ph_3P)_2Pt(CN)(C(CN)_2CH_3) \quad (6.68)$$

Oxidative additions of water, alcohols, phenols, thiophenols, and amines to low-valent transition metal complexes are known. As in other oxidative additions, attachment of basic ligands to the transition metal complexes enhances the nucleophilicity of the complex. Employment of bulky tertiary phosphines favors dissociation and helps to create a coordinatively unsaturated, reactive transition metal complex. As discussed in Section 6.1c, $Pt[P(i\text{-}Pr)_3]_3$ in solution gives a two-coordinate complex PtL_2 that oxidatively adds water and liberates OH^-.[109]

$$PtL_2 + H_2O \rightleftharpoons PtH(OH)L_2$$
$$PtH(OH)L_2 + S \rightleftharpoons PtH(S)L_2{}^+OH^- \rightleftharpoons PtH(S)L_2{}^+ + OH^- \quad (6.69)$$
$$S = \text{solvent}$$

The two-coordinate complex PtL_2 thus serves as a catalyst for the H–D exchange of active hydrogen compounds with D_2O, hydration of nitriles, and other reactions. The apparent pH of a water–THF solution containing $Pt(PEt_3)_3$ was estimated as 14.0, which is as strong as that of sodium hydroxide.

Methanol ($pK_a = 17.7$) oxidatively adds to the same complex giving *trans*-PtH_2L_2.[110]

Oxidative addition of phenol and pentafluorophenol to $Pt(PCy_3)_2$ gives *trans*-hydrido-phenoxo complexes.[112,113]

$$PtL_2 + HOAr \longrightarrow trans\text{-}PtH(OAr)L_2 \quad (6.70)$$
$$L = PCy_3 \qquad Ar = C_6H_5,\ C_6F_5$$

The same complex reacts with carboxylic acids to give *trans*-carboxylato-hydrido complexes.[112,113]

$$PtL_2 + RCOOH \longrightarrow trans\text{-}PtH(O_2CR)L_2 \quad (6.71)$$

Amines and imides oxidatively add to zero-valent group 10 metal complexes with cleavage of N–H bonds.[111,113]

$$PtL_2 + C_4H_4NH \text{ (pyrrole)} \longrightarrow \underline{trans}\text{-}PtH(NC_4H_4)L_2 \quad (6.72)$$

$$PtL_2 + C_6F_5NH_2 \longrightarrow trans\text{-}PtH(NHC_6F_5)L_2 \quad (6.73)$$
$$L = PCy_3$$

$$Ni(cod)_2 + 2L + HN(COCH_2)_2 \longrightarrow H\text{-}NiL_2\text{-}N(COCH_2)_2 \quad (6.74)$$
$$L = PCy_3$$

$$Pd(PCy_3)_2 + \text{succinimide} \longrightarrow H\text{-}PdL_2\text{-}N(\text{succinimidyl}) \tag{6.75}$$

Reaction of $Pt(PEt_3)_3$ with pentafluoroaniline did not give any oxidative addition product, presumably because the stronger coordinating ability of the PEt_3 ligands hinders the availability of the activation site for oxidative addition.

Substituted thiophenols HSC_6H_4Y oxidatively add to Vaska's complex.[114] The addition occurs in a cis manner, and the reaction rate is first order in the concentration of the iridium complex and the thiophenol.

$$\text{trans-}IrX(CO)L_2 + HSC_6H_4Y \longrightarrow (OC)(Cl)(H)L_2Ir\text{-}SC_6H_4Y \tag{6.76}$$

$L = PPh_3$, $X = Cl, Br$

Electron-withdrawing substituents Y increase the reaction rate. The relatively large negative entropies of activation, which range from -100 to -150 J mol^{-1} K^{-1}, and the relatively small enthalpies of activation (35–40 kJ mol^{-1}) suggest a rather constrained transition state.

Cleavage of the C–S bond in thioethers takes place on interaction of phenylalkenylthioethers with transition metal hydrides.[115] The reaction proceeds, however, not via oxidative addition but through a process involving insertion of the double bond into the Rh–H bond to give an alkyl complex and subsequent elimination of the PhS group at the β position.

$$CH_3CH{=}CHCH_2SPh + HRh(PPh_3)_4 \longrightarrow \text{olefin} + (PhS)Rh(PPh_3)_3 \tag{6.77}$$

The oxidative addition of hydrosilanes to transition metal complexes constitutes an important elementary step in catalytic hydrosilylation of olefins. Vaska's complex is known to react with R_3SiH with Si–H bond cleavage.[116]

$$IrCl(CO)(PPh_3)_2 + R_3SiH \longrightarrow IrCl(H)(SiR_3)(CO)(PPh_3)_2 \tag{6.78}$$

A special type of oxidative addition without bond cleavage involving addition of an ortho quinone to group 9 transition metal complexes has been reported.[117]

$$[M(P\text{-}P)_2]^+ + o\text{-}O_2C_6Cl_4 \longrightarrow [M(P\text{-}P)_2(O_2C_6Cl_4)]^+ \tag{6.79}$$

$P\text{-}P = Ph_2PCH{=}CHPPh_2$

Some olefins, acetylenes, and dienes add to transition metal complexes and form metallacycles. In the formation of metallacyclopentanes or metallacyclopentenes from two molecules of olefins or acetylenes, C–C bond formation, not C–C bond cleavage, is involved. We have seen examples of this in Section 4.2. Theoretical studies on the selectivity of products in these reactions have been reported.[118]

g. Reductive Elimination

Reductive elimination is the reverse of oxidative addition. In a typical reductive elimination (Eq. (6.80)] both the oxidation number and the coordination number decrease by 2.

$$L_nM(A)(B) \longrightarrow L_nM + A\text{–}B \tag{6.80}$$

When either or both A and B are bound to the metal with an M–C or an M–H bond, reductive elimination is one process by which an organometallic compound decomposes. In this process the bond can be formed between A and B. If both A and B are attached to the metal through a carbon atom, reductive elimination results in C–C bond formation. If one of the ligands is hydride, the reductive elimination causes C–H bond formation, and this occurs in many catalytic reactions such as hydrogenation and hydroformylation of olefins (see Chapters 7 and 8).

In spite of its obvious importance as an elementary step in metal-promoted organic syntheses,[118,119] the mechanism of reductive elimination has, until recently, received less attention than that of oxidative addition. This may largely be due to the limited availability of isolated alkyltransition metal complexes suitable for kinetic studies.

Reductive elimination of alkyl or aryl halides from $MR(X)L_n$ complexes has seldom been the subject of mechanistic studies, whereas there are numerous kinetic studies on oxidative addition of alkyl or aryl halides to low-valent transition metal complexes. The thermolysis of some monoalkyl–copper and monoalkyl–nickel complexes is complicated by intermolecular reactions.[120]

Reductive elimination of dialkyl transition metal complexes has received increasing attention. The coupling of the two alkyl groups generally proceeds in a concerted manner, although radical pathways may be involved in certain cases, particularly when the reactions are promoted by oxidants.[68,69,121]

An obvious but often overlooked requirement for concerted reductive elimination is that the two ligands to be reductively eliminated should be adjacent. In a planar cis dialkyl complex it follows that alkane may be reductively eliminated either with or without prior dissociation of an auxiliary ligand, whereas a trans dialkyl complex must first isomerize, either to a tetrahedral species or to the planar *cis*-dialkyl isomer. Reductive elimination can then occur from this species or from a three-coordinate complex formed by ligand dissociation.

For example, *trans*-$PdMe_2L_2$ (L = various tertiary phosphines) isomerizes to the cis isomer before elimination of ethane takes place, whereas *cis*-PdR_2L_2 (R = Me, Et, Pr)[16,17] liberates the reductive elimination products by a dissociative mechanism (Fig. 6.24).

Isotopic studies using CD_3 and CH_2CD_3 groups have established that reductive elimination from *cis*-PdR_2L_2 is an intramolecular process. Addition of excess phosphine hinders the trans → cis isomerization as well as reductive elimination from *cis*-PdR_2L_2. Stille found that dimethyl (transphos)palladium is quite resistant to reductive elimination, and CH_3CD_3 is produced at room temperature on addition of CD_3I to the complex.[16]

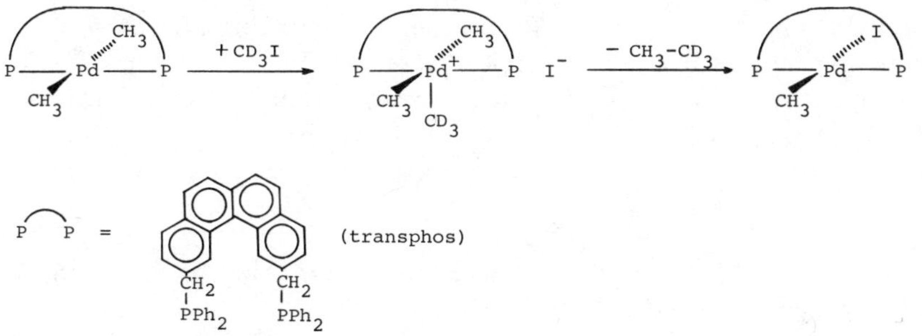

$$\text{P—Pd(CH}_3\text{)(CH}_3\text{)—P} \xrightarrow{+\,CD_3I} [\text{P—Pd}^{+}\text{(CH}_3\text{)(CH}_3\text{)(CD}_3\text{)—P}]\ I^- \xrightarrow{-\,CH_3\text{-}CD_3} \text{P—Pd(I)(CH}_3\text{)—P} \tag{6.81}$$

P⌒P = (transphos) [phenanthrene-type aromatic backbone bearing two CH_2PPh_2 groups]

The transphos ligand bridges trans coordination sites and does not allow the methyl groups to occupy adjacent positions in $PdMe_2$(transphos), thus inhibiting reductive elimination. Formation of $CH_3\text{–}CD_3$ by treatment of $PdMe_2$-(transphos) with CD_3I is thought to occur by reductive elimination from a cationic five-coordinate Pd(IV) intermediate. A similar reductive elimination of ethane from a Pd(IV) intermediate formed on trans addition of CD_3I to *cis*-$Pd(CH_3)_2L_2$ has been proposed to account for the proportions of CH_3CH_3 and CH_3CD_3 formed.[16b]

Although the trialkylpalladium(IV) species is thermally unstable and has not hitherto been detected experimentally, the corresponding trialkylplatinum(IV) complexes are thermally more stable and amenable to kinetic studies. It has been proposed that they decompose thermally by a reductive elimination mechanism involving dissociation of a tertiary phosphine ligand.[122]

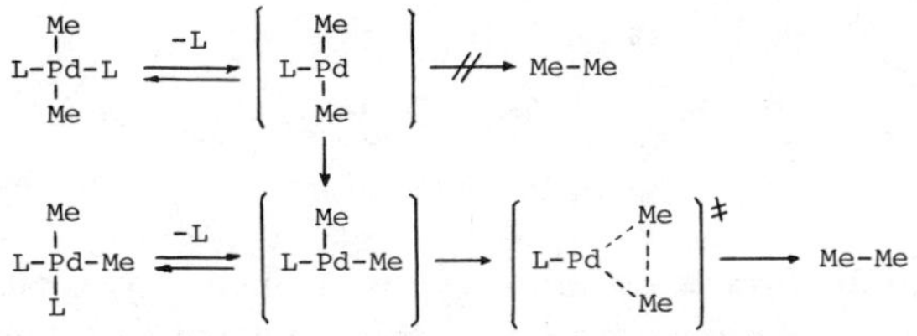

$$\text{L–Pd(Me)(Me)–L (trans)} \underset{}{\overset{-L}{\rightleftharpoons}} [\text{L–Pd(Me)(Me)}] \not\longrightarrow \text{Me–Me}$$

$$\downarrow$$

$$\text{L–Pd(Me)(Me)–L (cis)} \overset{-L}{\rightleftharpoons} [\text{L–Pd–Me(Me)}] \longrightarrow [\text{L–Pd}\cdots(\text{Me})(\text{Me})]^{\ddagger} \longrightarrow \text{Me–Me}$$

FIGURE 6.24. Mechanism of reductive elimination of ethane from *trans*- and *cis*-$PdMe_2L_2$ (L = tertiary phosphine).

$$\text{Me}_3\text{PtIL}_2 \xrightarrow{-\text{L}} [\text{Me}_3\text{PtIL}] \xrightarrow{-\text{Me-Me}} \text{trans-PtMeIL}_2 \qquad (6.82)$$

$L = PMe_3$, $PMePh_2$, PMe_2Ph

In diarylplatinum(II) complexes the trans isomer is observed to be thermally more stable than the cis isomer and trans → cis isomerization has been implicated in thermolysis of the solid trans isomer. It has been established that reductive elimination from *cis*-$PtAr_2L_2$ proceeds by a concerted mechanism.[123]

Addition of triphenylphosphine to *cis*-Pt(*p*-tolyl)$_2$(PPh_3)$_2$ enhances the rate of reductive elimination of bitolyl in toluene. The enthalpy and entropy of activation have been estimated as 74 ± 4 kJ mol^{-1} and -99 ± 16 J mol^{-1} K^{-1}. A three-center transition state in which the aryl orbitals interact with the metal has been proposed.[124]

$$\text{PtL}_2(\text{tol})_2 \xrightarrow{\text{L}} [\text{L}_3\text{Pt}(\text{tol})_2]^{\ddagger} \longrightarrow \text{Me-C}_6\text{H}_4\text{-C}_6\text{H}_4\text{-Me} + \text{PtL}_3 \qquad (6.83)$$

Stereochemical retention at the benzylic carbon atom has been established in the reductive elimination of a deuterium-labeled benzylmethylpalladium complex produced by *in situ* methylation of *trans*-Pd(CHDPh)X(PPh_3)$_2$, which supports a concerted mechanism for reductive elimination.[125]

$$\text{X-CH(D)Ph} + \text{PdL}_4 \xrightarrow[\text{inversion}]{-2\text{L}} \text{PhCH(D)-PdXL}_2 \xrightarrow[-\text{mX}]{\text{Me-m}} [\text{PhCH(D)-PdMeL}_2] \longrightarrow \text{PhCH(D)-Me} \qquad (6.84)$$

Me-m = $SnMe_4$, MeMgBr

Stereochemical retention has been confirmed also for the reductive elimination of alkyl and hydrido ligands from a molybdenum complex.[126]

$$\text{Cp}_2\text{Mo(D)-CH(CO}_2\text{CH}_3\text{)(CHDCO}_2\text{CH}_3\text{)} \xrightarrow[\text{toluene}]{\text{at room temp.}} [\text{Cp}_2\text{Mo}] + \text{HDC(CO}_2\text{CH}_3\text{)(CHDCO}_2\text{CH}_3\text{)} \qquad (6.85)$$

Reductive elimination of two alkyl groups from a square planar d^8 complex proceeding through a dissociative mechanism related to Figure 6.24 was first observed for a trialkylgold(III) triphenylphosphine complex.[15] In this process it is proposed that a T-shaped trialkylgold is first formed by dissociation of PPh_3 in solution and that this then undergoes reductive elimination through a Y-shaped activated complex.

$$\text{Et-Au(Me)}_2\text{-PPh}_3 \underset{}{\overset{-\text{PPh}_3}{\rightleftharpoons}} [\text{Et-AuMe}_2] \longrightarrow [\text{Et, Me} \cdots \text{Au-Me}]^{\ddagger} \longrightarrow \text{Me-Et} \quad (6.86)$$

The trialkylgold complex AuR_3L differs from the dialkylpalladium complex in its facile polytopal rearrangement behavior. The T-shaped Me_2AuEt species can readily undergo cis–trans isomerization by a windshield-wiper-like movement, whereas the similar isomerization is a high-energy process for $PdMe_2L$.

$$\text{Et—Au(Me)(Me) (trans Me)} \rightleftharpoons \text{Et—Au—Me (cis Me)}$$

Theoretical calculations by the extended Hückel MO method on reductive elimination from AuR_3L and PdR_2L_2 support this proposal.[15,18]

Comparison of the quite different behavior on thermolysis of trans and cis methyl–arylnickel complexes illustrates the important influence of the initial configuration on the reductive elimination pathway.[121,127] *Trans*-NiMe(Ph)$(PEt_3)_2$ is thermally stable, but it can be induced to decompose thermally by addition of aryl halide. Addition of PEt_3 to the complex hinders the thermolysis. Addition of $Me_2PCH_2CH_2PMe_2$(dmpe) at low temperature replaces the PEt_3 ligands from *trans*-NiMePh$(PEt_3)_2$ to give NiMe(Ph)(dmpe), in which the methyl and phenyl groups are forced to be in cis positions by the chelating dmpe ligand. This complex is thermally much less stable than *trans*-NiMe(Ph)$(PEt_3)_2$, and thermolysis to produce toluene can now be greatly accelerated by addition of sterically less demanding tertiary phosphines.

$$\text{L-Ni(Me)(Ph)-L} \xrightarrow[\;\not\;]{+\text{L}} \text{Me-Ph}; \quad \text{L-Ni(Me)(Ph)-L} \xrightarrow{\text{dmpe}} (\text{P}\frown\text{P})\text{Ni(Me)(Ph)} \xrightarrow[\text{acceleration}]{+\text{L}} \text{Me-Ph} \quad (6.87)$$

This striking difference in the reductive elimination of toluene from two types of nickel complexes that have the methyl and phenyl groups in trans and cis positions, respectively, has been accounted for by assuming five-coordinate trigonal bipyramidal intermediates (Fig. 6.25).[127a]

Coordination of the tertiary phosphine ligand L to *trans*-NiMe(Ph)L_2 gives either the intermediate [A], which has the methyl and phenyl groups at mutually trans positions, an obviously unfavorable configuration for the reductive elimination of Ph–Me, or intermediate [B], which has both the methyl and the phenyl groups at equatorial positions. Reductive elimination from configura-

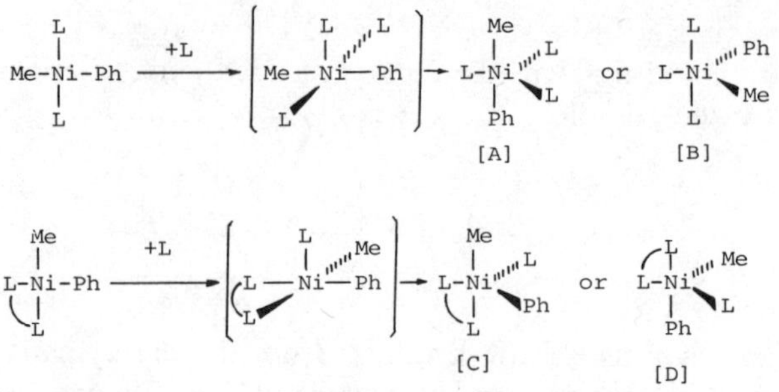

FIGURE 6.25. Five-coordinate intermediates in reductive eliminatin from *cis-* and *trans-*phenylmethylnickel complexes.

tion [B] is forbidden by orbital symmetry.[127b] On the other hand, coordination of L to *cis*-NiMe(Ph)L_2 gives intermediates [C] or [D] (Fig. 6.25), which have either of the methyl and phenyl groups at an apical position while the other is at an equatorial position. Reductive elimination from [C] or [D] is allowed by orbital symmetry and is indeed a low-energy process.[127b]

Reductive elimination from some transition metal dialkyls is accelerated by addition of an electron-withdrawing ligand. As mentioned previously, addition of electron-deficient olefins such as maleic anhydride and acrylonitrile to the thermally very stable complexes R_2Ni(bipy) markedly accelerates thermolysis and causes ethane to be readily lost from $NiMe_2$(bipy) and butane from $NiEt_2$(bipy).

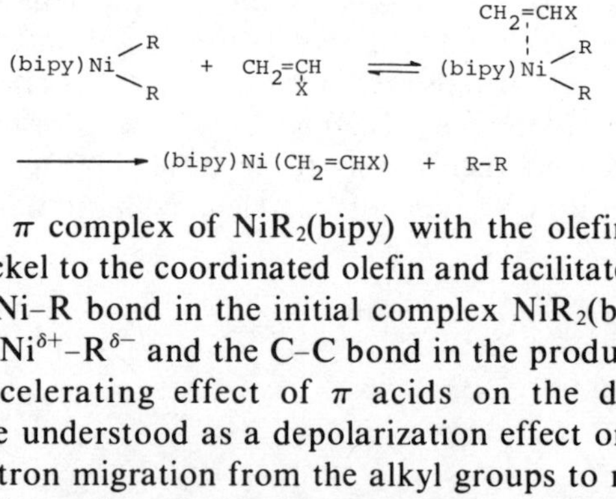

$$(\text{bipy})\text{NiR}_2 + \text{CH}_2{=}\text{CHX} \rightleftharpoons (\text{bipy})\text{NiR}_2(\text{CH}_2{=}\text{CHX}) \longrightarrow (\text{bipy})\text{Ni}(\text{CH}_2{=}\text{CHX}) + \text{R–R} \tag{6.88}$$

Formation of a π complex of NiR_2(bipy) with the olefins causes electron migration from nickel to the coordinated olefin and facilitates reductive elimination. Since the Ni–R bond in the initial complex NiR_2(bipy) is considered to be polarized as $Ni^{\delta+}$–$R^{\delta-}$ and the C–C bond in the product R–R should be nonpolar, the accelerating effect of π acids on the decomposition of NiR_2(bipy) may be understood as a depolarization effect on the Ni–C bonds by facilitating electron migration from the alkyl groups to nickel. The planar form of NiR_2(bipy) may be suitable for the complex formation with π acids. Similar acceleration effects of added π acids on reductive elimination have been observed in formation of carbocyclic compounds from metallacycles[128] and C–O bond formation to give an ester from an acyl–phenoxo bipyridine complex.[129]

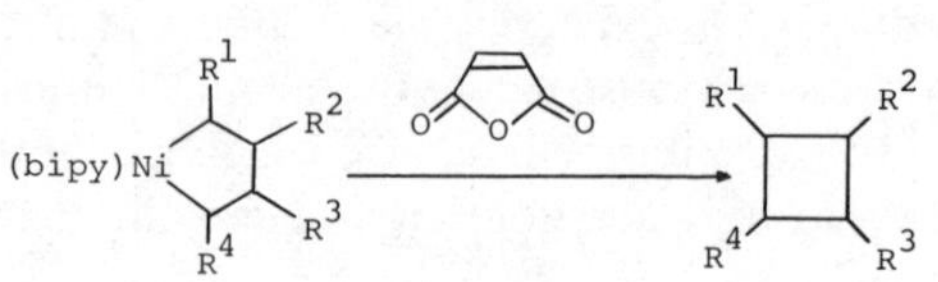

$$(\text{bipy})\text{Ni}[\text{C}(\text{R}^1)\text{–C}(\text{R}^2)\text{–C}(\text{R}^3)\text{–C}(\text{R}^4)] \xrightarrow{\text{maleic anhydride}} \text{cyclobutane}(\text{R}^1,\text{R}^2,\text{R}^3,\text{R}^4) \tag{6.89}$$

$$(\text{bipy})\text{Ni}(\text{OC}_6\text{H}_5)(\text{C(O)R}) \xrightarrow{\text{maleic anhydride}} \text{Ni(bipy)(maleic anhydride)}_2 + \text{RC(O)OC}_6\text{H}_5 \quad (6.90)$$

The reductive elimination of acid anhydrides from acyl-carboxylato-nickel(II) complex promoted by CO is another example of promotion of C–O bond formation aided by the presence of a π acid.[97]

Reductive elimination with C–C bond formation also constitutes an important elementary step in certain types of oligomerization of alkenes, alkynes, and dienes. This will be covered in detail in Section 7.2.

In thermolysis of transition metal complexes having two alkyl groups with β hydrogens, β-hydrogen elimination takes place first and is followed by reductive elimination of an alkane from the hydrido-alkyl complex thus formed. This reaction is called reductive disproportionation since disproportionation of the two alkyl groups to an alkane and an alkene is involved. Sometimes the term reductive β elimination is used to describe this process. The following are typical examples [Eqs. (6.91)[130-132] and (6.92)[133]].

$$(\text{R}_3\text{P})_2\text{Pt}(\text{CH}_2\text{CH}_3)_2 \xrightarrow{\Delta} [\text{Pt(PR}_3)_2] + \text{C}_2\text{H}_6 + \text{CH}_2\text{=CH}_2 \quad (6.91)$$

$$\text{cis-Fe(CH}_2\text{CH}_3)(\text{bipy})_2 \xrightarrow{\Delta} [\text{Fe(bipy)}_2] + \text{C}_2\text{H}_6 + \text{CH}_2\text{=CH}_2 \quad (6.92)$$

A similar reductive β-elimination process involving a metallacycle gives a 1-olefin.[134]

$$\text{L}_n\text{M}(\text{CH}_2\text{CH}_2\text{CH}_2\text{CH}_2) \xrightarrow{\Delta} [\text{L}_n\text{M}] + \text{CH}_3\text{CH}_2\text{CH=CH}_2 \quad (6.93)$$

Reductive β elimination will be dealt with in more detail in Section 6.3.

Certain types of reductive elimination proceed by a bimolecular mechanism as proved by isotopic labeling studies.[135]

Two mechanisms are possible for the hydrolysis of transition metal alkyls. One involves external attack on an alkyl group having anionic character by a proton, whereas the other mechanism involves the reductive elimination of the alkyl group with a hydrido ligand formed by attack of the proton on the transition metal. The latter pathway has been proposed for hydrolysis of an iron alkyl on the basis of the stereochemical retention of configuration at the α-carbon atom of the alkyl group.[136]

$$\text{Me-C}_6\text{H}_{10}\text{-FeCp(CO)}_2\ (\text{trans}) + \text{DX} \xrightarrow{\text{CH}_2\text{Cl}_2} \text{Me-C}_6\text{H}_{10}\text{-D}\ (\text{trans}) \quad (6.94)$$

$\text{X} = \text{Cl, CF}_3\text{CO}_2$

6.3. INSERTION AND DEINSERTION (EXTRUSION)

Insertion of an unsaturated compound into an M–C or M–H bond is a very important elementary reaction in many catalytic and stoichiometric reactions. The reaction may be expressed as follows.

$$\text{M-R} \begin{cases} \xrightarrow{\text{A=B}} \text{M-A-B-R} & \text{(a)} \\ \xrightarrow{\text{:A-B}} \underset{\substack{| \\ \text{B}}}{\text{M-A-R}} & \text{(b)} \end{cases} \tag{6.95}$$

The R stands for an alkyl, aryl, or hydrido ligand; A=B stands for C=C, C=O, or C=N; and :A—B stands for :CO, isocyanide, :C≡NR, or carbene $:CR_2$. The most important substrates among these are carbon monoxide and olefins.

a. CO Insertion and Decarbonylation

While most organometallic compounds of the main-group elements are unreactive toward carbon monoxide, many transition metal alkyls are known to undergo insertion with CO to give transition metal acyls.[137-140] In many cases this reaction is reversible. The reverse process is called decarbonylation, CO deinsertion, or extrusion reaction.

The forward and reverse reactions probably proceed in a concerted manner through a three-centered transition state.

$$L_nM\langle^{R}_{CO} \rightleftharpoons \left[L_nM\langle^{R}_{CO} \right]^{\ddagger} \rightleftharpoons L_nM\text{-}\underset{\substack{\| \\ O}}{C}\text{-}R \tag{6.96}$$

An obvious requirement for the reaction to proceed is that the R and CO groups are situated in mutually cis positions; if they are not, an isomerization to bring them to the cis positions must take place first.

The term *CO insertion* is used here without implying a particular mechanistic pathway. However, it has been established that CO "insertion" proceeds by *alkyl migration* to the coordinated CO ligand in most cases. For this reason the term *migratory insertion* is preferred by some authors. The alkyl migration mechanism has been confirmed for the reaction of methylpentacarbonylmanganese with CO labeled with ^{13}C and ^{14}C.[141]

$$\text{MeMn(CO)}_5 + {}^{*}\text{CO} \longrightarrow \text{(MeCO)Mn(CO)}_4({}^{*}\text{CO}) \tag{6.97}$$

$^{*}CO = {}^{13}CO$ or ^{14}CO

The added labeled CO was found to occupy the position cis to the unlabeled acetyl group in the product acetylmanganese complex. Furthermore, in the decarbonylation of $CH_3{*}COMn(CO)_5$ the methylmangenese complex formed was found to have the *CO ligand only at positions cis to the methyl group.

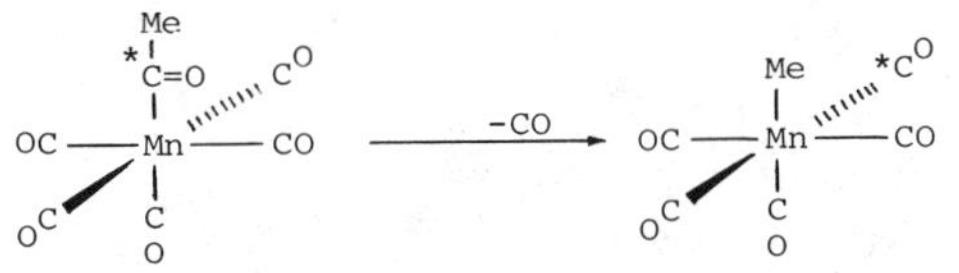

(6.98)

In the reaction of $CH_3Mn(CO)_5$ with a ligand such as PPh_3, the first step is methyl migration, which gives a coordinatively unsaturated acetyl complex $CH_3COMn(CO)_4$, and this is followed by coordination of PPh_3 to the vacant site in the complex.[141d]

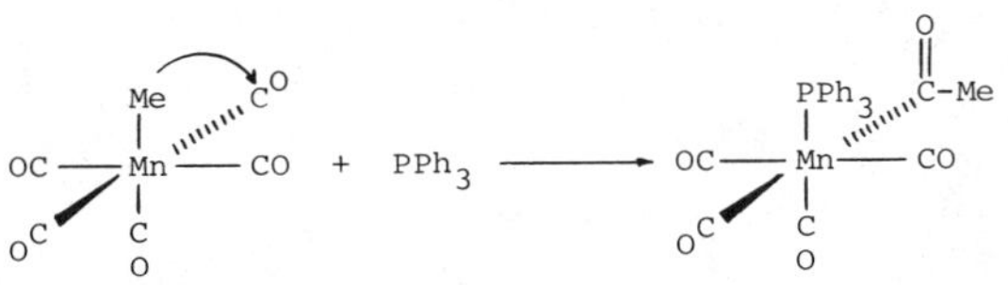

(6.99)

The rate of reaction is enhanced in polar solvents.

In other octahedral complexes trans effects can become important and can complicate the assignment of stereochemistry at the metal atom. However, for the following "pseudotetrahedral" systems[142] trans effects and trans–cis isomerization are absent.

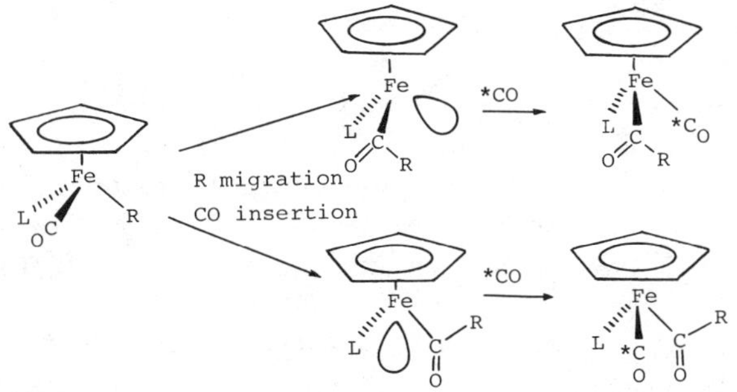

Detailed studies of these systems show that the stereochemistry at the metal depends on various factors, including solvent effects, the nature of the ligand, and the presence or absence of a catalyst. Both formal alkyl migration and CO migration were observed. An equilibrium between a coordinately unsaturated acyl species and an η^2-acyl-coordinated species is implicated.

In square planar complexes trans effects and cis–trans isomerization problems may also be involved. However, complexes that are configurationally stable can exhibit well-behaved reaction patterns. The CO insertion reactions of dialkylpalladium complexes PdR_2L_2 can be explained in terms of alkyl migration mechanisms.[143] The reactions of *cis*- and *trans*-$PdEt_2L_2$ (L = tertiary phosphine) with CO give totally different products depending on the configurations of the starting complexes.

$$cis\text{-}PdEt_2L_2 + CO \longrightarrow EtCHO + C_2H_4 + Pd(CO)_nL_m$$

$$trans\text{-}PdEt_2L_2 + CO \longrightarrow EtCOEt + Pd(CO)_nL_m$$

These reactions are hindered by the addition of L. The above results can be accounted for most consistently by assuming the displacement of L by CO and the subsequent ethyl group migration to the coordinated CO under the configurational constraint of the square planar geometry of the palladium complexes.

```
  Et                        [   Et     ]    [ CH2-CH2  ]
  |         +CO, -L         [   |      ]    [ |   |    ]      C2H4
L-Pd-Et   ==========>       [ L-Pd-Et  ] <=> [ L-Pd  H  ] --->  +
  |                         [   |      ]    [ |        ]      EtCHO
  L                         [   C      ]    [ C        ]
                            [   O      ]    [O   Et    ]

                                (A)             (B)

  Et                        [   Et     ]    [   Et      ]
  |         +CO, -L         [   |      ]    [   |       ]
L-Pd-L    ==========>       [ L-Pd-CO  ] <=> [ L-Pd-COEt ] ---> EtCOEt
  |                         [   |      ]    [           ]
  Et                        [   Et     ]    [           ]

                                (C)             (D)
```

In the reaction of *cis*-$PdEt_2L_2$ with CO the ethyl group migration in the CO-coordinated intermediate (A) gives a T-shaped intermediate (B) that has the ethyl and propionyl groups in "trans" positions. The "trans" configuration of (B) is not suitable for direct reductive elimination of EtCOEt. However, β-hydrogen elimination to liberate ethylene can occur, and the ensuing reductive elimination of the EtCO group and the hydrido ligand produces propionaldehyde. On the other hand, the reaction of *trans*-$PdEt_2L_2$ with CO gives (C) by displacement of L by CO. Subsequent ethyl group migration to the coordinated CO generates a T-shaped intermediate (D) that has the ethyl and propionyl groups in mutually "cis" positions, a configuration that is favorable for direct reductive elimination to liberate EtCOEt.

The reaction products of *cis*- and *trans*-$PdMe_2L_2$ with CO can also be interpreted reasonably by assuming the methyl group migration to the coordinated CO ligand. In these reactions only acetone is formed on treatment of *trans*-$PdMe_2L_2$ with CO, whereas biacetyl(2,3-butanedione) is obtained in addition to acetone in the reaction of *cis*-$PdMe_2L_2$ with CO.

$$\mathrm{L\text{-}Pd(Me)_2\text{-}L}\ \xrightleftharpoons{+CO,\ -L}\ \left[\mathrm{L\text{-}Pd(Me)(Me)(CO)}\right]_{(E)}\ \rightleftharpoons\ \left[\mathrm{L\text{-}Pd(Me)(C(=O)Me)}\right]_{(F)}\ \longrightarrow\ \mathrm{MeCOMe};\quad (F)\xrightarrow{CO}\mathrm{MeC(=O)C(=O)Me}$$

$$\mathrm{L\text{-}Pd(Me)(L)(Me)}\ \xrightleftharpoons{+CO,\ -L}\ \left[\mathrm{L\text{-}Pd(Me)(CO)(Me)}\right]_{(G)}\ \rightleftharpoons\ \left[\mathrm{L\text{-}Pd(Me)\text{-}C(=O)Me}\right]_{(H)}\ \longrightarrow\ \mathrm{MeCOMe}$$

In these reactions the T-shaped intermediate (H) formed by the reaction of *trans*-$PdMe_2L_2$ with CO through (G) is suitable for the facile reductive elimination of acetone, whereas the other T-shaped intermediate (F) formed from *cis*-$PdMe_2L_2$ through (E) has a "trans" configuration. Thus, (F) is susceptible to further coordination and insertion of CO to give MeCOCOMe. In this way all the reaction pathways can be interpreted in terms of alkyl migration under the constraint of square planar geometry. Other conceivable pathways do not satisfactorily account for the reaction products.

The reported evidence so far presented is consistent with the stereochemical retention at the carbon atom bonded to metal in the CO insertion process. The most direct evidence has been provided by NMR examination of the reaction of CpFe(CO)R (R = $CHDCHDCMe_3$) with CO.[144] Evidence for the stereochemical retention at the metal-bonded alkyl group has also been obtained in combination with stereochemical inversion at the metal-bonded carbon in oxidative additions of acyl halide.[63a,145] The acyl group formed by CO insertion into an M–C bond is sometimes η^2 bonded with the metal through the carbonyl carbon and oxygen atoms. The molecular structure of such an η^2-acyl metal complex formed by reaction of $Cp_2Ti(CO)_2$ with MeCOCl[146] is shown in Figure 6.26.

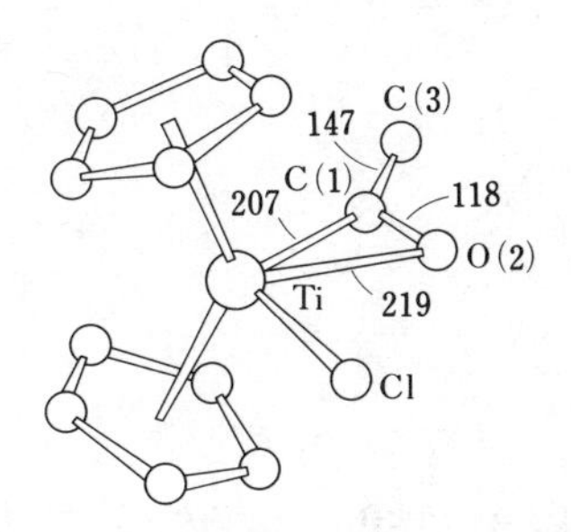

FIGURE 6.26. Molecular structure of $Cp_2Tie(COMe)Cl$; in picometers. (From Fachinetti et al.[146])

The carbonyl carbon in the complex has carbenelike character, and its behavior can be accounted for by assuming the following resonance contributions to the structure.

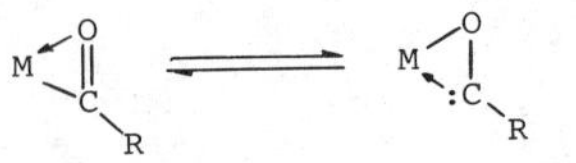

(6.100)

The acyl group in this bonding mode may be regarded as a three-electron ligand in terms of the definition that regards a simple η^1-acyl group as a one-electron ligand. The carbenelike character of η^2-acyl complexes sometimes confers unique properties on them. The following are examples of the chemical behavior of the η^2-acyl complexes formed by CO insertion reactions into M–C bonds of pentamethylcyclopentadienyl(C_5Me_5)zirconium complexes.[147,148]

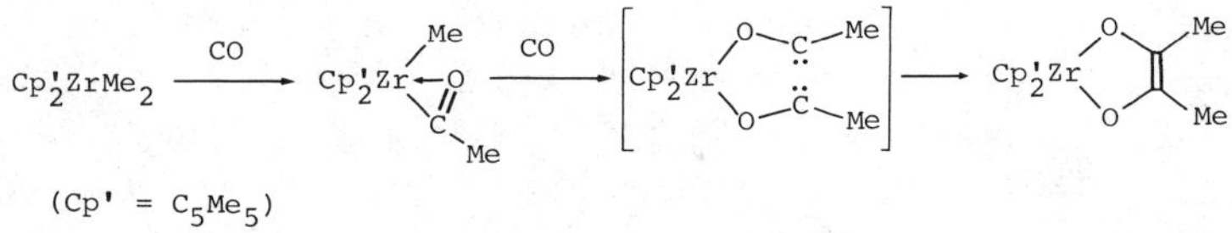

$$Cp'_2ZrMe_2 \xrightarrow{CO} Cp'_2Zr(\eta^2\text{-}COMe)Me \xrightarrow{CO} [\ldots] \longrightarrow \ldots \quad (6.101)$$

$(Cp' = C_5Me_5)$

$$(C_5Me_5)_2Zr(C_4H_8) + CO \longrightarrow \ldots \quad (6.102)$$

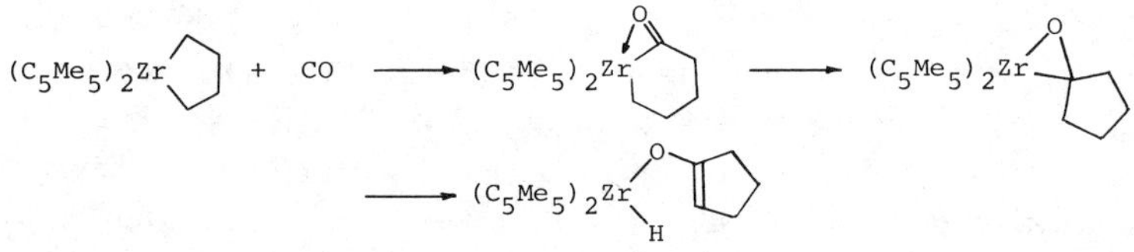

Alkyl migration to coordinated CO is greatly accelerated by addition of a Lewis acid, probably as a result of prior coordination of the Lewis acid. An adduct formed between an acetylmanganese complex and the Lewis acid has been structurally characterized.[149]

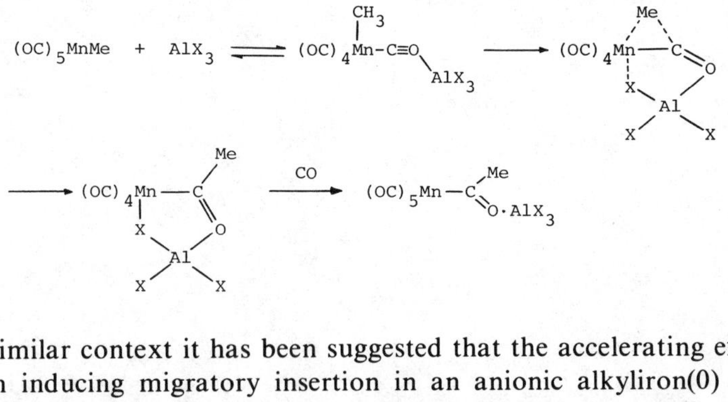

$$(OC)_5MnMe + AlX_3 \rightleftharpoons (OC)_4Mn\text{-}C(CH_3){=}O\text{-}AlX_3 \longrightarrow \ldots \xrightarrow{CO} (OC)_5Mn\text{-}C(Me){=}O\cdot AlX_3 \quad (6.103)$$

In a similar context it has been suggested that the accelerating effect of a cation in inducing migratory insertion in an anionic alkyliron(0) carbonyl complex in nonpolar solvents, where a tight ion pair is formed, is caused by coordination of the countercation to the oxygen atom of the carbonyl group.[150]

$$\left[(OC)_2(R)Fe^{-}(CO)-CO\cdots Na^{+} \right] \rightleftharpoons \left[(OC)_3Fe\text{---}C(R)\text{---}O^{-}\,Na^{+} \right] \xrightarrow{L} (OC)_3LFe\text{---}C(R)\text{---}O^{-}\,Na^{+} \quad (6.104)$$

To understand and design further synthetic reactions using CO, a basic knowledge of the thermodynamics and kinetics of CO insertion and decarbonylation is required. The first such study on the insertion of CO into methylpentacarbonylmangenese was reported by Calderazzo and Cotton.[151]

$$CH_3Mn(CO)_5 + CO \underset{k_{-1}}{\overset{k_1}{\rightleftharpoons}} CH_3C(=O)Mn(CO)_5$$

The following thermodynamic and kinetic parameters for the forward and reverse reactions have been obtained.

$$\Delta H^\circ = -52.9 \qquad \Delta H_1^\ddagger = 59.4 \qquad \Delta H_{-1}^\ddagger = 115.5 \text{ kJ mol}^{-1}$$

$$\Delta G^\circ = -14.9 \qquad \Delta G_1^\ddagger = 86.2 \qquad \Delta G_{-1}^\ddagger = 100.8 \text{ kJ mol}^{-1}$$

$$\Delta S^\circ = -125.5 \qquad \Delta S_1^\ddagger = -88.3 \qquad \Delta S_{-1}^\ddagger = 49.0 \text{ J K}^{-1} \text{ mol}^{-1}$$

For a series of $RMn(CO)_5$ complexes the enthalpies of formation and the Mn–C bond dissociation energies have been estimated (Table 3.3).[152] The enthalpies for the following reactions involving different R groups can be derived from the data.

$$RMn(CO)_5(c) + CO(g) \longrightarrow RCOMn(CO)_5(c) \quad (6.105)$$

The $-\Delta H$ values for reaction (6.105) are 63 ± 8 for phenyl, 54 ± 8 for methyl, and 12 ± 7 kJ mol^{-1} for CF_3, respectively. If one takes into account $-\Delta S^\circ$ (≈ 150 kJ^{-1} mol^{-1}) for the incorporation of gaseous CO into the solid acylmanganese complex, ΔG° for reaction (6.105) becomes positive for the CO insertion into the CF_3–Mn bond, whereas the ΔG° values for the phenyl and methyl complexes remain negative. Thus, the experimental fact that $CH_3Mn(CO)_5$ can be easily carbonylated, whereas $CF_3Mn(CO)_5$ is resistant to carbonylation under pressure, and that $CF_3COMn(CO)_5$ is readily decarbonylated to $CF_3Mn(CO)_5$ can be understood.

Other thermodynamic data available to date[153-156] clearly show that carbonylation becomes less favorable when the metal-bonded alkyl or aryl group becomes more electronegative. Figure 6.27 shows the variation of enthalpy for the following carbonylation reactions.[153]

$$2\ PtL(CO)(Cl)(R) \overset{K}{\rightleftharpoons} [L(RCO)Pt(\mu\text{-}Cl)_2Pt(COR)L] \quad (6.106)$$

L = $PMePh_2$ R = C_6H_4Y

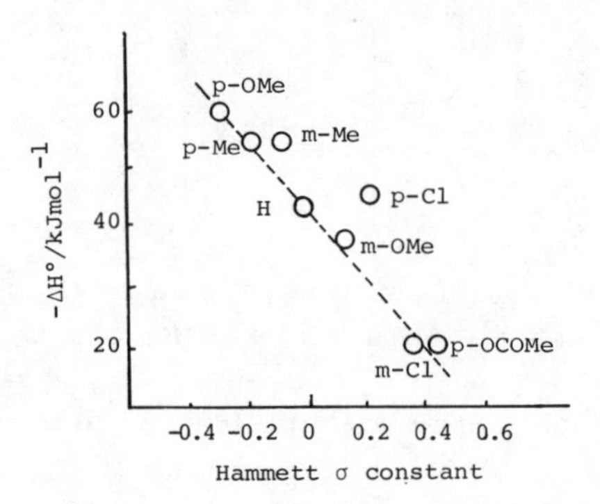

FIGURE 6.27. Variation of enthalpy on change of substituent Y in substituted phenylplatinum complexes in reaction (6.106).

Similarly, the equilibrium constant for the following decarbonylation reaction decreases in the order aryl > Me > primary alkyl.[154]

$$RhCl_2L_2(C(O)R) \overset{K}{\rightleftharpoons} RhCl_2L_2(R)(CO) \qquad (6.107)$$

These results suggest that transition metal alkyls with strong M–C bonds are thermodynamically unfavorable for carbonylation. This can be understood as follows.[139] In the carbonylation reaction

$$RM(CO)_n \overset{K}{\rightleftharpoons} RCOM(CO)_{n-1}$$

an M–R bond and an M–CO bond are cleaved, whereas an R–CO bond and an M–COR bond are formed. Therefore, the enthalpy of carbonylation can be expressed as

$$\Delta H^{\circ}_{solid} = E(M{-}R) + E(M{-}CO) + E(C{\equiv}O) - \{E(C(O){-}R) + E(M{-}C(O)R) + E(C{=}O)\}$$

where E denotes the bond energy contribution of the respective bond. The enthalpy of carbonylation in the solid state, ΔH°_{solid}, is approximately equal to the enthalpy of carbonylation in solution, ΔH°_{soln}, because the enthalpies of solvation for the acyl and alkyl complexes will be approximately equal. Since $E(C{\equiv}O)$ and $E(C{=}O)$ and $E(M{-}CO)$ are invariant with the nature of R, the enthalpy of carbonylation can be written

$$\Delta H^{\circ}_{soln} \approx \Delta H^{\circ}_{solid} \approx [E(M{-}R) - E(C(O){-}R) - E(M{-}C(O)R)] + \text{const}$$

Compared to the variation of E(M–R) with different R's, the variation of E[M–C(O)R] would be generally smaller since the effect of R on the M–C bond in the metal acyl complex would be smaller. The variation of E(R–CO) with R would not be too great either. Thus, in carbonylation of closely related transition metal alkyl complexes, the bond energy contribution to the metal alkyl will be the dominant factor in determining the equilibrium; the stronger the metal alkyl bond, the less favorable will be carbonylation.[139] This statement should be taken with caution when comparing the reactions of metal alkyls of widely different natures, but it is applicable to reactions of a series of closely related alkyls.

In most kinetic studies it has been found that the reaction pathway can be described by the two-step mechanisms shown in Eq. (6.108).[141]

$$L_nM(R)\text{-}CO \underset{k_{-1}}{\overset{k_1}{\rightleftharpoons}} \left[L_nM\text{-}C(=O)R \right] \underset{k_{-2},\ -L}{\overset{k_2,\ +L}{\rightleftharpoons}} L_{n+1}M\text{-}COR \qquad (6.108)$$

In the first step a coordinatively unsaturated species is formed by intramolecular migratory CO insertion, and the second step involves the addition of the external ligand L (tertiary phosphine or CO), affording the coordinatively saturated acyl complex. When pseudo-first-order reaction conditions are attained in the presence of a high concentration of L, the reaction rate can be expressed as

$$-\frac{d[RM(CO)L_n]}{dt} = k_{obs}[RM(CO)L_n] \quad \text{where } k_{obs} = \frac{k_1 k_2 [L]}{k_{-1} + k_2 [L]} \qquad (6.109)$$

From the dependence of k_{obs} on [L] the k_1 and k_{-1}/k_1k_2 values can be obtained experimentally from a plot of $1/k_{obs}$ against $1/[L]$.

The rates of carbonylation have been reported to decrease as the electron-withdrawing power of the substituent attached to the metal alkyl increases. A linear free energy relationship (LFER) between the rate of carbonylation and Taft's σ^* constant has been observed for the following reactions.[157]

$$RCH_2Mn(CO)_5 + CO \longrightarrow RCH_2COMn(CO)_5$$

$$RCH_2Fe(CO)_4^- + PPh_3 \longrightarrow RCH_2COFe(CO)_3(PPh_3)^-$$

The kinetics of carbonylation of metal alkyls that do not contain coordinated CO are different from those of alkyl metal carbonyls. Nevertheless, in the reaction of *trans*-$MRI(PPh_3)_2$ (M = Ni, Pd, Pt) with CO, the influence of the R group on the carbonylation rate is similar to that discussed above, namely, that as R is made more electron withdrawing, the rate of CO insertion decreases. In this case there are two parallel routes for carbonylation, one involving direct alkyl migration to the coordinated CO in a five-coordinate alkyl species and the other involving a ligand dissociation route. For both routes the more electron-releasing alkyl group acts to enhance the reaction rate.[158]

In the forward process of carbonylation (6.108) the coordinatively unsaturated species is formed by alkyl migration. The situation for the reverse process, decarbonylation of the coordinated saturated acyl complex, is somewhat different in that a vacant site has to be produced at the position cis to the acyl ligand in order to accommodate the alkyl ligand. This may be achieved by dissociation of a ligand, and the rate dependence on the nature of the R group in the acyl complex may be associated with the rate of dissociation of L. Thus, the kinetic parameters for decarbonylation should be interpreted with caution because they may not directly reflect the ease of alkyl migration.

Some rate studies of the decarbonylation of coordinatively unsaturated acyl complexes have been reported.[159]

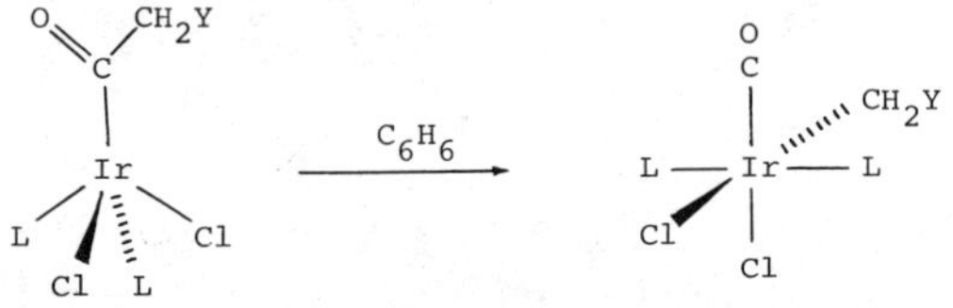

$Ir(COCH_2Y)Cl_2L_2 \xrightarrow{C_6H_6} Ir(CO)(CH_2Y)Cl_2L_2$ (6.110)

The rates of decarbonylation of the acyliridium(III) complexes decrease as the substituent Y becomes more electron withdrawing, the order being $p\text{-}CH_3OC_6H_4 > p\text{-}CH_3C_6H_4 > C_6H_5 > p\text{-}NO_2\text{-}C_6H_4 >> C_6F_5$. Although the starting acyliridium complex is coordinatively unsaturated, all the cis positions are occupied, thus apparently hindering alkyl migration. Clearly, a rearrangement must occur before decarbonylation can proceed, and the effect of the alkyl group might be associated with the rearrangement process. Further kinetic studies are needed to understand these decarbonylation processes.

When different acyl ligands are attached to a transition metal atom, the relative rates of decarbonylation can be directly compared. Casey has compared the decarbonylation products from manganese and rhenium complexes that have both acetyl and benzoyl groups.[160]

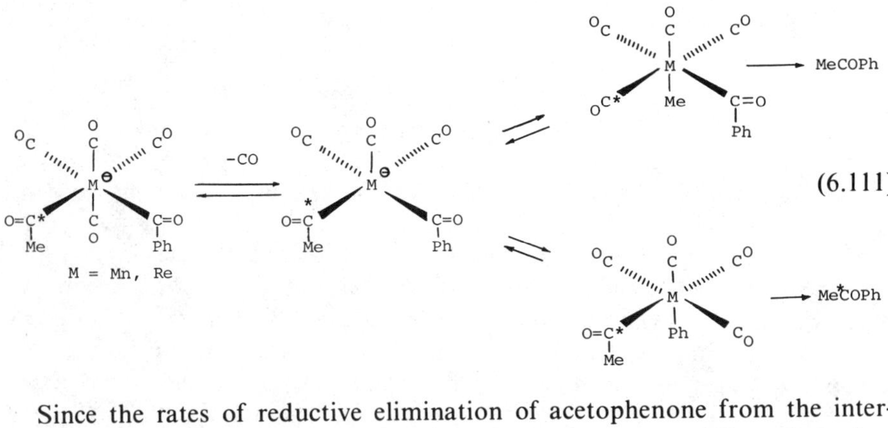

(6.111)

M = Mn, Re

$[M(CO)_4(C^*OMe)(COPh)]^- \underset{}{\overset{-CO}{\rightleftharpoons}} [M(CO)_3(C^*OMe)(COPh)]^- \rightleftharpoons M(CO)_3(C^*O)(Me)(COPh) \longrightarrow MeCOPh$

$[M(CO)_3(C^*OMe)(COPh)]^- \rightleftharpoons M(CO)_4(C^*OMe)(Ph) \longrightarrow MeC^*OPh$

Since the rates of reductive elimination of acetophenone from the intermediate acyl–alkyl complexes are rapid for manganese, studies of the less

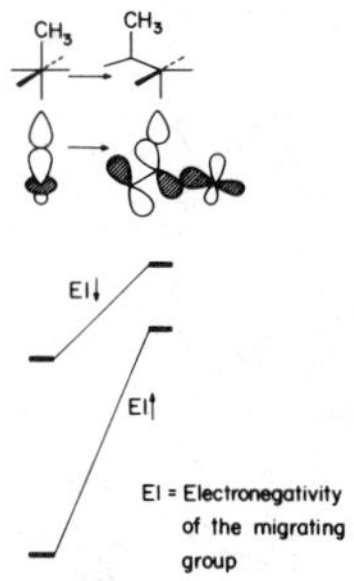

FIGURE 6.28. Effect of electronegativity change of migrating group in migratory CO insertion reaction of $RMn(CO)_5$. (Reprinted with permission from T. Berke and R. Hoffmann, *J. Am. Chem. Soc.* **100,** 7224. Copyright 1978 American Chemical Society.

reactive rhenium system are more informative. Detailed examination of the reaction products with the labeled CO ligand revealed that decarbonylation from MeCO–Re is *kinetically* favored over decarbonylation from PhCO–Re, whereas decarbonylation from PhCO–Re is *thermodynamically* favored over that from the MeCO–Re. It was also found that CO insertion into the Me–Re bond is favored over CO insertion into the Ph–Re bond.

Theoretical studies on CO insertion into $MeMn(CO)_5$ by the extended Hückel MO method have been carried out.[161] The results are consistent with the alkyl migration mechanism, but a process in which the CO group bends to meet the alkyl half way was implicated. The activation energy for methyl migration was calculated to be 80 kJ mol^{-1}, and an increase in the activation energy with an increase in the electronegativity of the alkyl group was found, in agreement with experimental results. The essential feature of the shape of the activated complex is shown in Figure 6.28. The electronegativity change affects the metal–alkyl σ bond more than the acyl–metal bond, thus lowering the energy of the starting complex and increasing the activation energy.

Recently, a more detailed ab initio study has been carried out on the insertion of CO into $PtCH_3(CO)(PH_3)F$.[162] The most probable processes were calculated to be (*a*) methyl migration from (A) to a T-shaped species (B) or (*b*) concerted movement of the methyl and CO groups with concomitant opening of the FPtP angle in (A) to give (C).

(*a*) Methyl migration to CO:

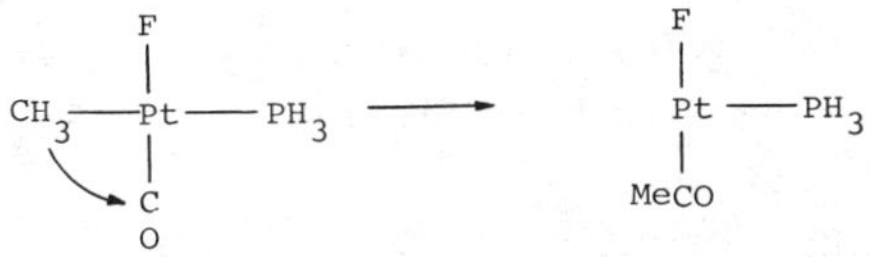

F
|
CH_3—Pt—PH_3 ⟶ Pt—PH_3 (with F above Pt, MeCO below)
|
C
O

(A) (B)

(*b*) Concerted movement of the methyl and CO groups:

$$\underset{\text{(A)}}{CH_3-Pt(F)(CO)-PH_3} \longrightarrow \underset{\text{(C)}}{F-Pt(COMe)-PH_3}$$

The insertion of CO into an M–H bond to give a formyl complex has been proposed as an important elementary step in the hydrogenation of carbon monoxide to ethylene glycol.[163] Direct evidence for CO insertion into an M–H bond is extremely limited and only a few cases are known, such as the reactions of an octaethylporphyrin-rhodium hydride[164] and a thorium hydride[165] with CO.

The enthalpy for CO insertion for the reaction

$$[Mn(CO)_5H](g) + CO(g) \longrightarrow [Mn(CO)_5CHO](g)$$

has been estimated to be 20 kJ mol^{-1} based on the known values of enthalpies of formation (Table 3.3). Taking additional account of a negative entropy of 45 kJ mol^{-1} at 300 K for incorporating gaseous CO into the formyl complex, it is clear that CO insertion into an M–H bond is a thermodynamically unfavorable process.

Although η^1-formyl complexes can be prepared by other routes (e.g., by hydrogenation of coordinated CO with borohydride), they readily undergo decarbonylation.[166] However, when the formyl complex thus obtained is coordinatively unsaturated and when the metal has a high affinity for the oxygen atom, thus favoring η^2-(C,O) coordination of the formyl group, CO insertion into an M–H bond may become thermodynamically favorable. The case of CO insertion into the Th–H bond provides an example of this behavior. The η^2-acyl complex has carbenelike character, and insertion of CO into the thorium hydride complex gives an ethylene glycol derivative by the following route.

$$(C_5Me_5)_2Th(H)(OR) + CO \longrightarrow \left[(C_5Me_5)_2Th(OR)(\eta^2\text{-}:O{=}CH) \longleftrightarrow (C_5Me_5)_2Th(OR)(\eta^2\text{-}O\text{-}:CH)\right]$$

$$\longrightarrow 1/2\ (C_5Me_5)_2Th(OR)-O-CH{=}CH-O-Th(C_5Me_5)_2(OR)$$

The insertion of CO into the porphyrin-rhodium hydride seems to be a special case, and the delocalized porphyrin ring may play a particular role in labilizing the Rh–H bond.

Another interesting problem is the feasibility of double CO insertion into an M–R bond. Studies on decarbonylation of a pyruvoyl complexes,

$MeCOCOMn(CO)_5$, indicated that CO insertion into the CH_3CO–Mn bond is thermodynamically unfavorable.[167] *Trans*-$PhCOCOPdCl(PMePh_2)_2$ is also readily decarbonylated to give *trans*-$Pd(COPh)Cl(PMePh_2)_2$.[168]

An example of double CO insertion has been reported in the reaction of benzyl chloride with cobalt carbonyl under phase transfer reaction conditions.[169]

The following double CO insertion mechanism has been proposed.

$$ArCH_2Co(CO)_4 \underset{}{\overset{CO}{\rightleftharpoons}} ArCH_2C(=O)Co(CO)_4 \underset{}{\overset{CO}{\rightleftharpoons}} ArCH_2C(=O)C(=O)Co(CO)_4$$

However, it has been pointed out that benzyl chloride may give thermodynamically unfavorable double-carbonylation products because the initially formed acyl might be converted into an enol form. This vinylcobalt intermediate could then undergo further CO insertion[169b,170]

$$Ph{-}CH_2{-}C(=O){-}Co(CO)_4 \rightleftharpoons Ph{-}CH{=}C(OH){-}Co(CO)_4 \xrightarrow{CO} PhCH{=}C(OH){-}C(=O){-}Co(CO)_4$$

Dihapto-acyl lanthanoid complexes are also involved in a special type of double carbonylation reaction.[171] Carbon monoxide reacts with $Cp_2Lu(CMe_3)(thf)$ at room temperature to form the diphaptoacyl complex $Cp_2Lu(\eta^2\text{-}COCMe_3)$, which reacts with additional CO to form a bimetallic complex where the two metal atoms are bridged by an enedione diolate ligand.

$$Cp_2LuCMe_3(thf) \xrightarrow{{}^*CO} [Cp_2Lu{-}\overset{*}{C}(=O){-}CMe_3 \longleftrightarrow Cp_2Lu(\eta^2\text{-}O{-}\overset{*}{C}CMe_3)]$$

$$\xrightarrow{{}^*CO} [Cp_2Lu(\mu\text{-}OC(=C{-}O)\ldots)]\ \text{(dimer: two } Cp_2Lu \text{ units bridged by } Me_3C{-}C(=O){-}C(O){=}C(O){-}C(=O){-}CMe_3\text{)}$$

Reactions involving dinuclear η^2-acyl complexes are attracting increasing attention.[147,172,173]

b. Insertion Reactions of Substrates Other Than CO

Insertion of SO_2 into a transition metal–carbon bond gives an *S*—sulfinate complex.[137,174]

$$CpFe(CO)_2R + SO_2 \xrightarrow{-40\ ^\circ C} Cp(CO)_2Fe{-}S(=O)_2{-}R$$

Other possible SO_2 insertion products are

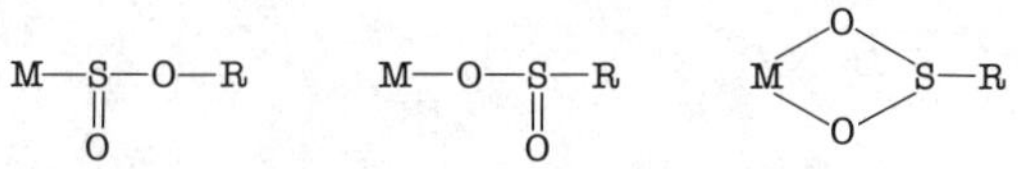

M—S(=O)—O—R   M—O—S(=O)—R   M(O)(O)S—R

The insertion products formed by insertion of SO_2 into main-group metal--carbon bonds are *O*-sulfinates.

There seems to be several reaction pathways for SO_2 insertion reactions into a transition metal–alkyl bond. In the following reaction the stereochemistry of the SO_2 insertion has been studied.[175,176]

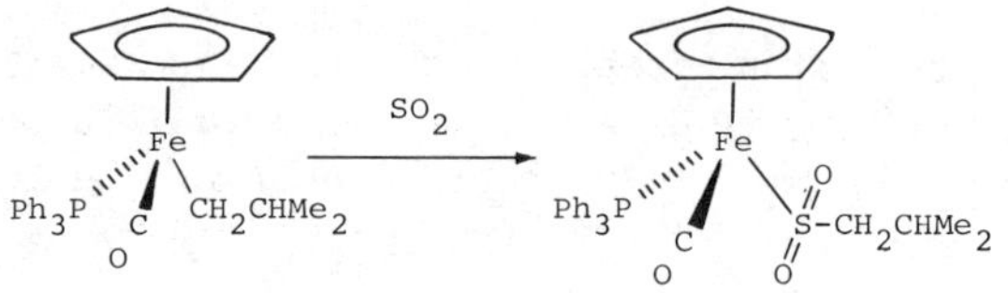

CpFe(PPh$_3$)(CO)CH$_2$CHMe$_2$ $\xrightarrow{SO_2}$ CpFe(PPh$_3$)(CO)S(O)$_2$-CH$_2$CHMe$_2$

The stereochemistry about the pseudotetrahedral, central iron atom is retained, whereas inversion occurs at the α carbon atom bound to iron. An S_N2 attack of the incoming SO_2 molecule on the alkyl group followed by rearrangement of the R–SO_2 entity to give the sulfinate ligand has been proposed.

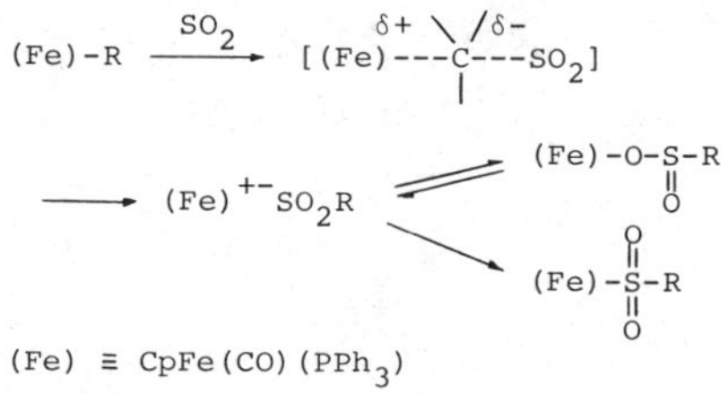

$$\text{(Fe)-R} \xrightarrow{SO_2} [\text{(Fe)}^{\delta+}\text{---C---}SO_2^{\delta-}] \longrightarrow \text{(Fe)}^{+-}SO_2R \rightleftharpoons \text{(Fe)-O-S(=O)-R}$$

$$\text{(Fe)}^{+-}SO_2R \longrightarrow \text{(Fe)-S(=O)}_2\text{-R}$$

(Fe) ≡ CpFe(CO)(PPh$_3$)

Isonitriles (isocyanide) have the following resonance formulas and show somewhat similar behavior to CO.[177,178]

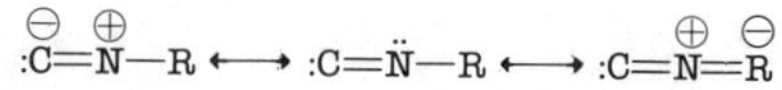

$$:\overset{\ominus}{C}{=}\overset{\oplus}{N}{-}R \longleftrightarrow :C{=}\ddot{N}{-}R \longleftrightarrow :C{=}\overset{\oplus}{N}{=}\overset{\ominus}{R}$$

In contrast to carbon monoxide, however, for which no established example of a double-insertion process has been reported, isonitriles are known to undergo multiple-insertion reactions.

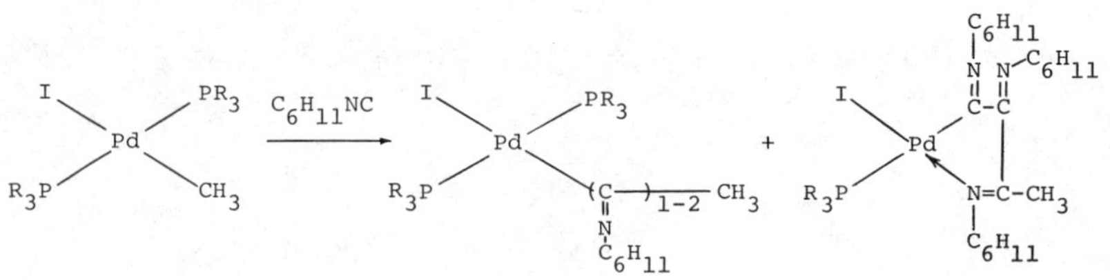

$$\text{Pd(I)(PR}_3\text{)(R}_3\text{P)(CH}_3) \xrightarrow{C_6H_{11}NC} \text{Pd(I)(PR}_3\text{)(R}_3\text{P)}(C(=NC_6H_{11}))_{1-2}CH_3 + \text{Pd(I)(R}_3\text{P)}[C(=NC_6H_{11})\text{-}C(=NC_6H_{11})\text{-}C(CH_3)=NC_6H_{11}]$$

The thiocarbonyl ligand is also very similar to the carbonyl ligand, but it differs from the carbonyl ligand in undergoing ready insertion into an M–H bond, affording a thioformyl complex.[179]

$$OsHCl(CS)(CO)(PPh_3)_2 \xrightarrow{CO} OsCl(CHS)(CO)_2(PPh_3)_2$$

Insertion of NO into metal–alkyl bonds received little attention until recently. A kinetic study of the reaction of PPh_3 with CpCo(NO)R has revealed that NO migratory insertion follows a similar kinetic pattern to that of CO insertion into the Mn–R bond in the reaction of $RMn(CO)_5$ with L (CO and PR_3).[180]

$$CpCo(NO)R \underset{k_{-1}}{\overset{k_1}{\rightleftharpoons}} CpCo{-}N(=O)R \xrightarrow[k_2]{L} CpCo(L){-}N(=O)R$$

fast

R = Me, Et, L = PPh_3

The product formed by insertion of NO into the Co–Et bond has been shown to contain an η^1-bonded nitrosoethane ligand.

Another type of reaction involving NO is the addition of olefins to coordinated NO to form chelate complexes, which can be hydrogenated to give diamines.[181]

$$CpCo(NO)_2 + R^1R^2C{=}CR^3R^4 \longrightarrow CpCo[N(O)CR^1R^2{-}CR^3R^4N(O)] \xrightarrow{LiAlH_4} H_2N{-}CR^1R^2{-}CR^3R^4{-}NH_2$$

(cis addition)

Insertion reactions of CO_2 and CS_2 were described in Section 4.7 and those of carbenes will be discussed later.

Although insertions of ketones, aldehydes, and esters into an M–C bond are well known for nontransition metal alkyls, examples of corresponding insertions into transition metal–carbon bonds are limited to some early transition metals.[182] Insertions of a C=O group into an M–H or M–C bond in group 8-10 transition metal complexes is virtually unknown. This is probably due to the kinetic instability of alkoxides of the later transition metals.[183] When the alkoxide has an electronegative substituent, a thermally stable alkoxide can be isolated.[215]

$$RuH_2L_4 + CF_3COPh \xrightarrow{-L} L_3Ru(H)(O{-}CH(Ph)CF_3)$$

$(L = PPh_3)$

c. Insertion and Deinsertion of Olefins

The insertion of olefins into M–C or M–H bonds constitutes one of the most important elementary steps in catalytic reactions of olefins, including hydrogenation, polymerization, and hydroformylation. Insertion of an olefin into an M–H bond is reversible; the reverse process is called β elimination, β-hydrogen elimination, or β-hydride transfer–alkene elimination.

In the olefin insertion reaction into an M–H bond, the olefin is thought to be activated on coordination to a transition metal; this activates the M–H bond as well to form the insertion product, a metal alkyl. The process may be regarded as a migratory transfer of the hydride ligand to the coordinated olefin, which is similar to the alkyl migration to coordinated CO observed in the CO insertion reaction.

$$M{-}H + C{=}C \rightleftharpoons H{-}M\cdots\|(C{=}C) \rightleftharpoons \left[\begin{matrix} H\cdots C \\ \vdots \quad | \\ M\cdots C \end{matrix} \right]^{\ddagger} \rightleftharpoons M{-}{}^{\alpha}C{-}{}^{\beta}C{-}H \quad (6.112)$$

As the CO insertion reaction, two cases are conceivable: one is the olefin migration to the M–H (or M–C) bond and the other is the hydride (or alkyl) transfer, as shown in Eq. (6.112). No established evidence to support either of the two cases has been provided. Thus, disregarding the mechanistic implications, we employ the term *insertion.*

In olefin insertion into an M–H bond or in the addition of a metal hydride to an olefin, *cis addition* has been established, a result consistent with an intramolecular process such as that shown in Eq. (6.112).

Unequivocal examples where transition metal alkyls have been obtained from olefin insertion into a metal–hydride bond are still limited. This may be due to the kinetic instability of transition metal alkyls, which have β-hydrogen atoms. These alkyls are susceptible to β-hydrogen elimination to give the initial hydride complex or to other decomposition processes. The previously described ethylplatinum complex *trans*-$PtEt(Br)(PEt_3)_2$ (Section 4.2) is one of the few transition metal alkyls prepared by insertion of ethylene into an M–H bond.[184] This process takes place more readily when a cationic platinum hydride is employed.[185]

$$[PtH(PEt_3)(Et_3P)(O{=}CMe_2)]^+ + C_2H_4 \longrightarrow [PtEt(PEt_3)(Et_3P)(O{=}CMe_2)]^+ \quad (6.113)$$

In this insertion process an intermediate *trans*-hydrido-ethylene complex has been isolated. In reaction (6.113) ethylene displaces acetone to give *trans*-$[PtH(C_2H_4)(PEt_3)_2]^+$, which is thought to rearrange to a cis geometry to allow ethylene insertion to proceed.[186] In the reverse reaction ethylene is extruded from *trans*-$PtEt(Br)(PEt_3)_2$ on heating and *trans*-$PtH(Br)(PEt_3)_2$ is formed. In both the forward and reverse processes, dissociative mechanisms seem to be involved. When the dissociative paths are blocked, the reactions are forced to go through five-coordinate intermediates, which require higher activation energies.

A hydrido-ethylene complex has been prepared by protonation of an ethylene complex, and a σ–π rearrangement has been observed.[187]

$$CpRh(C_2H_4)(PMe_3) + HBF_4 \longrightarrow [CpRhH(C_2H_4)(PMe_3)]^+BF_4^- \rightleftharpoons [CpRh(C_2H_5)(PMe_3)]^+BF_4^- \quad (6.114)$$

In the corresponding hydrido-ethylene complex of ruthenium, equilibrium can be shifted to the more stable ethyl complex by adding a tertiary phosphine.[188]

$$(C_6H_6)Ru(PMe_3)(C_2H_4) \xrightarrow[(ii)\ NH_4PF_6]{(i)\ CF_3COOH} [(C_6H_6)RuH(PMe_3)(C_2H_4)]^+PF_6^- \rightleftharpoons [(C_6H_6)Ru(PMe_3)(Et)]^+PF_6^- \xrightarrow{PMe_3} [(C_6H_6)Ru(PMe_3)_2(Et)]^+PF_6^-$$

When the olefin has an electronegative substituent, the alkyl complex becomes more stable and can be easily isolated.[186,189]

$$trans\text{-}RhH(CO)(PPh_3)_2 + C_2F_4 \longrightarrow trans\text{-}Rh(CF_2CF_2H)(CO)(PPh_3)_2 \quad (6.115)$$

$$trans\text{-}PtH(Cl)(PEt_3)_2 + C_2F_4 \longrightarrow trans\text{-}Pt(CF_2CF_2H)Cl(PEt_3)_2$$

Alkylzirconium complexes formed by alkene insertion into the Zr–H bond in $Cp_2ZrH(Cl)$ are also stable.[190]

In the insertion of monosubstituted olefin into an M–H bond, two insertion modes are possible. One gives a branched alkyl (Markownikoff addition, B) and the other gives a straight chain alkyl (anti-Markownikoff addition, A).

$$L_nM\text{-}H + RCH_2CH{=}CH_2 \overset{A}{\rightleftharpoons} \left[L_nM\cdots CH_2{=}CH(CH_2R)\cdots H\right]^{\ddagger} \rightleftharpoons L_nM\text{-}CH_2\text{-}CH_2\text{-}CH_2R \quad (6.116)$$

$$L_nM\text{-}H + RCH_2CH{=}CH_2 \overset{B}{\rightleftharpoons} \left[L_nM\cdots CH(CH_2R){=}CH_2\cdots H\right]^{\ddagger} \rightleftharpoons L_nM\text{-}CH(CH_3)(CH_2R)$$

If olefin insertion into an M–H bond and β-hydrogen elimination take place reversibly, reaction pathways for catalytic H–D exchange and isomerization of olefins are provided. For example, β-hydrogen elimination from a metal alkyl formed by olefin insertion into an M–D bond will give an olefin labeled with deuterium and a metal hydride. Also, elimination of a β-hydrogen atom from the RCH_2 group in the alkyl group produced by route B in Eq. (6.116) will liberate a 2-alkene, so that an internal olefin is formed by isomerization of a terminal olefin.

There are numerous studies reporting H–D scrambling,[191] skeletal alkyl isomerization,[192] and double-bond isomerization.[190] In these isomerizations involving insertion–elimination mechanisms, the steric effects of other ligands play important roles in determining the direction of isomerization. A representative example is the formation of the same straight-chain zirconium alkyl by insertion of isomeric terminal or internal olefins into a Zr–H bond.[190,193]

$$Cp_2Zr(Cl)(H) + [\text{isomeric octenes}] \longrightarrow Cp_2Zr(Cl)(n\text{-octyl}) \quad (6.117)$$

Although in most cases the direction of skeletal isomerization of the alkyl chain is from branched to straight chain, isomerization in the reverse direction has been observed when the alkyl group has an electronegative substituent.[194]

$$CpFe(CO)(PPh_3)-CH_2CH_2CN \xrightarrow[\text{toluene}]{\Delta} CpFe(CO)(PPh_3)-CH(CN)CH_3 \quad (6.118)$$

In this case the electronic stabilizing effect of the substituent on the alkyl group is considered to surpass the negative steric effect.

Alkynes generally give cis insertion into M–H bonds, as would be expected from a four-centered insertion mechanism similar to that shown in Eq. (6.112).[186]

$$trans\text{-}[PtH(Cl)(PEt_3)_2] + CF_3C{\equiv}CCF_3 \longrightarrow Cl-Pt(PEt_3)_2-C(CF_3){=}C(H)CF_3 \quad (6.119)$$

Occasionally, however, trans addition products are observed.[195] In certain cases the reaction proceeds by a radical mechanism.[196] It is often not easy to confirm the stereochemistry of the kinetic product if subsequent cis–trans isomerization occurs and the stereochemistry of insertion has been the subject of debate. When the mode of coordination of the alkyne is controlled by the arrangement of transition metal atoms in a cluster complex, trans addition can be forced to take place.[197]

The insertion of olefins into metal–carbon bonds is an important elementary step in various catalytic reactions such as polymerization and oligomerization, but direct evidence for this is still scarce. This may arise from the kinetic instability of the metal alkyl formed by the olefin insertion. After the first insertion of the olefin, the metal alkyl formed may readily undergo further reactions, such as successive olefin insertion and β-hydrogen elimination. It is not easy to hinder these subsequent reactions by adding other stabilizing ligands because coordinative unsaturation is required for the first olefin insertion into the M–C bond. With olefins that have electronegative substituents, the insertion products have been isolated, presumably because of the unavailability of the low-energy pathways for thermolysis of the metal alkyls produced.[198,199]

$$\text{trans-PtCH}_3\text{(X)(PEt}_3)_2 + \text{CF}_2\text{=CF}_2 \longrightarrow \text{trans-CH}_3\text{CF}_2\text{CF}_2\text{PtX(PEt}_3)_2$$

X = halogen

$$\text{CpFe(CO)}_2\text{Et} + \text{(NC)}_2\text{C=C(CN)}_2 \longrightarrow \text{CpFe(CO)}_2\text{C(CN)}_2\text{C(CN)}_2\text{Et}$$

The usually difficult insertion of ethylene into a metal–alkyl bond has recently been achieved by using ethylene under pressure.

$$\text{(acac)Ni(Me)(PCy}_3) \xrightarrow[\text{30 atm}]{\text{C}_2\text{H}_4} \text{(acac)Ni(R)(PCy}_3)$$

R = Et, Pr, Bu

The product was found to be a mixture of nickel ethyl, propyl, and butyl complexes.[200] The ethyl complex is considered to be formed by ethylene insertion into a nickel hydride species produced by β-hydrogen elimination from the Ni–Pr species generated by ethylene insertion into the Ni–Me bond. The butyl complex may be produced by ethylene insertion into the Ni–Et bond. Another example of this process has also been reported recently.[200b]

$$\text{CpNi(R)(C}_2\text{H}_4) \longrightarrow \{\text{CpNi-CH}_2\text{CH}_2\text{R}\} \xrightarrow[\text{50 bar}]{\text{C}_2\text{H}_4} \text{CpNi(CH}_2\text{CH}_2\text{R)(C}_2\text{H}_4)$$

Indirect evidence supporting ethylene insertion into a Co–methyl bond has been reported.[201] The reaction of $CpCo(CD_3)_2(PPh_3)$ with ethylene releases CD_3H and $CD_3CH{=}CH_2$. The result is consistent with ethylene insertion, β-hydrogen elimination, and the succeeding CD_3–H reductive elimination pathways.

$$CpCo(CD_3)_2(PPh_3) + CH_2{=}CH_2 \xrightarrow{-PPh_3} \underset{\displaystyle CpCo(CD_3)_2}{\overset{CH_2{=}CH_2}{|}} \longrightarrow CpCo\begin{smallmatrix}CH_2CH_2CD_3\\ CD_3\end{smallmatrix}$$

$$\longrightarrow \underset{\displaystyle \underset{CD_3}{|}}{\overset{CH_2{=}CH{-}CD_3}{|}}CpCo{-}H \xrightarrow[+PPh_3]{+C_2H_4} CD_3H + CD_3CH{=}CH_2 + CpCo(CH_2{=}CH_2)(PPh_3)$$

In the insertion of an alkyne into an M–C bond, cis addition takes place,[201] although the initial cis product isomerizes to a trans product.

$$(acac)Ni\begin{smallmatrix}L\\ Me\end{smallmatrix} + RC{\equiv}CR \xrightarrow{-L} (acac)Ni{-}C(R){=}C(R)Me$$

There are many examples of multiple alkyne insertions into metal–carbon[203] and metal–hydrogen bonds.[204] The multiple insertion of acetylene at low temperatures to give poly(acetylene) having conjugated cis double bonds[205] will be discussed in Section 7.2.

A representative example of β elimination that has been studied kinetically is thermolysis of the iron alkyl complex shown below.[206]

$$\begin{aligned} &CpFe(CO)(PPh_3)R \underset{}{\overset{-PPh_3}{\rightleftarrows}} CpFe(CO)H(olefin) \\ &\xrightarrow[+PPh_3]{-olefin} CpFe(CO)(PPh_3)H \end{aligned} \qquad (6.120)$$

Decomposition of the iron alkyl is completely blocked by addition of PPh_3, which suggests that a vacant site for β-hydrogen elimination must be created for thermolysis of such coordinatively saturated metal alkyls to occur. In reaction (6.120) no kinetic isotope effect was observed on thermolysis of an iron alkyl whose β position was labeled with deuterium atoms and H–D scrambling occurred prior to the thermolysis. Skeletal isomerization from a branched alkyl to the straight-chain alkyl also precedes thermal decomposition. These experimental results imply that olefin liberation from the hydrido-olefin complex generated by β elimination is the rate-determining step of thermolysis.

Thermolysis by β elimination to produce a metal hydride as in reaction (6.120) represents the simplest route for the decomposition of transition metal alkyls. The course of decomposition is often complicated by intermolecular reactions between two monoalkyl complexes or between the monoalkyl and the hydride complex produced from it.[120]

Thermolysis of transition metal dialkyls occurs by β elimination of one alkyl group to give alkene and reductive elimination of the resulting hydride ligand with the remaining alkyl group to liberate alkane. The β elimination may occur directly from the starting transition metal dialkyls or it may be preceded by dissociation of a ligand to create a vacant site for β elimination to occur. Ligand dissociation is a prerequisite for a coordinatively saturated complex to decompose by a β-elimination pathway. Even when the transition metal dialkyls are coordinatively unsaturated, a low-energy β-elimination pathway may sometimes be provided by ligand dissociation.

The thermolysis mechanisms of *cis*-$Pt(PPh_3)_2R_2$ have been studied in detail.[130-132] In the thermolysis alkane and alkene are evolved in a 1 : 1 ratio [cf. Eq. (6.91)], and the thermolysis is severely hindered by addition of PPh_3. In the absence of added PPh_3 thermolysis of platinum complexes with different alkyl groups (R_a and R_b) gives an alkane–alkene ratio [$R_aH/R_a(-H)$ or $R_b(-H)/R_bH$ ratio] that is proportional to the number of hydrogen atoms attached to the β-carbon atoms. For example, in the thermolysis of *cis*-$Pt(PPh_3)_2Et(n\text{-}Pr)$ the ethane–ethylene (or propylene–propane) ratio was 2 : 3. When one of the alkyl groups (R_b) has no β-hydrogens (e.g., Me or Ph), only R_bH and $R_a(-H)$ are formed. Although $CHD{=}CD_2$ and $CHD{=}CHD$ were formed in addition to $CD_2{=}CH_2$ on thermolysis of $Pt(CH_2CD_3)Pr^n(PPh_3)_2$, in the absence of added PPh_3 no H–D scrambling occurred in the remaining CH_2CD_3 group attached to platinum, and no isotope effect was observed in the thermolysis of platinum dialkyls labeled with deuteriums at the β-carbon atom. The thermolysis was first order in the platinum dialkyl complex. These results indicate that thermolysis of *cis*-$Pt(PPh_3)_2R_2$ in the absence of added tertiary phosphine proceeds by a dissociative mechanism, the rate-determining step being the formation of a T-shaped dialkyl intermediate. This undergoes β elimination and reductive elimination, producing the alkane and alkene.

$$L_2Pt(R)(CH_2CH_3) \xrightarrow{-L,\,rds} LPt(R)(CH_2CH_3) \longrightarrow LPt(R)(H\cdots CH_2CH_2) \underset{}{\overset{fast}{\rightleftharpoons}} LPt(R)(H)(CH_2{=}CH_2) \xrightarrow{-C_2H_4} LPt(R)(H) \longrightarrow C_2H_6$$

In the presence of added tertiary phosphine the dissociative pathway is blocked, and the platinum dialkyl complex is compelled to decompose through five-coordinate intermediates.

$$L_2Pt(R)(CH_2CH_3) \rightleftharpoons L_2Pt(R)(H\cdots CH_2CH_2) \xrightarrow{fast} L_2Pt(R)(H)(CH_2{=}CH_2) \xrightarrow{rds} C_2H_4 + RH$$

In this case H–D scrambling is observed in the ethyl group labeled with deuterium prior to the rate-determining loss of olefin. The absence of an isotope effect is also compatible with this scheme.

In contrast to the platinum dialkyls, thermolysis of *trans*-$PdEt_2(PR_3)_2$ is

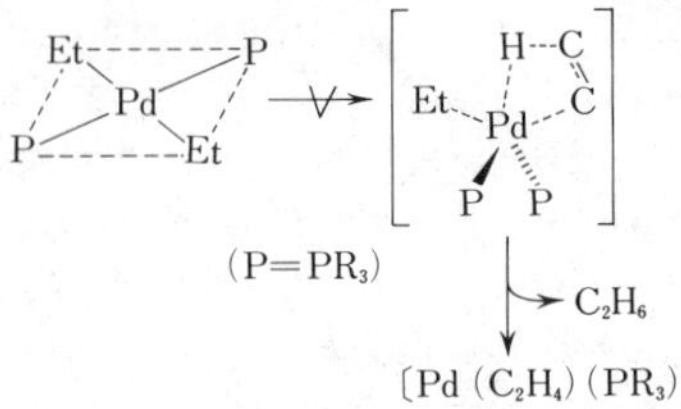

FIGURE 6.29. Thermolysis mechanism of *trans*-$PdEt_2(PR_3)_2$.

only slightly retarded by added tertiary phosphine.[207] No H–D scrambling was observed on thermolysis of *trans*-$Pd(CH_2CD_3)_2(PR_3)_2$, which liberated $CH_2{=}CD_2$ and CH_2DCD_3 in a 1 : 1 ratio with a small isotope effect of $k_H/k_D = 1.4 \pm 0.1$. The thermolysis rate increased with an increase in the cone angle of the tertiary phosphine. Based on these experimental results, the thermolysis mechanism shown in Figure 6.29 has been proposed.

In the thermolysis transition state, which may be close to a trigonal bipyramid, a β-hydrogen atom is abstracted from one ethyl group and transferred to the other, liberating ethane. Absence of H–D scrambling indicates that the β-hydrogen abstraction is an irreversible process. The ethylene produced from the ethyl group is coordinated to Pd(0) and is displaced by other more electronegative olefins or by heating.

A transition state for β-hydrogen atom abstraction has been suggested on the basis of the extended Hückel MO calculations.[208] It involves bending of the phosphine ligands away from planarity, as shown in Figure 6.29. For a β-hydrogen atom to be abstracted from an ethyl group, the two carbon atoms in the ethyl group and the metal atom must be coplanar. Furthermore, a bending of the M–C–C angle from its normal tetrahedral value of 109° may be required to bring the β-hydrogen atom into a region where it can interact with the metal d orbitals. The molecular structure of $Ti(dmpe)EtCl_3$ (dmpe = $Me_2PCH_2CH_2PMe_2$) is interesting in this respect because it shows exactly this type of distortion of the M–C–C angle from the normal tetrahedral value of 109.3° to 85.9°, as shown in Fig. 6.30.[209]

The complex may be viewed as being ready to have a β-hydrogen atom abstracted by the transition metal. On the basis of ab initio MO calculations, the bending of the M–C–C angle from the normal tetrahedral angle has been

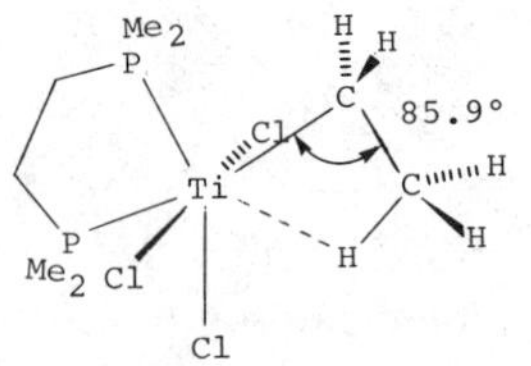

FIGURE 6.30. Molecular structure of $Ti(Me_2PCH_2CH_2PMe_2)EtCl_3$ (from ref. 209).

found for a Ti–Et complex[209b] and for a transition state in insertion of ethylene into a Rh–H bond in $RhH_2(Cl)(PH_3)_2$.[209c]

In certain cases β-hydrogen abstraction constitutes the rate-determining step in the thermolysis of transition metal alkyls. The kinetic isotope effect (k_H/k_D) in the thermolysis of an octyliridium complex $C_8H_{17}Ir(CO)_2(PPh_3)$ where the β-hydrogens are labeled with deuterium was found to be 2.28 ± 0.20.[210] A similar value of 2.30 ± 0.5 was observed in the thermolysis of $Et_2Co(acac)(PMe_2Ph)_2$.[211] This complex differs from the octyliridium complex in being coordinatively saturated, and phosphine dissociation is required for thermolysis by β elimination to occur. The result of the kinetic study, however, indicates that the rate-determining step for thermolysis of $Et_2Co(acac)(PR_3)_2$ is β-hydrogen elimination, not ligand dissociation.

The requirement of coplanarity of the M–C–C–H atoms for cis β-hydrogen elimination is usually fulfilled in simple metal alkyls whose M–C and C–C bonds can freely rotate. However, the situation is different in metallacycles. Platinum metallacycles $Pt(PPh_3)_2(CH_2)_n$ (n = 4–6) decompose by a β-elimination mechanism that liberates alkenes. The five- and six-membered metallacycles are 10^4 times more stable than the seven-membered one.[212]

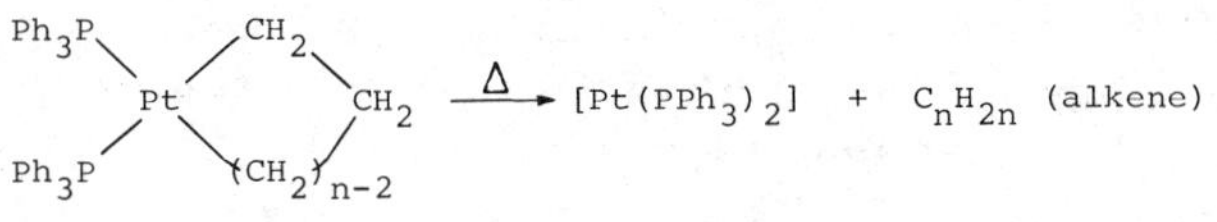

$$(Ph_3P)_2\overline{Pt-CH_2-CH_2-(CH_2)_{n-2}} \xrightarrow{\Delta} [Pt(PPh_3)_2] + C_nH_{2n} \text{ (alkene)}$$

This is because in metallacycles, in contrast to simple alkyls, it is difficult for the dihedral angle M–C–C–H shown below to become 0°; this is particularly true for the rather rigid five- and six-membered metallacycles. In a seven-membered metallacycle the constraint is largely relieved and its thermolysis rate is about two-thirds that of the simple dialkyls.

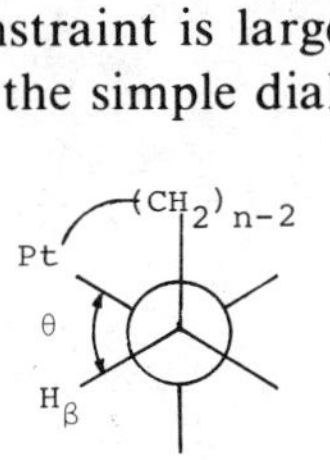

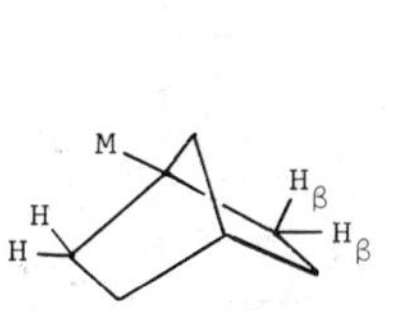

The stability of metal alkyls that have the norbornyl groups attached to the metal at the bridgehead position can be explained similarly by the difficulty of bringing the dihedral angle M–C–C–H close to zero.

In many substitution reactions of olefins catalyzed by palladium complexes, the stereochemistry of the products can be explained by *cis addition* of a Pd–Me entity to the olefin and subsequent *cis elimination* of a Pd–H entity from the palladium alkyl formed. For example, in the methylation of *E*-deuterated styrene with methyllithium catalyzed by palladium acetylacetonate, inversion of stereochemistry at the olefin is observed.

$$\text{(H)(Ph)C=C(D)(H)} + \text{MeLi} \xrightarrow[\text{THF}]{\text{Pd(acac)}_2} \text{(H)(Ph)C=C(Me)(D)}$$

In this system a methylpalladium complex may be formed by methylation of $Pd(acac)_2$ with MeLi. Coordination of styrene to this complex, insertion, and β-hydrogen elimination after rotation around the C–C bond can give the product, as shown below.

$$\text{(H)(Ph)C=C(D)(H)} \cdots \text{Pd–Me} \longrightarrow \text{(H)(Ph)C(Pd)–C(D)(H)(Me)} \longrightarrow \text{(H)(Ph)C(Pd)–C(Me)(D)(H)} \xrightarrow{-\text{Pd–H}} \text{(H)(Ph)C=C(Me)(D)}$$

On the other hand, the stereochemical course is different when a more carbanionic nucleophile such as sodium diethyl malonate is employed. The stereochemistry of the product is more consistent with a mechanism involving external ***trans attack*** of the palladium-coordinated olefin by the nucleophile, as will be described in Section 6.4.[213]

The β elimination reaction is not limited to β-hydrogen elimination, and other atoms or groups may be eliminated, although established examples are still limited. When a group attached to the β-carbon of a metal alkyl is a good leaving group or the metal has a strong affinity with the leaving group, the group may be eliminated from the metal alkyl.

$$\text{–C(M)–C(Y)–R} \longrightarrow \text{M–Y} + \text{C=C(R)} \tag{6.121}$$

$Y = HCO_2,\ CH_3CO_2,\ Br,\ OH,\ OPh,\ OSO_2Me,\ OCH_2CH{=}CH_2,\ SPh,\ OSiMe_3$

$R = Me,\ H$

Reaction products consistent with olefin insertion into a metal–carbon bond and the subsequent β-elimination reactions shown above have been identified in reactions of substituted olefins with isolated transition metal phenyl complexes[214] and other transition metal aryl complexes generated *in situ.*[214c]

β Elimination is probably involved in some thermolysis reactions of transition metal alkoxides, particularly of the later transition metals. As described in Section 4.5, β-hydrogen elimination of a metal alkoxide is probably involved in the formation of Vaska's complex, $IrCl(CO)(PPh_3)_2$, by the reaction of alcohols with $IrCl_3$ in the presence of PPh_3. The course of the reaction has not been established with certainty because group 8–10 transition metal alkoxides with β-hydrogens are relatively rare. When the alkoxo group is substituted by fluorine atoms, stable group 8–10 metal alkoxides are isolable.[215]

β Elimination involving abstraction of an atom or group other than hydrogen can take place in transition metal alkoxides as in transition metal alkyls. In a reaction of a cobalt hydride with $CH_3CO_2CH_2CF_3$, an alkoxocobalt complex is produced by insertion of the ester into the Co–H bond. In a related reaction of the cobalt hydride complex with CH_3CO_2Ph, a cobalt phenoxide is produced by phenoxy group abstraction from the putative intermediate alkoxide complex. The reaction is reminiscent of the reaction of Grignard reagents with esters.[96]

$$CoH(N_2)(PPh_3)_3 + CH_3COOCH_2CF_3 \longrightarrow (Ph_3P)_3Co{-}O{-}CH(CH_3)(OCH_2CF_3)$$

$$CoH(N_2)(PPh_3)_3 + CH_3CO_2Ph \longrightarrow \left[(Ph_3P)_3Co{-}OCH(CH_3)(OPh)\right] \xrightarrow{-CH_3CHO} Co(OPh)(PPh_3)_3$$

γ-Hydrogen elimination is sometimes observed on thermolysis of transition metal alkyls. An important consequence of this reaction is the formation of metallacyclobutanes from transition metal neopentyls.[216]

$$L_2M(CH_2CMe_3)_2 \longrightarrow L_2M(CH_2)_2CMe_2 + CMe_4 \qquad (6.122)$$

Similar γ-hydrogen elimination is involved in the internal metallation of coordinated triarylphosphines, triarylphosphites, and trialkylphosphines.[57]

Metallacyclobutanes are important intermediates in catalytic olefin metathesis reactions. The metallacyclobutanes can be prepared, among other methods, by olefin insertion into metal carbenes[217]:

$$M{=}C + C{=}C \rightleftharpoons \text{metallacyclobutane} \qquad (6.123)$$

The reaction is reversible, and the metallacyclobutanes can sometimes be regarded as precursors for metal–carbene complexes. Carbene complexes can be stabilized by forming a four-membered ring with a Lewis acid, such as alkyl aluminum compounds. Tebbe isolated a binuclear complex from the reaction of Cp_2TiCl_2 and $AlMe_3$.[218] Tebbe's complex reacts with other olefins to form titanacyclobutanes.[219]

$$Cp_2Ti(\mu\text{-}CH_2)(\mu\text{-}Cl)AlMe_2 + RCH{=}CH_2 \xrightarrow{Py} Cp_2Ti(CH_2)_2C(R)(H) + AlMe_2Cl{\cdot}py. \quad (6.124)$$

The isolated titanacyclobutane complexes undergo exchange reactions with other olefins, probably through intermediate carbene complexes.

$$Cp_2Ti(CH_2)_2C(R)(H) + R'CH{=}CH_2 \longrightarrow Cp_2Ti(CH_2)_2C(R')(H) + RCH{=}CH_2$$

Crystallographic studies have revealed that some metallacyclobutanes have puckered rings, whereas others, particularly unsubstituted ones, have flat rings with a dihedral angle α close to 180°.[219,220]

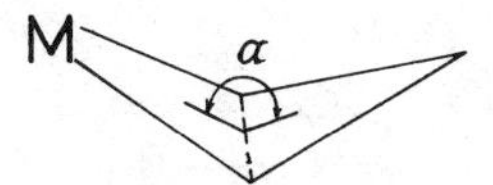

The puckering may have an important effect on the behavior of these metallacyclobutanes and consequently on the course of some olefin metathesis reactions.

Schrock alkylidene complexes can be prepared by α-hydrogen elimination of transition metal alkyls, as described in Section 4.3.

$$M{-}CH_2R \rightleftharpoons H{-}M{=}C(R)(H)$$

In olefin metathesis reactions catalyzed by mixtures of transition metal complexes and methylaluminum compounds, the catalytically active carbene complex may be formed by α-hydrogen elimination. A carbene complex can also be formed on thermolysis of dimethyltransition metal complexes with evolution of methane.[221]

$$M(CH_3)_2 \rightleftharpoons CH_2{=}M(H)CH_3 \xrightarrow{-CH_4} CH_2{=}M$$

The alkylidene entity can serve as a bridging ligand connecting two metal atoms.[222] A μ-methylene complex is formed by thermolysis of a dimethyl complex with a direct metal–metal bond.[223]

$$Na_2[Os_2(CO)_8] \xrightarrow{CH_3I} (OC)_4(Me)Os{-}Os(Me)(CO)_4 \xrightarrow{-CH_4} (OC)_4Os(\mu\text{-}CH_2)Os(CO)_4$$

The μ-methylene complex undergoes an ethylene insertion reaction to give a metallacycle complex containing two metal atoms.

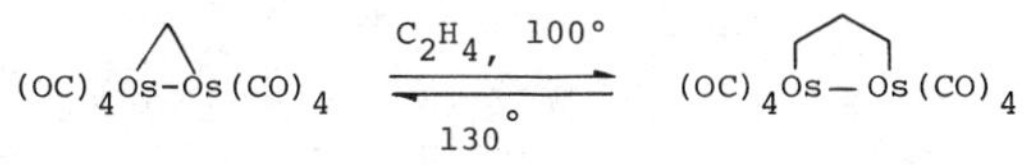

$$(OC)_4Os\text{–}Os(CO)_4 \;\underset{130^\circ}{\overset{C_2H_4,\ 100^\circ}{\rightleftharpoons}}\; (OC)_4Os\text{—}Os(CO)_4$$

The reaction is reversible and the μ-CH_2 complex can be regenerated with evolution of C_2H_4, $CH_3CH{=}CH_2$, and CH_4.

A unique type of insertion of an olefin into the C–H bond of a μ-methine complex has been reported.[224]

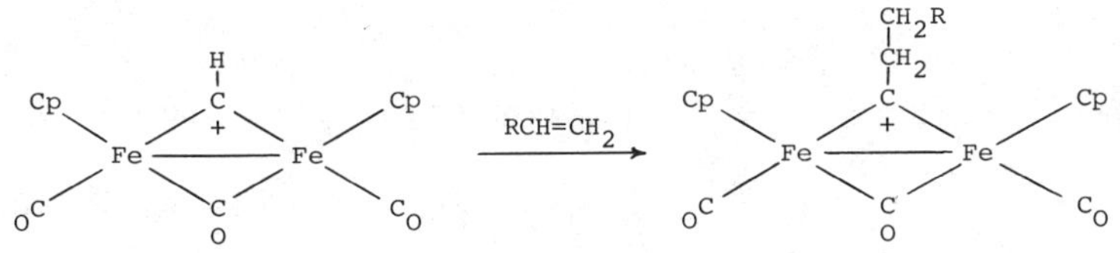

$$\text{RCH=CH}_2$$

d. Insertion of Dienes

Insertion of a diene molecule into an M–C or M–H bond is an important elementary step in the polymerization and oligomerization of dienes such as butadiene and isoprene.

Two types of coordination with transition metals through the two double bonds of dienes are known. One is the cisoid (s-cis) type of coordination observed in many diene complexes and the other is the transoid (s-trans) type of coordination, which has been recently found for zirconium complexes.[182,225]

On insertion of butadiene into an M–H bond, two types of η^3-allyl complexes, the syn form and the anti form shown in Figures 3.15*a,b* may be formed. The insertion of butadiene into the M–H or M–R bond may be viewed as migration of the hydrido or alkyl ligand to the coordinated butadiene to form an alkenyl–metal complex that may be delocalized to an η^3-allyl complex. Addition of the M–H bond to the butadiene ligand coordinated to the metal in the s-cis form would give the anti form of the η^3-allyl complex.

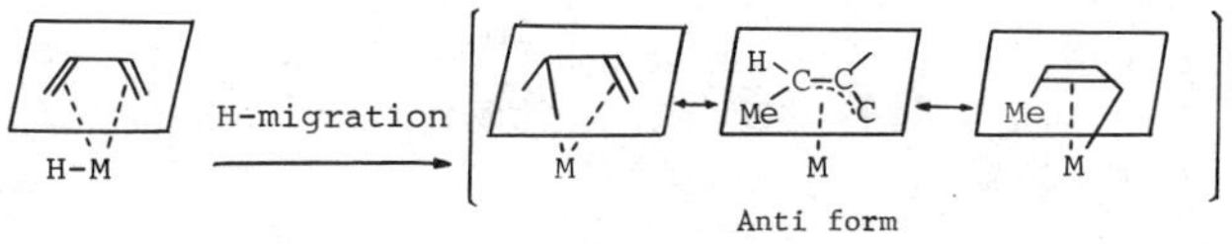

H–M —H-migration→ [M ↔ Me M ↔ Me M] (6.125)

Anti form

On the other hand, addition of the M–H bond to the coordinated butadiene in the s-trans form would give the syn form of the η^3-allyl complex. Butadiene may also coordinate to a metal through only one double bond. In this case the s-trans form is considered to be more stable than the s-cis form. Insertion of butadiene coordinated through one double bond into the M–H bond would lead to an alkenyl complex that on delocalization would rearrange to the syn η^3-allyl complex.

H–M $\xrightarrow{\text{H-migration}}$ [M′ ⟷ M ⟷ M′] (Syn form) (6.126)

From reaction systems for the oligomerization of butadiene catalyzed by transition metal complexes, many η^3-allyl complexes both in the syn and anti forms have been isolated and characterized by X-ray crystallography.[226] It has been found in many cases that the syn form is thermodynamically favored[227] and that the syn and anti forms are interconvertible above certain temperatures. The anti–syn isomerization may proceed through a σ-allyl intermediate, as shown in Figure 5.12.

It can be seen from Eqs. (6.125) and (6.126) that further butadiene insertion into the anti and syn forms of η^3-allyl complexes would yield polymers having *cis*-1,4 and *trans*-1,4 repeating units, respectively. These insertion reactions are relevant to the stereospecific polymerization of butadiene and isoprene catalyzed by transition metal complexes, as will be discussed in Section 7.2.

In some octahedral complexes, which are less labile to ligand substitution, the availability of coordination sites is restricted. When octahedral complexes react with dienes, η^1-allyl complexes are formed. For example, $MnH(CO)_5$ reacts with butadiene to give a mixture of *cis*- and *trans*-η^1-crotyl manganese complexes, and these rearrange to anti and syn η^3-allyl complexes, respectively.[228]

$$MnH(CO)_5 \xrightarrow{CH_2{=}CHCH{=}CH_2} (OC)_5Mn{-}CH_2CH{=}CHCH_3 \ (\text{cis } 85\%) + (OC)_5Mn{-}CH_2CH{=}CHCH_3 \ (\text{trans } 15\%)$$

$$\text{cis} \xrightarrow{-CO,\ Ir(dppe)_2Cl} \eta^3\text{-}CH_2CHCHCH_3Mn(CO)_4 \ \text{anti}(85\%) \qquad \text{trans} \xrightarrow{-CO,\ Ir(dppe)_2Cl} \eta^3\text{-}CH_2CHCHCH_3Mn(CO)_4 \ \text{syn}(15\%)$$

On treatment with $Ir(dppe)_2Cl$, a strong CO acceptor, the η^1 complexes can be converted at room temperature into η^3-allyl complexes with retention of configuration of the 1-crotyl ligand.

6.4. REACTIONS OF COORDINATED LIGANDS

a. Reactions of Coordinated Olefins and Acetylenes

A coordinated olefin shows a different reactivity from that of an uncoordinated olefin. The most important reaction revelant to catalytic reactions of olefins is external attack on the coordinated olefin by a nucleophile. The reverse of this process is expulsion of the attached nucleophile.

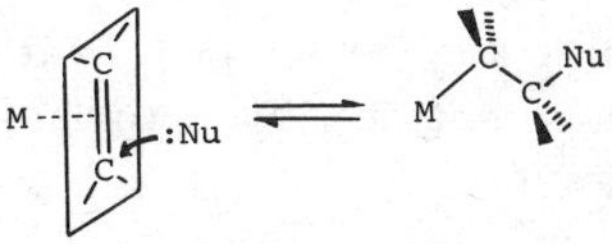

M---C=C ←:Nu ⇌ M–C–C–Nu (6.127)

The nucleophilic attack on the coordinated olefin resembles the olefin insertion reaction but differs from it in stereochemistry. It is not easy to distinguish this type of intermolecular reaction from intramolecular "migratory insertion" involving prior coordination of the nucleophile. In the migratory insertion of an olefin into a metal–hydride bond as represented by Eq. (6.112), cis attack takes place, while in reaction (6.127) the nucleophile directly attacks the coordinated olefin from the outside (anti side) without prior coordination. Thus, the double bond opens in a trans fashion. The mode of double-bond opening can be identified by using a 1,2-disubstituted olefin. A problem that remained unsolved until recently concerning the important Wacker reaction involving nucleophilic attack on coordinated ethylene by OH^- (or H_2O) to give acetaldehyde (cf. Section 7.4) is whether coordinated OH^- attacks the coordinated ethylene internally or uncoordinated OH^- attacks the coordinated olefin externally. The question was solved by employing 1,2-d_2-ethylene and examining the configuration of the product.[229] It has been established that *cis*-CHD=CHD gives the threo isomer (A) on reaction with Pd(II) in a system similar to the Wacker catalyst, whereas *trans*-CHD=CHD gives the erythro isomer (B).

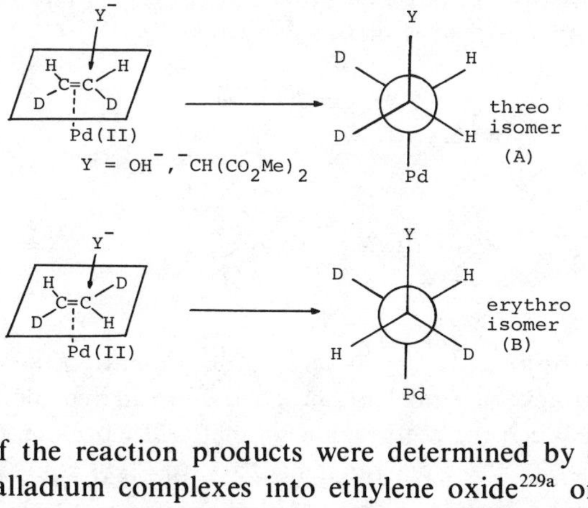

Y^-; H, H, C=C, D, D; Pd(II); Y = OH^-, $^-CH(CO_2Me)_2$ → Y, D, H, D, H, Pd — threo isomer (A)

Y^-; H, D, C=C, D, H; Pd(II) → Y, D, H, H, D, Pd — erythro isomer (B) (6.128)

Configurations of the reaction products were determined by converting the β-hydroxyethylpalladium complexes into ethylene oxide[229a] or β-lactone.[229b] More direct evidence for trans attack on the coordinated ethylene by a nucleophile has been obtained by Kurosawa,[229c–e] who made an NMR examination of the σ-alkylpalladium complexes derived by treatment of $[CpPd(PPh_3)(CHD{=}CHD)]^+ClO_4^-$ with $CH(COMe)_2^-$ or CH_3O^-. Addition of acetate ion to an olefin coordinated to a palladium complex was also established to be trans.[229f]

Addition of an amine to a coordinated olefin has also been shown to occur by trans attack. The configuration of carbon-2 in butene-1 coordinated to a platinum complex containing a chiral amine [(*S*)-α-methylbenzylamine] is *S*, and attack on the coordinated butene-1 by diethylamine after hydrolysis yields (*S*)-*N,N*-diethyl-*sec*-butylamine.

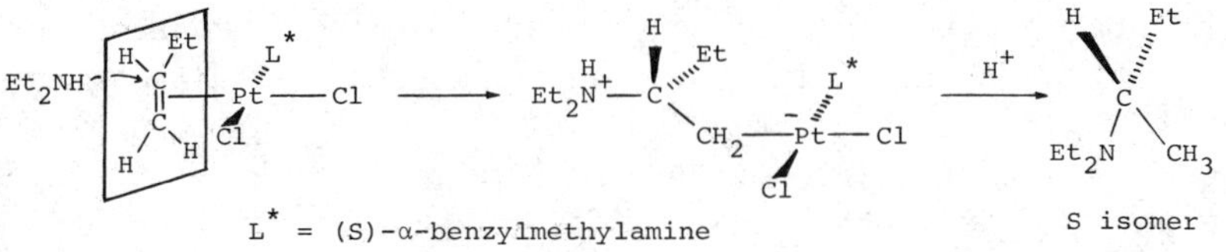

L^* = (S)-α-benzylmethylamine

S isomer

The result is consistent with trans attack of the amine on the coordinated olefin.[230]

An olefin bound to a cationic Fe(II) complex undergoes attack by a variety of nucleophiles to produce stable, neutral σ-alkyliron complexes.

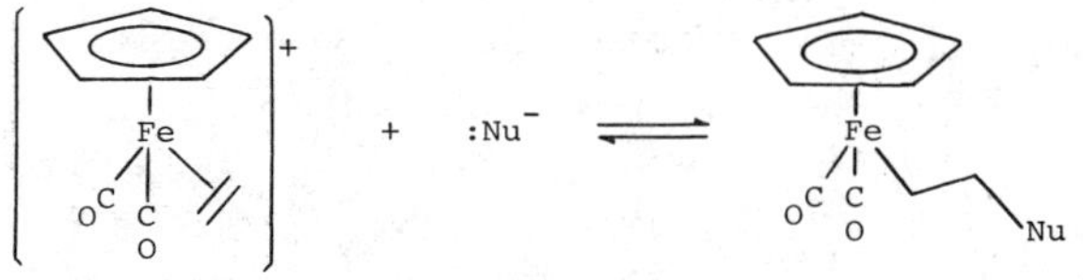

Nucleophilic attack occurs primarily at the most substituted olefinic position in the absence of excessive steric demands.[231]

Attack of an amine on an olefin coordinated to a Pt(II) complex also proceeds at the more substituted olefinic position.[231d]

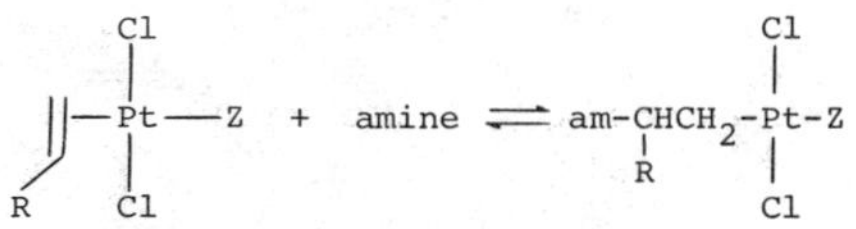

Z = $NHMe_2$

am = amino group

main product (80-90%)

On the basis of an extended Hückel MO calculation, it has been proposed that external attack of a nucleophile on a coordinated olefin is energetically more favorable when the olefin is unsymmetrically bonded to the metal, as in (B) below. This facilitates rearrangement to a σ-alkyl complex (C).[232a]

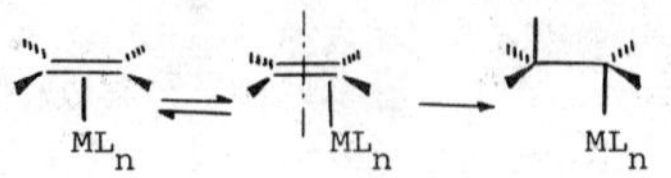

(A) (B) (C) (6.129)

Some experimental evidence in support of the proposal has been provided.[232b–d]

Coordinated acetylene also undergoes nucleophilic attack. The mode of addition varies depending on the nature of the nucleophile, and careful examination is required to determine whether the product under examination is a kinetic or thermodynamic product. Trans addition products have been reported in reactions of coordinated acetylenes with nucleophiles such as organocuprates.[233]

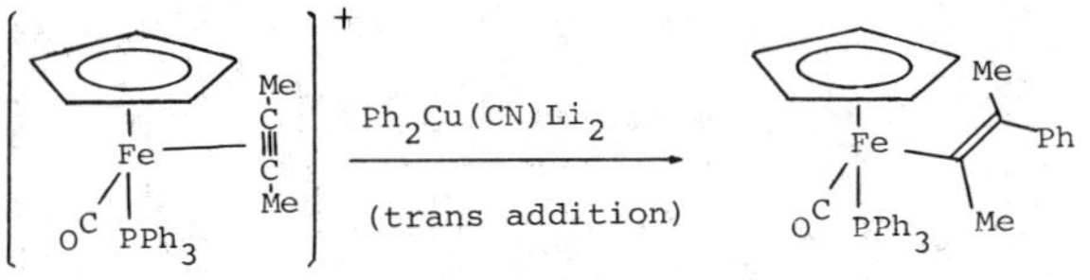

$Ph_2Cu(CN)Li_2$ (trans addition)

On the other hand, cis addition has been found in reactions of coordinated acetylenes with hydrides.[234]

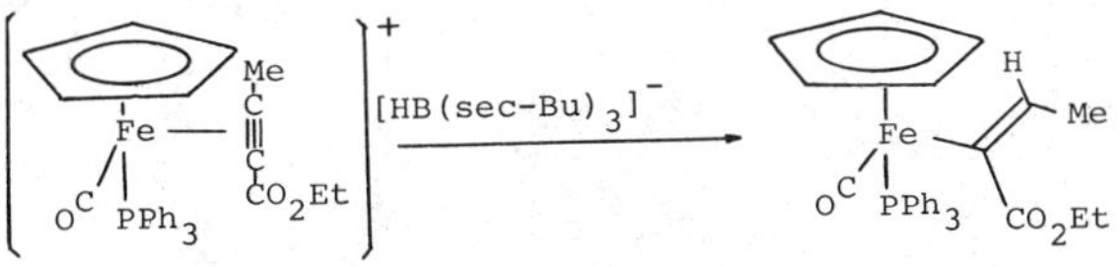

$[HB(sec\text{-}Bu)_3]^-$

Interestingly, a detailed study using borodeuteride revealed that the first step in the above reaction is external attack on the cyclopentadienyl group, and this is followed by migration of the endo hydrogen to give the cis product.

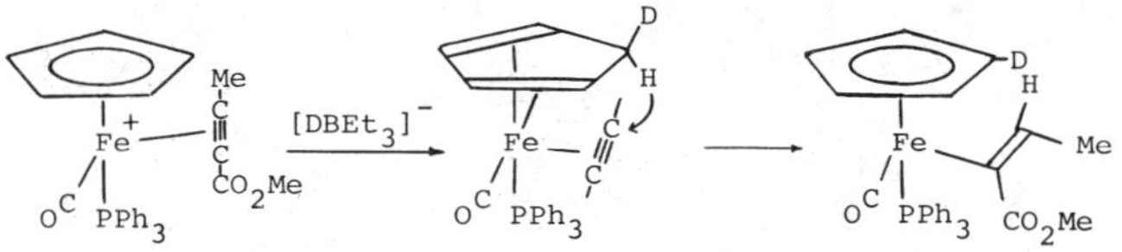

$[DBEt_3]^-$

Acetylene coordinated to palladium also undergoes nucleophilic attack by alcohol to give a β-alkoxyvinyl derivative that undergoes further insertions of acetylene and subsequent ring closure.[235]

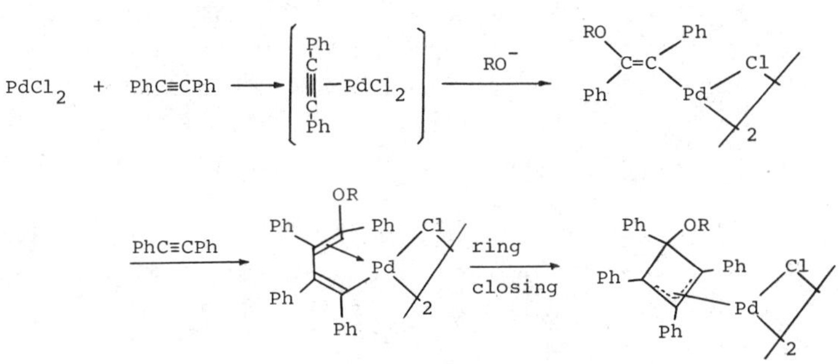

$PdCl_2 + PhC{\equiv}CPh \longrightarrow [PhC{\equiv}CPh{-}PdCl_2] \xrightarrow{RO^-}$

$\xrightarrow{PhC{\equiv}CPh}$ $\xrightarrow{\text{ring closing}}$

Nucleophilic attack of alkoxide or hydroxide on a coordinated chelated diene, such as cyclooctadiene-1,5 proceeds by an external trans attack to give a σ-alkyl-π-olefin complex.[236]

Allylamine can form a chelate complex with Pd(II) through the nitrogen atom and the double bond, and this undergoes the external nucleophilic attack by methanol.[237]

Nucleophilic attack of amine on an olefin coordinated to palladium followed by treatment with CO leads to an acyl palladium complex.[238]

$$R\text{-}CH{=}CH_2 + R'_2NH + PdCl_2(PhCN)_2$$

b. Reactions of η^3-Allyl Ligands

η^3-Allyl complexes can be prepared by various methods as described in Sections 4.1 and 6.2, including treatment of a transition metal compound with an allyl Grignard reagent, insertion of butadiene into a metal hydride, electrophilic or nucleophilic attack on a coordinated diene ligand, and oxidative addition of allylic compounds. The η^3-allyl ligand thus formed may be attacked by nucleophiles. Nucleophilic attack on the η^3-allyl palladium complexes has found particular synthetic utility. The synthetic applications will be dealt with in Section 8.2d. External and internal attacks on the coordinated η^3-allyl ligand have been reported, the mode of attack depending on the complex and the nucleophile involved. Several representative examples dealing with the stereochemistry in the nucleophilic attack of the η^3-allyl ligands are described here.

In catalytic alkylation of allylic acetates, which involves oxidative addition of allylic acetates to Pd(0) followed by nucleophilic addition to the resulting η^3-allyl palladium complexes, the reaction was found to proceed with an overall net retention of configuration.[239,240]

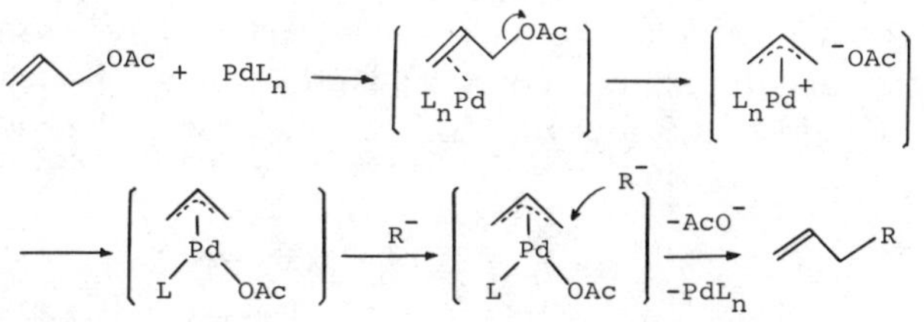

This result can be explained by trans elimination of the acetato group to form an η^3-allyl palladium complex followed by trans attack on the η^3-allyl ligand by the nucleophile. Allylic alkylation of an optically active allylic acetate with a malonate anion catalyzed by a palladium complex was established to proceed with 100% stereochemical retention at the α-allyl carbon atom.[241]

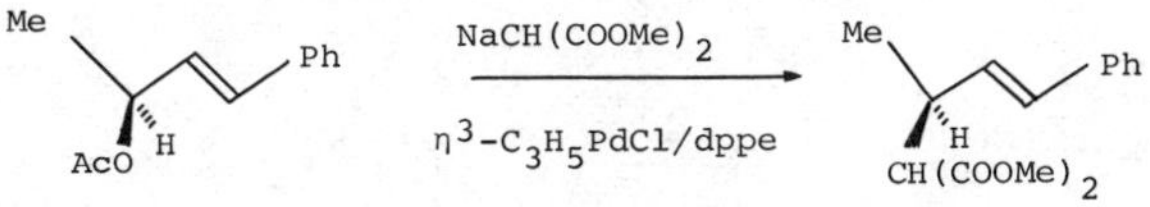

Furthermore, the individual steps of oxidative addition of allylic acetate to Pd(0) and nucleophilic attack of the malonate on the resulting η^3-allyl palladium complex were confirmed to proceed with stereochemical inversion.[241]

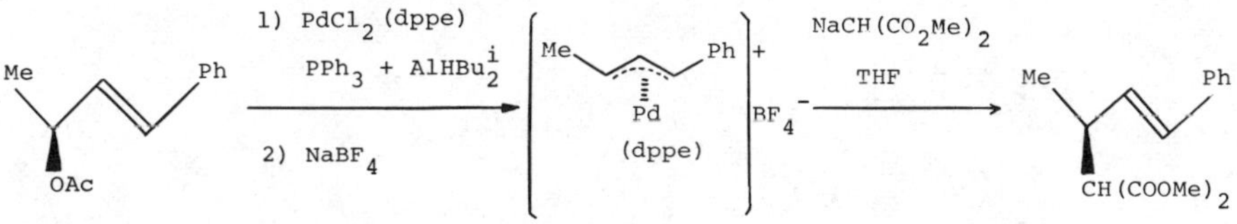

A similar external attack of nucleophiles on allyl molybdenum complexes has been confirmed.[242]

c. Reactions of Coordinated Arenes

The reactivity of arenes can sometimes be significantly altered by complexation to a transition metal. There are many arene-coordinated transition metal carbonyls prepared in various ways. Thus, on complexation with a metal carbonyl moiety, which acts as a π acid, the electron density on the arene is decreased and the arene becomes susceptible to *nucleophilic* attack, a mode of reaction rarely observed in free aromatic compounds. For example, a range of η^6-arene complexes can be prepared by heating $Cr(CO)_6$ with arenes.

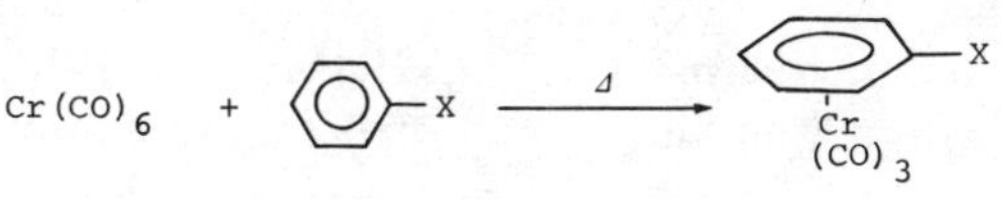

X = Me, NH_2, OMe, Cl, I, MeCO

The $Cr(CO)_3$ entity attached to the arene attracts electrons as strongly as a *p*-nitro group in uncoordinated arenes. Thus, the following nucleophilic substitution can be induced without the aid of an electron-withdrawing substituent attached to the arene ring.

$$(C_6H_5Cl)Cr(CO)_3 + NaOMe \xrightarrow{MeOH} (C_6H_5OMe)Cr(CO)_3 \qquad (6.130)$$

Similarly, chlorobenzene coordinated to $Cr(CO)_3$ can be alkylated with carbanions.[243]

$$(C_6H_5Cl)Cr(CO)_3 + R^- \longrightarrow (C_6H_5R)Cr(CO)_3 \xrightarrow{I_2} C_6H_5R + CO + Cr(III)$$

$R^- = CH(COOMe)_2^-, CMe_2CN^-$

The alkylated arene can be freed by oxidation with iodine.

Arenes coordinated to a metal in a high oxidation state can also be attacked by nucleophiles. Thus, double nucleophilic attack on benzene coordinated to a dicationic cobalt complex has been reported.[244]

$$[(C_6H_6)Co(C_5H_5)]^{2+}[BF_4^-]_2 \xrightarrow{2NaY} (C_6H_6Y_2)Co(C_5H_5) \qquad Y = C_5H_5^-, {}^-OMe$$

The nucleophilic attack on the coordinated arene can take a different course depending on the nucleophile and the metal complex. Addition of $LiCuMe_2$ to $[(C_6H_6)Mn(CO)_3]^+$ proceeds by initial methylation at manganese and subsequent internal cis migration to afford an endo-methylated η^5-methylcyclohexadienylmanganese complex.[245]

$$[(C_6H_6)Mn(CO)_3]^+ \xrightarrow[-CO]{Me_2CuLi} (C_6H_6)Mn(CO)_2(CH_3) \xrightarrow{PPh_3} (C_6H_6Me)Mn(CO)_2(PPh_3) \ \text{(endo)}$$

On the other hand, treatment of $[(C_6H_6)Mn(CO)_3]^+$ with lithium aluminum hydride or with alkyllithium is reported to give exo products.[246]

R
H
(exo)
RLi
Mn(CO)3 +
Mn(CO)3

The cyclopentadienyl ligand is also susceptible to nucleophilic attack. The reaction course varies depending on the complex involved, the reagent, and temperature. Thus, depending on temperature, either exo or endo attack may be observed.[247]

H
D
(endo)
+
-78°
LiAlD4
Fe
P
P
CO
Fe— CO
P
P
D
90°
toluene
Fe
P
P
CO
D
H
LiAlD4
70°
(exo)
Fe
P
P
CO

A reaction that is similar to nucleophilic attack on benzene coordinated to the $Mn(CO)_3$ entity is the external attack of MeLi on the cyclopentadienyl ligand in $[C_5H_5Mn(CO)(NO)PPh_3)]^+$.[248]

+
Me
(exo)
H
+ LiMe
Mn
ON C PPh3
O
Mn
ON C PPh3
O

d. Reactions of Coordinated CO and Carbene Ligands

Despite its unsaturated nature, free carbon monoxide is a rather unreactive compound. The reactivity develops on coordination to a metal. This causes back donation of electrons from the metal to CO, increasing electron density on oxygen and making the carbon relatively electron deficient. Thus, the oxygen and carbon of the coordinated CO ligand become susceptible to electrophilic and nucleophilic attack, respectively.

$$\overset{\delta+}{\text{M-C}}\equiv\overset{\delta-}{\text{O}} \quad (\text{Nu}^- \rightarrow \text{C};\ \text{O} \rightarrow \text{E}^+)$$

A hydride ion in reagents such as $NaBH_4$ and $KHB(OR)_3$ attacks the carbonyl carbon to give a formyl complex. Further reaction of the hydride with the formyl complex eventually gives a methyl complex through a hydroxymethyl intermediate.[249-251]

$$\text{M-CO} \xrightarrow{H^-} \left[\text{M-}\overset{O}{\overset{\|}{C}}\text{-H}\right] \xrightarrow{2H^-} \left[\text{M-}\overset{OH}{\overset{|}{C}}H_2\right] \xrightarrow{2H^-} \left[\text{M-CH}_3\right] \tag{6.131}$$

The reaction sequence has been established for a rhenium complex.[249]

$$[(C_5H_5)Re^+(NO)(CO)_2]BF_4^- \xrightarrow[THF/H_2O]{NaBH_4} (C_5H_5)Re(NO)(CO)(CHO) \xrightarrow[THF/H_2O]{NaBH_4} (C_5H_5)Re(NO)(CO)(CH_2OH) \xrightarrow[THF/H_2O]{NaBH_4} (C_5H_5)Re(NO)(CO)(CH_3)$$

The individual steps of the reaction are observed in THF–H_2O mixtures, whereas reaction of the starting carbonyl complex with $NaBH_4$ in anhydrous THF directly gives the final methylrhenium complex. The effect of solvent on the reduction of CO with $NaBH_4$ has also been observed with an iron complex, $[(C_5H_5)Fe(CO)_3]^+PF_6^-$.[252]

The reaction of the rhenium carbonyl complex with $NaBH_4$ was explained as follows.[249]

$$Re^+\text{-CO} + HBH_3^- \longrightarrow \left[\text{Re-C}(\cdots O \cdots BH_3 \cdots H \cdots)\right] \longrightarrow \text{Re-C}(=O:BH_3)\text{H}$$

The carbonyl carbon is also susceptible to attack by OH^- to give a hydroxycarbonyl complex, which is readily decarboxylated to afford a hydride complex.

$$Fe(CO)_5 + OH^- \longrightarrow \left[(OC)_4Fe\text{-C}(=O)\text{O-H}\right] \longrightarrow (OC)_4FeH^- + CO_2 \tag{6.132}$$

The reaction is relevant to the water gas shift reaction and will be discussed in Section 7.3.

Attack of an alkoxide ion on the carbonyl carbon yields an alkoxycarbonyl complex.[253]

$$[L_nM\text{-C}\equiv O]^+ + OR^- \longrightarrow L_nM\text{-}\overset{O}{\overset{\|}{C}}\text{-OR} \tag{6.133}$$

Attack of ammonia or amines gives a carbamoyl complex.

$$[L_nM\text{-}C{\equiv}O]^+ + 2HNRR' \longrightarrow L_nM\text{-}\overset{O}{\overset{\|}{C}}\text{-}NRR' + H_2\overset{+}{N}RR' \qquad (6.134)$$

Formation of a Fischer carbene complex on treatment of a metal carbonyl by alkyllithium has already been described in Section 4.3. The carbonyl carbon is first attacked by R^- to give an anionic acyl complex. Further reaction of the acyl complex with an alkyl cation yields the carbene complex.

$$M\text{-}CO + RLi \longrightarrow Li^+[M\text{-}\overset{O}{\overset{\|}{C}}\text{-}R]^- \xrightarrow{R'^+} \left[M{=}C\begin{matrix}O\text{-}R'\\ R\end{matrix}\right]$$

On the other hand, a Schrock carbene complex can be generated by hydride abstraction from an alkyl complex by triphenylmethyl cation. The carbene complex can be converted into an alkyl complex on treatment with an alkyllithium or alkyl Grignard reagent.[254]

$$CpRe(NO)(L)(CH_3) \xrightarrow[CH_2Cl_2]{Ph_3C^+PF_6^-} [CpRe(NO)(L)(=CH_2)]^+ \xrightarrow[\text{or } RMgX]{RLi} CpRe(NO)(L)(CH_2R)$$

Attack of a deuteride ion on an ethylidene rhenium complex takes place at thc anti side of the triphenylphosphine ligand.

$$[CpRe(NO)(PPh_3)(=CHCH_3)]^+ \underset{Ph_3C^+PF_6^-}{\overset{LiBEt_3D}{\rightleftharpoons}} CpRe(NO)(PPh_3)(CHDCH_3)$$

β-Hydrogen (deuterium) abstraction from the resulting ethylrhenium complex by a triphenylmethyl cation also occurs trans to the coordinated triphenylphosphine ligand.

Reactions of carbene complexes with olefins lead either to formation of cyclopropane or to exchange of the carbene moiety with olefins.[217]

$$(OC)_5W{=}CPh_2 \xrightarrow{CH_2{=}CMe_2} \text{1,1-diphenyl-2,2-dimethylcyclopropane} + CH_2{=}CPh_2 \qquad (6.135)$$

$$(OC)_5W{=}CPh_2 \xrightarrow{CH_2{=}C(Ph)OMe} (OC)_5W{=}C(Ph)(OMe) + CH_2{=}CPh_2$$

These reactions probably proceed through metallacyclobutane complexes.

(6.136)

Other properties of carbene complexes have been described in Section 4.3 in relation to their synthesis. The catalytic properties of carbene complexes that are relevant to olefin metathesis will be dealt with in Section 7.5. The rapid increase in interest in transition metal–carbene complexes is continually leading to the discovery of new reactions so that any review is soon likely to be incomplete and obsolete.

Isocyanide coordinated to a transition metal shows a reactivity similar to that of CO. Attack of alcohol on the carbon atom of coordinated isocyanide affords a carbene complex.[255]

$$PtCl_2(PEt_3)(CNPh) \xrightarrow{EtOH} (EtO)(PhNH)C{=}PtCl_2(PEt_3)$$

e. Nucleophilic Reactions of Transition Metal Alkyls

In the previous sections of this chapter we dealt with the various fundamental reactions of transition metal alkyls; in the next section we shall discuss the thermal stabilities and thermolysis pathways of transition metal alkyls. The other reactions of transition metal alkyls, particularly with electrophiles, will be discussed in this section.

Two pathways are possible in reactions of electron-rich transition metal alkyls with electrophiles such as protic acids, halogens, and other metal ions. One is the oxidative addition of the electrophiles to the transition metal alkyls followed by internal reactions of the oxidative addition products. The other is the direct attack of the electrophile on the alkyl carbon or hydrogen atoms on the alkyl. The criteria for distinguishing between these pathways are provided by stereochemical studies of the reaction courses.[140,256,259]

In the direct S_E2 reactions of transition metal alkyls with electrophiles two stereochemical possibilities arise, inversion or retention at the α carbon atom attached to the metal. Possible transition states in these reactions are shown below.

$$M\text{-}R + E\text{-}X \nearrow [X\text{---}E\text{---}R\text{---}M]^{\ddagger} \xrightarrow[\text{inversion}]{S_E2} M\text{-}X + E\text{-}R \quad (6.137a)$$

$$M\text{-}R + E\text{-}X \searrow [X\text{----}E(R)(M)]^{\ddagger} \longrightarrow M\text{-}X + E\text{-}R \quad (6.137b)$$

Cleavage of the iron–alkyl bond in the complex $CpFe(CO)_2(CHDCHDBu^t)$ with halogens has been established to proceed with inversion at the α-carbon atom.[144]

CMe3, H, D, D, H, Fe(CO)2Cp — erythro $\xrightarrow[CS_2,\ C_5H_{12}]{Br_2}$ CMe3, D, H, D, H, Br — threo + $CpFe(CO)_2Br$

The inversion of stereochemistry from the initial erythro to the threo configuration was deduced on the basis of the ^1H–^1H coupling constants of the vicinal protons. Attack on the α-carbon atom bound to iron from the anti side of the metal explains the stereochemical inversion.[144]

$\delta-$ Br-Br + (R, H, D)C ⊂⊃ Fe(Cp)(CO)(CO) ⟶ Br—C(R)(D)(H) + Cp—Fe—Br (CO)(CO)

This result is in contrast to the stereochemical retention observed on insertion of CO into the same complex, as previously discussed.

The complicated nature of the reaction pathway, however, was revealed in the reaction of a deuterium-labeled *threo*-phenethyliron complex with halogens. An experiment using an alkyl labeled with ^{13}C at the α-carbon indicated that the threo product was formed but that scrambling of the ^{13}C-labeled carbon had occurred.[259]

Ph, H, D, *C, H, D, Fe(CO)2Cp — threo $\xrightarrow[X=Br\ or\ Cl]{X_2}$ { threo-$PhCHD^{13}CHDX$ + threo-$Ph^{13}CHDCHDX$ }

$^*C = {}^{13}C$

A related experiment in which $Cp(CO)_2Fe(CD_2CH_2Ph)$ was treated with halogens gave $PhCH_2CD_2X$ and $PhCD_2CH_2X$ in a 1 : 1 ratio.[260]

These results may be explained by assuming an intermediate as shown below.

$$\text{[phenonium-type cation: } C_6H_5^+ \text{ bridging CD}_2\text{–CH}_2\text{]} \quad X\text{–Fe(CO)}_2\text{Cp} \quad X^-$$

A halide ion X^- formed by the reaction of X_2 with the electron-rich $CpFe(CO)_2R$ complex would have an equal opportunity to attack either of the two methylene carbons in the symmetrical phenethyl cation intermediate.

Stereochemical retention was observed on transfer of the *t*-BuCHDCHD group from the $FeCp(CO)_2$ entity to mercury ion.

$$\text{CMe}_3(\text{H,D})\text{C–C(D,H)Fe(CO)}_2\text{Cp (erythro)} \xrightarrow[C_6H_6]{HgCl_2} \text{CMe}_3(\text{H,D})\text{C–C(D,H)HgCl (erythro)} + \text{CpFe(CO)}_2\text{Cl}$$

The reaction course, however, may differ depending on the transition metal alkyl involved. In contrast to the above example, *erythro*-(*t*-BuCHDCHD)-$Co(dmgH)_2py$ (dmg = dimethylglyoxymato, cf. Section 9.1) transfers its alkyl group to Hg^{2+} in water to give *threo*-(*t*-BuCHDCHD)Hg^+ with stereochemical inversion.[257,261]

Another example involving the inversion of stereochemistry at the α-carbon has been reported in the process of alkyl transfer from an alkyltin compound to palladium.[262]

$$\text{Ph(H)(D)C–SnBu}_3 \xrightarrow{PhCOPdCl(PPh_3)_2} \text{Ph(D)(H)C–C(=O)–Ph}$$

The mechanism of this alkyl transfer reaction in polar media is thought to involve an S_E2 transition state.

$$\text{PhCO–Pd(L)(Cl}^{\delta-}\text{)}^{\delta+}\text{----C(H)(D)(Ph)----SnBu}_3^{\delta+}$$

The ensuing reductive elimination of the benzyl and benzoyl group is likely to proceed with stereochemical retention.

The other type of reaction involving cleavage of a transition metal–alkyl bond is its reaction with electrophiles. This proceeds by prior oxidative addition followed by reductive elimination.

$$L_nM{-}R + E{-}X \xrightarrow{\text{Oxidative Addition}} L_nM(X)(E){-}R \longrightarrow E{-}R$$

$$\text{or} \quad [L_nM(E){-}R]^+ X^- \longrightarrow E{-}R \qquad (6.138)$$

The iron–alkyl bond in a methylcyclohexyliron complex is cleaved with stereochemical retention at the α-carbon [cf. Eq. (6.94)].

The *trans*-methylcyclohexyliron complex gives *trans*-1-methyl-4-deuteriocyclohexane, whereas the *cis*-methylcyclohexyliron complex yields *cis*-1-methyl-4-deuteriocyclohexane.[136] This result is consistent with protonation of the starting complex and subsequent reductive elimination of the alkane.

$$CpFe(CO)_2R + DX \longrightarrow [CpFe(CO)_2(R)(D)]^+ X^- \longrightarrow CpFe(CO)_2X + RD$$

Alkylation of a copper compound with a cyclopropyl group followed by hydrolysis proceeds with retention of stereochemistry at the α-carbon.[263] The process probably involves the direct attack of a deuteron at the alkyl group attached to copper.

$$\text{Ph}_2\text{C}_3\text{H}_2(\text{Li})(\text{NC}) \xrightarrow{\text{CuX}} \xrightarrow{D_2O} \text{Ph}_2\text{C}_3\text{H}_2(\text{D})(\text{NC})$$

In certain transition metal alkyls a β-hydrogen atom or another group attached to the β-carbon atom is abstracted by an electrophile. The abstraction proceeds predominantly by trans elimination.[258]

$$PhCH_2CH_2Fe(CO)_2Cp \xrightarrow{Ph_3C^+BF_4^-} [\text{H}\cdots{}^+CPh_3] \longrightarrow Ph\overset{+}{C}HCH_2Fe(CO)_2Cp \longrightarrow \left[PhCH{=}CH_2 \cdots Fe(CO)_2Cp \right]^+$$

In the oxidation of transition metal alkyls with reagents such as Cu(II) and Ce(IV), an electron transfer is involved, and often the reactions proceed through radical intermediates with loss of stereochemistry at the α-carbon atom.[264] Thus, the stereochemistry in the oxidation of metal alkyls may vary depending on the alkyl group and the oxidants, and the reaction course is often unpredictable.

η^1-Allyl ligands are known to act as nucleophiles toward strong electrophiles, giving cationic η^2-olefin complexes.[265]

$$M\text{-}CH_2CH{=}CH_2 + E^+ \longrightarrow [M]^+\text{—}(\eta^2\text{-}CH_2{=}CHCH_2E) \qquad (6.139)$$

The following are typical examples.

$$CpMo(CO)_3(\eta^1\text{-}C_3H_5) \xrightarrow{H^+} [CpMo(CO)_3(\eta^2\text{-}CH_2{=}CHCH_3)]^+$$

$$Cp(OC)_2Fe\text{-}CH_2CH{=}CH_2 \xrightarrow{H^+} [Cp(OC)_2Fe(\eta^2\text{-}CH_2{=}CHCH_3)]^+$$

An η^1-allylplatinum complex undergoes electrophilic attack by tetracyanoethylene to afford a cyclopentylplatinum complex.[266]

$$CH_2{=}CHCH_2PtClL_2 + (NC)_2C{=}C(CN)_2 \longrightarrow \text{3,3,4,4-tetracyanocyclopentyl-}PtClL_2$$

L = tertiary phosphine

In the case of L = PCy_3, electrophilic attack at the allylic double bond is assumed to occur.

$$L_2ClPt\text{-}CH_2CH{=}CH_2 + \text{TCNE} \longrightarrow \left[ClL_2Pt^+\text{--}\|\text{-}CH_2\text{-}C(CN)_2\text{-}{:}C^-(CN)_2\right] \longrightarrow ClL_2Pt\text{-}C_5H_5(CN)_4$$

6.5. STABILITY OF TRANSITION METAL ALKYLS

Because by mid-1950s so few transition metal alkyls were known, they were considered as intrinsically unstable compounds. The number of isolated transition metal alkyls has steadily increased since then, and they are now available for most transition metals, although their stability differs greatly from one transition metal alkyl to another. Why, then, were they considered intrinsically unstable and why do their stabilities differ so widely? And how can we activate some of the stable transition metal alkyls so that they can be utilized as catalysts for organic reactions?

These are fundamental problems in organometallic chemistry, and like most fundamental problems, they are difficult to answer. One of the reasons that transition metal alkyls were considered unstable is that they are usually very sensitive to oxygen and water, and the lack of proper techniques for handling these air-sensitive compounds hindered their isolation. The progress in experimental techniques described in Chapter 5 and the participation of an increasing number of chemists in organotransition metal chemistry spurred by the discoveries of ferrocene and Ziegler and Wacker catalysts revealed that some transition metal alkyls can be synthesized as thermally stable compounds even though they may be quite air sensitive. As described in Chapter 3, the dissociation energies of transition metal–carbon bonds are now known to be fairly large and are comparable in magnitude to those of main-group metal alkyls. Nevertheless, transition metal alkyls are often quite readily thermolyzed in the absence of suitable stabilizing ligands such as tertiary phosphines, cyclopentadienyl, CO, and bipyridine. What, then, is the role of these stabilizing ligands?

Some of the answers to these fundamental problems are being found by studying the thermolysis pathways of isolated transition metal alkyls. We can now state that most, if not all, of the reasons for the thermal instability of transition metal alkyls are kinetic. By providing a suitable energy barrier for blocking low-energy thermolysis pathways, most transition metal alkyls can be stabilized. Conceivable thermolysis pathways are (i) homolytic splitting of the M–C bond releasing radicals and (ii) concerted paths such as reductive elimination, β elimination, α elimination, and γ elimination.

Examples of thermolysis involving radical formation are relatively limited.[68,267-271] In view of the relatively large dissociation energies of transition metal–alkyl bonds, homolytic M–C splitting is not likely to occur around room temperature. It does occur under other conditions, for example, when other reagents are added or when the system is irradiated with light to promote homolysis. Addition of oxidizing reagents such as O_2, Ce(IV), and Ir(IV) or of electron acceptors such as alkyl halides and tetracyanoethylene is known to accelerate the thermolysis of transition metal alkyls by formation of radicals.[68,256] However, the majority of transition metal alkyls are thermolyzed via one of the concerted pathways.

The most frequently observed of the concerted thermolysis pathways are β-hydrogen elimination and reductive elimination. Wilkinson considered that the β-hydrogen elimination route is decisive in determining the stability of transition metal alkyls and that stabilizing ligands such as tertiary phosphines contribute to the stabilization of transition metal alkyls merely by occupying the vacant site required for β elimination.[272] Based on this view, various transition metal alkyls containing groups that lack a readily abstractable β-hydrogen atom have been successfully prepared.[272-274] Alkyl groups such as $PhCH_2$-, Me_3SiCH_2-, and 1-norbornyl have been used for this purpose. However, since β-hydrogen elimination does not constitute the sole thermolysis pathway, blocking of this route may not be enough to stabilize the transition metal alkyls. Permethyl compounds having no β-hydrogen atoms have been prepared, but they are found to be unstable; WMe_6 and $TaMe_5$ decompose explosively.[275] Whereas $TiMe_4$ is an unstable compound that decomposes readily above −80 °C, addition of 2,2′-bipyridine helps to stabilize the system, and $TiMe_4$(bipy) is thermally much more stable.

The other important thermolysis pathway is reductive elimination of two alkyl groups. If a transition metal compound having more than two alkyl groups is to decompose by this route, the two alkyl groups must be in *adjacent* positions, as discussed in Section 6.2. Thus, *trans*-$PdMe_2(PR_3)_2$ is not in an appropriate configuration and trans → cis isomerization must take place first for the dimethylpalladium complex to decompose by the reductive elimination pathway. Addition of a tertiary phosphine ligand blocks this isomerization route and increases the thermal stability of *trans*-$PdMe_2(PR_3)_2$.[16,17] It is interesting to note that thermolysis of *trans*-$PdEt_2(PR_3)_2$, which proceeds via β elimination with the liberation of ethane and ethylene, is not blocked by addition of tertiary phosphine, whereas reductive elimination of the ethyl groups of *cis*-$PdEt_2(PR_3)_2$ to give butane is hindered by added PR_3. For this reason, it has been proposed that the dissociative pathway for the thermolysis of *cis*-$PdEt_2(PR_3)_2$ involves a T-shaped intermediate. A similar dissociative pathway proceeding through a T-shaped trialkyl species has been proposed for the thermolysis of $MeEt_2Au(PPh_3)$.[15] These suggestions have been supported by MO studies.[15,18]

On the other hand, thermolysis of *cis*-$PtR_2(PR'_3)_2$ proceeding by β elimination via T-shaped dialkyl intermediates is hindered by added tertiary phosphine. Other examples show that thermolysis proceeds through dissociative pathways.[54] In these cases the supporting ligand blocks a coordination site and hinders the low-energy pathway, either by reductive elimination or by β elimination.

On the other hand, in the thermolysis of certain metal dialkyls addition of tertiary phosphine ligands serves to promote decomposition, not to stabilize the complexes. Examples of this behavior are *cis*-$PtPh_2(PPh_3)_2$ and NiMe(Ar)(dmpe) (Section 6.2). In these cases the configurations of the intermediate five-coordinate diorgano species seem to be critical.

Thus, it is clear that the role of stabilizing ligands is not restricted simply to blocking the site for β elimination.

Another factor in the stabilizing effect of ligands is electronic. Most of the empirically useful stabilizing ligands such as cyclopentadienyl, CO, tertiary phosphines, and 2,2′-bipyridine are capable of having π-bonding interactions with transition metals. The electronic factors seem to be associated with the stabilizing effect of these ligands. It is known that for some transition metal alkyls an organic nitrogen base such as 2,2′-bipyridine serves as an excellent stabilizing ligand, whereas for some metal alkyls it is not effective and tertiary phosphines are much better. The exact reasons for these differences in stabilizing effects are not known.

Thermolysis mechanisms of monoalkyltransition metal complexes can be quite different from those of dialkyl compounds. At least there is no direct path for reductive elimination. During thermolysis, however, dialkyls may be formed by an intermolecular alkyl exchange reaction, thus complicating the thermolysis mechanism. When such an intermolecular process is not involved, the monoalkyl complex may be stable because reductive elimination cannot take place.

Although one cannot provide an all-embracing discussion that covers the whole array of transition metal alkyls that decompose by various thermolysis mechanisms, a common factor or factors should be associated with the stability of a series of transition metal alkyls that decompose by a similar thermolysis pathway, for example, by reductive elimination. The most important factor in determining the stability of metal alkyls is the M–C bond energy. The following discussion is an attempt to relate the bond energy to the thermal stability of transition metal alkyls that decompose by reductive elimination.

The dialkylnickel complexes [NiR_2(bipy), R = Me and Et] decompose by a nondissociative pathway, the thermolysis rate of the ethyl complex being higher than that of the methyl complex and the decomposition temperature of the methyl complex being higher than that of the ethyl complex. When they are thermolyzed in solution, the following reactions occur.

$$(bipy)Ni\langle^{R}_{R} \xrightarrow{\Delta} \left[(bipy)Ni\langle^{R}_{R}\right]^{\ddagger} \longrightarrow (bipy)Ni(solvent) + R\text{–}R$$

In this reductive elimination process two Ni–R bonds are cleaved and one C–C bond is formed, and solvent molecules may be bound to nickel in the thermolysis product. The process envisaged is shown in Fig. 6.31.

In the thermolysis process the two Ni–R bonds are lengthened and couple with each other to generate the R–R bond. Data on the Ni–Et bond energy are not currently available. However, the trend in the bond energies of transition metal alkyls most probably parallels that in the main-group metal alkyls where a metal–ethyl bond is invariably weaker than a metal–methyl bond. The bond energy difference between the Me–Me and Et–Et bonds is very small, and other factors such as solvation energies cancel when one compares the methyl- and ethylnickel complexes. Thus, the diethylnickel complex is thermodynamically less stable than the dimethylnickel complex. Although the exact nature of the curve for bond dissociation and bond forming is not known, the methyl–nickel and ethyl–nickel bond dissociation energy curves are likely to be similar. Thus, it can be seen from Figure 6.31*a* that the activation energy for reductive elimination of nickel dimethyl is greater than that for nickel diethyl, in agreement with experimental results.[38]

Similar arguments can be extended to the thermolysis of nickel diphenyls, but one must take account of the different C–C single-bond energies of the $C(sp^2)$–$C(sp^2)$ bond (420 kJ mol^{-1}) and the $C(sp^3)$–$C(sp^3)$ bond (370 kJ mol^{-1}). Since the Ni–CH_3 and Ni–Ph bond energy data are not available, we tentatively use the available Mn–CH_3 and Mn–Ph bond energies in $CH_3Mn(CO)_5$ (153 kJ mol^{-1}) and $PhMn(CO)_5$ (170 kJ mol^{-1}).[152] The stability of NiR_2(dmpe) [dmpe = 1,2-bis(dimethylphosphino)ethane] complexes de-

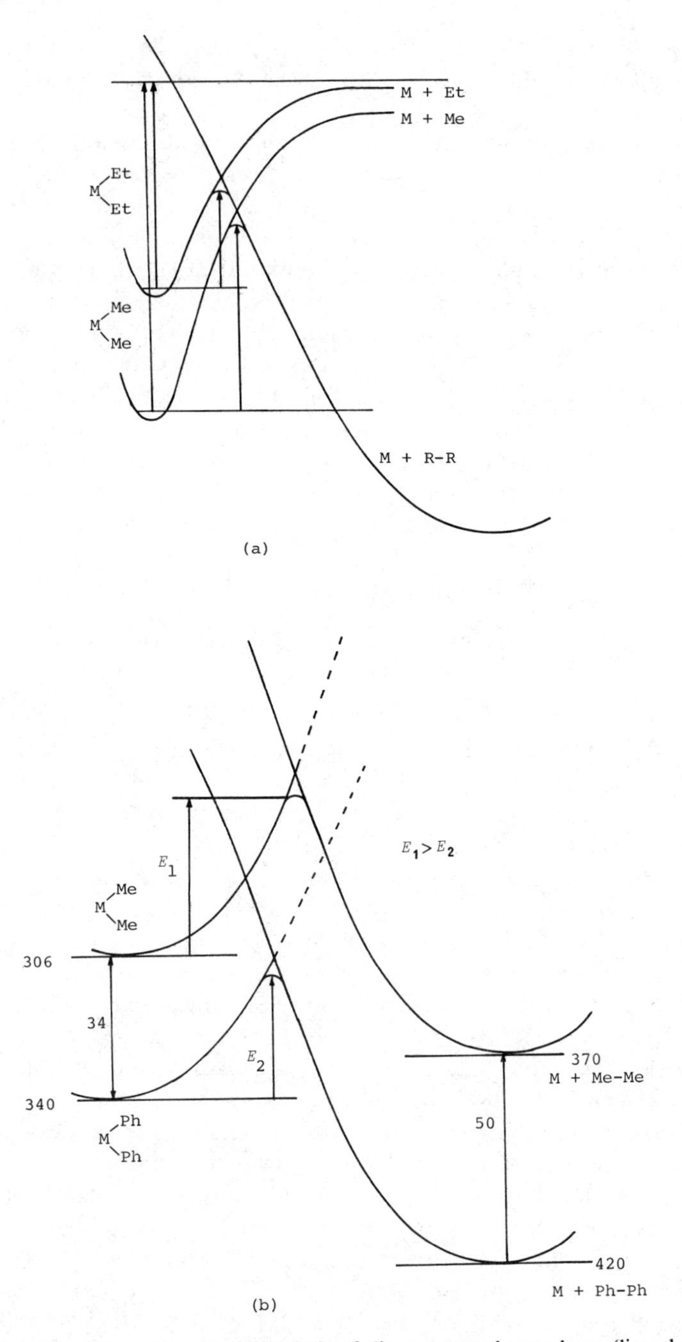

FIGURE 6.31. Reaction profiles in thermolysis of diorganometal complexes (ligands omitted). (*a*) Thermolysis of dimethyl- and diethylmetal compounds via reductive elimination pathway. (*b*) Thermolysis of dimethyl- and diphenylmetal complexes. Numbers are bond energies, in kJ mol^{-1}.

creases in the order $NiMe_2$(dmpe) > NiMePh(dmpe) > $NiPh_2$(dmpe).[127a] The higher thermal stability of the dimethylnickel complex relative to the methylphenyl- and diphenylnickel complexes may appear strange at first sight in view of the fact that the M–Ph bond energy is greater than that of M–Me bond. However, in the process of reductive elimination of R–R from $NiMe_2$(dmpe) and $NiPh_2$(dmpe), two Ni–Me and Ni–Ph bonds are cleaved and Me–Me and Ph–Ph bonds are formed. Since the Ph–Ph bond energy is greater than the Me–Me bond energy by 50 kJ mol^{-1}, more energy is gained by formation of the Ph–Ph bond than by formation of the Me–Me bond.

If the activation energy difference is associated with enthalpy change (in other words, if a linear free energy relationship holds), the activation energy for reductive elimination of Ph–Ph would be lower than that of Me–Me, as roughly sketched in Figure 6.31*b*.

Furthermore, as an extended Hückel MO calculation shows, interaction of the π orbitals of the phenyl group with the methyl group or other phenyl group in thermolysis of NiPhMe(dmpe) or $NiPh_2$(dmpe) may serve to lower the transition state.[127c] We must await further progress in theoretical studies before we can sensibly compare the transition states for reductive elimination of transition metal alkyls and aryls.

Another aspect that awaits theoretical clarification is the activation of transition metal–carbon bonds on interaction with π acids such as olefins and carbon monoxide. As mentioned in Section 5.2, thermolysis of thermally stable NiR_2(bipy) complexes is greatly accelerated on interaction with electron acceptors such as olefins bearing electron-withdrawing substituents. The thermolysis rate increases with the π-acceptor ability of the olefin.[38] This result indicates that the transition state for reductive elimination of R–R has been lowered on interaction of NiR_2(bipy) with the acceptor [Eq. (6.88)]. An interpretation of the cause of the acceleration of the reductive elimination has been presented,[38] and earlier proposals[276,277] have been modified. Activation of the M–R bond was correlated with the energy difference between the σ(M–C) orbital and the vacant d orbital of nickel (LUMO), as shown in Figure 6.32.

It was hypothesized that the metal–alkyl bond is activated by promoting an electron from the bonding σ(M–C) orbital to the LUMO orbital of the metal and that the greater the energy difference, the greater the stability of the metal alkyl. On interaction of the metal alkyl with a π-acceptor olefin, the metal d orbitals are decreased in energy as revealed by the spectroscopic study described in Section 5.2. This makes electronic excitation from the bonding σ(M–C) orbital easier as the energy required for electronic promotion is decreased from (*a*) to (*b*) in Figure 6.32. The energy level of the σ(M–C) orbital is in the order Ni–Me < Ni–Et < Ni–Pr < Ni–Bu. Using this hypothesis, the decreasing stability order Ni–Me > Ni–Et > Ni–Pr > Ni–Bu can be reasonably explained by correlating it with the electron promotion energy from the σ(M–C) orbital to the LUMO, a consequence of the activation effect of the olefin.

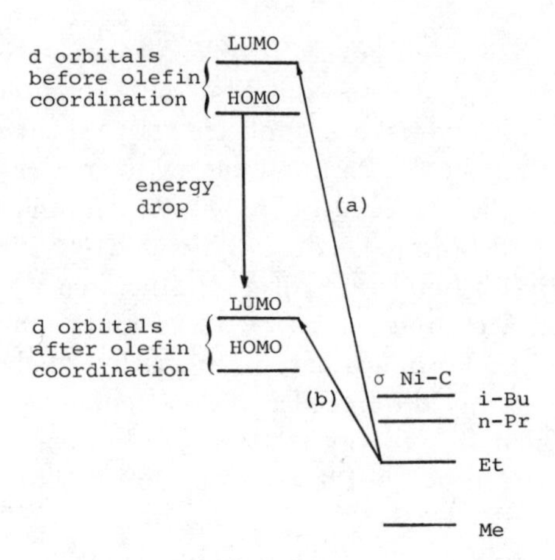

FIGURE 6.32. Interpretation of M–R bond activation in NiR_2(bipy) by coordination of π-acceptor olefins.

A criticism of this type of hypothesis is that the thermal process of the M–C bond cleavage cannot be correlated with the energy difference for electronic excitation.[119a–c] The stability of the M–C bond may be correlated, however, with the energy difference between the bonding σ(M–C) orbital and the antibonding σ^*(M–C) orbital as follows. Figure 6.33 shows the splitting of the M–C bond by a thermal process. The potential energy of the σ(M–C) orbital will increase with the lengthening of the M–C bond brought about thermally. If the potential energy curve of the σ(M–C) orbital crosses that of the

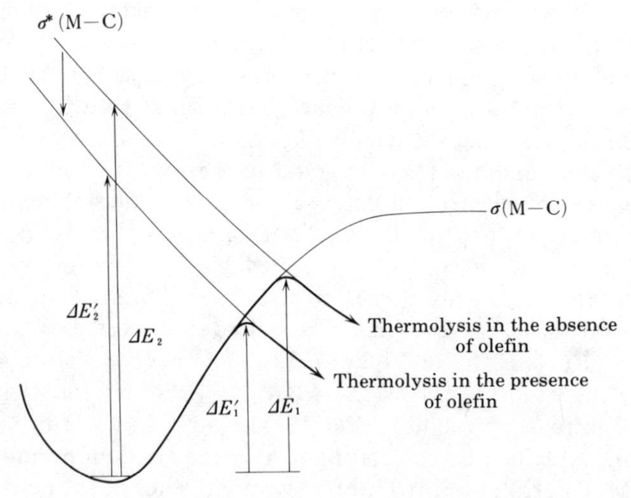

FIGURE 6.33. Thermolysis pathways of metal alkyls in the presence or absence of an added olefin.

σ^*(M–C) orbital, the complex will decompose, following the potential energy curve of the σ^*(M–C) orbital. The activation energy ΔE_1 required for thermolysis is considered to be proportional to the electron excitation energy ΔE_2. Thus, it is possible that the electronic excitation energy is related to the activation energy in thermolysis. In the presence of electronegative olefins the energy of the σ^*(M–C) orbital will be lowered from ΔE_2 to $\Delta E_2'$; correspondingly, the activation energy for thermolysis will be lowered from ΔE_1 to $\Delta E_1'$ in Figure 6.33. This is in agreement with experimental observation. A similar argument may apply to the activating effect of the π-acceptor CO ligand in weakening the M–C bond, a process that occurs in CO insertion.

On the other hand, factors influencing the stabilities of complexes decomposing by dissociative mechanisms involving the rate-determining dissociation of a supporting ligand may differ from the factors influencing thermolysis mechanisms involving nondissociative pathways. In this case the nature of the alkyl group may indirectly affect the rate of dissociation of the supporting ligand.

$$L_nMR_2 \underset{\text{rds}}{\overset{-L}{\rightleftharpoons}} L_{n-1}MR_2 \xrightarrow{\text{fast}} L_{n-1}M + R\text{–}R$$
$$L_{n-1}M \xrightarrow{+L} L_nM$$

In complexes having the tertiary phosphine as the supporting ligand, the ligand–metal interaction may be viewed as a Lewis base–Lewis acid interaction. The more basic the ligand, the more reluctant it is to leave the metal that acts as the Lewis acid. Conversely, decrease of the Lewis acidity on the metal will enhance the dissociation rate of the ligand as a Lewis base. The higher rate of thermolysis of transition metal ethyl complexes relative to their methyl counterparts, therefore, may be ascribed to the higher electron-donating ability of the ethyl group, which causes dissociated, unstable metal alkyls to be formed more rapidly by loss of a ligand.

Factors influencing β elimination are also not well understood. The critical step in this process is interaction of the metal d orbital with the C–H bond attached to the β-carbon atom. Two types of interaction are possible, as shown in Figure 6.34. One is the interaction between the bonding σ(C–H) orbital with a vacant d orbital (Fig. 6.34*a*). The other is a π-type interaction between the antibonding σ^*(C–H) orbital and a filled metal d orbital. The relationship is reminiscent of metal–olefin π bonding. In the former type of interaction the lowering of the vacant metal d orbital (LUMO) is expected to increase the interaction, whereas in the latter case the rise in the energy of the filled metal d orbital (HOMO) will enhance the π-type interaction. In at least one case, namely $TiCl_3Et(PH_3)_2$ an ab initio study revealed that a lowering of and d orbital energy of $TiCl_3Et(PH_3)_2$ causes a decrease in the Ti–C–C bond angle, indicative of increased interaction between the metal and the C–H bond.[209b] The result of this calculation is in agreement with the experimental observation by X-ray crystallography of a conspicuous distortion of the nor-

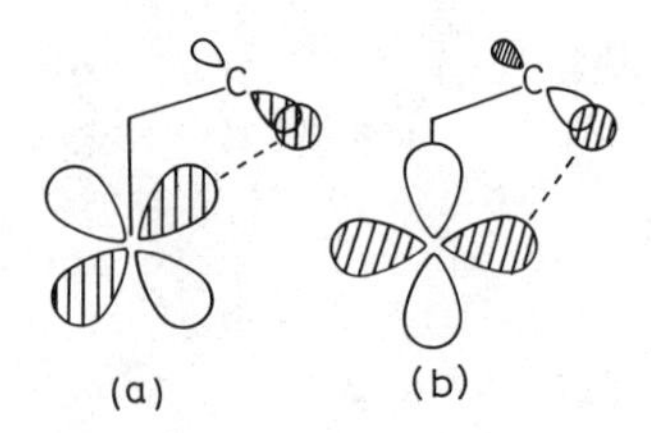

FIGURE 6.34. Interaction of C–H bonding and antibonding orbitals with metal d orbitals.

mal M–C–C bond angle to approximately 90° from its expected value of about 110° for sp^3 hybridization.[209a] A similar interaction may be operative in the activation of γ-C–H bonds.

Thermolysis pathways of transition metal alkyls vary from one complex to another. The thermolysis behavior of a particular type of metal alkyl can be quite different from that of its congener belonging to the same column in the periodic table. For example, *cis*-$PdEt_2(PR_3)_2$ and *cis*-$PtEt_2(PR_3)_2$ are thermolyzed by dissociative mechanisms involving the rate-determining liberation of the coordinated tertiary phosphine. However, the palladium complex decomposes by reductive elimination, whereas the platinum analog is thermolyzed by β-hydrogen elimination and evolves ethylene and ethane. We must await more refined theoretical studies and further kinetic studies before we fully understand these variations in thermolysis behavior and can control the stability and reactivity of transition metal alkyls.

REFERENCES

1. For other classifications, see (a) C. Tolman, *Chem. Rev.*, **1,** 337 (1972); (b) K. F. Purcell and J. C. Kotz, *An Introduction to Inorganic Chemistry,* Holt-Saunders, Philadelphia, 1980, p. 539.
2. J. K. Kochi, *Organometallic Mechanism and Catalysis,* Academic Press, New York, 1978.
3. F. Basolo and R. G. Pearson, *Mechanisms of Inorganic Reactions,* 2nd ed., Wiley, New York, 1967.
4. C. H. Langford and H. B. Gray, *Ligand Substitution Processes,* W. A. Benjamin, New York, 1965.
5. H. Taube, *Chem. Rev.*, **50,** 69 (1952).
6. There are fine distinctions between these extreme mechanisms, such as concerted interchange mechanisms, I_d and I_a, but they need not concern us here.
7. R. E. Dessy and F. Paulik, *J. Chem. Educ.*, **40,** 185 (1963).
8. J. E. Huheey, *Inorganic Chemistry: Principles of Structure and Reactivity,* 2nd ed., Harper & Row, New York, 1978, p. 489.
9. B. Rosenberg, J. E. Trosko, and V. H. Mansour, *Nature,* **222,** 358 (1969); T. A. Connors and J. J. Roberts, eds., *Platinum Coordination Complexes in Cancer Chemotherapy,* Springer-Verlag, Berlin, 1974.
10. (a) F. Basolo, J. Chatt, H. B. Gray, R. G. Pearson, and B. L. Shaw, *J. Chem. Soc.*, 2207 (1961); (b) G. Farone, V. Ricevuto, R. Romeo, and M. Trozzi, *Inorg. Chem.*, **8,** 2207 (1969); **9,** 1525 (1970); (c) *J. Chem. Soc. A,* 1877 (1971).

11. (a) G. K. Anderson and R. J. Cross, *Chem. Soc. Rev.*, **9**, 185 (1980); (b) R. Favez, R. Roulet, A. A. Pinkerton, and D. Schwarzenbach, *Inorg. Chem.*, **19**, 1356 (1980); (c) A. W. Verstuyft and J. H. Belson, *Inorg. Chem.*, **14**, 1501 (1975); **16**, 1317 (1977).

12. W. J. Louw, *Inorg. Chem.*, **16**, 2147 (1977); R. van Eldik, D. A. Palmer, H. Kelm, and W. J. Louw, *Inorg. Chem.*, **19**, 3551 (1980); W. J. Louw, R. van Eldik, and H. Kelm, *Inorg. Chem.*, **19**, 2878 (1980).

13. R. Romeo, *Inorg. Chem.*, **17**, 2040 (1978); R. Romeo, D. Minniti, and M. Trozzi, *Inorg. Chem.*, **15**, 1134 (1976).

14. J. H. Price, J. P. Birk, and B. B. Wayland, *Inorg. Chem.*, **17**, 2245 (1978).

15. S. Komiya, T. A. Albright, R. Hoffmann, and J. K. Kochi, *J. Am. Chem. Soc.*, **98**, 7255 (1976).

16. (a) A. Gillie and J. K. Stille, *J. Am. Chem. Soc.*, **102**, 4933 (1980); (b) A. Moravskiy and J. K. Stille, *J. Am. Chem. Soc.*, **103**, 4182 (1981); (c) M. Loar and J. K. Stille, *J. Am. Chem. Soc.*, **103**, 4174 (1981).

17. (a) F. Ozawa, T. Ito, Y. Nakamura, and A. Yamamoto, *Bull. Chem. Soc. Jpn.*, **54**, 1868 (1981); (b) R. S. Paonessa and W. C. Trogler, *J. Am. Chem. Soc.*, **104**, 3529 (1982).

18. K. Tatsumi, R. Hoffmann, A. Yamamoto, and J. K. Stille, *Bull. Chem. Soc. Jpn.*, **54**, 1857 (1981).

19. Y. W. Yared, S. L. Miles, R. Bau, and C. A. Reed, *J. Am. Chem. Soc.*, **99**, 7076 (1977).

20. H. Nakazawa, F. Ozawa, and A. Yamamoto, *Organometallics*, **2**, 241 (1983).

21. J. Chatt and R. G. Wilkins, *J. Chem. Soc.*, 273, 4300 (1952); 525 (1956).

22. G. R. Dobson, *Acc. Chem. Res.*, **9**, 300 (1976).

23. C. A. Tolman, *Chem. Rev.*, **77**, 313 (1975).

24. (a) J. D. Smith and J. O. Oliver, *Inorg. Chem.*, **17**, 2585 (1978); (b) E. Alyea, S. Dias, G. Fergusson, and R. J. Restivo, *Inorg. Chem.*, **16**, 2329 (1977).

25. K. W. Barnett and T. G. Pollmann, *J. Organometal. Chem.*, **69**, 413 (1974).

26. R. Heck, *J. Am. Chem. Soc.*, **85**, 651, 655, 657 (1963).

27. C. A. Tolman, *J. Am. Chem. Soc.*, **92**, 2953 (1970).

28. W. Strohmeier and F. J. Müller, *Chem. Ber.*, **100**, 2812 (1967).

29. (a) W. Partenheimer, *Inorg. Chem.*, **11**, 743 (1972); (b) W. Partenheimer and E. F. Hoy, *Inorg. Chem.*, **12**, 2805 (1973).

30. T. Yoshida, T. Matsuda, T. Okano, T. Kitami, and S. Otsuka, *J. Am. Chem. Soc.*, **101**, 2027 (1979).

31. (a) M. E. Rerek and F. Bassolo, *Organometallics*, **2**, 372 (1983); (b) C. P. Casey, J. M. O'Conner, W. D. Jones, and K. J. Haller, *Organometallics*, **2**, 535 (1983); (c) C. P. Casey and J. M. O'Conner, *J. Am. Chem. Soc.*, **105**, 2919 (1983).

32. T. Fueno, O. Kajimoto, and J. Furukawa, *Bull. Chem. Soc. Jpn.*, **41**, 782 (1968).

33. T. Fueno, T. Okuyama, T. Deguchi, and J. Furukawa, *J. Am. Chem. Soc.*, **87**, 170 (1965).

34. (a) H. Kurosawa, T. Majima, and N. Asada, *J. Am. Chem. Soc.*, **102**, 6996 (1980); (b) H. Kurosawa and N. Asada, *J. Organometal. Chem.*, **217**, 259 (1981): (c) H. Kurosawa and N. Asada, *Organometallics*, **2**, 251 (1983).

35. (a) K. Miki, M. Yama, Y. Kai, and N. Kasai, *J. Organometal. Chem.*, **239**, 417 (1982); (b) K. Miki, Y. Kai, N. Kasai, and H. Kurosawa, *J. Am. Chem. Soc.*, **105**, 2482 (1983).

36. (a) J. G. Traynham and J. R. Olechowski, *J. Am. Chem. Soc.*, **81**, 571 (1959); (b) W. Partenheimer and E. H. Johnson, *Inorg. Chem.*, **12**, 1274 (1973); (c) W. Partenheimer and B. Durham, *J. Am. Chem. Soc.*, **96**, 3800 (1974).

37. R. Cramer, *J. Am. Chem. Soc.*, **86**, 217 (1964); **89**, 4621 (1967).

38. T. Yamamoto, A. Yamamoto, and S. Ikeda, *J. Am. Chem. Soc.*, **93**, 3350, 3360 (1971).

39. C. A. Tolman, *J. Am. Chem. Soc.*, **96**, 2780 (1974).

40. S. D. Ittel, *Inorg. Chem.*, **16,** 2589 (1977).

41. (a) S. Otsuka, T. Yoshida, and Y. Tatsuno, *J. Am. Chem. Soc.*, **93,** 6462 (1971); (b) T. Yoshida, Y. Tatsuno, and S. Otsuka, *Pure Appl. Chem.*, **52,** 713 (1980).

42. C. A. Tolman and W. L. Seidel, *J. Am. Chem. Soc.*, **96,** 2774 (1974).

43. C. E. Moore, *Atomic Energy Levels,* NBS Circular No. 467, Vol. I, 1949, Vol. II, 1952, Vol. III, 1958.

44. (a) J. R. Joy and M. Orchin, *J. Am. Chem. Soc.*, **81,** 305 (1959); (b) S. I. Shupak and M. Orchin, *J. Am. Chem. Soc.*, **86,** 586 (1964).

45. (a) S. J. Bryan, P. G. Huggett, K. Wade, J. A. Daniels, and J. R. Jennings, *Coord. Chem. Rev.*, **44,** 149 (1982); (b) B. N. S. Storhoff and H. C. Lewis, Jr., *Coord. Chem. Rev.*, **23,** 1 (1977).

46. S. Komiya, J. Ishizu, A. Yamamoto, T. Yamamoto, A. Takenaka, and Y. Sasada, *Bull. Chem. Soc. Jpn.*, **53,** 1283 (1980).

47. (a) T. A. Albright, *Acc. Chem. Res.*, **15,** 149 (1982); (b) T. A. Albright, R. Hoffmann, J. C. Thibeault, and D. L. Thorn, *J. Am. Chem. Soc.*, **101,** 3801 (1979); (c) N. Rösch and R. Hoffmann, *Inorg. Chem.*, **13,** 2656 (1974).

48. R. M. Pitzer and H. F. Schaefer III, *J. Am. Chem. Soc.*, **101,** 7176 (1979).

49. (a) P. J. Hay, *J. Am. Chem. Soc.*, **103,** 1390 (1981); (b) T. Ziegler and A. Rauk, *Inorg. Chem.*, **18,** 1558 (1979).

50. K. Kitaura, S. Sakaki, and K. Morokuma, *Inorg. Chem.*, **20,** 2292 (1981).

51. (a) W. C. Hamilton, K. A. Klandermann, and R. Spratley, *Acta Crystallogr.*, **A25,** S172 (1969); (b) W. Dreissig and H. Dietrich, *Acta Crystallogr.*, **B37,** 931 (1981); (c) J. K. Stalick and J. A. Ibers, *J. Am. Chem. Soc.*, **92,** 5333 (1970); (d) A. J. Schulz, R. K. Brown, J. M. Williams, and R. R. Schrock, *J. Am. Chem. Soc.*, **103,** 169 (1981).

52. D. M. P. Mingos, *Adv. Organometal. Chem.*, **15,** 1 (1977).

53. J. Ashley-Smith, M. Green, and D. C. Wood, *J. Chem. Soc. A,* 1847 (1970).

54. T. Yamamoto, A. Yamamoto, and S. Ikeda, *Bull. Chem. Soc. Jpn.*, **45,** 1104 (1972).

55. (a) W. E. Buhro, A.-H. Patton, C. E. Strouse, J. A. Gladysz, F. B. McCormick, and M. C. Etter, *J. Am. Chem. Soc.*, **105,** 1056 (1983); (b) W. Kropp, V. Skibbe, G. Erker, and C. Krüger, *J. Am. Chem. Soc.*, **105,** 3353 (1983); (c) S. Gambarotta, C. Floriani, A. Chiesi-Villa, and C. Guastini, *J. Am. Chem. Soc.*, **104,** 2019 (1982); (d) G. R. Clark, C. E. L. Headford, K. Marsden, and W. R. Roper, *J. Organometal. Chem.*, **231,** 335 (1982).

56. (a) J. P. Collman, *Acc. Chem. Res.*, **1,** 136 (1968); (b) L. Vaska, *Acc. Chem. Res.*, **1,** 335 (1968); (c) J. Halpern, *Acc. Chem. Res.*, **3,** 386 (1970); (d) A. J. Deeming, *MIP Int. Rev. Sci. Inorg. Chem. Ser. One,* Vol. **9,** 117 (1972); (e) J. P. Collman and W. R. Roper, *Adv. Organometal. Chem.*, **7,** 53 (1968).

57. H.-P. Abicht and K. Issleib, *Z. Chem.*, **17,** 1 (1977); G. W. Parshall, *Acc. Chem. Res.*, **3,** 139 (1970); M. I. Bruce, *Angew. Chem. Int. Ed. Engl.*, **16,** 73 (1977); I. Omae, *Coord. Chem. Rev.*, **32,** 235 (1980).

58. J. M. Guss and R. Mason, *Chem. Commun.*, 58, (1971).

59. S. Hietkamp, D. J. Stufkens, and K. Vrieze, *J. Organometal. Chem.*, **139,** 189 (1977).

60. N. Barr and S. F. Dyke, *J. Organometal. Chem.*, **243,** 223 (1983).

61. A. Dedieu and A. Strich, *Inorg. Chem.*, **18,** 2940 (1979).

62. K. Kitaura, S. Obara, and K. Morokuma, *J. Am. Chem. Soc.*, **103,** 2891 (1981).

63. (a) J. K. Stille and K. S. Y. Lau, *Acc. Chem. Res.*, **10,** 434 (1977); (b) R. J. Mureinik, M. Weitzberg, and J. Blum, *Inorg. Chem.*, **18,** 915 (1979).

64. (a) Y. Becker and J. K. Stille, *J. Am. Chem. Soc.*, **100,** 838 (1978); (b) K. S. Y. Lau, R. W. Fries, and J. K. Stille, *J. Am. Chem. Soc.*, **96,** 4983 (1974); (c) K. S. Y. Lau, P. K. Wong, and J. K. Stille, *J. Am. Chem. Soc.*, **98,** 5832 (1976); (d) P. K. Wong, K. S. Y. Lau, and J. K. Stille, *J. Am. Chem. Soc.*, **96,** 5956 (1974).

65. (a) W. H. Thompson and C. T. Sears, Jr., *Inorg. Chem.*, **16,** 769 (1977); (b) R. Ugo, A. Pasini, A. Fusi, and S. Cenini, *J. Am. Chem. Soc.*, **94,** 7364 (1972); (c) P. B. Chock and J. Halpern, *J. Am. Chem. Soc.*, **88,** 3511 (1966).
66. J. Rajaram, R. G. Pearson, and J. A. Ibers, *J. Am. Chem. Soc.*, **96,** 2103 (1974).
67. J. F. Fauvarque, F. Pflüger, and M. Troupel, *J. Organometal. Chem.*, **208,** 419 (1981).
68. J. K. Kochi, *Organometallic Mechanisms and Catalysis,* Academic Press, New York, 1979.
69. T. T. Tsou and J. K. Kochi, *J. Am. Chem. Soc.*, **101,** 6319 (1979).
70. J. A. Osborn, in *Organotransition-Metal Chemistry,* Y. Ishii and M. Tsutsui, eds., Plenum Press, New York, 1976.
71. R. G. Pearson and P. E. Figdore, *J. Am. Chem. Soc.*, **102,** 1541 (1980).
72. J. Burgess, M. J. Hacker, and R. D. Kemmitt, *J. Organometal. Chem.*, **72,** 121 (1974).
73. (a) J. Chatt and M. Davidson, *J. Chem. Soc.*, 843 (1965); (b) C. A. Tolman, S. D. Ittel, A. D. English, and J. P. Jesson, *J. Am. Chem. Soc.*, **101,** 1742 (1979); (c) S. D. Ittel, C. A. Tolman, A. D. English, and J. P. Jesson, *J. Am. Chem. Soc.*, **100,** 7577 (1978).
74. (a) R. Werner and H. Werner, *Angew. Chem., Int. Ed. Engl.*, **20,** 793 (1981); (b) W. D. Jones and F. J. Feher, *J. Am. Chem. Soc.*, **104,** 4240 (1982).
75. (a) G. W. Parshall, *Acc. Chem. Res.*, **8,** 113 (1975); (b) *Catalysis,* **1,** 335 (1977).
76. M. L. H. Green, *Ann. N. Y. Acad. Sci.*, **333,** 229 (1980); N. J. Cooper, M. L. H. Green, and R. Mahtab, *J. Chem. Soc. Dalton Trans.*, 1557 (1979), 1950 (1979).
77. A. H. Janowicz and R. G. Bergman, *J. Am. Chem. Soc.*, **105,** 3929 (1983).
78. J. Fornies, M. Green, J. L. Spencer, and F. G. A. Stone, *J. Chem. Soc., Dalton Trans.*, 1006 (1977).
79. J. K. Hoyano and W. A. Graham, *J. Am. Chem. Soc.*, **104,** 3723 (1982).
80. W. D. Jones and F. J. Feher, *Organometallics,* **2,** 562 (1983).
81. (a) D. Baudry, M. Ephritikhime, H. Felkin, and J. Zakrzewski, *J. Chem. Soc. Chem. Commun.*, 1235 (1982) and references cited therein; (b) R. H. Crabtree, M. F. Mellea, J. M. Mihelcic, and J. M. Quirk, *J. Am. Chem. Soc.*, **104,** 107 (1982).
82. D. E. Webster, *Adv. Organometal. Chem.*, **15,** 147 (1979).
83. (a) G. Carturan, A. Scrivanti, and F. Morandini, *Angew. Chem. Int. Ed. Engl.*, **20,** 112 (1981); (b) H. Bönnemann, *Angew. Chem.*, **82,** 699 (1970); (c) D. J. Cole-Hamilton and G. Wilkinson, *J. Chem. Soc. Chem. Commun.*, 59 (1977); (d) T. H. Tulip and J. A. Ibers, *J. Am. Chem. Soc.*, **101,** 4201 (1979).
84. S. Komiya, T. Ito, M. Cowie, A. Yamamoto, and J. A. Ibers, *J. Am. Chem. Soc.*, **98,** 3874 (1976).
85. M. Kubota, A. Miyashita, S. Komiya, and A. Yamamoto, *J. Organometal. Chem.*, **139,** 111 (1977).
86. (a) A. I. Deeming and M. Underhill, *J. Chem. Soc., Chem. Commun.*, 277 (1973); (b) J. B. Keister and J. R. Shapley, *J. Organometal. Chem.*, **85,** C29 (1975).
87. (a) S. D. Ittel, C. A. Tolman, A. D. English, and J. P. Jesson, *Adv. Chem. Ser.*, Vol. **173,** Am. Chem. Soc. 1979, p. 67; (b) J. W. Suggs, *J. Am. Chem. Soc.*, **100,** 640 (1978); (c) D. Milstein, *Organometallics,* **1,** 1549 (1982).
88. (a) J. Tsuji, *Organic Synthesis via Metal Carbonyls,* Vol. II, I. Wender and P. Pino, eds., Wiley-Interscience, New York, 1977; (b) D. Milstein, *Acc. Chem. Res.*, **17,** 221 (1984).
89. (a) J. A. Gladysz, *Adv. Organometal. Chem.*, **20,** 1 (1982); (b) D. L. Thorn, *J. Mol. Cat.*, **17,** 279 (1982); (c) K. L. Brown, G. R. Clark, C. E. L. Headford, K. Marsden, and W. Roper, *J. Am. Chem. Soc.*, **101,** 503 (1979); (d) D. R. Fahey, *J. Am. Chem. Soc.*, **103,** 136 (1981).
90. (a) A. Yamamoto, T. Yamamoto, T. Ito, S. Komiya, J. Ishizu, T. Kohara, and K. Igarashi, in *Fundamental Research in Organometallic Chemistry,* van Nostrand Reinhold, New York, and Science Press, Beijing, Y. Ishii, M. Tsutsui, and Y. Z. Huang, eds., 1982, p. 327;

(b) B. M. Trost, *Tetrahedron,* **33,** 2615 (1977); (c) G. P. Chiusoli, *Pure Appl. Chem.,* **52,** 635 (1980).

91. (a) J. Ishizu, T. Yamamoto, and A. Yamamoto, *Chem. Lett.,* 1091 (1976); (b) T. Yamamoto, J. Ishizu, and A. Yamamoto, *J. Am. Chem. Soc.,* **103,** 6863 (1981).
92. T. Yamamoto, O. Saito, and A. Yamamoto, *J. Am. Chem. Soc.,* **103,** 5600 (1981).
93. (a) J. Tsuji, *Bull. Chem. Soc. Jpn.,* **46,** 1896; (b) J. Tsuji and T. Yamakawa, *Tetrahedron Lett.,* 613 (1979); (c) J. Tsuji, T. Yamakawa, M. Kaito, and T. Mandai, Tetrahedron Lett, 2075 (1978); (d) T. Tsuda, Y. Chujo, S. Nishi, K. Tawara, and T. Saegusa, *J. Am. Chem. Soc.,* **102,** 6381 (1980).
94. T. Yamamoto, J. Ishizu, T. Kohara, S. Komiya, and A. Yamamoto, *J. Am. Chem. Soc.,* **102,** 3758 (1980).
95. T. Yamamoto, S. Miyashita, Y. Naito, S. Komiya, T. Ito, and A. Yamamoto, *Organometallics,* **1,** 808 (1982).
96. Y. Hayashi, T. Yamamoto, A. Yamamoto, S. Komiya, and Y. Kushi, *J. Am. Chem. Soc.* in press.
97. (a) S. Komiya, A. Yamamoto, and T. Yamamoto, *Chem. Lett.,* 193 (1981); (b) J. P. Collman and S. R. Winter, *J. Am. Chem. Soc.,* **95,** 4089 (1973).
98. K. Sano, T. Yamamoto, and A. Yamamoto, *Chem. Lett.,* 115 (1983).
99. J. Tsuji, I. Shimizu, I. Minami, and Y. Ohashi, *Tetrahedron Lett.,* **23,** 4809 (1982); J. Tsuji, K. Sato, and H. Okamoto, *J. Org. Chem.,* **49,** 1341 (1984).
100. S. Komiya and A. Yamamoto, *J. Organometal. Chem.,* **87,** 333 (1970).
101. T. Yamamoto, J. Ishizu, and A. Yamamoto, *Bull. Chem. Soc. Jpn.,* **55,** 623 (1982).
102. (a) D. Milstein and J. C. Calabrese, *J. Am. Chem. Soc.,* **104,** 3773 (1982); (b) R. Schlodder, J. A. Ibers, M. Lenarda, and M. Graziani, *J. Am. Chem. Soc.,* **96,** 6893 (1974).
103. (a) J. Rajaram and J. A. Ibers, *J. Am. Chem. Soc.,* **100,** 829 (1978); (b) J. Burgess, R. I. Haines, E. R. Hammer, R. D. W. Kemmitt, and M. A. R. Smith, *J. Chem. Soc. Dalton Trans.,* 2579 (1975).
104. K. P. C. Vollhardt and T. W. Weidman, *J. Am. Chem. Soc.,* **105,** 1676 (1983).
105. D. H. Gerlach, A. R. Kane, G. W. Parshall, J. P. Jesson, and E. L. Muetterties, *J. Am. Chem. Soc.,* **93,** 3543 (1971); G. W. Parshall, *J. Am. Chem. Soc.,* **96,** 2360 (1974).
106. A. Morvillo and A. Turco, *J. Organometal. Chem.,* **208,** 103 (1981).
107. J. L. Burmeister and L. M. Edwards, *J. Chem. Soc., (A),* 1663 (1971).
108. F. Ozawa, K. Iri, and A. Yamamoto, *Chem. Lett.,* 1707 (1982).
109. T. Yoshida, T. Matsuda, T. Okano, T. Kitami, and S. Otsuka, *J. Am. Chem. Soc.,* **101,** 2027 (1979).
110. T. Yoshida and S. Otsuka, *J. Am. Chem. Soc.,* **99,** 2134 (1977).
111. J. Fornies, M. Green, J. L. Spencer, and F. G. A. Stone, *J. Chem. Soc. Dalton Trans.,* 1006 (1977).
112. P. W. Atkins, J. C. Green, and M. L. H. Green, *J. Chem. Soc. (A),* 3350 (1971).
113. T. Yamamoto, K. Sano, and A. Yamamoto, *Chem. Lett.,* 907 (1982).
114. J. R. Gaylor and C. V. Senoff, *Can. J. Chem.,* **50,** 1868 (1972).
115. K. Osakada, K. Matsumoto, T. Yamamoto, and A. Yamamoto, *Organometallics,* **4,** 857 (1985).
116. (a) A. J. Chalk and J. F. Harrod, *J. Am. Chem. Soc.,* **87,** 16 (1965); (b) K. M. Mackay and B. K. Nicholson, Comprehensive Organometallic Chemistry, Vol. 6, Pergamon Press, Oxford, 1982, p. 1043; (c) J. L. Speier, *Adv. Organometal. Chem.,* **17,** 407 (1979).
117. J. U. Mondal, R. Bullo, and D. M. Blake, *Inorg. Chem.,* **21,** 1668 (1982); A. L. Balch and Y. S. Sohn, *J. Organometal. Chem.,* **30,** C31 (1971).
118. A. Stockis and R. Hoffmann, *J. Am. Chem. Soc.,* **102,** 2952 (1980).
119. (a) P. S. Braterman and R. J. Cross, *Chem. Soc. Rev.,* **2,** 271 (1973); (b) *J. Chem. Soc. Dalton Trans.,* 657 (1972); (c) P. S. Braterman, *Top. Cur. Chem.,* **92,** 149 (1980); (d) M. C.

Baird, *J. Organometal. Chem.*, **64,** 289 (1974); (e) J. R. Norton, *Acc. Chem. Res.*, **12,** 139 (1979).

120. (a) G. M. Whitesides, E. J. Panek, and E. R. Stedronsky, *J. Am. Chem. Soc.*, **94,** 232 (1972); (b) A. Miyashita, T. Yamamoto, and A. Yamamoto, *Bull. Chem. Soc. Jpn.*, **50,** 1109 (1977); (c) T. Yamamoto, T. Saruyama, Y. Nakamura, and A. Yamamoto, *Bull. Chem. Soc. Jpn.*,**49,** 589 (1976); (d) T. Yamamoto, T. Kohara, and A. Yamamoto, *Bull. Chem. Soc. Jpn.*, **54,** 1720, 2010 (1961).
121. (a) D. G. Morrel and J. K. Kochi, *J. Am. Chem. Soc.*, **97,** 7262 (1975); (b) G. Smith and J. K. Kochi, *J. Organometal. Chem.*, **198,** 199 (1980).
122. M. P. Brown, J. R. Puddephatt, and C. E. E. Upton, *J. Chem. Soc. Dalton Trans.*, 2457 (1974); M. P. Brown, R. J. Puddephatt, C. E. E. Upton, and S. W. Lavington, *J. Chem. Soc. Dalton Trans.*, 1613 (1974).
123. (a) T. Debaerdemaeker, H. P. Klein, M. Wiege, and H. A. Brune, *Z. Naturforsch.*, **36b,** 958 (1981); (b) U. Bayer and H. A. Brune, *Z. Naturforsch.*, **38b,** 226, 621, 632 (1983) and references therein.
124. P. S. Braterman, R. J. Cross, and G. B. Young, *J. Chem. Soc. Dalton Trans.*, 1892 (1977).
125. D. Milstein and J. K. Stille, *J. Am. Chem. Soc.*, **101,** 4981 (1979).
126. A. Nakamura and S. Otsuka, *J. Am. Chem. Soc.*, **95,** 7262 (1973).
127. (a) S. Komiya, Y. Abe, T. Yamamoto, and A. Yamamoto, *Organometallics*, **2,** 1466 (1983); (b) K. Tatsumi, A. Nakamura, S. Komiya, A. Yamamoto, and T. Yamamoto, *J. Am. Chem. Soc.*, **106,** 8181 (1984); (c) K. Tatsumi, unpublished results.
128. (a) M. J. Doyle, J. McMeeking, and P. Binger, *J. Chem. Soc. Chem. Commun.*, 376 (1976); (b) P. Binger, J. H. Doyle, C. Krüger, and Y. H. Tsay, *Z. Naturforsch.*, **34b,** 1289 (1979).
129. (a) T. Kohara, S. Komiya, T. Yamamoto, A. Yamamoto, *Chem. Lett.*, 1513 (1979); (b) S. Komiya, Y. Akai, K. Tanaka, T. Yamamoto, and A. Yamamoto, *Organometallics*, **4,** 1130 (1985).
130. G. M. Whitesides, *Pure Appl. Chem.*, **53,** 287 (1981).
131. T. T. McCarthy, R. G. Nuzzo, and G. M. Whitesides, *J. Am. Chem. Soc.*, **103,** 3396, 3404 (1981).
132. S. Komiya, Y. Morimoto, T. Yamamoto, and A. Yamamoto, *Organometallics*, **1,** 1528 (1982).
133. (a) T. Yamamoto, A. Yamamoto, and S. Ikeda, *Bull. Chem. Soc. Jpn.*, **45,** 1104 (1972); (b) W. Lau, J. C. Huffman, and J. K. Kochi, *Organometallics*, **1,** 155 (1982).
134. (a) R. H. Grubbs, A. Miyashita, M. Liu, and P. Burk, *J. Am. Chem. Soc.*, **100,** 2418 (1978); (b) J. McDermitt, J. F. White, and G. M. Whitesides, *J. Am. Chem. Soc.*, **98,** 6521 (1976).
135. (a) J. R. Norton, *Acc. Chem. Res.*, **12,** 139 (1979); (b) J. Halpern, *Acc. Chem. Res.*, **15,** 332 (1982); (c) A. H. Janowicz and R. G. Bergman, *J. Am. Chem. Soc.*, **103,** 2488 (1981).
136. W. N. Rogers and M. Baird, *J. Organometal. Chem.*, **182,** C65 (1979).
137. A. Wojcicki, *Adv. Organometal. Chem.*, **11,** 87 (1973); **12,** 33 (1974).
138. F. Calderazzo, *Angew. Chem. Int. Ed. Engl.*, **16,** 299 (1977).
139. E. J. Kuhlmann and J. J. Alexander, *Coord. Chem. Rev.*, **33,** 195 (1980).
140. T. C. Flood, *Top Stereochem.*, **12,** 37 (1980).
141. (a) K. Noack and F. Calderazzo, *J. Organometal. Chem.*, **10,** 101 (1967); (b) T. C. Flood, J. E. Jensen, and J. A. Statler, *J. Am. Chem. Soc.*, **103,** 4410 (1981); (c) R. W. Glyde and R. J. Mawby, *Inorg. Chem.*, **10,** 854, (1971); (d) R. J. Mawby, F. Basolo, and R. G. Pearson, *J. Am. Chem. Soc.*, **86,** 3994 (1964).
142. (a) H. Brunner, B. Hammer, I. Bernal, and M. Draux, *Organometallics*, **2,** 1595 (1983). (b) T. C. Flood, K. D. Campbell, H. H. Downs, and S. Nakanishi, *Organometallics*, **2,** 1590 (1983).

143. F. Ozawa and A. Yamamoto, *Chem. Lett.*, 289 (1981).

144. P. L. Bock, D. J. Boschetto, J. R. Rasmussen, J. P. Demers, and G. M. Whitesides, *J. Am. Chem. Soc.*, **96,** 2814 (1974).

145. M. A. Dunham and M. C. Baird, *J. Chem. Soc. Dalton Trans.*, 774 (1975).

146. G. Fachinetti, C. Floriani, and H. Stoeckli-Evans, *J. Chem. Soc., Dalton Trans.*, 2297 (1977).

147. P. T. Wolczanski and J. E. Bercaw, *Acc. Chem. Res.*, **13,** 121 (1980).

148. J. M. Manriquez, D. R. McAlister, R. D. Sanner, and J. E. Bercaw, *J. Am. Chem. Soc.*, **100,** 2716 (1978).

149. T. G. Richmond, F. Basolo, and D. F. Schriver, *Inorg. Chem.*, **21,** 1272 (1982).

150. J. P. Collman, R. Finke, J. N. Cawse, and J. I. Brauman, *J. Am. Chem. Soc.*, **100,** 4766 (1978).

151. F. Calderazzo and F. A. Cotton, *Inorg. Chem.*, **1,** 30 (1962).

152. J. A. Connor, M. T. Zafurani-Moattar, J. Bickerton, N. I. ElSaied, S. Suradi, R. Carson, G. AlTakhin, and H. A. Skinner, *Organometallics,* **1,** 1166 (1982).

153. R. Cross and J. Gemmill, *J. Chem. Soc. Dalton,* 2317 (1981).

154. D. L. Egglestone, M. C. Baird, C. J. L. Lock, and G. Turner, *J. Chem. Soc., Dalton Trans.*, 1576 (1977); D. Egglestone and M. C. Baird, *J. Organometal. Chem.*, **113,** C25 (1976); D. S. Slack, D. L. Egglestone, and M. C. Baird, *J. Organometal. Chem.*, **146,** 71 (1978).

155. (a) D. M. Blake, J. de Faller, Y. L. Chung, and A. Winkelman, *J. Am. Chem. Soc.*, **96,** 5568 (1974); (b) D. M. Blake, A. Winkelman, and Y. L. Chung, *Inorg. Chem.*, **14,** 1326 (1975).

156. J. A. Connor, *Top. Cur. Chem.*, **71,** 71 (1977).

157. J. N. Cawse, R. A. Fiato, and R. L. Pruett, *J. Organometal. Chem.*, **172,** 405 (1979).

158. (a) P. E. Garrou and R. F. Heck, *J. Am. Chem. Soc.*, **98,** 4115 (1976); (b) W. Sugita, J. V. Minkiewicz, and R. F. Heck, *Inorg. Chem.*, **17,** 2809 (1978).

159. (a) M. Kubota, D. M. Blake, and S. A. Smith, *Inorg. Chem.*, **10,** 1430 (1971); (b) D. M. Blake, A. Vinson, and R. Dye, *J. Organometal. Chem.*, **204,** 257 (1981); (c) M. A. Bennett and J. C. Jeffery, *Inorg. Chem.*, **19,** 3763 (1980); (d) M. A. Bennett, J. C. Jeffery, and G. B. Robertson, *Inorg. Chem.*, **20,** 323 (1981).

160. C. P. Casey and D. M. Scheck, *J. Am. Chem. Soc.*, **102,** 2723, 2728 (1980).

161. H. Berke and R. Hoffmann, *J. Am. Chem. Soc.*, **100,** 7224 (1978).

162. S. Sakaki, K. Kitaura, K. Morokuma, K. Ohkubo, *J. Am. Chem. Soc.*, **105,** 2280 (1983).

163. W. Keim, ed., *Catalysis in* C_1 *Chemistry,* D. Reidel, Dordrecht, 1983.

164. B. B. Wayland and B. A. Woods, *J. Chem. Soc., Chem. Commun.*, 700 (1981); B. B. Wayland, B. A. Woods, and R. Pierce, *J. Am. Chem. Soc.*, **104,** 302 (1982); B. B. Wayland, A. Duttahmed, and B. A. Woods, *J. Chem. Soc. Chem. Commun.*, 142 (1983).

165. P. J. Fagan, K. J. Moloy, and T. J. Marks, *J. Am. Chem. Soc.*, **103,** 6959 (1981).

166. C. P. Casey, S. M. Neumann, M. A. Andrews, and D. R. McAlister, *Pure Appl. Chem.*, **52,** 625 (1980).

167. C. P. Casey, C. A. Bunnel, and J. C. Calabrese, *J. Am. Chem. Soc.*, **98,** 1166 (1976).

168. F. Ozawa, T. Sugimoto, T. Yamamoto, and A. Yamamoto, *Organometallics,* **3,** 692 (1984).

169. (a) H. des Abbayes and A. Buloup, *J. Chem. Soc. Chem. Commun.*, 1090 (1978); (b) L. Cassar, *Ann. N.Y. Acad. Sci.*, **333,** 208 (1980).

170. F. Francalanci, A. Gardono, L. Abis, T. Fiorani, and M. Foà, *J. Organometal. Chem.*, **243,** 87 (1983).

171. W. J. Evans, A. L. Wayda, W. H. Hunter, and J. L. Atwood, *J. Chem. Soc., Chem. Commun.*, 706 (1981).

172. (a) C. Floriani, *Pure Appl. Chem.*, **55,** 1 (1983); (b) S. Gambarotta, C. Floriani, A. Chiesi-Villa, and C. Guastini, *J. Am. Chem. Soc.*, **105,** 1690 (1983).

173. K. G. Caulton, *J. Mol. Cat.*, **13,** 71 (1981).

174. A. Wojcicki, *Acc. Chem. Res.*, **4,** 344 (1971).

175. S. L. Mils, D. L. Mils, R. Bau, and T. C. Flood, *J. Am. Chem. Soc.*, **100,** 7278 (1978).

176. T. G. Attig, R. G. Teller, S.-M. Wu, R. Bau, and A. Wojcicki, *J. Am. Chem. Soc.*, **101,** 619 (1979).

177. Y. Yamamoto and H. Yamazaki, *Coord. Chem. Rev.*, **8,** 225 (1972); P. M. Treichel, *Adv. Organometal. Chem.*, **11,** 21 (1973).

178. Y. Yamamoto and H. Yamazaki, *Bull. Chem. Soc. Jpn.*, **43,** 2653, 3634 (1970).

179. T. J. Collins and W. R. Roper, *J. Chem. Soc. Chem. Commun.*, 1044 (1976).

180. W. P. Weiner, M. A. White, and R. G. Bergman, *J. Am. Chem. Soc.*, **103,** 3612 (1981); W. P. Weiner and R. G. Bergman, *J. Am. Chem. Soc.*, **105,** 3922 (1983).

181. P. N. Becker and R. G. Bergman, *Organometallics*, **2,** 787 (1983); *J. Am. Chem. Soc.*, **105,** 2985 (1983).

182. (a) H. Yasuda, Y. Kajihara, K. Mashima, K. Nagasuna, and A. Nakamura, *Chem. Lett.*, 671 (1981); (b) M. Akita, H. Yasuda, and A. Nakamura, *Chem. Lett.*, 217 (1983).

183. M. A. Bennett and T. Yoshida, *J. Am. Chem. Soc.*, **100,** 1750 (1978); D. P. Arnold, M. A. Bennett, M. S. Bilton, and G. B. Robertson, *J. Chem. Soc. Chem. Commun.*, 115 (1982); D. P. Arnold and M. A. Bennett, *J. Organometal. Chem.*, **199,** C17 (1980).

184. J. Chatt and B. L. Shaw, *J. Chem. Soc.*, 5075 (1962).

185. H. C. Clark and C. R. Jablonski, *Inorg. Chem.*, **13,** 2213 (1974).

186. H. C. Clark, *J. Organometal. Chem.*, **200,** 63 (1980) and references therein.

187. (a) H. Werner and R. Fraser, *Angew. Chem. Int. Ed. Engl.*, **18,** 157 (1979); (b) *J. Organometal Chem.*, **232,** 351 (1982); (c) J. W. Byrne, H. J. Blaser, and J. A. Osborn, *J. Am. Chem. Soc.*, **97,** 3871 (1975); (d) J. W. Byrne, J. R. Kress, J. A. Osborn, L. Picard, and R. E. Weiss, *J. Chem. Soc. Chem. Commun.*, 662 (1977).

188. H. Werner and R. Werner, *J. Organometal. Chem.*, **174,** C63 (1979).

189. G. Yagupsky, C. K. Brown, and G. Wilkinson, *J. Chem. Soc. Chem. Commun.*, 1244 (1969).

190. J. Schwarz, *Pure Appl. Chem.*, **52,** 733 (1980).

191. (a) T. Yamamoto, T. Saruyama, and A. Yamamoto, *Bull. Chem. Soc. Jpn.*, **49,** 589 (1976); (b) T. Yamamoto and A. Yamamoto, *J. Organometal. Chem.*, **117,** 365 (1976).

192. (a) G. M. Whitesides, J. F. Gaash, and E. R. Stedronsky, *J. Am. Chem. Soc.*, **94,** 5258 (1972); (b) S. Komiya, Y. Morimoto, T. Yamamoto, and A. Yamamoto, *Organometallics*, **2,** 1528 (1983); (c) M. A. Bennett and R. Charles, *J. Am. Chem. Soc.*, **94,** 666 (1972); (d) A. Tamaki and J. K. Kochi, *J. Chem. Soc. Chem. Commun.*, 423 (1973); (e) K. Tamao, Y. Kiso, K. Sumitani, and M. Kumada, *J. Am. Chem. Soc.*, **94,** 9268 (1972); (f) C. P. Casey and C. R. Cyr, *J. Am. Chem. Soc.*, **95,** 2240 (1973); (g) C. P. Casey, C. R. Cyr, and J. A. Grant, *Inorg. Chem.*, **13,** 910 (1974).

193. (a) J. Schwarz and J. A. Labinger, *Angew. Chem., Int. Ed. Engl.*, **15,** 333 (1976); (b) D. W. Hart and J. Schwarz, *J. Am. Chem. Soc.*, **96,** 8115 (1974).

194. D. L. Reger and P. J. McElligott, *J. Organometal. Chem.*, **216,** C12 (1981).

195. S. Otsuka and A. Nakamura, *Adv. Organometal. Chem.*, **14,** 245 (1976).

196. H. C. Clark and C. S. Wong, *J. Am. Chem. Soc.*, **99,** 7073 (1977).

197. R. R. Burch, A. J. Shusterman, E. L. Muetterties, R. G. Teller, and J. M. Williams, *J. Am. Chem. Soc.*, **100,** 3546 (1983).

198. H. C. Clark and R. J. Puddephatt, *Inorg. Chem.*, **10,** 18 (1971).

199. S. R. Su and A. Wojcicki, *Inorg. Chem.*, **14,** 89 (1975).

200. (a) G. T. Crisp, S. Holle, and P. W. Jolly, *Z. Naturforsch.*, **37b**, 1667 (1982); (b) H. Lehmkuhl, C. Naydowski, R. Benn, A. Rufinska, and G. Schroth, *J. Organometal. Chem.*, **246**, C9 (1983).
201. E. R. Evitt and R. G. Bergman, *J. Am. Chem. Soc.*, **101**, 3973 (1979).
202. J. M. Huggins and R. G. Bergman, *J. Am. Chem. Soc.*, **101**, 4410 (1979).
203. J. L. Davidson, M. Green, F. G. A. Stone, and A. J. Welch, *J. Chem. Soc. Dalton Trans.*, 2044 (1976).
204. P. M. Maitlis, *J. Organometal. Chem.*, **200**, 161 (1980).
205. S.Ikeda, Kogyo Kagaku Zasshi, **70**, 1880 (1967); H. Shirakawa and S. Ikeda, *J. Polymer Sci., Polymer Chem. Ed.*, **12**, 1924 (1974).
206. D. L. Reger and E. C. Culbertson, *J. Am. Chem. Soc.*, **98**, 2789 (1976); *Inorg. Chem.*, **16**, 3104 (1977).
207. F. Ozawa, T. Ito, and A. Yamamoto, *J. Am. Chem. Soc.*, **102**, 6457 (1980).
208. D. L. Thorn and R. Hoffmann, *J. Am. Chem. Soc.*, **100**, 2079 (1978).
209. (a) Z. Dawoodi, M. L. H. Green, V. S. B. Mtetwa, and K. Prout, *J. Chem. Soc., Chem. Commun.*, 802 (1982); (b) N. Koga, S. Obara, and K. Morokuma, *J. Am. Chem. Soc.*, **106**, 4625 (1984); (c) A. Dedieu, *Inorg. Chem.*, **20**, 2803 (1981).
210. J. Evans, J. Schwarz, and P. W. Urquhart, *J. Organometal. Chem.*, **81**, C37 (1974).
211. T. Ikariya and A. Yamamoto, *J. Organometal. Chem.*, **120**, 257 (1976).
212. J. X. McDermott, J. F. White, G. M. Whitesides, *J. Am. Chem. Soc.*, **98**, 6521 (1976).
213. S.-I. Murahashi, M. Yamamura, and N. Mita, *J. Org. Chem.*, **42**, 2870 (1977).
214. (a) K. Maruyama, T. Ito, A. Yamamoto, *J. Organometal. Chem.*, **155**, 359 (1978); (b) *Bull. Chem. Soc. Jpn.*, **52**, 849 (1979); (c) T. Sugino, Y. Shiraiwa, M. Hasegawa, and K. Ichikawa, *Bull. Chem. Soc. Jpn.*, **52**, 3629 (1979).
215. Y. Hayashi, S. Komiya, T. Yamamoto, and A. Yamamoto, *Chem. Lett.*, 1363 (1984).
216. P. Foley and G. M. Whitesides, *J. Am. Chem. Soc.*, **101**, 2732 (1979).
217. (a) C. P. Casey, *Chemtech.*, 378 (1978); (b) R. H. Grubbs, *Prog. Inorg. Chem.*, **24**, 1 (1979); (c) T. J. Katz, *Adv. Organometal. Chem.*, **16**, 283 (1977); N. Calderon, J. P. Lawrence, E. A. Ofstead, *Adv. Organometal. Chem.*, **18**, 449 (1979); J. M. Basset and M. Leconte, *Chemtech.*, **10**, 762 (1980).
218. F. N. Tebbe, G. W. Parshall, and G. S. Reddy, *J. Am. Chem. Soc.*, **100**, 3611 (1978).
219. T. R. Howard, J. B. Lee, and R. H. Grubbs, *J. Am. Chem. Soc.*, **102**, 6876 (1980).
220. J. Rajaram and J. A. Ibers, *J. Am. Chem. Soc.*, **100**, 829 (1978).
221. R. R. Schrock, *Acc. Chem. Res.*, **12**, 98 (1979).
222. W. A. Herrmann, *Adv. Organometal. Chem.*, **20**, 159 (1982).
223. K. M. Motyl, J. R. Norton, C. K. Schrauzer, and O. P. Anderson, *J. Am. Chem. Soc.*, **104**, 7325 (1982).
224. C. P. Casey, P. J. Fagan, W. H. Miles, and S. R. Marder, *J. Mol. Catal.*, **21**, 173 (1983) and references cited therein.
225. (a) G. Erker, J. Wicher, K. Engel, F. Rosenfeldt, W. Dietrich, and C. Krüger, *J. Am. Chem. Soc.*, **102**, 6346 (1980); (b) G. Erker, J. Wicher, K. Engel, and C. Krüger, *Chem. Ber.*, **115**, 3300 (1982); (c) H. Yasuda, K. Kajihara, K. Mashima, K. Nagasuna, K. Lee, and A. Nakamura, *Organometallics*, **1**, 388 (1982).
226. F. Hartley and S. Patai, eds., *The Chemistry of the Metal–Carbon Bond*, Wiley-Interscience, New York, 1982, Chapters 1, 8, and 9; M. I. Lobach and V. A. Kormer, *Russ. Chem. Rev.*, **48**, 758 (1979).
227. C. A. Tolman, *J. Am. Chem. Soc.*, **92**, 6785 (1970).

228. N. N. Druz, V. I. Klepikova, M. I. Lobach, and V. A. Kormer, *J. Organometal. Chem.*, **162,** 343 (1978); V. A. Kormer, M. I. Lobach, and N. N. Druz, *Dokl. Akad. Nauk SSSR,* **246,** 1372 (1979).

229. (a) J. E. Bäckvall, B. Åkermark, and S. O. Ljunggren, *J. Am. Chem. Soc.*, **101,** 2411 (1979); (b) J. K. Stille and R. Divakaruni, *J. Organometal. Chem.*, **169,** 239 (1979); (c) H. Kurosawa, T. Majima, and N. Asada, *J. Am. Chem. Soc.*, **102,** 6996 (1980); (d) H. Kurosawa and N. Asada, *Tetrahedron Lett.*, 255 (1979); (e) T. Majima and H. Kurosawa, *J. Chem. Soc., Chem. Commun.*, 610 (1977); (f) O. S. Andell and J.-E. Bäckvall, *J. Organometal. Chem.*, **244,** 401 (1983).

230. A. Panunzi, A. DeRenzi, and G. Paiaro, *J. Am. Chem. Soc.*, **92,** 3488 (1970).

231. (a) P. Lennon, A. M. Rosan, and M. Rosenblam, *J. Am. Chem. Soc.*, **99,** 8426 (1977); (b) P. Lennon, M. Madhavarao, A. Rosan, and M. Rosenblum, *J. Organometal. Chem.*, **108,** 93 (1976); (c) *J. Am. Chem. Soc.*, **99,** 2823 (1977); (d) M. Green, J. K. K. Sarhan, and I. M. Al-Najjar, *J. Chem. Soc. Dalton Trans.*, 1565 (1981).

232. (a) O. Eisenstein and R. Hoffmann, *J. Am. Chem. Soc.*, **102,** 6149 (1980); (b) L. L. Wright, R. M. Wing, and M. F. Rettig, *J. Am. Chem. Soc.*, **104,** 610 (1982); (c) A. Tiripicchio, M. Tiripicchio-Camellini, L. Maresca, G. Natile, and G. Rizzardi, *Cryst. Struct. Commun.*, **8,** 689 (1979); (d) L. Maresca and G. Natile, *J. Chem. Soc. Chem. Commun.*, 40 (1983).

233. (a) D. L. Reger, P. J. McEligott, *J. Am. Chem. Soc.*, **102,** 5923 (1980); (b) D. L. Reger, K. A. Belmore, E. Mintz, N. G. Charles, E. A. H. Griffith, and E. L. Amma, *Organometallics,* **2,** 101 (1983).

234. (a) D. L. Reger, K. A. Balmore, J. L. Atwood, and W. E. Hunter, *J. Am. Chem. Soc.*, **105,** 5710 (1981); (b) S. R. Allen, P. K. Baker, S. G. Barmes, M. Bottrill, M. Green, and A. G. Orpen, *J. Chem. Soc. Dalton Trans.*, 927 (1983).

235. (a) P. M. Maitlis, *Acc. Chem. Res.*, **9,** 93 (1976); (b) *Pure Appl. Chem.*, **30,** 427 (1972); (c) S. H. Taylor and P. M. Maitlis, *J. Am. Chem. Soc.*, **100,** 4700 (1978); (d) T. R. Jack, C. J. May, and J. Powell, *J. Am. Chem. Soc.*, **100,** 5057 (1978).

236. (a) J. Chatt, L. M. Vallarino, and L. M. Venanzi, *J. Chem. Soc.*, 2496 (1957); (b) J. K. Stille and D. E. James, *J. Organometal. Chem.*, **108,** 401 (1976); (c) J. E. Bäckvall, B. Åkermark, and S. O. Ljunggren, *J. Chem. Soc., Chem. Commun.*, 264 (1977); (d) J. K. Stille and R. Divakaruni, *J. Am. Chem. Soc.*, **100,** 1303 (1978).

237. (a) A. C. Cope, J. M. Kliegman, and F. C. Friedrich, *J. Am. Chem. Soc.*, **89,** 287 (1967).

238. (a) L. S. Hegedus and K. Siirala-Hansen, *J. Am. Chem. Soc.*, **97,** 1184 (1975); (b) L. S. Hegedus, O. P. Anderson, K. Zetterberg, G. Allen, K. Siirala-Hansen, P. J. Olsen, and A. B. Packard, *Inorg. Chem.*, **16,** 1887 (1977).

239. (a) B. M. Trost, *Tetrahedron,* **33,** 2615 (1977); (b) B. M. Trost and T. R. Verhoeven, *Comprehensive Organometallic Chemistry,* Vol. 8, Pergamon Press, Oxford, 1982, p. 799.

240. B. M. Trost and T. R. Verhoeven, *J. Am. Chem. Soc.*, **98,** 630 (1976); **100,** 3435 (1978).

241. T. Hayashi, T. Hagihara, M. Konishi, and M. Kumada, *J. Am. Chem. Soc.*, **105,** 7767 (1983).

242. J. W. Faller and K.-H. Chao, *J. Am. Chem. Soc.*, **105,** 3893 (1983).

243. (a) M. F. Semmelhack, *J. Organometal. Chem. Libr.*, **1,** 361 (1976); (b) *Ann. N. Y. Acad. Sci.*, 36 (1977); (c) M. F. Semmelhack and H. T. Hall, Jr., *J. Am. Chem. Soc.*, **96,** 7091, 7092 (1974).

244. Y. H. Lai, W. Tam, and K. P. C. Vollhardt, *J. Organometal. Chem.*, **216,** 97 (1981).

245. H. Brookhardt, A. R. Pinhas, and A. Lukacs, *Organometallics,* **1,** 1730 (1982).

246. (a) D. Jones, L. Pratt, and G. Wilkinson, *J. Chem. Soc.*, 4458 (1962); 2479 (1964); (b) M. R. Churchill and R. Mason, *Proc. Chem. Soc.*, 112 (1963).

247. S. G. Davies, J. Hibberd, and S. J. Simpson, *J. Organometal. Chem.*, **246,** C16 (1983).

248. G. Evard, R. Thomas, B. R. Davis, and I. Bernal, *Inorg. Chem.*, **15,** 52 (1976).

249. J. R. Sweet and W. A. G. Graham, *J. Am. Chem. Soc.*, **104,** 2811 (1982).

250. W. Tom, W.-K. Wong, and J. A. Gladysz, *J. Am. Chem. Soc.*, **101,** 1589 (1979); W. Tom, G.-Y. Lin, W.-K. Wong, W. A. Kiel, V. K. Wong, and J. A. Gladysz, *J. Am. Chem. Soc.*, **104,** 141 (1982).

251. C. P. Casey, M. A. Andrews, and J. E. Rinz, *J. Am. Chem. Soc.*, **101,** 741 (1979).

252. C. Lapinte and D. Astruc, *J. Chem. Soc. Chem. Commun.*, 430 (1983).

253. R. Angelici, *Acc. Chem. Res.*, **5,** 335 (1972).

254. W. A. Kiol, G.-Y. Lin, G. S. Bodner, and J. A. Gladysz, *J. Am. Chem. Soc.*, **105,** 4958 (1983).

255. E. M. Bradley, J. Chatt, R. L. Richards, and G. A. Sim, *Chem. Commun.*, 1322 (1969).

256. G. W. Daub, *Prog. Inorg. Chem.*, **22,** 409 (1977).

257. M. D. Johnson, *Acc. Chem. Res.*, **11,** 57 (1978).

258. D. Slack and M. C. Baird, *J. Chem. Soc., Chem. Commun.*, 701 (1974).

259. D. Slack and M. C. Baird, *J. Am. Chem. Soc.*, **98,** 5539 (1976).

260. T. C. Flood and F. J. DiSanti, *J. Chem. Soc., Chem. Commun.*, 18 (1975).

261. H. L. Fritz, J. H. Espenson, D. A. Williams, and G. A. Molander, *J. Am. Chem. Soc.*, **96,** 2378 (1974).

262. J. W. Labadie and J. K. Stille, *J. Am. Chem. Soc.*, **105,** 669, 6129 (1983).

263. M. P. Periasamy and H. M. Walborsky, *J. Am. Chem. Soc.*, **97,** 5930 (1975).

264. W. N. Rogers, J. A. Page, and M. C. Baird, *Inorg. Chem.*, **20,** 3521 (1981).

265. (a) M. L. H. Green and P. L. In Nagy, *Z. Naturforsch.*, **18b,** 162 (1963); (b) *J. Chem. Soc.*, 189 (1963); (c) M. Rosenblum, *Acc. Chem. Res.*, **7,** 125 (1974).

266. M. Calligaris, G. Carturan, G. Nardin, A. Scrivanti, and A. Wojcicki, *Organometallics*, **2,** 865 (1983) and references cited therein.

267. J. Halpern, *Pure Appl. Chem.*, **55,** 1059 (1983).

268. J. Halpern, *Acc. Chem. Res.*, **15,** 238 (1982).

269. J. Halpern, *Acc. Chem. Res.*, **15,** 332 (1982).

270. M. J. Nappa, R. Santi, S. P. Diefenbach, and J. Halpern, *J. Am. Chem. Soc.*, **104,** 619 (1982).

271. (a) F. T. T. Ng, C. L. Rempel, and J. Halpern, *J. Am. Chem. Soc.*, **104,** 621 (1982); (b) T. T. Tsou, M. Loots, and J. Halpern, *J. Am. Chem. Soc.*, **104,** 623 (1982).

272. (a) W. Mowat, S. G. Yagupsky, N. J. Hill, M. Yagupsky, and G. Wilkinson, *J. Chem. Soc., Dalton Trans.*, 533 (1972); (b) G. Wilkinson, *Chimia,* **27,** 165 (1973); (c) *Science,* **185,** 109 (1974); (d) W. Mowat and G. Wilkinson, *J. Organometal. Chem.*, **38,** C35 (1972).

273. (a) J. Davidson, M. M. Lappert, and R. Pearce, *Acc. Chem. Res.*, **7,** 209 (1974); (b) G. Barker, M. F. Lappart, and J. A. K. Howard, *J. Chem. Soc., Dalton Trans.*, 734 (1978).

274. (a) R. R. Schrock and G. W. Parshall, *Chem. Rev.*, **76,** 243 (1976); (b) P. J. Davidson, M. F. Lappert, and R. Pearce, *Chem. Rev.*, **76,** 219 (1976); (c) R. Taube, H. Drevs, and D. Steinborn, *Z. Chem.*, **18,** 425 (1978).

275. A. L. Galyer and G. Wilkinson, *J. Chem. Soc., Dalton Trans.*, 2235 (1976); R. R. Schrock, *J. Organometal. Chem.*, **122,** 209 (1976).

276. J. Chatt and B. L. Shaw, *J. Chem. Soc.*, 705 (1959); B. L. Shaw and N. I. Tucker, in *Comprehensive Inorganic Chemistry,* Vol. 4, A. F. Trotman-Dickenson, ed., 1973, p. 781.

277. P. Cossee, *J. Catalysis,* **3,** 80 (1964).

CHAPTER 7

Application of Transition Metal Complexes to Catalysis

Transition metal complexes have been used extensively in industrial and laboratory processes to catalyze a variety of chemical reactions. In this and the following chapters we try to understand these processes as combinations of the elementary steps discussed in Chapter 6. Processes that are mainly related to industrial applications constitute the subject of this chapter,[1] whereas laboratory scale syntheses will be dealt with in the next chapter.

7.1. CHARACTERISTICS OF TRANSITION METAL COMPLEX CATALYSTS: MILD CONDITIONS AND HIGH SELECTIVITY

In this chapter we adopt a rather broad definition of a transition metal complex catalyst. It is a catalyst in which a transition metal complex serves as an active site for the catalytic reaction. The catalyst is usually homogeneous, but heterogeneous systems such as the Ziegler catalysts are also included in this category because they share common features with other soluble transition metal complex catalysts.

Transition metal complex catalysts have advantages as well as disadvantages in comparison with heterogeneous catalysts such as metals and metal oxides. A heterogeneous catalyst activates a substrate as a result of the high surface reactivity of the solid, thus allowing a specific reaction to proceed.

One great advantage of heterogeneous catalysts is that they are easily separated from reaction products. Another advantage is that they are usually robust. Industrial processes using heterogeneous catalysts can therefore be operated over a wide range of conditions and especially at high temperature;

moreover, the catalysts are relatively insensitive to poisons that may readily deactivate transition metal complex catalysts.

However, the fact that heterogeneous catalysis, by its very nature, involves a surface reaction makes study of the catalytic reaction extremely difficult. Despite the tremendous advances that have occurred recently in techniques for studying surface chemistry, an in-depth analysis of the nature of the active site in heterogeneous catalysts is fraught with enormous difficulties. Our understanding of the interaction of the active site with a substrate and of reaction mechanisms in heterogeneous catalysis is based mostly on circumstantial evidence, as we are hindered by the lack of a more direct approach to unravelling the "mystery" of what is occurring on the surface of a catalyst.

A homogeneous transition metal complex catalyst consisting of discrete molecules stands in sharp contrast to the heterogeneous catalysts in this respect. We can utilize a wide range of research methods and obtain detailed information about the nature of catalytically active transition metal complexes. The time of development in homogeneous catalysis using transition metal complexes coincided with a period of remarkable progress in modern experimental techniques such as NMR and IR spectroscopy. Kinetic studies on homogeneous catalysis also provide us with much more dependable data than those available from studies of heterogeneous systems. In certain cases it is even possible to isolate a transition metal complex catalyst itself and to study in detail its interaction with a particular substrate, a feat that has never been performed with a heterogeneous catalyst.

Particularly impressive advances have been made by making use of the progress in X-ray crystallography combined with computer technology. Single-crystal X-ray structural analysis of a complex catalyst enables us to investigate the arrangement of ligands around the metal and to examine how the complex can interact with a substrate, thus allowing very plausible arguments regarding the reaction mechanism to be advanced.

However, one must be careful of the pitfall inherent in a reaction mechanism that is based only on information derived from a transition metal complex isolated from a catalyst system. Although such a complex may be close to a real catalyst, it may nevertheless be a species in a blind alley, separate from the mainstream of a catalytic cycle. The real catalyst action may be somewhat different from what one concludes based on the study of such an isolated model complex.

The main features of transition metal complex catalysts are selectivity and catalytic activity under mild conditions. One can control the selectivities of homogeneous catalysts by choosing combinations of the transition metal complexes with suitable ligands. Such a control or modification of selectivity is difficult to achieve in heterogeneous catalysts because of the very nature of heterogeneity in active sites. The most outstanding illustration of selectivity in homogeneous catalysts is the almost 100% optical yield obtained in asymmetric hydrogenation of olefins by employing suitably chosen metal complexes containing specially designed asymmetric tertiary phosphine ligands.

Attempts to overcome the disadvantage of homogeneous catalysis, that is, the difficulty in separation of the catalyst from products, are being made by supporting transition metal complex catalysts on solid surfaces, inorganic or organic, as will be discussed in Section 9.2. Some improvements have been also reported concerning the thermal instability and sensitivity to air by attaching transition metal complex catalysts to solid surfaces.

Another important aspect of catalysis is the function of transition metal clusters. Metal clusters can be regarded as a bridge between homogeneous mononuclear catalysts and heterogeneous catalysts that are difficult to characterize. These topics will be discussed in Chapter 9.

7.2. POLYMERIZATION AND OLIGOMERIZATION OF OLEFINS AND DIENES

In the period from 1950 to the mid-1970s petrochemical industries in the world expanded enormously. Particularly outstanding was the development of technology utilizing olefins and dienes containing C_2–C_5 carbon units as feedstocks. This growth of the petrochemical industry matched the stable and abundant supply of inexpensive oil and important discoveries such as the Ziegler process and the Wacker process. Polyolefins and polydienes are the largest volume polymer materials produced in the world. The total quantities of the polymers of ethylene, propylene, butadiene, and isoprene produced using Ziegler catalysts in the United States alone was well over 3 million tons in 1977.[1a] Thus, it is appropriate to begin this chapter with a discussion of Ziegler catalysts.[2]

Discovery of Ziegler Catalysts

Karl Ziegler and his co-workers were studying the insertion of ethylene into an Al–C bond to give long-chain aluminum alkyls [Eq. (7.1)] by heating triethylaluminum with ethylene in an autoclave. One day they found that only 1-butene was produced in the autoclave and that the long-chain aluminum alkyls were not formed [Eq. (7.2)].

$$AlEt_3 + C_2H_4 \longrightarrow Al\!\lessdot\!(CH_2CH_2)_nEt \qquad (7.1)$$

$$C_2H_4 + C_2H_4 \xrightarrow{\text{"Ni," } AlEt_3} CH_3CH_2CH{=}CH_2 \qquad (7.2)$$

Investigation revealed that the cause of this divergent result was a trace of nickel compound that had been used in the autoclave for a previous hydrogenation experiment and had not been removed. When he discovered the "nickel effect," Ziegler reasoned: nickel is a heavy metal, so let us examine the effect of adding heavy metals to the system. At the end of 1953, while examining the effect of other transition metal compounds on the insertion of ethylene into the Al–C bond, Ziegler's group was astonished to find that addition of zirconium acetylacetonate to the system led to the formation of high-molecular-weight polyethylene instead of the expected 1-butene.

It was soon found that a combination of $TiCl_4$ and $AlEt_3$ was the most effective ethylene polymerization catalyst. This was a most surprising discovery at the time, since hitherto ethylene had only been polymerized under extremely drastic conditions, employing pressures of more than 1000 atm at temperatures above 200 °C. In his lectures Ziegler vividly describes his satisfaction in showing the process to visitors and his pleasure at their amazement.[3]

The information on the mixed catalyst system was conveyed to the Montecatini company and to G. Natta at the Milan Polytechnic Institute.[4a] This spurred the tremendous development of the new polymerization technique and led to a variety of discoveries, the most outstanding being the preparation of stereoregular polymers of propylene. Other remarkable discoveries include stereospecific polymerization of butadiene and isoprene, copolymerization of ethylene and propylene, polymerizations of alkylene oxides, cyclopentene, and acetylene (cf. "The Golden Airplane" in Section 7.1b).

Ten years later Ziegler and Natta jointly received the Nobel prize in chemistry.

The Ziegler catalyst is unique and marvelous.[2a] It consists of a mixture of a transition metal compound and an alkyl or hydride of a main-group element. By choosing proper combinations of these components and changing the proportions of the components and the time for which the mixture is allowed to stand before polymerization, a variety of monomers can be specifically polymerized under mild conditions.

a. Polymerization of Ethylene with Ziegler Catalysts

Ethylene can be polymerized by catalyst systems composed of $TiCl_4$ or $TiCl_3$ mixed with organoaluminum compounds such as $AlEt_3$ and $AlEt_2Cl$. Ziegler initially called this polymerization the Mülheim process, after the town where the Max-Planck Institute for Coal Research, of which Ziegler was director, was situated. However, the catalyst mixture is now called a Ziegler catalyst, following a proposal by Natta. Catalysts containing transition metal compounds other than titanium compounds in combination with the alkyl compounds of the main-group elements are called Ziegler-type catalysts.

Extensive studies have been made in an effort to unravel the nature of the active sites in Ziegler catalysts. Nevertheless, ambiguities still remain regarding the identity of the active center and the polymerization mechanism. This is unavoidable to some extent because of the heterogeneous nature of the catalyst systems and because the catalytic activity depends on the conditions under which the mixed catalysts are prepared. The following is a description of processes that probably occur during catalyst preparation.

When $TiCl_4$ is mixed with $AlEt_3$, an alkyl exchange reaction probably occurs through bridged dinuclear species to give an ethyltitanium compound.

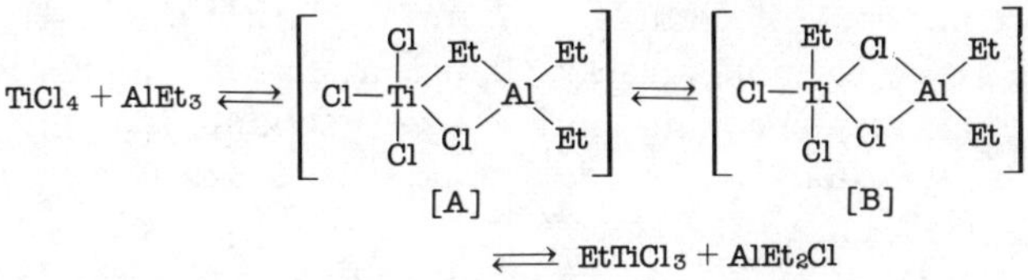

$$TiCl_4 + AlEt_3 \rightleftharpoons [A] \rightleftharpoons [B] \rightleftharpoons EtTiCl_3 + AlEt_2Cl \quad (7.3)$$

The suggestion that $EtTiCl_3$ is formed is supported by the fact that its congener $MeTiCl_3$ can be isolated from the reaction of $TiCl_4$ with $AlMe_3$. The exchange between the alkyl group and Cl atom is likely to proceed through bridged species [A] and [B]. The ethyltitanium compound is thermally less stable than $MeTiCl_3$ and decomposes at about room temperature to give a Ti(III) compound together with ethylene and ethane. Since $TiCl_3$ is insoluble in hydrocarbon solvents in which the Ziegler catalysts are prepared, the system becomes heterogeneous. Further interaction of $TiCl_3$ with $AlEt_3$ or $AlEt_2Cl$ gives an ethylated product, $EtTiCl_2$ or its adduct with an aluminum component.

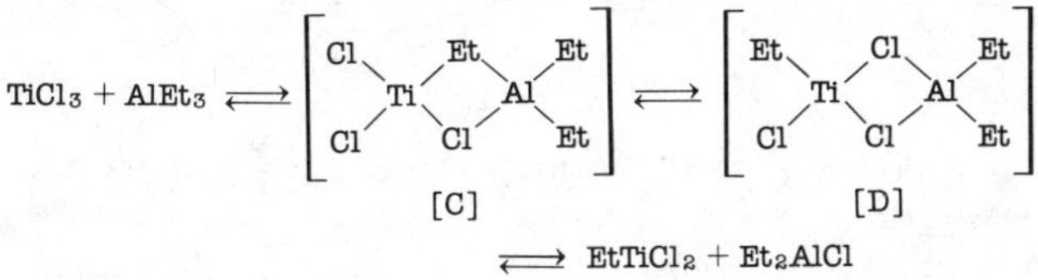

$$TiCl_3 + AlEt_3 \rightleftharpoons [C] \rightleftharpoons [D] \rightleftharpoons EtTiCl_2 + Et_2AlCl \quad (7.4)$$

Although evidence supporting the formation of bridged species such as [C] and [D] is difficult to obtain in the actual Ziegler catalyst systems, the formation of such binuclear complexes is considered likely in view of the isolation of the binuclear complexes **7-1** and **7-2** from mixtures containing Cp_2TiCl_2 and $AlEt_3$, $AlEt_2Cl$, or $AlEtCl_2$.[5]

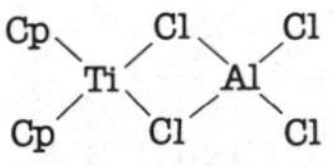

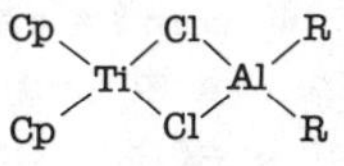

R = Cl or Et

7-1 **7-2**

A methyl-bridged complex of titanium **7-3**, however, can be obtained by treatment of $\{Cp_2TiCl\}_2$ with $LiAlMe_4$.[5d,e]

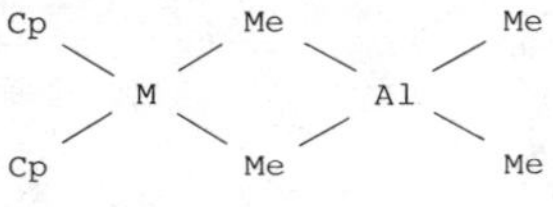

M = Ti, Y, Yb

7-3

Reduction of $EtTiCl_2$ under ambient conditions can lead to formation of Ti(II) species. Furthermore, there is some evidence for the formation of an ionic species containing an aluminate anion such as $[Cp_2TiR]^+[AlRCl_3]^-$ in a mixture of Cp_2TiCl_2 and AlR_2Cl.[4b] Thus, the actual Ziegler catalysts are likely to be composed of complex mixtures of Ti(IV), Ti(III), and Ti(II) species, which may be mononuclear, dinuclear, or ionic and whose relative proportions may vary depending on the "aging" conditions and the period for which the catalyst system is allowed to stand before a monomer is added.

Despite their complexity, Ziegler catalyst systems probably contain ethyltitanium compounds. Cossee has proposed the following mechanism in which an ethyltitanium complex serves as the active center for polymerization.[6] Triethylaluminum is supposed to act simply as an alkylating agent so that the ethyltitanium compound can be formed from $TiCl_4$. Coordination of ethylene to an available vacant site on the ethyltitanium active center gives a pseudo-octahedral ethylene-coordinated ethyltitanium complex. Subsequent ethylene insertion into the Ti–Et bond is the initiation step of the polymerization.

Initiation process:

(a) π Coordination of ethylene → (b) Activated state → (c) Insertion completed (7.5)

After the insertion step, which may be viewed as the migration of the ethyl group to the coordinated ethylene ligand, a butyltitanium species with a vacant coordination site is formed. Further ethylene coordination at this site and insertion of ethylene into the Ti–C bond lead to growth of the polymer chain attached to the titanium active center, and this constitutes the propagation process. If β-hydrogen elimination takes place during the propagation process, a titanium hydride and a polymer having a terminal double bond are produced. The titanium hydride may further react with ethylene to regenerate the ethyltitanium species, which will further react with ethylene to produce a polymer.

Propagation process:

$$[Ti]\text{-}CH_2CH_2Et \xrightarrow[\text{coordination}]{C_2H_4} \overset{CH_2=CH_2}{[Ti]}\text{-}CH_2CH_2Et \xrightarrow{\text{insertion}} [Ti]\text{-}(CH_2CH_2)_2Et$$

$$\xrightarrow[\text{insertion}]{C_2H_4} [Ti]\text{-}(CH_2CH_2)_nEt \qquad (7.6)$$

Termination or chain transfer processes:

$$[Ti]-(CH_2CH_2)_nEt \xrightarrow{\beta\text{-elimination}} [Ti]-H + H_2C{=}CH(CH_2CH_2)_{n-1}Et \quad (7.7)$$

$$[Ti]-H \xrightarrow[\text{insertion}]{C_2H_4} [Ti]-Et \xrightarrow{n\,C_2H_4} [Ti]-(CH_2CH_2)_nEt \quad (7.8)$$

The precise details of the termination process have not been established. In addition to the termination mechanism involving β elimination as expressed by Eq. (7.7), a bimolecular termination mechanism has been proposed. If the chain transfer process as expressed by Eq. (7.7) takes place frequently, the molecular weight will decrease.

Hydrogen is sometimes added to control the molecular weight. The reaction with H_2 is considered to proceed through a four-center mechanism to give a titanium hydride and a polymer with a saturated terminus.

$$[Ti]-(CH_2CH_2)_nEt + H_2 \longrightarrow \begin{matrix}[Ti]\text{---}CH_2CH_2(CH_2CH_2)_{n-1}Et \\ \vdots \qquad\quad \vdots \\ H\text{----}H\end{matrix} \longrightarrow [Ti]-H + Et(CH_2CH_2)_{n-1}Et \quad (7.9)$$

Another process for the polymerization of ethylene was developed by the Phillips Company at about the same time as the discovery of the Ziegler catalyst.[7] The Phillips catalysts are prepared by deposition of chromium oxides on silica. They have an advantage over Ziegler catalysts in that the catalysts are more easily separated from the polymer. The mechanism of formation of the active species for ethylene polymerization in the Phillips process or its modification, the Union Carbide process, is not clearly understood. It is believed that an active species with a Cr–H bond may be formed somehow, possibly by a reaction of the chromium compound with an OH group on silica or by interaction of the chromium with ethylene in the initial stage of the reaction. Insertion of ethylene into the Cr–H bond followed by repeated ethylene insertions into the Cr–C bond may lead to a high-molecular-weight polymer.

$$\gt Cr-H + C_2H_4 \longrightarrow \gt Cr-C_2H_5 \xrightarrow{nC_2H_4} \gt Cr-(CH_2CH_2)_nEt \quad (7.10)$$

Supporting of a transition metal catalyst on a solid surface can sometimes enhance its catalytic activity in addition to making separation easier. Use of $MgCl_2$ as a solid support in combination with other additives has brought about a marked enhancement in the activity of conventional Ziegler catalysts. By increasing catalyst activity so that 10^6 g of polyethylene are obtained per gram of a titanium-based catalyst, the amount of the catalyst residue in the polymer can be reduced to the order of parts per million. At this level the costly process of removing the catalyst residue from the polymer, known as *deashing,* can be omitted, thus affording considerable saving in construction and operational costs.

Another recent development is gas phase polymerization without solvent. With the advent of these new processes it has been said that the Ziegler process is now in its second era of development.

Polyethylene produced by Ziegler processes has fewer branches on its polymer skeleton than polyethylene produced by a high-pressure radical process. The density of the polyethylene produced by the low-pressure process, typically 0.94–0.97 g cm^{-3}, is higher than the density of the polyethylene produced by the high-pressure process (0.92–0.93 g cm^{-3}). The polyethylene produced by the low-pressure process is referred to as high-density polyethylene (HDPE) while that using a high-pressure process is referred to as low-density polyethylene (LDPE).

The HDPE is suitable for production of moulded materials but not for making thin films. Thus, the high-pressure process has maintained the market for producing films despite its operational disadvantages. The LDPE can be produced, however, by a low-pressure process in which 1-butene is used as a comonomer to give a polymer that has a lower density owing to branching. Another approach for decreasing the polymer density is to increase the β-hydrogen elimination rate for producing α-olefins, which may be copolymerized into polyethylene to produce highly branched polyethylene.

b. Polymerization of Propylene by Ziegler–Natta Catalysts

The discovery by G. Natta of the process for the polymerization of propylene is significant in two respects. The first is its commercial development. Before Natta's discovery propylene was considered to be not polymerizable to macromolecules useful for plastics or fibers. Therefore, the new process enormously enhanced the value of propylene and helped to bring about a tremendous expansion in the petrochemical industry. The second is the impact on polymer chemistry. As a result of studies of the mechanism of stereospecific polymerization, our understanding of stereoregular polymers and their synthesis advanced quite significantly. Furthermore, the mechanistic studies on stereospecific polymerization by transition metal complexes deepened our understanding of the coordination and reactivity of olefins in transition metal complexes. The fact should not be overlooked that the conceptual advances in stereospecific coordination polymerization paved the way for later development in asymmetric hydrogenation by transition metal complexes.

After their initial work on polymerization of ethylene with a Ziegler catalyst consisting of $TiCl_4$ and $AlEt_3$, Natta and his group turned to propylene to see if it could be polymerized with the same catalyst system. They soon discovered that use of solid $TiCl_3$ instead of $TiCl_4$ in combination with $AlEt_3$ or $AlEt_2Cl$ gave a crystalline polymer having the structure shown below.

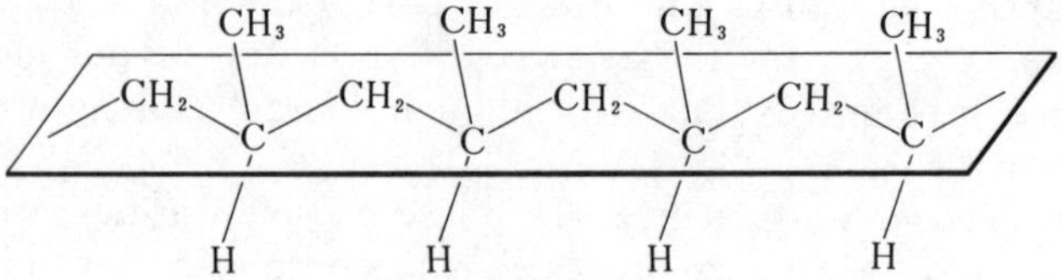

CH_3 CH_3 CH_3 CH_3
CH_2 C CH_2 C CH_2 C CH_2 C
H H H H

Structure of isotactic propylene

The polymer is available in a crystalline form that consists of helical macromolecules containing three monomer units in one turn of the spiral chain. The configuration of the tertiary carbon atom to which the methyl group is attached is the same throughout the polymer chain. Thus, if one stretches the spiral polymer chain in a trans zig-zag conformation as shown above, the methyl groups are lined up on the same side of the plane formed by the trans zig-zag chain, whereas the hydrogen atoms all oriented on the other side of the plane. At the suggestion of his wife Natta called this type of polymer *isotactic* (from Greek, *iso,* meaning same, *tacticos,* meaning arrangement).

On the other hand, polymerization of propylene by a homogeneous catalyst system composed of VCl_4–$AlEt_2Cl$ at low temperatures gives another type of polymer with the following structure.

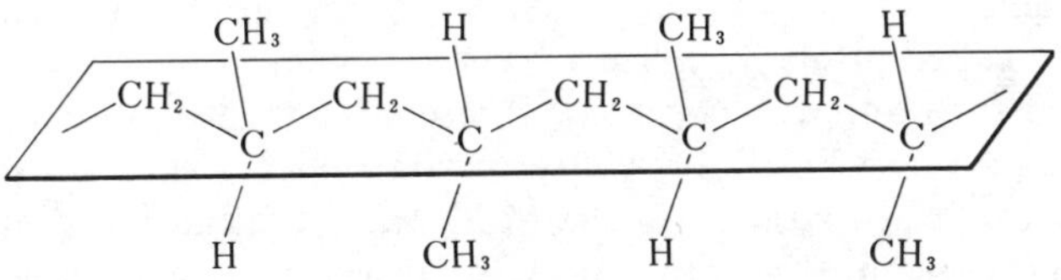

Structure of syndiotactic propylene

In this polymer the methyl groups and the hydrogen atoms attached to the tertiary carbon atom in each propylene unit alternate along the polymer chain. Natta called this type of polymer *syndiotactic* (from Greek, *syndio,* meaning alternating).

In addition to these two types of stereoregular polymers, another type of polymer lacks stereoregularity so far as the configuration at the tertiary carbon in the propylene unit is concerned. This polymer is called *atactic,* and the degree of stereoregularity is sometimes called *tacticity*. Commercially, isotactic polypropylene is the desired product and can be used for many purposes. On the other hand, atactic polypropylene, which has a waxy appearance, is of little use and has an adverse effect on the properties of isotactic polypropylene. Therefore, a process to remove it is required. Because a huge volume of polypropylene is now being produced, the amount of unwanted atactic polymer can be very large. Therefore, intensive effort has been devoted to increasing the proportion of isotactic polypropylene by raising the stereoregularity of the polymer and to enhancing catalyst activity so that the dashing process can be dispensed with.

To produce stereoregular polymers with a coordination catalyst, two factors must be controlled. One is the mode of coordination of propylene to the active center of the alkyltitanium complex, and the other is the mode of double-bond opening. If we assume that the polymerization proceeds by insertion of coordinated propylene into the titanium–alkyl bond, two modes of insertion are possible.

$$\underset{\substack{\delta- \;\; \vdots \;\; \delta+ \\ H_2C{=}CH \\ | \\ Me}}{M{-}R^{\delta-}} \xrightarrow{\text{insertion}} M{-}CH_2{-}\underset{\substack{| \\ Me}}{CH}{-}R \xrightarrow{C_3H_6} \longrightarrow \cdots M{-}(CH_2{-}\underset{\substack{| \\ Me}}{CH})_n{-}R \quad (7.11)$$

$$\underset{\substack{\vdots \\ HC{=}CH_2 \\ | \\ Me}}{M{-}R} \xrightarrow{\text{insertion}} M{-}\underset{\substack{| \\ Me}}{CH}{-}CH_2{-}R \xrightarrow{C_3H_6} \longrightarrow \cdots M{-}(\underset{\substack{| \\ Me}}{CH}{-}CH)_n{-}R \quad (7.12)$$

The first type of reaction [Eq. (7.11)] is addition of the titanium alkyl to propylene following Markownikoff's rule to give a primary alkyl group attached to the metal. The second mode of insertion proceeds by anti-Markownikoff addition of the metal alkyl to the coordinated propylene to give a secondary alkyl group attached to the metal, Eq. (7.12). These two processes are called primary and secondary insertion, respectively. If the alkyl group bound to titanium has carbanionic character, the propylene molecule polarized as shown in Eq. (7.11) will prefer to undergo the primary insertion. However, because polarization of the M–C bond and the coordinated propylene is not very pronounced, the mode of insertion is determined by subtle factors and secondary insertion also takes place. Consecutive occurrence of one mode of insertion gives a polymer containing propylene units linked head to tail. Change of the insertion mode from primary to secondary or vice versa can lead to lowering of the tacticity by forming head-to-head or tail-to-tail linkages.

There is evidence supporting the occurrence of primary insertion in the polymerization of propylene to isotactic polypropylene initiated by the $TiCl_3$–$AlEt_3$ system. On the other hand, secondary insertion has been inferred in the polymerization of propylene initiated by homogeneous catalyst mixtures such as VCl_4–$AlEt_2Cl$ and VCl_4–$AlEt_2Cl$–anisole.[8]

To produce stereoregular polymers by coordination polymerization, steric control of the mode of coordination of propylene to the metal is also required. A propylene monomer π bonded to a metal alkyl can adopt four different orientations, as shown in Figure 7.1.

In the polymerization of specifically deuterated propylene, the formation of both isotactic and syndiotactic polypropylenes has been shown to involve a cis opening of the propylene double bond.[9] This result is consistent with the mechanism involving coordination of propylene to a transition metal alkyl followed by propylene insertion into the metal–carbon bond, as represented in Figure 7.1.

In polymerization to give isotactic polypropylene, primary insertion takes place. The two modes of propylene coordination are shown in Figures 7.1*a,b*.[10] One mode will be favored over the other to an extent dependent on the steric influence of the ligands around the metal. The propylene molecule will coordinate to the metal so as to minimize steric interaction between the methyl group and the ligand. The reason highly isotactic polypropylene is

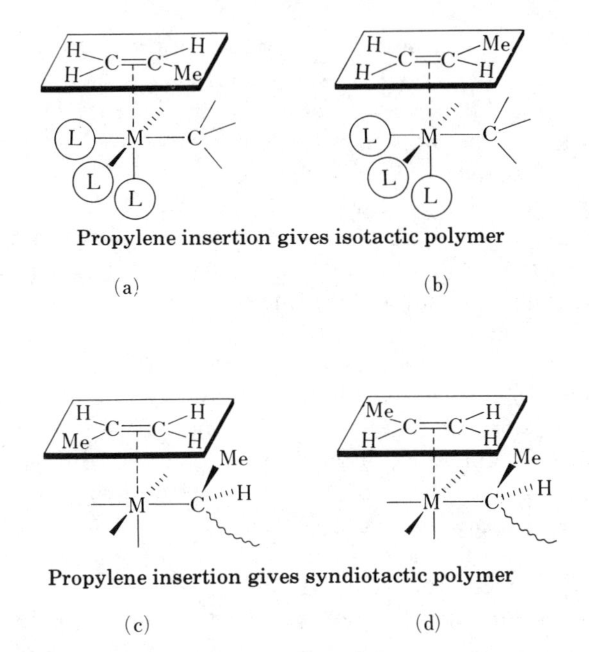

Propylene insertion gives isotactic polymer
(a)
(b)
Propylene insertion gives syndiotactic polymer
(c)
(d)

FIGURE 7.1. Orientation of propylene in coordination to transition metal alkyls.

obtained when crystalline $TiCl_3$ is used in combination with $AlEt_3$ is believed to be due to the nature of the stereochemical environment on the surface of the crystalline $TiCl_3$, which allows the coordination of propylcnc only in onc direction.[11]

On the other hand, in homogeneously catalyzed propylene polymerization to give the syndiotactic polymer at low temperature, which proceeds by the secondary insertion process [Eq. (7.12)], steric repulsion between the ligands and the methyl group attached to the α-carbon atom is less important, and steric repulsion between the methyl group on the α-carbon atom of the alkyl chain and the methyl group in the coordinating propylene becomes the dominant factor, as shown in Figures 7.1*c,d*. In this case the propylene molecule will coordinate to the metal in such a way as to avoid the repulsion caused by the methyl group in the polymer chain. Thus, coordination mode (*d*) is favored over (*c*) because in the former repulsion between the two methyl groups is smaller than in the latter. These consecutive coordination and insertion processes are thought to give the syndiotactic polypropylene, which has a polymer chain containing tertiary carbon atoms of alternating configurations.

We have treated the mechanisms of polymerization of ethylene and propylene according to the hypothesis proposed by Cossee. Although the Cossee mechanism accommodates most of the experimental facts relating to olefin polymerization by Ziegler–Natta catalysts, the crucial part of the mechanism concerning olefin insertion into the Ti–C bond still remains to be established despite extensive mechanistic studies by both academic and industrial chemists. Attachment of the terminal ethyl group remnant arising from $AlEt_3$ to the polymer produced by a $TiCl_4$–$AlEt_3$ catalyst mixture was established by an experiment using $AlEt_3$ labeled with ^{14}C. The fact that $Ti(CH_2Ph)_4$ that is free from the aluminum component can polymerize ethylene provides further support for a mechanism proceeding through ethylene insertion into the Ti–C bond, although the activity is lower than that with a Ziegler catalyst.[12] This evidence is in agreement with Cossee's mechanism, but examples of olefin insertion into metal–carbon σ bonds have been established only for later transition metal alkyls[13] and lanthanoid methyls.[14] Green, Ivin, and their coworkers raised a fundamental question regarding the validity of the olefin insertion mechanism and proposed a novel mechanism involving α-hydrogen abstraction followed by metallacycle formation, as shown in Figure 7.2.[15a–c]

In this mechanism a carbene species is formed by α-hydrogen elimination from the terminal carbon atom of the polymer chain. Coordination of propylene and subsequent cyclization of the π-bonded propylene with the carbene species gives a metallacyclobutane. If the abstracted hydride subsequently returns to the α-carbon atom of the metallacycle to regenerate the metal alkyl, one step of the propylene "insertion" is completed without invoking the direct insertion of the coordinated propylene into the metal alkyl. This mechanism includes the extra steps of α-hydrogen elimination and return of the hydrogen atom to the α-carbon atom but accommodates most of the other features of the Cossee mechanism to explain stereoregulation, except that less attention is paid to the question of primary or secondary insertion. The advantage of the Green mechanism is that it accommodates both Ziegler polymerization and olefin metathesis by assuming common carbene and metallacyclobutane intermediates. It also explains olefin dimerization as described later. The proposal opened up a new controversy about the mechanism of olefin polymerization by Ziegler–Natta catalysts. Still another proposal involving a metallacycle intermediate has been made based on semiempirical MO calculations.[15d] Because a recently reported study[16a] using a niobium alkyl provides a convincing argument in support of the Green mechanism, the polymerization process involving the carbene intermediate should not be excluded, at least for certain types of catalyst systems. However, a few lines of evidence against the Green mechanism have been presented.[16b,c]

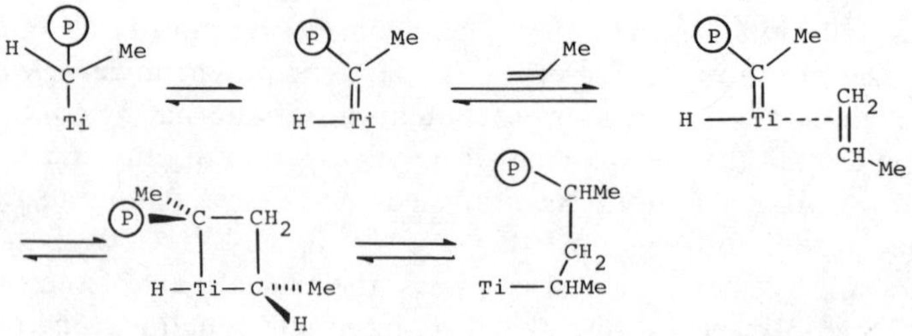

FIGURE 7.2. Green–Ivin mechanism for propylene polymerization.

Furthermore, Grubbs quite recently presented evidence for olefin insertion into the Ti–C bond in the following reaction.[16d]

$$Cp_2Ti(Cl)(CH_2CH_2CH_2CH_2CH=CH_2) \xrightarrow{EtAlCl_2} Cp_2Ti(Cl)(CH_2\text{-}c\text{-}C_5H_9)$$

The isolated cyclopentylmethyltitanium complex proved to initiate ethylene polymerization. It was further demonstrated that a titanacyclobutane initiates the polymerization of norbornene.

These results indicate that there are two types of polymerization mechanisms, one involving the olefin insertion into the Ti–C σ bond and the other comprising the metallacyclobutane–carbene intermediates (cf. Section 7.1d). It seems quite likely that olefin polymerization by Ziegler catalysts proceeds by the former mechanism.

Ethylene and propylene can be copolymerized by Ziegler–Natta catalysts. By using a third component consisting of a nonconjugated diene in combination with ethylene and propylene, a small amount of unsaturation can be introduced into the copolymer. Thus, elastomers can be produced from the ethylene–propylene copolymer by causing cross-linking at the site of unsaturation using the conventional technique of "curing."

Intermezzo — *The Golden Airplane*

After Ziegler's discovery of ethylene polymerization catalyzed by a mixture of $TiCl_4$–$AlEt_3$ at the end of 1953, a remarkable development occurred in Italy. Using the same catalyst, Natta and his co-workers in Italy discovered the stereospecific polymerization of propylene and expanded the scope of coordination polymerization at a truly astounding pace. One can see in the two volumes in which the work of Natta's group is compiled that almost 200 quite important papers were published in a period of five years, that is, almost one paper per week.

One can only wonder how the information was conveyed from Ziegler to Natta and how it was possible to conduct the research so quickly. The story I heard in Germany was as follows. Shortly after Ziegler's discovery Natta visited Ziegler's institute and was shown the newly developed ethylene polymerization process. He asked a co-worker of Ziegler if they had tried to polymerize propylene with the catalyst mixture. The answer was yes, but the results were not too exciting. After Natta had returned to Italy, a patent claiming that propylene could be polymerized to give crystalline polymers was filed by Natta's group on June 8, 1954. This aroused some suspicion among Ziegler's co-workers. One of them told me that Natta should bring a gold-plated airplane for the information. This is the German version of the discovery.

The Italian version of the story is different.[17] In 1952 Natta attended Ziegler's lecture in Frankfurt on the growth reaction of triethylaluminum with ethylene. Natta immediately saw the importance of the process and persuaded executives of the Montecatini chemical company to negotiate a contract with Ziegler so that information on the trialkylaluminum process could be supplied to the company, which would in turn support Ziegler's group with a grant. Natta started a kinetic study of the reaction of ethylene with $AlEt_3$ and sent his co-workers to Ziegler's institute so that they could learn the technique of handling air-sensitive compounds. When the information on the Ziegler catalyst was brought to Montecatini *according to the agreement*, Natta was in a position to know and utilize all the information on the new process. He also had a large research group including 6 talented research assistants and 10–20 young chemists from Montecatini who were being trained under him. And it would be difficult to find anyone more suitable than Natta to develop the project. He had an excellent knowledge of the use of X-ray crystallography to analyze crystalline polymers, of heterogeneous catalysis as well as homogeneous cobalt carbonyl catalysts for the oxo process, and of polymer chemistry. He acted swiftly when the information on the Ziegler process was brought to him. He reorganized his group in order to concentrate on the new polymerization, and they had already found that propylene could be polymerized to a crystalline polymer before Natta visited Ziegler.

Although the shift in focus in coordination polymerization from Germany to Italy may not have been too pleasant for Ziegler's co-workers, the transfer of information had important implications for the scientific community. It would be fair to say that without Natta's contribution such a remarkable development in polymer chemistry would not have been achieved in such a short time span. The development must have also enhanced the importance of Ziegler's original discovery, without which Natta's activity would not have existed. One could say that Natta eventually brought the golden airplane, transformed into a Nobel prize, to Mülheim. Ziegler remarked[4a]: "I have never tried to exhaust the possibilities of my newly won knowledge. I was satisfied to open the door."

c. Oligomerization of Olefins

If β-hydrogen elimination takes place after a few steps of olefin insertion into an M–C bond, a chain transfer reaction will result and further growth of the polymer chain will be interrupted to give oligomers (from Greek *oligo,* meaning small).[18] The nickel effect discovered by Ziegler during his investigation of the growth reaction of triethylaluminum with ethylene is related to the dimerization of ethylene promoted by a catalytically active nickel species formed by the reaction of a nickel compound with $AlEt_3$.

Nickel compounds such as $NiCl_2$ and $Ni(acac)_2$ give on reaction with $AlEt_3$ an unstable ethylnickel compound.[19] Ethylnickel compounds can be isolated in the presence of stabilizing ligands such as tertiary phosphines, but when these ligands are absent, the ethylnickel compounds are converted into a nickel hydride species by a β-hydrogen elimination reaction. Alternatively, the nickel hydride species may be formed after ethylene insertion into the $Ni{-}C_2H_5$ bond and subsequent β elimination. The nickel hydride species is believed to be the catalytically active species responsible for olefin dimerization (Fig. 7.3).

Insertion of ethylene into the metal–hydride bond gives an ethyl complex that on further reaction with ethylene forms a butyl complex. The β elimination from the butyl complex regenerates the metal hydride, which carries the catalytic cycle and liberates 1-butene, the dimerization product of ethylene. It is possible that an organoaluminum component is bound to the transition metal active center, but coligands are omitted from Figure 7.3 for simplicity. If the butyl complex reacts further with ethylene, the process may give ethylene oligomers higher than trimers.

Dimerization of propylene is more complicated than that of ethylene because there are two possible modes of insertion, namely, primary and secondary insertions (Fig. 7.4).

The isomer distribution of the products of propylene dimerization can be varied by adding tertiary phosphines to the catalyst system. These ligands control the direction of propylene insertion and β-hydrogen elimination. Addition of fairly bulky phosphines is observed to give a higher proportion of

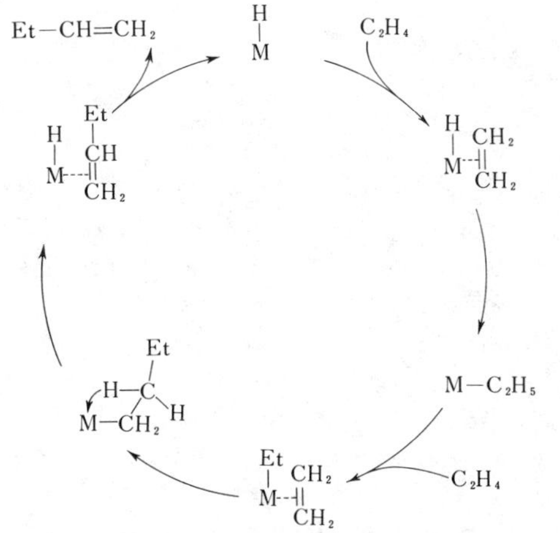

FIGURE 7.3. Mechanism of ethylene dimerization by insertion–β-elimination process.

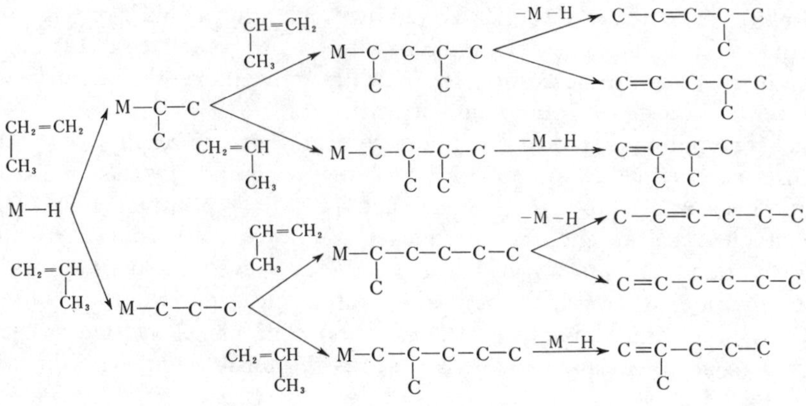

M—H + CH_2=CH_2—CH_3 → M—C(C)—C
→ (CH=CH_2—CH_3) M—C(C)—C—C(C)—C —M—H→ C—C=C—C(C)—C ; C=C—C—C(C)—C
→ (CH_2=CH—CH_3) M—C—C(C)—C(C)—C —M—H→ C=C(C)—C(C)—C

M—H + CH_2=CH—CH_3 → M—C—C—C
→ (CH=CH_2—CH_3) M—C(C)—C—C—C—C —M—H→ C—C=C—C—C—C ; C=C—C—C—C—C
→ (CH_2=CH—CH_3) M—C—C(C)—C—C—C —M—H→ C=C(C)—C—C—C

FIGURE 7.4. Mechanism of propylene dimerization by insertion–β-elimination process.

branched olefins.[20] The presence of the phosphines influences the mode of propylene insertion into the M–H and M–C bonds. The production of more branched olefins by addition of bulky phosphines may be correlated with an increased tendency to undergo primary insertion into the M–C bond because β-hydrogen elimination from a species having an M–CH(Me)–CH_2 entity produced by secondary insertion would give more linear olefins.

Developments in metallacycle chemistry have led to the proposal of an entirely different mechanism for selective olefin dimerization. Two molecules of an olefin can oxidatively add to a low-valent transition metal complex to form a five-membered metallacycle, a metallacyclopentane. β-Hydrogen elimination from the metallacycle gives butene derivatives.[21] Formation of a metallacyclopentane derivative by the reaction of a tantalum carbene complex with propylene is known.[22]

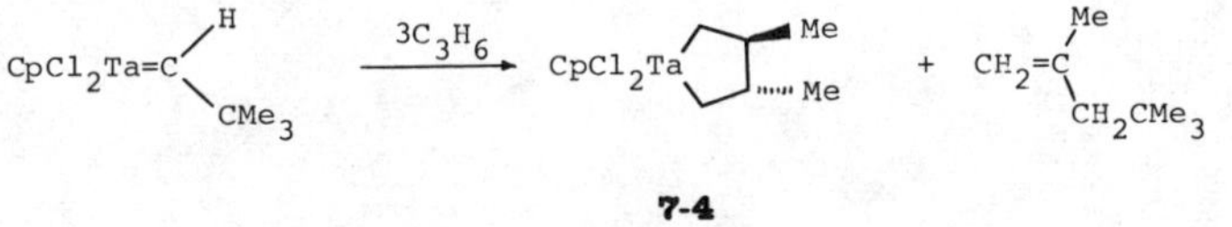

$$CpCl_2Ta{=}C(H)(CMe_3) \xrightarrow{3C_3H_6} CpCl_2Ta(\text{cyclo-}C_4H_6Me_2) + CH_2{=}C(Me)CH_2CMe_3 \qquad (7.13)$$

7-4

Complex **7-4** acts as a propylene dimerization catalyst and affords 2,3-dimethyl-1-butene with a selectivity of over 95% at 35 °C. In Figure 7.5 is shown the proposed mechanism for propylene dimerization involving a metallacycle intermediate.

The β elimination from a metallacyclopentane (**7-4**) will generate a hydrido-alkyl species at 3 o'clock in the catalytic cycle, and its reductive elimination liberates 2,3-dimethyl-1-butene. Coordination of two molecules of propylene to the zero-valent species at 9 o'clock followed by cyclization regenerates complex **7-4**, which continues the catalytic cycle. In Figure 7.5 the ligands attached to the metal and a concomitant process in which 2,3-

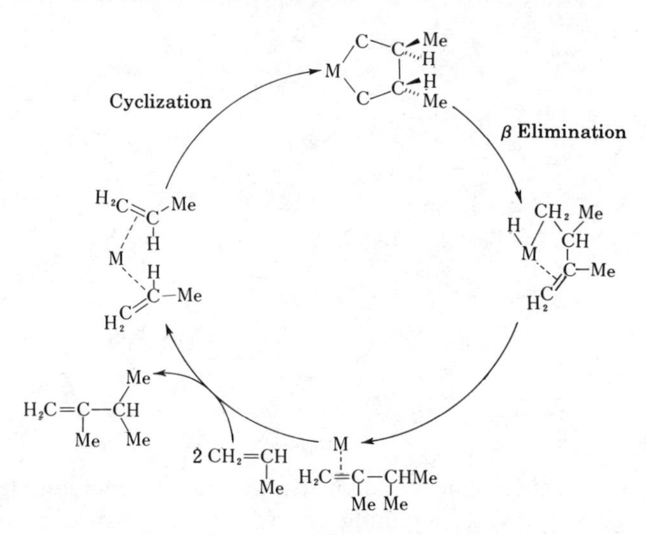

FIGURE 7.5. Propylene dimerization mechanism involving the metallacycle formation.

dimethyl-1-butene is formed via a metallacyclobutane intermediate are omitted for simplicity.[22]

This type of mechanism may be fairly common for olefin dimerization, but its significance has been recognized only recently. The fact that bulky phosphines favor 2,3-dimethyl-1-butene can also be accommodated by this mechanism. A metallacycle such as **7-4**, which has its two methyl groups away from the ligands, may be favored by bulky tertiary phosphine ligands because steric interactions exerted by the bulky ligand are minimized.

d. Polymerization of Cyclic Olefins

Cycloolefins such as cyclopentene can be polymerized to macromolecules having internal double bonds by Ziegler-type catalysts such as $MoCl_5$–$AlEt_3$ and WCl_5–$AlEt_3$[23] as well as by the solid catalyst MoO_3/Al_2O_3.[24]

$$n \text{ (cyclopentene)} \longrightarrow -(CH_2CH_2CH_2CH{=}CH-)_n$$

Studies using isotope-labeled cycloolefins revealed that the polymerization proceeds with the opening of the C–C double bond, and a ring expansion mechanism involving the formation of a pseudocyclobutane ring on the transition metal catalyst was first proposed.[25] As the chemistry of olefin metathesis promoted by transition metal complexes has progressed, this mechanism has become less popular, and a mechanism whereby a carbene intermediate is generated and a metallacyclobutane is formed has been generally accepted.[26] Polymerization of cycloolefins is likely to proceed by a mechanism similar to that proposed for olefin metathesis.

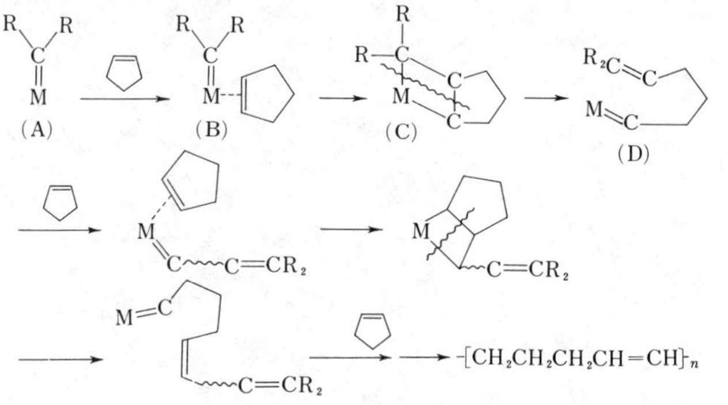

FIGURE 7.6. Mechanism of cycloolefin polymerization.

In fact, cyclopentene can be polymerized by a carbene complex, $Ph_2C{=}W(CO)_5$, to give a polymer retaining the cis stereochemistry.[27] The polymerization probably proceeds as shown in Figure 7.6.

The polymerization is initiated by coordination of the cycloolefin to the carbene complex (A) to form (B), which then forms the metallacyclobutane (C) and undergoes rearrangement to give the new carbene complex (D). Repetition of these processes enables cycloolefins to be inserted into the M–C double bond, leading to the cis cyclopentenamer. Because the polymer has unsaturated carbon units, it can be vulcanized to give a synthetic rubber.

e. Polymerization and Oligomerization of Alkynes

Acetylene is a very reactive compound and once served as an important feedstock for the organic chemical industry. Because of its high price, however, it has been replaced by olefins and other less expensive feedstocks. However, a useful process based on acetylene that exploits the reactivity of the carbon–carbon triple bond could still be developed.

The oligomerization of acetylene was first discovered by Reppe.[28] Acetylene can be converted to cyclooctatetraene when it is heated at 60–70 °C at 15–20 atm in the presence of $Ni(CN)_2$ or $Ni(CO)_4$.

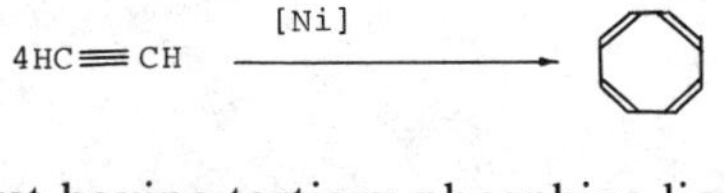

$$4HC{\equiv}CH \xrightarrow{[Ni]} \text{cyclooctatetraene} \quad (7.14)$$

When a nickel catalyst having tertiary phosphine ligands is used under similar conditions, acetylene is cyclotrimerized to benzene. The $Ni(CO)_2(PPh_3)_2$ acts as a cyclotrimerization catalyst for a variety of monosubstituted acetylenes, and the catalyst tolerates functional groups such as esters, ketones, hydroxyls, ethers, olefins, and nitro.

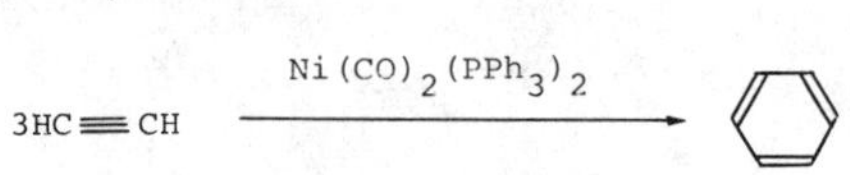

$$3HC{\equiv}CH \xrightarrow{Ni(CO)_2(PPh_3)_2} \text{benzene} \quad (7.15)$$

An elegant application of cyclotrimerization to organic synthesis is the one-step synthesis of a steroid analog as shown below.[29]

$Me_3Si{-}C{\equiv}C{-}SiMe_3$ + HC≡C / HC≡C (diyne) $\xrightarrow{CpCo(CO)_2}$ Me_3Si, Me_3Si (H, H)

Hagihara[30] and Schrauzer[31] initially proposed that the catalytic cyclooligomerization of acetylene proceeds by simultaneous coordination of three or four acetylene molecules and subsequent concerted cyclization of the coordinated acetylenes. However, stepwise processes involving metallacyclopentadiene intermediates are now favored for the cyclooligomerization mechanism.[22,32-36] The reaction probably proceeds by initial coordination of one alkyne to give a monoalkyne complex, coordination of a second alkyne to form a bisalkyne complex, and rearrangement of this to a metallacyclopentadiene complex. In the reaction of $CpCo(PPh_3)_2$ with alkynes the monoalkyne complexes and metallacyclopentadiene complexes have been isolated.[34]

$$CpCo(PPh_3)_2 \xrightarrow[-PPh_3]{RC\equiv CR} Cp(PPh_3)Co{\cdots}(RC\equiv CR) \xrightarrow{RC\equiv CR} Cp(PPh_3)Co(C_4R_4) \quad (7.16)$$

7-5

Coordination of a third molecule of alkyne and its insertion into the metallacycle followed by reductive elimination may lead to a cyclotrimer of the alkyne. When a dienophile such as dialkyl acetylenedicarboxylate is used, the metallacyclopentadiene **7-5** is likely to react with it in Diels–Alder fashion to give the cyclotrimer.

The cyclization process can be applied to the synthesis of heterocycles. Insertion of the C≡N group in nitriles into the Co—C bond in **7-5** followed by reductive elimination gives pyridine derivatives.[35,36]

$$2HC\equiv CH + RCN \xrightarrow{CpCoL_2} \left[CpCo(C_4H_4) \cdots N\equiv CR \right] \longrightarrow \text{2-R-pyridine} \quad (7.17)$$

As an application of the catalytic cyclization process, 2,2′-bipyridine can be prepared from acetylene and 2-cyanopyridine by using CpCo(cod).

$$2HC\equiv CH + \text{2-cyanopyridine (CN)} \xrightarrow{CpCo(cod)} \text{2,2'-bipyridine}$$

Ziegler systems also catalyze the polymerization and oligomerization of acetylene. When the catalyst mixture contains halide, either in the transition metal compound or in the alkylaluminum compound, cyclotrimerization of acetylene to benzene is the main process, whereas a catalyst mixture containing no halide, such as $Ti(OBu)_4/AlEt_3$, gives polyacetylene as an insoluble black solid.[37] Since it has been established that an alkyl group derived from the alkylaluminum compound becomes attached to the end of the polymer, the polymerization is likely to proceed by acetylene insertion into a metal–alkyl bond.[37,38] It has further been shown that poly-*cis*-acetylene is formed by polymerization at low temperatures and that it subsequently isomerizes at higher temperatures to poly-*trans*-acetylene. These results are consistent with acetylene initially coordinating side-on and then inserting into the M–C bond (the cis addition mechanism). Since cyclotrimerization of acetylene with $TiCl_4$–AlR_3 gives benzene containing some alkylbenzene, this cyclization is also considered to proceed by an insertion mechanism into the M–C bond. The processes that account most reasonably for the polymerization and cyclotrimerization of acetylene are shown in Figure 7.7.

When three acetylene molecules insert consecutively into an M–C bond, two possibilities arise. Cyclization at this stage gives alkylbenzene, as confirmed experimentally, and a hydrido complex that on further acetylene insertion produces benzene catalytically. If the cyclization route is somehow hindered and acetylene insertion continues, polyacetylene having long conjugated double bonds is formed. The polyacetylene thus prepared does not conduct electricity, having an electric conductivity of 10^{-5}–10^{-9} S cm^{-1}, but on doping with I_2 or AsF_5, a remarkable enhancement in conductivity is achieved to give a polymer with a conductivity as high as 10^3 S cm^{-1}. Thus, because of its potential application in batteries and other electronic devices, polyacetylene is attracting attention.

Acetylenes are also known to insert into a Pd–Cl bond. Further consecutive insertion of acetylenes into the resulting Pd–C bond followed by cyclization gives cyclobutadiene and benzene derivatives.[39]

$$PdCl_2 \xrightarrow{RC\equiv CR} Pd(Cl)(CR{=}CRCl) \xrightarrow{RC\equiv CR} Pd(Cl)(CR{=}CR{-}CR{=}CRCl) \xrightarrow{RC\equiv CR} Pd(Cl)(CR{=}CR{-}CR{=}CR{-}CR{=}CRCl)$$

$$\downarrow \quad Cl_2Pd(C_4R_4) \qquad\qquad \downarrow \quad C_6R_6$$

f. Polymerization of Dienes and Oligomerization of Dienes

Studies of the polymerization and oligomerization of dienes by transition metal complex catalysts were started immediately after the discovery of the Ziegler catalyst. The Goodrich–Gulf Chemical Company, having been in-

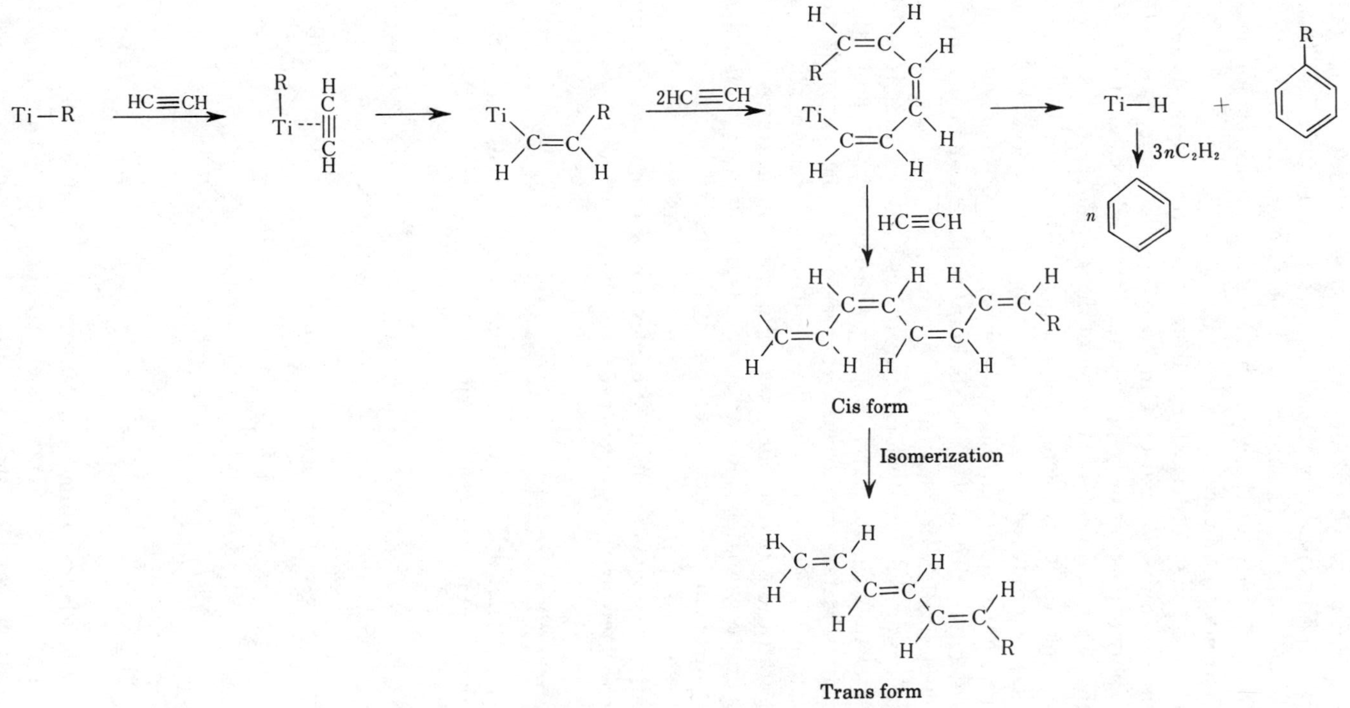

$Ti{-}R \xrightarrow{HC\equiv CH} $ R–Ti···HC≡CH $\longrightarrow$ Ti(H)C=C(R)H $\xrightarrow{2HC\equiv CH}$ Ti–CH=CH–CH=CH–CH=CHR $\longrightarrow$ Ti–H + R–C$_6$H$_5$

Ti–H $\xrightarrow{3nC_2H_2}$ n C$_6$H$_6$

$\downarrow HC\equiv CH$

Cis form

$\downarrow$ Isomerization

Trans form

FIGURE 7.7. Oligomerization and polymerization mechanisms of acetylene catalyzed by Ziegler catalysts.

formed by Ziegler of his discovery before it became public, first examined the Ziegler catalyst for polymerization of dienes.[2a] Because of the industrial importance of polydienes as materials for synthetic rubbers and because of academic interest in the mechanisms of diene polymerization, a great effort has since been devoted to the research and development of polymerization and oligomerization of dienes.

Natural rubber made of latex taken from *Hevea* trees is constituted of *cis*-1,4-polyisoprene units. Goodrich–Gulf workers, and later Natta's group, found that polymerization of isoprene with the Ziegler catalyst gave *cis*-1,4-polyisoprene. With another combination of the Ziegler catalyst mixture *trans*-1,4-polyisoprene, the constituent of gutta percha used for making chewing gum, can also be produced.

While the controversy regarding the active site for olefin polymerization by the Ziegler catalyst, that is, whether the transition metal is solely responsible for the polymerization or whether the organoaluminum component also is playing a role, was still unsettled, elegant studies by Wilke and his co-workers in Ziegler's institute in the early 1960s established beyond doubt that low-valent transition metal complexes are responsible for the polymerization and oligomerization of dienes.[40a,41] They have isolated a variety of aluminum-free low-valent transition metal complexes from reaction mixtures of transition metal salts and organoaluminum or organomagnesium compounds and have shown that these isolated complexes can initiate the polymerization or oligomerization of dienes. They have also shown for the first time that the course of polymerization or oligomerization of dienes can be controlled by addition of ligands such as tertiary phosphines and phosphites.

(1) Dimerization of Dienes

The following dimerization products of dienes can be produced by employing transition metal complexes as catalysts.

1,5-cyclooctadiene (COD) | 1,2-divinylcyclobutane (DVC) | octatrienes (OT)

2-methylene-1-vinylcyclopentane | 4-vinyl-1-cyclohexene | 3-methyl-1,4,6-heptatriene

The first five of these isomeric butadiene dimers can be selectively synthesized by use of nickel-based catalysts containing suitably chosen ligands and under the correct reaction conditions, a fact that demonstrates the ability of ligands to control the course of butadiene oligomerization by modifying the nickel active center. Cobalt complexes are more suitable for production of 3-methyl-1,4,6-heptatriene. By use of appropriate combinations of transition metals and the modifying ligands, the mode of butadiene coordination can be controlled. The structures of the intermediate species determine the nature of the dimerization products.

For example, complex **7-9** in Figure 7.8 has been isolated, its structure characterized by X-ray analysis, and its activity in butadiene cyclodimerization established. The catalytically active nickel complex can be prepared either by reduction of a nickel salt (e.g., nickel acetylacetonate) with an alkylaluminum compound in the presence of butadiene and the ligand or by ligand exchange of a zero-valent nickel complex such as $Ni(cod)_2$ with butadiene and the ligand. The process of coordination of two butadiene molecules to a nickel complex (**7-6**) having one ligand to give **7-7** and its subsequent rearrangement to give either a bis(π-allyl) intermediate (**7-8**) or a π-allyl σ-allyl intermediate (**7-9**) are included in Figure 7.8. A complex of type **7-8** has been isolated by use of a tertiary phosphite, $P(OC_6H_4\text{-}o\text{-}C_6H_5)_3$, and a type **7-9** complex has been isolated with a tertiary phosphine ligand, $P(cyclohexyl)_3$. The π-allyl and σ-allyl forms are interconvertible depending on the experimental conditions. Reductive elimination from a bis(σ-allyl)nickel intermediate may give cyclic butadiene dimers such as cyclooctadiene, 4-vinyl-1-cyclohexene, and 1,2-

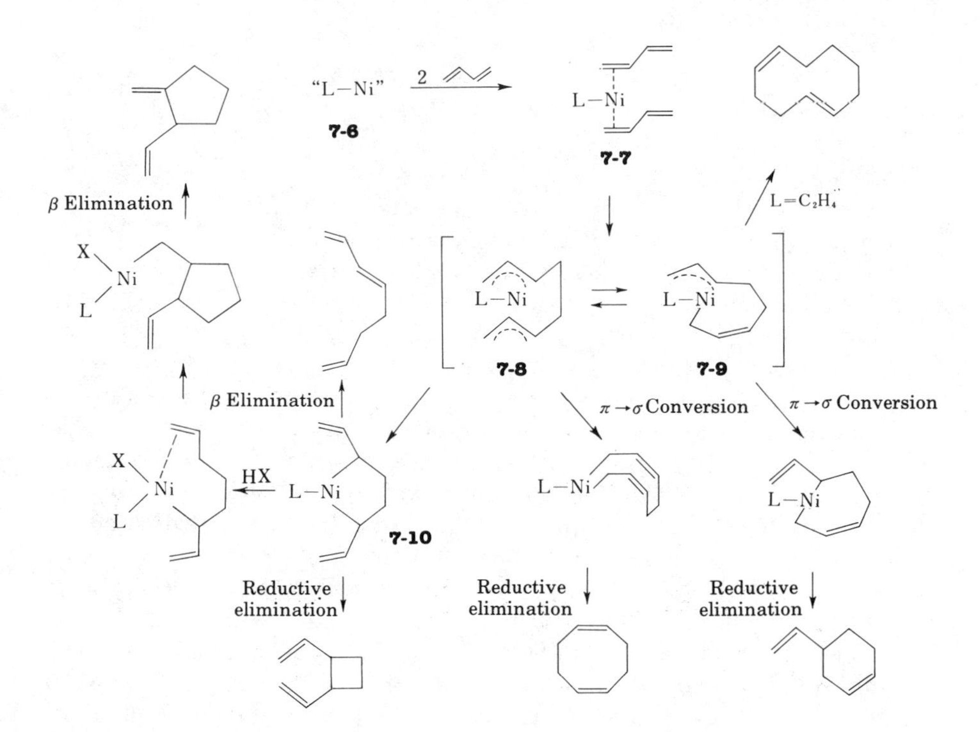

FIGURE 7.8. Mechanism of butadiene cyclodimerization catalyzed by nickel complexes.

divinylcyclobutane. In the catalytic process reductive elimination may be assisted by coordination of other butadiene molecule(s). Insertion of ethylene into the Ni–C bond in **7-9** followed by reductive elimination yields also 1,5-cyclodecadiene.

Linear dimers of butadiene may be formed from a metallacycle (**7-10**) by β-hydrogen elimination followed by reductive elimination. In the presence of a protonic reagent the course of dimerization may be modified to produce 2-methylene-1-vinylcyclopentane, as shown in Figure 7.8.

In certain cases the protonic reagent may be incorporated in the butadiene dimers. Palladium complexes are known to be more suitable than nickel complexes for this type of reaction,[42] which has found application in organic syntheses.[43,44]

$$2C_4H_6 + HY \xrightarrow{Pd[P(OPh)_3]_4} \text{Y-substituted octadienes (linear + branched)}$$

HY = HOPh, H_2NR, H_2O, ROH, RCO_2H, $HSiR_3$, HCN

Linear dimerization of butadiene to 3-methyl-1,4,6-heptatriene is catalyzed by various cobalt complexes. Studies of this process using deuterium-labeled butadiene have revealed that the terminal hydrogen in butadiene is transferred to the other terminal group.[45]

$$CD_2{=}CHCH{=}CD_2 + D_2C{=}CHCH{=}CD_2 \xrightarrow{CoEt(bipy)_2} CD_2{=}CHCH(CD_3)CD{=}CHCH{=}CD_2$$

A complex relevant to the linear dimerization has been isolated from a system containing $CoCl_2$, $NaBH_4$, and butadiene, and its molecular structure has been determined by X-ray crystallographic analysis.[46] Complex **7-11** has both butadiene and methylheptatriene coordinated to the central cobalt atom, as shown in Figure 7.9.

The butadiene is in its η^4-bonded, s-cis form, and methylheptatriene is bound to cobalt through an η^3-allyl group and an η^2-vinyl group. Based on this information, the following mechanism for butadiene dimerization has been proposed.[46] The hydrogen atom attached to carbon atom 4 in complex **7-11** may be abstracted by cobalt and transferred to the coordinated butadiene ligand, as depicted in Figure 7.9*a,* thus converting the coordinated butadiene ligand into an η^3-1-methylallyl ligand and giving 3-methyl-1,4,6-heptatriene bonded to cobalt (Fig. 7.9*b*). Attack of an incoming butadiene molecule at the allylic ligand as shown in (*c*) gives the complex in (*d*). Displacement of the 3-methyl-1,4,6-heptatriene in (*d*) by a new butadiene molecule again leads to the initial complex in (*a*), now upside down. The proposal is somewhat speculative but is quite reasonable in view of the X-ray data as well as the experimental results using isotopically labeled butadiene.

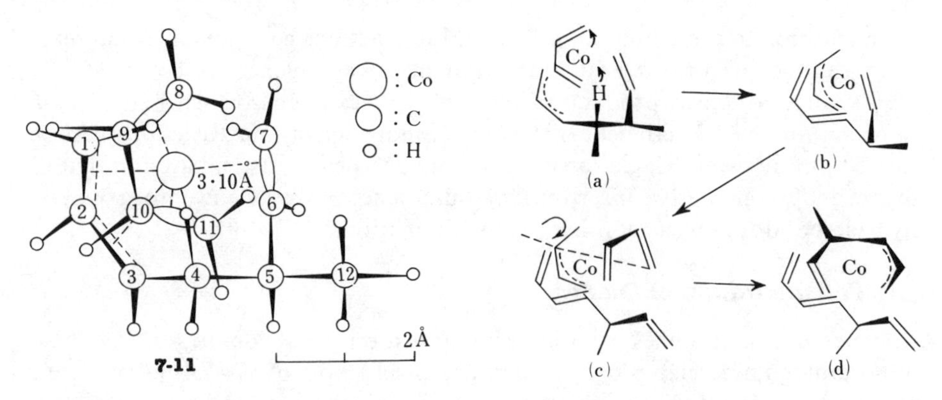

FIGURE 7.9. Mechanism of butadiene dimerization catalyzed by cobalt complex **7-11**. (From Allegra et al.[46])

Use of isoprene in place of butadiene gives a greater variety of oligomerization products. The oligomers of isoprene are especially interesting for the synthesis of terpenoid and sesquiterpenoid compounds of biological interest. They may also be used as intermediates in the synthesis of perfumes.

(2) Trimerization of Dienes

Butadiene can be cyclotrimerized by a mixture of $TiCl_4$ and an alkylaluminum chloride to give a trans, trans, cis isomer of cyclododecatriene (CDT). Zero-valent nickel catalysts without phosphine or phosphite ligands produce an all-trans isomer.

ttc ttt

The cyclododecatriene can be partially hydrogenated and then transformed into lactam. 12-Nylon is available by ring-opening polymerization of the lactam and is used to produce plastics of excellent dimensional stability. A nickel complex of formula **7-12** (shown below) has been isolated and characterized by Wilke's group. The complex contains a C_{12} entity in which cyclization is incomplete. Cyclization to cyclododecatriene is induced by addition of a tertiary phosphine or by introduction of butadiene. Further reaction with butadiene gives the cyclotrimer of butadiene catalytically.

Ni $\xrightarrow{C_4H_6}$

7-12 ttt-CDT

In addition to these homooligomerization reactions cooligomerization reactions can be performed with transition metal complexes. A commercially important example is production of *trans*-1,4-hexadiene, which is used as a third comonomer in ethylene–propylene copolymerization. Reactions can be carried out by using Ni, Co, or Rh compounds as catalysts and are thought to proceed by consecutive insertions of butadiene and ethylene into a metal–hydride bond followed by β-hydrogen elimination.[47]

(3) Polymerization of Dienes

Polymerization of dienes is initiated by transition metal complexes. The most important commercial processes are the production of *cis*-1,4-polyisoprene and *cis*-1,4-polybutadiene, which are used as synthetic elastomers.

The growth reaction in diene polymerization is generally thought to be insertion of the diene into a π-allyl complex. However, a growth reaction involving insertion of the diene into a σ-allyl complex is also a likely process, particularly in diene polymerization by very active catalysts. In Section 6.3 we discussed the formation of π-allyl complexes by interaction of butadiene with a metal hydride. The π-allyl complexes thus formed are tautomeric with σ-allyl complexes, as shown below:

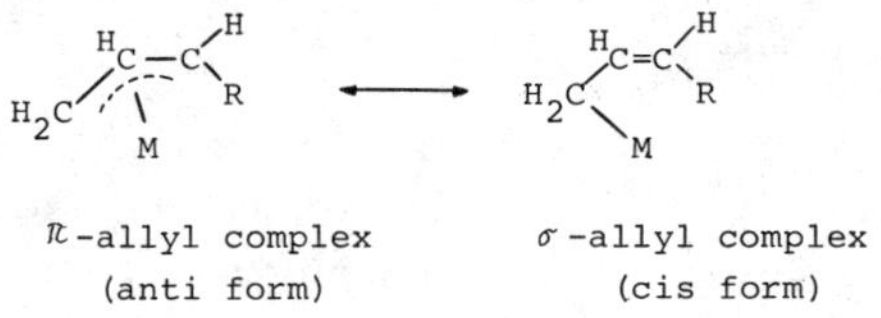

$$H_2C\cdots CH\cdots C(H)(R)\ (\eta^3\text{-}M) \longleftrightarrow H_2C(M){-}CH{=}C(H)(R) \qquad (7.18)$$

π-allyl complex (anti form) $\qquad$ σ-allyl complex (cis form)

$$H_2C\cdots CH\cdots C(R)(H)\ (\eta^3\text{-}M) \longleftrightarrow H_2C(M){-}CH{=}C(R)(H) \qquad (7.19)$$

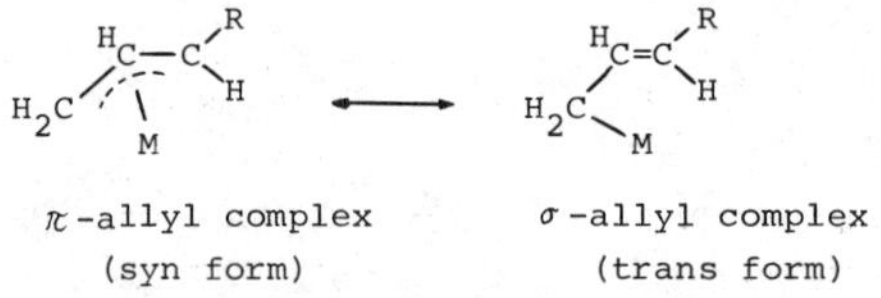

π-allyl complex (syn form) $\qquad$ σ-allyl complex (trans form)

Consecutive insertion of butadiene into the π-allyl complex in its anti form or into the corresponding σ-allyl complex in its cis form thus leads to *cis*-1,4-polybutadiene, whereas similar butadiene insertion into the π-allyl complex in its syn form or into the corresponding σ-allyl complex in its trans form leads to *trans*-1,4-polybutadiene. If syn–anti isomerization is slow, the stereoregularity of the polybutadiene can be enhanced.

Another important factor controlling stereoregularity is the mode of coordination of butadiene to the transition metal complex. As discussed in Section 6.3 [see Eqs. (6.125) and (6.126)], butadiene coordination through two double bonds in an s-cis manner is likely to give a π-allyl complex in the anti form, whereas coordination through one of the double bonds of butadiene favors the s-trans conformation and thus promotes formation of a syn π-allyl complex. If the modes of coordination and subsequent insertion of butadiene are maintained throughout the polymerization, the stereoregularity of the polybutadiene may be determined by the mode of butadiene coordination. Although it is difficult to establish experimentally which factor is decisive, the following result suggests the importance of the mode of monomer coordination in determining the stereoregularity of polybutadiene.

Polymerization of butadiene by the π-crotyl complexes [A] and [B] yields *cis*-1,4-polybutadiene from [A] and *trans*-1,4-polybutadiene from [B]. However, NMR studies of [A] and [B] indicate that both the chloride and iodide complexes are in the syn form.[48]

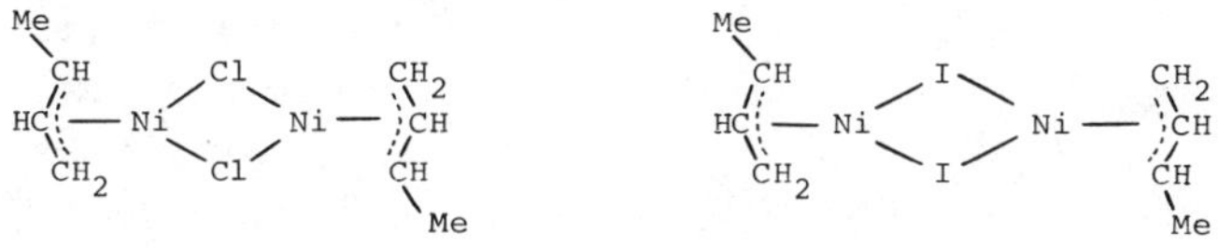

[A] [B]

Therefore, the steric structures of the polymers are not dictated by the structure of the π-allyl group in the starting complexes. The following example illustrates how *cis*-1,4-polybutadiene can be produced by insertion of butadiene, coordinated to nickel through its two double bonds, into an allyl–nickel bond.[50]

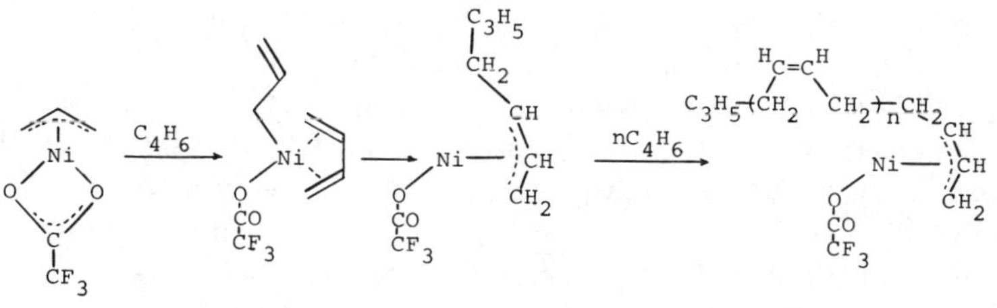

(7.20)

If one adds a strongly bonding ligand such as $P(OR)_3$ to a system that is polymerizing butadiene to give *cis*-1,4-polybutadiene as shown above, the nature of the catalyst center may be modified so that *trans*-1,4-polybutadiene is produced instead. Thus, one can obtain a block copolymer containing a *cis*-1,4-polybutadiene sequence and a *trans*-1,4-polybutadiene sequence or equibinary mixtures of these units. The effect of the additive may be understood as the blocking of a coordination site for incoming butadiene so that the monomer is forced to coordinate to the metal through one double bond in a s-trans manner, thus giving a *trans*-1,4-polybutadiene sequence.

Another type of polymer obtained from butadiene is 1,2-polybutadiene, which has pendant vinyl groups. If one assumes a π-allyl complex to be the propagating species for butadiene polymerization, it can be seen that the site at which the diene enters decides whether a 1,4- or a 1,2-polymer is formed. Thus, attack at the terminal methylene carbon atom (site A) gives 1,4-polymer (Eq. (7.21)], whereas attack at the methine carbon (site B) leads to 1,2-polymer (Eq. (7.22)].

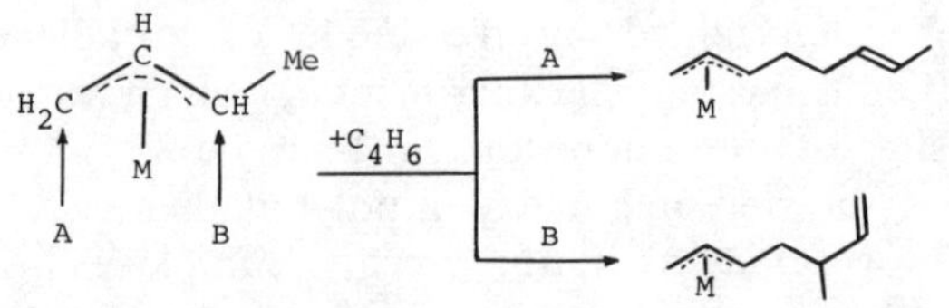

(7.21)

(7.22)

NMR studies using specifically deuterated π-crotyl nickel iodide in combination with $CD_2{=}CHCH{=}CD_2$ were compatible with butadiene insertion into the π-allyl complex.[49] However, this catalyst system has rather low activity for butadiene polymerization, a feature that makes the system amenable to the NMR study. One could argue that in more active catalyst systems polymerization may proceed by butadiene insertion into the M–C bond of a σ-allyl species. Formation of 1,2-polybutadiene, then, can be explained by assuming the insertion of s-trans-bonded butadiene, as shown below.

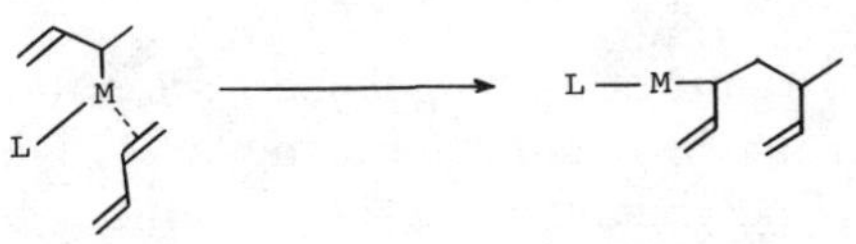

(7.23)

In addition to butadiene, isoprene polymerization is commercially very important. The mechanism of polymerization is probably similar to that of butadiene, with more variations in the modes of coordination and insertion.

Because of the industrial importance of diene polymers, an enormous body of information has accumulated about catalyst systems that polymerize butadiene and isoprene. It is now possible, at least conceptually, to explain how the polymerization occurs by making assumptions about how butadiene coordinates to the transition metal active site with its associated ligands. However, the optimum conditions for polymerizing dienes to polymers of desirable physical properties have to be established largely by painstaking experimental work.[51]

Polymerization of another type of diene, 1,2-propadiene, or allene, has been studied. By means a rhodium-based catalyst it can be converted into a polymer having methylene groups attached to the polymer backbone.[52]

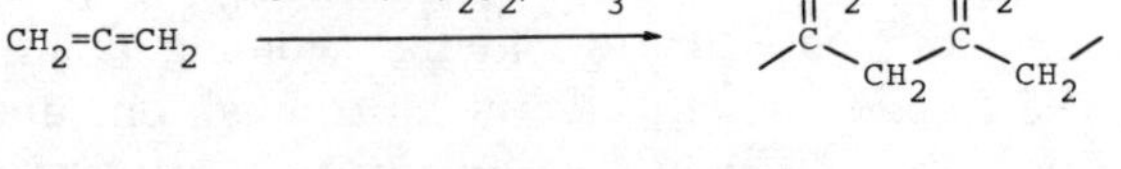

$$CH_2{=}C{=}CH_2 \xrightarrow{[RhCl(CO)_2]_2/PPh_3} \sim C(=CH_2)-CH_2-C(=CH_2)-CH_2\sim$$

7.3. SYNTHESES WITH CARBON MONOXIDE

Carbon monoxide is one of the most important feedstocks in chemical industry. Despite its unsaturated nature, it is unreactive under usual reaction conditions and requires activation by a transition metal to be converted into useful organic products. A variety of products can be produced by catalytic reactions using carbon monoxide as the unique source of carbon or as a component used in combination with other compounds.[53] The importance of carbon monoxide as an industrial feedstock has been greatly enhanced by the realization that the world's supply of petroleum is dwindling and by the sharp increase in the price of oil. The recent success in commercialization of methanol carbonylation encouraged the search for processes to produce hydrocarbons and other products using carbon monoxide. The Fischer–Tropsch process, which had already been discovered in 1936, is now attracting renewed attention as an alternative route to hydrocarbons. The term C_1 chemistry has recently been coined to encompass various chemical processes starting from carbon monoxide.

a. Hydroformylation of Olefins (Oxo Process)

The hydroformylation of olefins was discovered by Roelen in 1938 while studying the mechanism of Fischer–Tropsch synthesis.[54] The total world capacity of the oxo process in 1979 exceeded 6,500,000 tons/year. In this process olefins are converted into aldehydes, which are versatile chemical intermediates.

$$RCH{=}CH_2 + CO + H_2 \xrightarrow{[Co]\text{ or }[Rh]} RCH_2CH_2CHO + RCH(CH_3)CHO \qquad (7.24)$$

A cobalt catalyst was used in the older processes, but a process utilizing a rhodium catalyst is now attracting attention because of its higher activity and selectivity in producing linear aldehyde relative to branched aldehyde. The linear aldehydes (*n*-aldehydes) are more useful than the branched aldehydes (isoaldehydes), and a process that produces aldehydes in a high *n*–iso ratio is preferred. For example, *n*-butanal obtained by hydroformylation of propylene can be converted into 1-butanol by hydrogenation or into 2-ethylhexanol after aldol condensation of the 1-butanal and subsequent dehydration and hydrogenation. The alcohol thus obtained is called oxo alcohol and is used for the manufacture of esters, which are useful as plasticizers.

2 (CO + H2) ; OH ; -H2O ; 2H2 ; OH

The mechanism of hydroformylation of olefins is based on a combination of the fundamental processes discussed in Chapter 6. The following is an account of the mechanism of propylene hydroformylation using $Co_2(CO)_8$ as a catalyst. It is based on the original proposal of Heck and Breslow.[55] The $Co_2(CO)_8$ is converted into a hydridocobalt tetracarbonyl under H_2.

$$Co_2(CO)_8 + H_2 \rightleftharpoons HCo(CO)_4 \qquad (7.25)$$

The hydridocobalt complex is coordinatively saturated and one CO molecule must dissociate in order to create a coordinatively unsaturated species that can react with the olefin (Fig. 7.10).

Propylene coordinates to the coordinatively unsaturated species $HCo(CO)_3$ and undergoes insertion into the Co–H bond. Depending on its direction, the insertion can give either a linear alkylcobalt or a branched alkylcobalt complex.

$$\begin{matrix} CH_3CH{=}CH_2 \\ \vdots \\ H{-}Co(CO)_3 \end{matrix} \rightleftharpoons CH_3CH_2CH_2{-}Co(CO)_3 \tag{7.26}$$

$$\begin{matrix} CH_2{=}CHCH_3 \\ \vdots \\ H{-}Co(CO)_3 \end{matrix} \rightleftharpoons (CH_3)_2CH{-}Co(CO)_3 \tag{7.27}$$

Further coordination of the CO molecule to the propylcobalt intermediates is followed by CO insertion into the Co–propyl bonds to give an acylcobalt complex. Oxidative addition of H_2 to the acylcobalt complex and subsequent reductive elimination of the acyl group and the hydrido ligand liberates the aldehyde. This regenerates $HCo(CO)_3$, which is the carrier of the catalytic cycle. If the acylcobalt complex $C_3H_7COCo(CO)_3$ reacts with CO, a coordinatively saturated species $C_3H_7COCo(CO)_4$ is produced. This complex is observable in the reaction system, but it seems to occupy a blind alley apart from the main catalytic cycle. The scheme shown in Figure 7.10 satisfies the results of kinetic studies. An alternative route to the aldehyde has been proposed in which there is a bimolecular reaction between the acylcobalt species and the hydridocobalt complex $HCo(CO)_4$.[56]

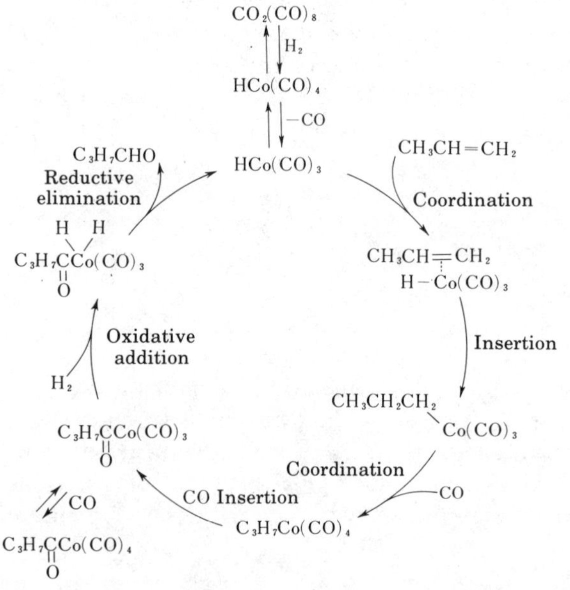

FIGURE 7.10. Mechanism of hydroformylation of propylene with cobalt carbonyl catalyst.

Addition of a ligand such as a tertiary phosphine increases the stability of the catalysts but decreases the catalytic activity. An important consequence of addition of a tertiary phosphine is an increase in the *n*–iso ratio of the product aldehyde.[57] Combinations of electronic and steric effects as quantified by Tolman's χ_i and cone angle values[58] seem to correlate with the ability to produce aldehydes of different *n*–iso ratios. Addition of tertiary phosphine ligands also enhances the hydrogenation of the olefins used and the aldehydes produced.

Catalyst systems based on rhodium complexes have the advantages of very mild reaction conditions, high conversion rates of olefin, and high selectivity for *n*-aldehyde. The addition of phosphine ligands to these catalysts leads to excellent selectivity for the formation of the desired *n*-aldehydes without seriously sacrificing catalytic activity. Another advantage of the rhodium system is that it is more amenable to mechanistic studies,[53h] as exemplified by the pioneering work of Wilkinson using $RhH(CO)(PPh_3)_3$.[59] The essential features of the mechanism of rhodium-catalyzed hydroformylation of olefins resemble those shown in Figure 7.10. In the rhodium-catalyzed system the conversion of olefin to aldehyde has been shown to proceed by cis addition of the formyl group and hydrogen atom to the double bond.[60] Asymmetric hydroformylation using chiral tertiary phosphine ligands is also being studied.[61] 1,3-Butadiene can be partly converted into dialdehyde by rhodium complexes having tertiary phosphine ligands.[62]

b. Carboxylation of Olefins

If hydrogen donors are used in place of H_2 in the hydroformylation of olefins, derivatives of carboxylic acids can be produced. This process, which gives the —C(O)O— moiety is sometimes called oxacarbonylation and includes processes for converting olefins, CO, and alcohols into carboxylic acid esters. The processes that give carboxylic acids and esters are called hydroxycarbonylation and alkoxycarbonylation, respectively. The original synthetic method was discovered by Reppe during his systematic study of the reaction of unsaturated compounds with CO and nucleophiles in the presence of metal carbonyls.[53,63] Part of the reaction mechanism is similar to that of hydroformylation of olefins. It involves the insertion of olefin into an M–H bond followed by a further insertion of CO to give an acyl metal derivative. It is generally assumed that the acyl complex then undergoes attack by a nucleophile such as water, alcohol, or amine; in the last case carboxylic amide can be produced. When ethylene is used, propionic acid or its derivatives may be produced.

$$\text{H–M} \xrightarrow{C_2H_4} C_2H_5\text{–M} \xrightarrow{CO} C_2H_5C(=O)\text{–M} \begin{cases} \xrightarrow{H_2O} C_2H_5COOH \\ \xrightarrow{ROH} C_2H_5COOR \\ \xrightarrow{R_2NH} C_2H_5CONR_2 \end{cases} \quad (7.28)$$

After attack of the nucleophile to give the carboxylic acid or its derivatives, the M–H bond may be regenerated to carry the catalytic cycle. Propionic acid is made commercially by this hydroxycarbonylation route from ethylene, CO, and water, whereas higher carboxylic acids are produced as methyl or ethyl esters.

When acetylene is used, hydroxycarbonylation or alkoxycarbonylation catalyzed by nickel carbonyl provides acrylic acid or its ester. This is another variation of the Reppe processes.

$$HC{\equiv}CH + CO + H_2O(ROH) \xrightarrow{Ni(CO)_4} \begin{array}{l} H_2C{=}CHCOOH \\ (H_2C{=}CHCOOR) \end{array} \tag{7.29}$$

For the manufacture of acrylic acid and acrylates, these processes have been commercialized, and production capacity worldwide is estimated to be about 350,000 tons/year.[53a]

Although attack on the acyl metal species by nucleophiles is generally postulated as in Eq. (7.28), an alternative route may exist. The carbonyl ligand attached to a transition metal is susceptible to nucleophilic attack, for example, by water, alcohol, or amine, to yield a hydroxycarbonyl, alkoxycarbonyl, or carbamoyl species,[64] as discussed in Section 6.4.

$$M{-}CO \begin{cases} \xrightarrow{H_2O} M{-}\underset{\underset{O}{\|}}{C}{-}OH + H^+ \\ \xrightarrow{ROH} M{-}\underset{\underset{O}{\|}}{C}{-}OR + H^+ \\ \xrightarrow{R_2NH} M{-}\underset{\underset{O}{\|}}{C}{-}NR_2 + H^+ \end{cases} \tag{7.30}$$

These complexes, having the metal–carbon σ bond, may react with the alkyl group to reductively eliminate the carboxylic acid, ester, or amide. It is also possible that the olefins or alkynes insert into the metal–carbon σ bond shown in Eq. (7.30), and the resultant alkyl or alkenyl complex is trapped by a proton or reacts with a hydride ligand to release the products shown in Eq. (7.28). In fact, insertion of butadiene into a MeOCO–Co bond has been substantiated.[65]

It is yet to be determined whether these hydroxycarbonylation, alkoxycarbonylation, or amidocarbonylation reactions proceed by a mechanism involving the nucleophilic attack on the coordinated carbonyl ligand or by the conventional mechanism involving nucleophilic attack on the acyl metal species. It is also possible that the reaction mechanism depends on the metal, the ligand, and the nucleophile.

Another route for the production of carboxylic acids from olefins and CO is copper-catalyzed hydroxycarbonylation in concentrated sulfuric acid. The course of the reaction is not completely clear, but it is believed that a carbenium ion may be formed first. The carbenium ion is trapped by carbon monoxide in a reaction mediated by a Cu(I) compound, presumably

$[Cu(CO)_3]^+$, which can exist only in a high concentration of sulfuric acid.[53a,53f,66]

$$=\!\!\!\langle^{R} \xrightarrow{H_2SO_4} -\!\!\langle^{R}_{+} \xrightarrow{[Cu(CO)_3]^+} -\!\!\langle^{R}-CO^+ \xrightarrow{H_2O} -\!\!\langle^{R}-COOH$$

This is called a Koch reaction, and it is suitable for making a highly branched carboxylic acid. The reaction is attracting renewed interest as a route to carboxylic acids that are not otherwise accessible. Of particular recent interest is the normal pressure synthesis of glycolic acid and its esters from CO and formaldehyde.[66d]

$$HCHO + CO \xrightarrow[acid]{Cu(CO)_n^+} HOCH_2CO^+ \begin{cases} \xrightarrow{ROH} HOCH_2COOR \\ \xrightarrow{H_2O} HOCH_2COOH \end{cases}$$

c. Synthesis of Acetic Acid by Carbonylation of Methanol

Acetic acid is one of the most important industrial products and organic intermediates in the chemical industry. The world capacity of acetic acid production in 1983, which exceeded 4.5 million tons, testifies to its importance. Acetic acid is used for the manufacture of vinyl acetate, cellulose acetate, terephthalic acid, acetic acid esters, and chloroacetic acid, the demand decreasing in this order. Acetic acid was first produced by fermentation and oxidation of the resulting ethanol and later by oxidation of acetaldehyde prepared by mercuric-ion-catalyzed hydration of acetylene. The latter process was replaced by the Wacker synthesis of acetaldehyde produced from petroleum-based ethylene. It is ironic that the Wacker process, the once glamorous new technology that helped to boost the growth of the petrochemical industry, is now being challenged by another new technology based on the utilization of coal. Since methanol and CO, the two components for production of acetic acid, can be derived by routes utilizing coal, the process should become economically more favorable as the oil becomes more expensive.

There are two commercial processes for making acetic acid from methanol and carbon monoxide. One is the cobalt-based high-pressure process commercialized in the early 1960s by BASF. The BASF process operates under severe conditions, that is, 200–250 °C and 500–700 atm, and the reaction mixture is corrosive. The other process, which operates at low pressure, was commercialized by Monsanto Company in the late 1960s and early 1970s. The Monsanto process uses a rhodium-based catalyst that is much more reactive than the cobalt-based catalyst, and it can be operated at much lower temperatures and pressures (150–200 °C, 1–40 atm). The selectivity to acetic acid is on the order of 99% based on methanol and 90% based on carbon monoxide. The reaction is first order in rhodium and iodine added to the system. Figure 7.11 depicts the probable reaction mechanism.[67,68]

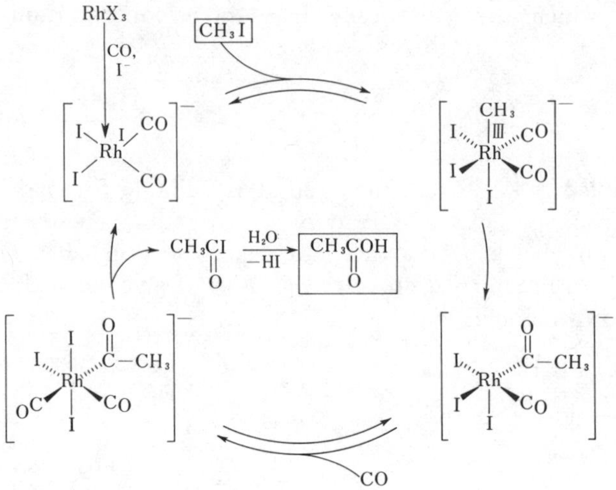

FIGURE 7.11. Mechanism of carbonylation of methanol catalyzed by a rhodium catalyst.

The catalyst precursor, a Rh(III) compound, is converted into the Rh(I) anion $[RhI_2(CO)_2]^-$ in the reaction system, which contains CO and an iodide ion. This active species oxidatively adds methyl iodide to form a methyl–Rh(III) species. The high activity of the rhodium-based catalytic system is due to the high rate of oxidative addition, which constitutes the rate-determining step in the whole catalytic cycle. The methyl iodide for the reaction is formed *in situ* according to the following reaction.

$$CH_3OH + HI \rightleftharpoons CH_3I + H_2O \tag{7.31}$$

The methylrhodium species thus formed undergoes CO insertion to afford an acetylrhodium(III) species. Reductive elimination and subsequent hydrolysis of acetyl iodide gives acetic acid. The $[RhI_2(CO)_2]^-$ complex is regenerated to carry the catalytic cycle, and the hydrogen iodide thus produced gives methyl iodide, according to Eq. (7.31). An alternative to the reductive elimination of acetyl iodide is direct hydrolysis of the acetylrhodium species to liberate acetic acid. Although rhodium is more expensive than cobalt, its excellent catalytic activity more than compensates for the difference in cost of the catalysts, and the rhodium-based process has been rapidly adopted worldwide. Iridium also serves as an excellent catalyst and has attracted some mechanistic studies,[67] although the behavior of this system is somewhat more complicated.

Intermezzo — *Important Timing*

The invention of the Monsanto process, which converts methanol to acetic acid, seems to have been made possible by right timing. James Roth and his co-workers at the Monsanto Company were working on the oxo process using a tertiary phosphine-modified rhodium catalyst. At a weekly technical conference a research director in the petrochemical division who was aware of the then impending cost reduction of methanol asked whether the process might be used for carbonylation of methanol. Experiments with the same rhodium catalyst proved to be totally negative. Then they rationalized that the problem might reside in the difficulty of forming a metal–carbon bond. Recalling a lecture of Professor Jack Halpern, who served as consultant to the company, on oxidative addition of methyl iodide to metal complexes, they then decided to add an iodine promoter.

In the first experiment with added iodide they observed with a mixture of disbelief and delight the catalytic carbonylation of methanol to acetic acid. Roth later told me that he was not aware at that time of the high-pressure process reported by BASF in 1965 for the carbonylation of methanol using an iodide-promoted cobalt catalyst. They were also not aware that Union Carbide had made a similar discovery and was but a few months behind them.

A project team was promptly formed, and after only five months of intensive effort the project was transferred to the operating division for development of a commercial process, which produced 150,000 tons of acetic acid in 1970.

The process became the forerunner of what is now called C_1 chemistry and marked a turning point from a totally petroleum-based technology to the process utilizing carbon monoxide.[68b]

d. Reactions of Synthesis Gas

The mixture of carbon monoxide and hydrogen produced by burning coal in the presence of oxygen and steam is called synthesis gas, or *syngas*. It is also produced by gasification of natural gas and petroleum. Since known coal reserves are estimated to have a lifetime of over 250 years, syngas is regarded as a quite stable feedstock for the foreseeable future. Its competitiveness with the petroleum as a feedstock will increase with further increases in petroleum prices.

There are broadly two types of hydrogenation of carbon monoxide,[53,69] one in which the C–O bond is cleaved and the other in which the C–O bond is retained and oxygen-containing compounds are formed. The former is related to Fischer–Tropsch synthesis and the latter to processes such as the synthesis of ethylene glycol.

The Fischer–Tropsch process was developed in Germany in the period 1923–1925 by Franz Fischer and Hans Tropsch. Fischer was the first director of the Kaiser Wilhelm Institute for Coal Research, now called the Max-Planck Institute for Coal Research in Mülheim/Ruhr. The process supplied liquid fuel during World War II and had a peak capacity of 600,000 tons/year in 1944. Although the abundant supply of cheap oil after the end of the war made the process uneconomic, the process has been used by the Republic of South Africa for political reasons since 1957. The Sasol process uses a heterogeneous catalyst and produces over 2 million tons of primary product, ranging from methane to paraffin wax.[53g] Although heterogeneous catalysts are used in the Fischer–Tropsch process, considerable efforts have been made to unravel the mechanism using homogeneous catalysts and model systems.

In the process of converting CO into hydrocarbons a process of dehydration must occur. This may take place on a single metal atom or it may proceed by cooperation of more than two metal atoms. A model for the first of these possibilities involves the formation of formyl, hydroxymethyl, and carbene complexes:

$$\overset{\displaystyle H}{\underset{|}{}}\!\!\!\!M\text{—CO} \longrightarrow M\text{—CHO} \longrightarrow M\text{—CH}_2\text{OH} \longrightarrow M{=}CH_2 \qquad (7.32)$$

We dealt with the processes represented in Eq. (7.32) in Section 6.4. The key step is the formation of the formyl complex. The simple insertion of CO into the M–H bond to give an η^1-formyl complex seems to be thermodynamically unfavorable. The process may become more favorable when an η^2-formyl species is formed[70] or when CO insertion is aided by interaction with a Lewis acid.[53i] The reduction steps in Eq. (7.32) have been demonstrated to occur when a hydridic reducing agent is used. It seems probable that in catalytic systems the reactions also proceed by interaction with the transition metal hydride formed in the systems.

The C–O bond can be more easily activated when the CO molecule interacts with more than two metal atoms, as will be discussed in Section 9.3. It is known that a CO ligand interacting with four iron atoms has a considerably longer C–O bond than the usual type of terminal CO ligand. Multiple interaction with more than two metal atoms can also occur in the intermediate formyl complex. An illustrative case of C–O bond cleavage in a formyl complex has been reported.[71]

$$Cl_2(\eta^5\text{-}C_5Me_4Et)Ta(\mu\text{-}\eta^2\text{-HC–O})(\mu\text{-H})Ta(\eta^5\text{-}C_5Me_4Et)Cl_2 \xrightarrow{PMe_3} Cl_2(\eta^5\text{-}C_5Me_4Et)Ta(\mu\text{-}CH(PMe_3))(\mu\text{-O})(\mu\text{-H})Ta(\eta^5\text{-}C_5Me_4Et)Cl_2$$

Thus, there are several routes by which reactive carbene, μ-methylene, carbyne, and carbido species can be formed from the interaction of CO-coordinated species with H_2 or with hydride species formed from it. Hydrocarbon mixtures can be formed from surface CH_2 entities, as has been demonstrated by Pettit.[72] Analysis of the distribution of hydrocarbons formed on decomposition of a diazomethane-H_2 mixture on Ni, Pd, Fe, Co, Ru, and Cu surfaces as compared with the hydrocarbon distribution by the Fischer–Tropsch process led to the following conclusion. In the presence of hydrogen a surface methylene is transformed into a surface methyl that can initiate chain propagation through methylene insertion.

$$\overset{CH_2}{\wedge}\ \overset{CH_2}{\wedge}\ \overset{CH_3}{|} \longrightarrow \overset{CH_2CH_2CH_3}{|}$$

Because Fischer–Tropsch synthesis is a surface reaction, its mechanism is quite difficult to study, and it will remain a controversial subject of further research because of its obvious importance and complexity.[73]

Ethylene glycol can be prepared from a synthesis gas in a process in which the C–O bond is not ruptured. Because ethylene glycol is so useful, its selective synthesis has been the target of many industrial and governmental laboratories. Rh- and Ru-based catalysts have been extensively examined. The process developed by Union Carbide employs a rhodium cluster complex as a catalyst. A difficulty associated with the process is the severe operational conditions (210–250 °C and 500–3400 atm), but yields of ethylene glycol range up to 70%.

Spectroscopic studies of the catalytic system revealed the presence of $[Rh_{12}(CO)_{30}]^{2-}$. However, the possibility that a catalytically more reactive mononuclear or oligonuclear species is responsible for the reaction is difficult to exclude.

The mechanism of ethylene glycol synthesis is not clear, but it may be related to the reaction of a zirconium hydride with CO in which methanol and ethylene glycol are formed (See Chapter 6). In this reaction CO inserts into the Zr–H bond, and the resulting η^2-formyl group, which has a carbenelike character, undergoes dimerization.

$$2Zr\text{-}H \xrightarrow{CO} 2Zr\langle(\eta^2\text{-}C(H){=}O) \longrightarrow Zr\text{-}O(H)C{=}C(H)O\text{-}Zr \xrightarrow{H_2} HOCH_2CH_2OH$$

Another possibility is that formaldehyde is formed and is converted to a hydroxymethyl species. CO insertion into the M–C bond then produces a hydroxyacetyl species that liberates glycolaldehyde. Further reaction of the glycolaldehyde with the metal hydride in the presence of H_2 may lead to ethylene glycol.[74-76]

$$CH_2O + HM(CO)_n \longrightarrow HOCH_2M(CO)_n \xrightarrow{CO} HOCH_2C(=O)M(CO)_n$$

$$\xrightarrow{H_2} HOCH_2CHO \xrightarrow[H_2]{HM(CO)_n} HOCH_2CH_2OH \qquad (7.33)$$

In this connection a stoichiometric synthesis of glycolaldehyde from formaldehyde has been reported.[77]

$$CH_2O + HCo(CO)_4 \xrightarrow{CO} HOCH_2CHO \tag{7.34}$$

Although the formation of formaldehyde from synthetic gas is thermodynamically unfavorable, it has been argued that the concentration of formaldehyde permitted by thermodynamics is sufficient for it to exist as a transient intermediate.[76]

Alcohols also can be synthesized from synthesis gas.

$$CO + 2H_2 \longrightarrow CH_3OH \xrightarrow{CO/H_2} C_2H_5OH + H_2O \tag{7.35}$$

Although the process can be catalyzed by a homogeneous catalyst such as $Co_2(CO)_8$, copper-based heterogeneous catalysts produce methanol rapidly and cleanly at pressures below 100 atm. Homologation of methanol to mixtures of higher alcohols may be of more practical interest.[78]

At the present stage of development the selectivity in homologation is not particularly high. In homologation of methanol catalyzed by cobalt-based catalysts a variety of side products other than ethanol are formed. These include aldehydes, acetals, ethers, acetates, and acetic acid. However, the reaction mechanisms are not clear. In iodine-promoted catalyst systems, a likely process is the formation of methyl iodide by Eq. (7.31). Oxidative addition of CH_3I to Co(I) to give a methylcobalt(III) species and ensuing CO insertion into the Co–CH_3 bond provide an acetylcobalt species. If H_2 oxidatively adds after reductive elimination of I_2, an acetyl-hydrido species may be formed. Reductive elimination of acetyl and hydrido ligands liberates acetaldehyde as the primary product. Ethanol is considered to be formed by hydrogenation of acetaldehyde. Further studies are required to establish the mechanism.

Vinyl acetate is another target of industrial importance. A route to vinyl acetate starting from methyl acetate and synthesis gas has been developed by the Halcon Corporation. Since methyl acetate can be produced from CO and H_2, the vinyl acetate can be regarded as derived solely from synthesis gas. The process involves formation of ethylidene diacetate, which is converted to vinyl acetate and acetic acid in high yields by passage over a metal oxide catalyst at high temperature.

$$CH_3OAc \xrightarrow{CO/H_2} CH_3CH(OAc)_2 \xrightarrow[\Delta]{-HOAc} H_2C{=}CHOAc \tag{7.36}$$

The reaction of methyl acetate with synthesis gas is catalyzed by a rhodium or palladium salt in combination with an iodide.[79] For example, a 1 : 2 CO–H_2 mixture at 150 °C and about 140 atm reacts with methyl acetate to give $CH_3CH(OAc)_2$, the process being catalyzed by $RhCl_3 \cdot 3H_2O$ and methyl iodide in the presence of 3-picoline. The catalyst system resembles the one used for the Monsanto acetic acid process. Oxidative addition of methyl iodide to a Rh(I) species and subsequent CO insertion into the resulting CH_3–Rh bond may occur.[1a]

$$CH_3OAc \xrightarrow{HI} CH_3I \xrightarrow{[Rh]} CH_3RhI \xrightarrow{CO} CH_3CORhI$$

$$CH_3CORhI \begin{cases} \xrightarrow{AcOH} (Ac)_2O + [Rh] + HI \\ \xrightarrow{H_2} CH_3CHO + [Rh] + HI \end{cases} \tag{7.37}$$

$$CH_3CHO + (Ac)_2O \rightleftharpoons CH_3CH(OAc)_2$$

The mechanism in which oxidative addition of methyl iodide occurs to produce acetaldehyde as the primary product seems to have much in common with homologation of methanol by an iodide-promoted transition metal complex.

The water gas shift reaction is a process performed to increase the hydrogen content in a synthesis gas:

$$CO + H_2O \rightleftharpoons CO_2 + H_2 \tag{7.38}$$

The reaction is catalyzed heterogeneously, but mechanistic studies have been made using homogeneous transition metal complexes such as $[Rh(CO)_2I_2]^-$, $Ru_3(CO)_{12}$, $[HFe(CO)_4]^-$, and $Pt(PR_3)_3$.[80] The activity of these catalysts is too low to have any industrial significance, and the homogeneous catalyst systems suffer from the problem of separation; but some relevant information has been obtained about the key step of the reaction. The process involves nucleophilic attack on a coordinated CO ligand by water to give a hydroxycarbonyl complex that loses CO_2, thus causing reduction of the transition metal.

$$M^{III}{-}CO\ (\uparrow OH_2) \xrightarrow{-H^+} M{-}C(=O)OH \longrightarrow M^I + CO_2 + H^+ \tag{7.39}$$

The reduced species reacts with water to evolve H_2, and the metal is oxidized.

$$M^I + CO + 2H^+ \longrightarrow M^{III}(CO) + H_2 \tag{7.40}$$

The overall process of Eqs. (7.39) and (7.40) is equivalent to Eq. (7.38), which represents the catalytic water gas shift reaction. The reaction taking place on the surface of a solid catalyst most likely resembles the above processes.

e. Synthesis of Oxalates

The first process for the synthesis of oxalic acid and its esters in the petrochemical industry was the oxidation of ethylene glycol derived from ethylene. A new process to synthesize oxalates has been developed by Fenton[81,82] and its modified process has recently been successfully commercialized by Ube Chemical Industries in Japan.[83] The feasibility of further hydrogenation of oxalic acid to ethylene glycol is now being investigated. If successful, it would represent a complete reversal of the feedstock stream stimulated by the necessity to switch from petroleum-based feedstock to coal-based feedstock. The total process can be represented as follows:

$$2CO + 2ROH + \tfrac{1}{2}O_2 \xrightarrow{[Pd]} ROCOCOOR + H_2O \qquad (7.41)$$

The precise mechanism has not been established,[81] but it is likely that an important step is nucleophilic attack on a CO ligand coordinated to palladium to give an alkoxycarbonylpalladium complex. Subsequent reductive elimination of two alkoxycarbonyl ligands would afford oxalic ester.

$$PdCl_2(CO)_2 + 2RO^- \longrightarrow \left[Cl_2Pd(COOR)_2 \right]^{2-} \qquad (7.42)$$

$$\longrightarrow ROCOCOOR + Pd + 2Cl^-$$

Liberation of oxalic ester from $(Ph_3P)_2Pd(COOR)_2$ in the presence of CO has been established.[82]

A critical step in this reaction, as in the Wacker oxidation of ethylene to acetaldehyde catalyzed by a palladium compound, is reoxidation of Pd(0) to Pd(II). This can be performed with Cu(II) but is more conveniently done with alkyl nitrite in the Ube process.[83]

$$\begin{aligned} 2CO + 2BuONO &\xrightarrow{Pd(II)} (COOBu)_2 + 2NO \\ Pd(0) + RONO &\longrightarrow Pd(OR)_2 + 2NO \\ 2NO + \tfrac{1}{2}O_2 + 2BuOH &\longrightarrow 2BuONO + H_2O \end{aligned} \qquad (7.43)$$

Although the idea for the process stemmed from a homogeneous system, the commercialized Ube process utilizes a palladium catalyst supported on charcoal.

7.4. OXIDATION REACTIONS

Since many useful organic compounds contain oxygen atoms, finding ways to selectively produce oxygen-containing compounds is important.[84] Many known oxidation processes using air as an oxidant are not very selective and tend to produce a variety of side products that could cause environmental problems. Enhancement of selectivity is also desirable from the viewpoint of conserving the earth's energy and resources. The low selectivity in air oxidation processes arises from the nature of the radical reactions, which are difficult to control. If one could find appropriate means to control oxidation reactions by using transtion metal complexes, it would have enormous potential applications. Soluble metal complexes are used at present in many important catalytic oxidation reactions, such as the oxidation of *p*-xylene to terephthalic acid, which is the largest process using a homogeneous catalyst, producing over 2 million tons in 1977 in the United States alone. However, in most cases the metal complexes seem to be involved not in the *major* oxidation pathways but rather in the catalytic decomposition of hydroperoxides.[1a] Thus, in many oxidation processes organometallic complexes are not involved, and their treatment is outside the scope of this book. An outstanding exception is olefin oxidation catalyzed by Pd(II), which is discussed in this section. The other important process in which an organometallic intermediate may be involved is the synthesis of oxirans catalyzed by molybdenum compounds.

a. Oxidation of Olefins by the Wacker Process

The fact that ethylene can be oxidized to acetaldehyde by palladium chloride was observed as early as 1894 by Phillips.[85] The reaction is stoichiometric, and palladium chloride is reduced to palladium metal, which precipitates from the solution. In 1956 the process was combined with oxidation processes to catalytically convert ethylene to acetaldehyde.[86] The process, generally known as the Wacker process or the Hoechst-Wacker process, was invented in the laboratories of the Consortium für Elektrochemische Industrie, a branch of Wacker-Chemie. This was one of the major technological breakthroughs of the 1950s and led to the replacement of acetylene by ethylene as a major feedstock for the production of organic chemicals. Now the Wacker process is becoming obsolete, a reflection of the rapid pace of technological innovation, but mechanistic studies of palladium-catalyzed olefin oxidations have contributed greatly to our understanding of catalytic reactions promoted by palladium and its analogs. They have also stimulated the design of new processes based on concepts developed as a result of the studies.

The overall process can be represented as a combination of the following reactions.

$$
\begin{array}{rcl}
C_2H_4 + PdCl_2 + H_2O & \longrightarrow & CH_3CHO + Pd^0 + 2HCl \\
Pd^0 + 2CuCl_2 & \longrightarrow & PdCl_2 + 2CuCl \\
2CuCl + 2HCl + \frac{1}{2}O_2 & \longrightarrow & 2CuCl_2 + H_2O \\
\hline
C_2H_4 + \frac{1}{2}O_2 & \longrightarrow & CH_3CHO
\end{array}
\qquad (7.44)
$$

$$(\Delta H = -243 \text{ kJ mol}^{-1})$$

The success of this process depended on application of a known reoxidation process of the noble metal palladium by the cation of a less noble metal, Cu(II). The oxidation of Pd(0) to Pd(II) can be carried out at high Cl^- ion concentrations and allows the reusage of the precious metal in a catalytic sense.

There remains some disagreement regarding the details of the process, but an outline of the oxidation mechanism is represented in Figure 7.12.

The reaction is initiated by nucleophilic attack by water on ethylene coordinated to Pd(II). As discussed in Section 6.4, stereochemical studies employing deuterated ethylene have established that this attack by a water molecule or an OH^- ion takes place from outside (although internal attack may still occur under the usual operational conditions for the Wacker process). The external attack converts the π complex to a β-hydroxyethylpalladium species. A subsequent 1,2-hydrogen shift is followed by liberation of acetaldehyde and H^+ to give Pd(0), which is reoxidized and reenters the catalytic cycle.

The 1,2-shift probably proceeds by a β-hydrogen elimination mechanism. In reactions with D_2O, which give deuterated products, no H/D exchange with the solvent is observed. Thus, the possibility that vinyl alcohol is liberated from the $HOCH_2CH_2Pd$ species (**7-13**) and isomerizes to acetaldehyde can be excluded. If such a reaction did take place, the deuterium in D_2O should be found in acetaldehyde in the form of CH_2DCHO. Therefore, the 1,2-shift must occur rapidly in the coordination sphere of palladium. On the other hand, a vinyl alcohol–palladium complex corresponding to **7-14**, which may be formed by β-hydrogen elimination, was isolated and characterized in a different reaction.[87] A plausible reaction course is shown below.

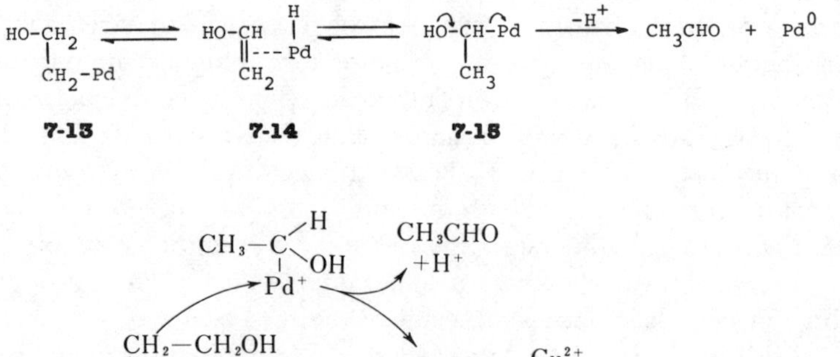

$$\underset{\textbf{7-13}}{HO\text{-}CH_2CH_2\text{-}Pd} \rightleftharpoons \underset{\textbf{7-14}}{(HO\text{-}CH{=}CH_2)\cdots Pd\text{-}H} \rightleftharpoons \underset{\textbf{7-15}}{HO\text{-}CH(CH_3)\text{-}Pd} \xrightarrow{-H^+} CH_3CHO + Pd^0 \qquad (7.45)$$

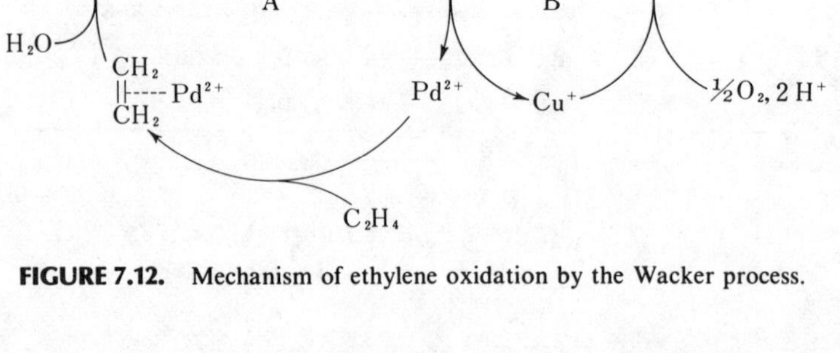

FIGURE 7.12. Mechanism of ethylene oxidation by the Wacker process.

After the 1,2-shift from **7-13** to the α-hydroxyethyl palladium complex **7-15**, deprotonation releases acetaldehyde and Pd(0). However, the exact course of the deprotonation in the actual aqueous system may not take the clear-cut route represented in Eq. (7.45). On the basis of kinetic and isotopic studies the formation of the hydroxyethylpalladium species **7-13** was found to be rate determining, and decomposition from the β-hydroxyethylpalladium species was suggested to proceed through a concerted process involving C–O double-bond formation and hydride shift.[88]

$$\left[\begin{array}{l} CH_2 — \overset{H}{C} = O \cdots H \\ ClPd \cdots H \end{array} \right] \longrightarrow CH_3CHO + HCl + Pd^\circ$$

The Pd(0) produced in cycle A (Fig. 7.12) is oxidized by Cu(II), which is reduced to Cu(I) in the redox process. The Cu(I) species is air oxidized to Cu(II) (cycle B in Fig. 7.12).

The Wacker process can be used to oxidize higher α-olefins, for example, propylene can be converted into acetone and 1-butene to methyl ethyl ketone. The rates and yields, however, decrease with increasing alkyl chain length.

The worldwide capacity of acetaldehyde production reached 2.6 million tons/year in 1978. The main commercial use of acetaldehyde is in the manufacture of acetic acid and acetic anhydride. However, the advent of the Monsanto process, which produces acetic acid by carbonylation of methanol, undermined the competitiveness of the Wacker process.

The ability to convert a terminal olefin into a ketone in one step has been applied to laboratory-scale organic synthesis of ketones.[89] Since internal olefins bind more weakly than terminal olefins to Pd(II), the rate of oxidation of terminal olefins is much greater than that of internal olefins. Therefore, one can selectively oxidize only a terminal double bond in a compound having both terminal and internal double bonds.[90]

COOR —86 %→ COOR

Cyclic olefins such as cyclohexene and cyclooctene are not oxidized by this method.

b. Acetoxylation of Olefins

Palladium compounds serve as catalysts for the production of vinyl acetate and allyl acetate. The reaction mechanisms in these processes seem to have features in common with that of the Wacker process. The process for vinyl acetate is based on the discovery by Moiseev that ethylene is stoichiometrically oxidized to vinyl acetate by $PdCl_2$ and sodium acetate in acetic acid.[91]

$$C_2H_4 + 2NaOAc + PdCl_2 \longrightarrow CH_2{=}CHOAc + 2NaCl + AcOH + Pd \qquad (7.46)$$

When coupled with the reoxidation of Pd(0) to Pd(II), this reaction can be made catalytic, like the Wacker process.

$$C_2H_4 + AcOH + \tfrac{1}{2}O_2 \xrightarrow{[Pd]} H_2C{=}CHOCOCH_3 + H_2O \quad (7.47)$$

$$(\Delta H = -176\ \text{kJ mol}^{-1})$$

Two companies built large plants using this principle but were forced to shut down because of corrosion problems. These problems were solved by the development of a heterogeneous catalyst that converts a gaseous mixture of ethylene, oxygen, and acetic acid vapor to vinyl acetate. The chemistry of this process seems to be similar to that in the liquid phase.

Ethylene coordinated to Pd(II) is considered to undergo external attack by AcO^- to give a β-acetoxyethylpalladium(II) complex (**7-16**) that on subsequent β-hydrogen elimination liberates vinyl acetate.

$$XPd\text{---}\|(CH_2{=}CH_2) + {}^-OAc \longrightarrow XPd\text{-}CH_2CH_2OAc\ (\textbf{7-16}) \xrightarrow{\beta\text{-H-elim.}} H(X)Pd\text{---}\|(CH(OAc){=}CH_2) \longrightarrow H_2C{=}CHOAc + [HPdX] \quad (7.48)$$

Stereochemical studies of nucleophilic attack on coordinated ethylene by nucleophiles such as MeO^- and AcO^- indicate that exo (anti) attack occurs[92] in agreement with the mechanism shown in Eq. (7.48). The unstable palladium hydride decomposes to metallic palladium unless an oxidant intercepts it.

Vinyl acetate is mainly used in the manufacture of poly(vinyl acetate). The world production capacity of vinyl acetate amounts to 2 million tons.

Use of propylene in place of ethylene in the above process affords allyl acetate.

$$H_2C{=}CHCH_3 + CH_3COOH + \tfrac{1}{2}O_2 \xrightarrow{[Pd]} H_2C{=}CHCH_2OOCCH_3 + H_2O \quad (7.49)$$

$$(\Delta H = -188\ \text{kJ mol}^{-1})$$

Depending on the reaction conditions, oxidation of butadiene by palladium salts gives a variety of products.[93]

c. Synthesis of Acrylates and Related Derivatives

When carried out in the presence of carbon monoxide, the palladium-promoted oxidative process of ethylene can be used to synthesize acrylic acid, an important material for the chemical industry. In the Union Oil process acrylic acid can be produced from ethylene and CO with a selectivity of 85% based on ethylene at 140–150 °C and 77 atm with a catalyst mixture similar to the one used in the Wacker process.[81c,94]

$$CH_2{=}CH_2 + CO + \tfrac{1}{2}O_2 \longrightarrow CH_2{=}CHCOOH \quad (7.50)$$

A possible mechanism involves nucleophilic attack on the CO ligand by OH^- and ethylene insertion to give a β-hydroxycarbonylethylpalladium complex that liberates acrylic acid.

$$Cl-\overset{Cl}{\underset{CO}{Pd}}-CO \xleftarrow{} OH^- \longrightarrow \left[Cl-\overset{Cl}{\underset{CO}{Pd}}-\overset{O}{\overset{\|}{C}}-OH \right]^- \underset{}{\overset{C_2H_4}{\rightleftharpoons}} \left[Cl-\overset{Cl}{\underset{CH_2=CH_2}{Pd}}-\overset{O}{\overset{\|}{C}}-OH \right]^-$$

$$\xrightarrow{CO} \left[Cl-\overset{Cl}{\underset{CH_2CH_2COOH}{Pd}}-CO \right]^- \longrightarrow CH_2=CHCOOH + Pd + CO + HCl + Cl^-$$

Further study is necessary to establish whether this mechanism is correct. Other products such as propionic acid, its ester, and succinic ester can be produced in this process by modification of the reaction conditions.[94]

Another route to acrylic acid is based on the Reppe process,[63] using acetylene, CO, and water:

$$HC{\equiv}CH + CO + H_2O \longrightarrow CH_2{=}CHCOOH$$

When alcohol is used, corresponding acrylic ester is formed. This process has been replaced by synthesis based on propylene oxidation in the United States but is still used elsewhere.

The preferred catalyst is based on nickel carbonyl. The reaction is believed to be initiated by formation of a nickel hydride. Once the hydride is formed, subsequent acetylene insertion into the Ni–H bond followed by CO insertion is a likely process:

$$H-\overset{|}{\underset{|}{Ni}}-CO + HC{\equiv}CH \longrightarrow H_2C{=}CH-\overset{|}{\underset{|}{Ni}}-CO \longrightarrow H_2C{=}CHCO\overset{|}{\underset{|}{Ni}}-$$

Nucleophilic attack on the acyl nickel complex by water gives acrylic acid. An alternative path is nucleophilic attack of water on the coordinated CO to afford a hydroxycarbonyl group, which may be reductively eliminated with the vinyl group to liberate acrylic acid. Another possibility is acetylene insertion into the hydroxycarbonyl–nickel bond to give a nickel–bonded hydroxycarbonylvinyl group that is protonated and liberates acrylic acid.

d. Olefin Epoxidation

The epoxidation of olefins[84] by hydroperoxides is an important nonradical process that is promoted homogeneously by a transition metal compound. The most important commercial process is the manufacture of propylene oxide developed by Halcon Corporation in 1967.[1a,95]

$$CH_3CH{=}CH_2 + ROOH \longrightarrow CH_3\underbrace{CH-CH_2}_{O} + ROH \qquad (7.51)$$

The process is likely to displace the older technology, which is based on dehydrochlorination of propylene chlorohydrin. In the Halcon process the propylene oxidation is homogeneously catalyzed by Mo(VI), W(VI), V(V), and Ti(IV) ions in decreasing activity, whereas a heterogeneous catalyst of Ti(IV) supported on SiO_2 is used in the Shell process.

The mechanism of the epoxidation of olefin still remains unsettled. Sharpless proposed the mechanism shown in Figure 7.13.[1a,96,97]

In this mechanism the molybdenum alkoxide [A] undergoes ester exchange with alkyl hydroperoxide to form the alkylperoxomolybdate [B]. Reaction of this complex with an olefin is assumed to take place via an intermediate or transition state [C], in which the Mo-bound peroxy oxygen atom transfers to the olefin. This transfer gives the epoxide, which is displaced by alcohol or hydroperoxide to complete the catalytic cycle.

Epoxide formation from an olefin may also be involved in monoxygenase activity of an enzyme called P-450 (cf. Section 9.1).

Concerning the intermediate participating in oxygen transfer to olefin, Mimoun[98] proposed a mechanism somewhat different from that shown in Figure 7.13. Rather than the intermediate [C] in the figure, Mimoun favors an intermediate in which the olefin coordinates to the metal and forms a peroxy pseudometallacycle. The metallacycle subsequently rearranges to liberate the epoxide and regenerates the alkoxide [A].

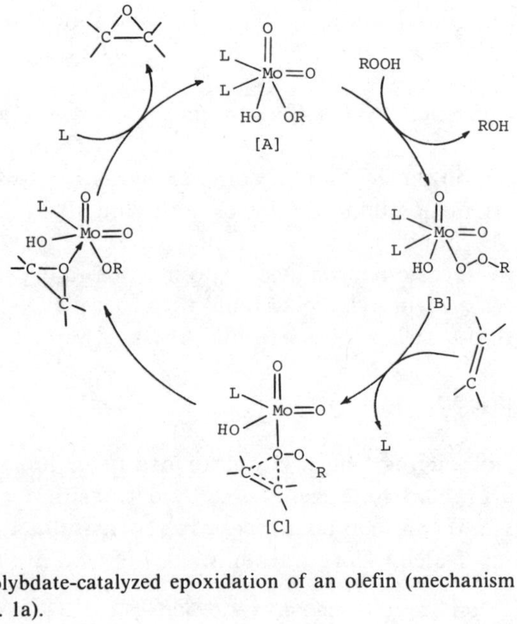

FIGURE 7.13. Molybdate-catalyzed epoxidation of an olefin (mechanism proposed by Sharpless, taken from ref. 1a).

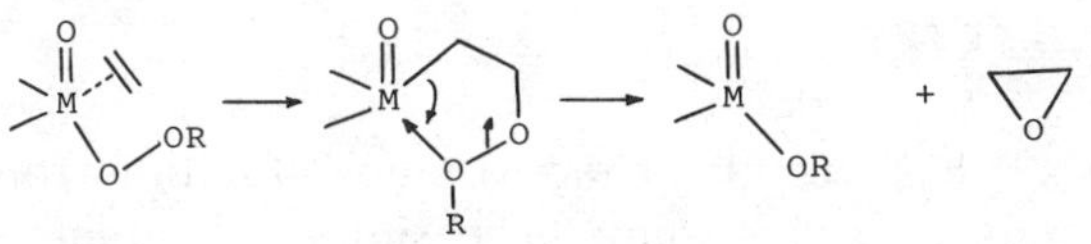

The pseudocyclic peroxymetallation assumed here resembles the formation of the five-membered dioxometallacycle observed in the reaction of the dioxygen–platinum complex with the electrophilic olefin (cf. complex **4-95**).

Another product in the oxidation of olefins catalyzed by transition metal complexes is ketone. Mimoun found the oxidation of terminal olefins to methyl ketones catalyzed by palladium and rhodium complexes. Palladium catalysts promote the oxidation of terminal olefins with *tert*-butyl hydroperoxide or H_2O_2 to methyl ketones, and rhodium catalysts serve for olefin oxidation utilizing molecular oxygen. Formation of the pseudocyclic five-membered intermediates is assumed as shown below for the palladium-catalyzed oxidation of olefins.

$ROO\text{-}Pd^{II} + R'CH{=}CH_2 \longrightarrow ROO\text{—}Pd \cdots R'CH{=}CH_2 \longrightarrow$

ROH, ROOH, RO-Pd, $R'COCH_3$

(R = tert-Bu or H)

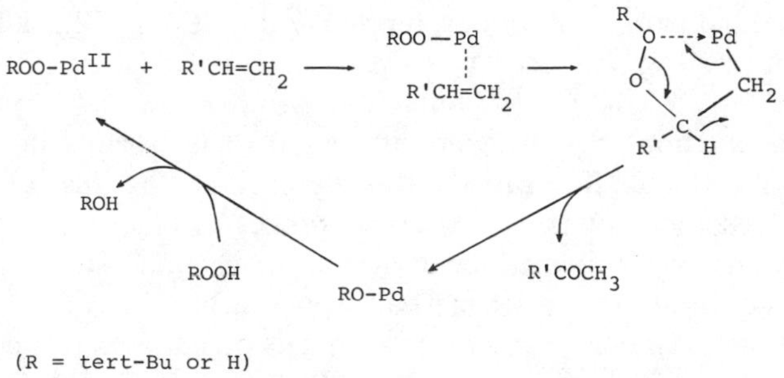

Collapse of the pseudocyclic peroxyalkyl intermediate via O–O bond rupture and β-hydrogen shift gives the methyl ketone and generates palladium alkoxide (or hydroxide). The reaction of the alkoxide with ROOH regenerates ROOPd(II), which carries the catalytic cycle. The rhodium-catalyzed oxidation of terminal olefins to methyl ketones by molecular oxygen is considered to involve the similar cyclic peroxymetallation but is somewhat more complicated by intervention of a second step of Wacker hydroxymetallation.

Although there remain some unsettled problems concerning the detailed mechanism, the metal-catalyzed oxidation of olefins has already found considerable synthetic utility.

Sharpless succeeded in developing a versatile synthetic method for stereospecific oxidation of various olefinic substrates.[97] Particularly useful is oxidation of allylic alcohols by employing the combination of *tert*-butyl hydroperoxide, titanium tetraisopropoxide, and chiral diethyl tartarate to give epoxides in 70–90% yields and over 90% optical yields.[99]

7.5. OLEFIN METATHESIS

Olefin metathesis[100] has recently attracted much attention from both the academic and the industrial viewpoints. The simplest example of olefin metathesis is shown below.

$$CH_3-CH{=}CH_2 + CH_2{=}CH-CH_3 \rightleftharpoons CH_3CH{=}CHCH_3 + CH_2{=}CH_2 \quad (7.52)$$

In olefin metathesis, or olefin disproportionation as it was called, the alkene carbon–carbon double bonds are broken and reformed to give a statistical distribution of the alkylidene entities. For this reason, it is sometimes called the transalkylidene reaction.

Both heterogeneous and homogeneous catalysts are used for olefin metathesis. The most frequently used metals in both homogeneous and heterogeneous catalysts are molybdenum, tungsten, and rhenium. Since olefin conversions can be performed in olefin metathesis, the process can be used in the petrochemical industry. The process represented by Eq. (7.52), called the Phillips Triolefin process, was used when propylene prices were low and butene was required as a feedstock for butadiene production.[101] The process is an equilibration reaction, but the volatility of ethylene permits its selective removal from the system. The process was discontinued because of the increasing cost of feedstock. In the commercial processes heterogeneous catalysts, commonly oxides of molybdenum, tungsten, or rhenium supported on silica or alumina, are used. The other method of preparing the supported catalysts is to deposit carbonyl complexes of these metals on surfaces of silica, alumina, or magnesia. For homogeneous catalysis Ziegler catalysts consisting of mixtures of transition metal compounds and alkylaluminum, magnesium, or lithium reagents are employed.

There has been much controversy about the mechanism of this intellectually challenging process. A once popular mechanism was a concerted process involving the quasicyclobutane intermediate shown below.

$$M + 2R-CH{=}CH_2 \rightleftharpoons \begin{matrix} R-CH-CH_2 \\ \boxed{M} \\ R-CH-CH_2 \end{matrix} \rightleftharpoons \begin{matrix} RCH \\ \| \\ RCH \end{matrix} + \begin{matrix} CH_2 \\ \| \\ CH_2 \end{matrix} + M$$

However, this clear-cut and intellectually appealing pairwise mechanism was abandoned because of its failure to explain several pieces of experimental evidence. Figure 7.14 shows the currently accepted mechanism based on the original proposal of Chauvin.[102]

This mechanism invokes the intermediacy of carbene and metallacycle complexes. A carbene complex (**7-17**) formed in the catalyst system binds an

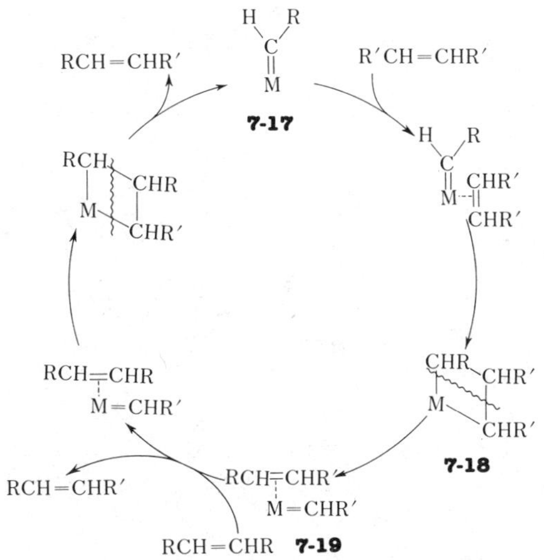

RCH=CHR′   H R C M 7-17   R′CH=CHR′

H R C M CHR′ CHR′

RCH CHR M CHR′   CHR CHR′ M CHR′ 7-18

RCH=CHR M=CHR′

RCH=CHR′   RCH=CHR′ M=CHR′   RCH=CHR 7-19

FIGURE 7.14. Mechanism of olefin metathesis.

olefin, R′CH=CHR′. Intramolecular cycloaddition yields a metallacyclobutane complex (**7-18**). Rupture of the metallacyclobutane ring either regenerates the starting complex or produces the new olefin-coordinated carbene complex **7-19**, as shown by the arrow. Displacement of the coordinated olefin by the new olefin RCH=CHR followed by subsequent metallacyclobutane formation and its rupture completes the catalytic cycle. If R=H and R′=CH_3 in Figure 7.14, the reaction corresponds to Eq. (7.52).

In a Ziegler catalyst the carbene species may be formed by α-hydrogen elimination in an alkyl.

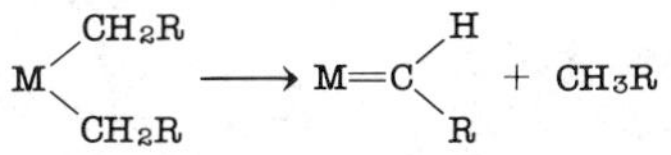

$$M(CH_2R)_2 \longrightarrow M{=}C(H)(R) + CH_3R$$

This type of mechanism, involving a carbene complex, may operate not only in olefin metathesis but also in the polymerization of cyclic alkenes, as discussed in Section 7.2d. In fact, the carbene complex $Ph_2C{=}W(CO)_5$ has been shown to act as a catalyst for the polymerization of cyclopentene, to undergo alkylidene exchange on interaction with the olefin $R_2C{=}CH_2$, and to afford a cyclopropane derivative, which probably arises by decomposition of a metallacyclobutane intermediate.

$$(OC)_5W{=}CPh_2 + R_2C{=}CH_2 \rightleftharpoons (OC)_4W({=}CPh_2)(\eta^2\text{-}R_2C{=}CH_2) \rightleftharpoons (OC)_4W{-}CPh_2{-}CH_2{-}CR_2 \text{ (metallacycle)} \longrightarrow \text{1,1-diphenyl-2,2-dialkylcyclopropane}$$

$$(OC)_4W({=}CPh_2{-}CH_2{-}CR_2) \rightleftharpoons (OC)_4W({=}CR_2)(\eta^2\text{-}Ph_2C{=}CH_2) \rightleftharpoons (OC)_5W{=}CR_2 + Ph_2C{=}CH_2 \quad (7.53)$$

Some carbene complexes are stabilized by forming a μ-methylene bridge with Lewis acids. This type of complex (e.g., Tebbe's complex, **7-20**) can be regarded as the precursor of a carbene complex. It has been established by studies using ^{13}C-labeled isobutylene that Tebbe's complex undergoes the alkylidene exchange reaction.[104]

$$Cp_2Ti(\mu\text{-}CH_2)(\mu\text{-}Cl)AlMe_2\ (\textbf{7-20}) + H_2C^*{=}CMe_2 \rightleftharpoons Cp_2Ti(\mu\text{-}C^*H_2)(\mu\text{-}Cl)AlMe_2 + H_2C{=}CMe_2 \quad (7.54)$$

Complex **7-20** also acts as an olefin metathesis catalyst, albeit of low activity.[105]

$$\text{(methylenecyclohexane)}\ CH_2{=}C_6H_{10} + {}^*CH_2{=}CMe_2 \xrightleftharpoons[51^\circ]{Cp_2Ti(\mu\text{-}CH_2)(\mu\text{-}Cl)AlMe_2} {}^*CH_2{=}C_6H_{10} + CH_2{=}CMe_2 \quad (7.55)$$

Complex **7-20** can be converted to a titanacyclobutane complex when it is treated with other olefins in the presence of pyridine [cf. Eq. (6.124)].

These reactions strongly support the idea that carbene and metallacyclobutane complexes are intermediates in olefin metathesis.

Acetylene metathesis can also be carried out by using a molybdenum- or tungsten-based catalyst.[106]

$$\text{p-tolyl}{-}C{\equiv}C{-}Ph \xrightarrow{[Mo]} PhC{\equiv}CPh + \text{p-tolyl}{-}C{\equiv}C{-}\text{tolyl-p}$$

A metallacyclobutadiene complex, which is a putative intermediate in acetylene metathesis, has been isolated by the reaction of a carbyne complex with an acetylene, and its structure has been established.[107]

$$(dme)Cl_3W{\equiv}CCMe_3 + RC{\equiv}CR \longrightarrow Cl_3W[C(CMe_3)C(R)C(R)]$$

dme = 1,2-dimethoxyethane R = Me

In a reaction that is closely related both to the Wittig reaction and to olefin metathesis, a carbene complex or its metallacyclobutane precursor reacts with a ketone to give an olefin in one step.[108]

$$(Me_3CCH_2)Ta{=}CHCMe_3 + \text{cyclohexanone} \longrightarrow \text{cyclohexylidene}{=}C(H)CMe_3$$

Olefin metathesis and related reactions involving carbene or metallacycle intermediates are a rapidly expanding field. Undoubtedly, more exciting novel reactions unique to organotransition metal complexes will be developed.

REFERENCES

1. For reviews, see: (a) G. W. Parshall, *Homogeneous Catalysis. The Applications and Chemistry of Catalysis by Soluble Transition Metal Complexes,* Wiley, New York, 1980; (b) A. Nakamura and M. Tsutsui, *Principles and Applications of Homogeneous Catalysis,* Wiley, New York, 1980; (c) M. M. Taqui Khan and A. E. Martell, *Homogeneous Catalysis by Metal Complexes,* Academic Press, New York, 1974; (d) C. Masters, *Homogeneous Transition-metal Catalysis: A Gentle Art,* Chapman and Hall, London, 1981; (e) G. Henrici-Olivé and S. Olivé, *Coordination and Catalysis,* Verlag Chemie, Weinheim, 1976; (f) J. P. Collman and L. S. Hegedus, *Principles and Applications of Organotransition Metal Chemistry,* University Science Books, Mill Valley, Calif., 1980; (g) K. Weissermal and H.-J. Arpe, *Industrielle Organische Chemie, Bedeutende Vor- und Zwischenprodukte,* Verlag Chemie, Weinheim, 1976; (h) L. H. Pignolet, ed., *Homogeneous Catalysis with Metal Phosphine Complexes,* Plenum, New York, 1983.
2. For reviews on coordination polymerization by transition metal catalysts, see: (a) J. Boor, *Ziegler–Natta Catalysts and Polymerizations,* Academic Press, New York, 1978; (b) H. Sinn and W. Kaminsky, *Adv. Organometal. Chem.,* **18,** 99 (1980); (c) J. C. W. Chien, ed., *Coordination Polymerization,* Academic Press, New York, 1975; (d) T. Keii, *Kinetics of Ziegler–Natta Polymerization,* Kodansha, Tokyo, 1972; (e) A. Yamamoto and T. Yamamoto, *Macromol. Rev.,* **13,** 161 (1978); (f) C. C. Price and E. J. Vandenberg, eds., *Coordination Polymerization,* Plenum Press, New York, 1983; (g) A. D. Caunt, *Catal.,* **1,** 234 (1977).
3. (a) K. Ziegler, *Angew. Chem.,* **67,** 543 (1955); (b) K. Ziegler, E. Holzkamp, H. Breil, and H. Martin, *Angew. Chem.,* **67,** 541 (1955).
4. (a) G. Wilke, in *Coordination Polymerization,* J. C. W. Chien, ed., Academic Press, New York, 1975, p. 11; (b) F. S. Dyachkovsky, in *Coordination Polymerization,* J. C. W. Chien, ed., Academic Press, New York, 1975, p. 199.
5. (a) G. Natta, P. Pino, G. Mazzanti, U. Guannini, E. Mantica, and M. Perado, *J. Polymer Sci.,* **26,** 120 (1957); (b) G. Natta, P. Pino, G. Mazzanti, and U. Giannini, *J. Inorg. Nucl. Chem.,* **8,** 612 (1958); (c) G. Natta, P. Corradini, and I. W. Bassi, *J. Am. Chem. Soc.,* **80,** 755 (1958); (d) D. G. H. Ballard and R. Pearce, *J. Chem. Soc. Chem. Commun.,* 621 (1975); (e) J. Holton, M. F. Lappert, G. R. Scollary, D. G. H. Ballard, R. Pearce, J. L. Atwood, and W. E. Hunter, *J. Chem. Soc. Chem. Commun.,* 425 (1976).
6. P. Cossee, *J. Catal.,* **3,** 80 (1964).
7. A. Clark, *Catal. Rev.,* **3,** 145 (1969).
8. Y. Takegami and T. Suzuki, *Bull. Chem. Soc. Jpn.,* **42,** 848, 1060 (1969); **43,** 1484 (1970).
9. T. Miyazawa and T. Ideguchi, *J. Polymer Sci.,* **B1,** 389 (1963).

10. (a) A. Zambelli and C. Tosi, *Fortschr. Hochpolym. Forsch.*, **15,** 31 (1974); (b) A. Zambelli and G. Allegra, *Macromolecules,* **13,** 42 (1980).
11. (a) E. J. Arlman, *J. Catal.*, **3,** 89 (1964); (b) E. J. Arlman and P. Cossee, *J. Catal.*, **3,** 99 (1964).
12. D. G. H. Ballard, *Adv. Catal.*, **23,** 263 (1973).
13. (a) G. T. Crisp, S. Holle, and P. W. Jolly, *Z. Naturforsch.*, **37b,** 1667 (1982); (b) H. Lehmkuhl, C. Naydowski, R. Benn, A. Rufinska, and G. Schroth, *J. Organometal. Chem.*, **246,** C9 (1983); (c) E. R. Evitt and R. Bergmann, *J. Am. Chem. Soc.*, **101,** 3973 (1979).
14. P. L. Watson, *J. Am. Chem. Soc.*, **104,** 337 (1982).
15. (a) K. J. Ivin, J. J. Rooney, C. D. Stewart, M. L. H. Green, and R. Mahtab, *J. Chem. Soc. Chem. Commun.*, 604 (1978); (b) M. L. H. Green and A. Mahtab, *J. Chem. Soc. Dalton Trans.*, 262 (1979); (c) M. L. H. Green, *Pure Appl. Chem.*, **50,** 27 (1978); (d) R. J. McKinney, *J. Chem. Soc. Chem. Commun.*, 490 (1980); (e) K. J. Ivin, *Olefin Metathesis,* Academic Press, New York, 1983.
16. (a) H. W. Turner, R. R. Schrock, J. D. Fellmann, and S. J. Holmes, *J. Am. Chem. Soc.*, **105,** 4942 (1983); (b) A. Zambelli, P. Locatelli, M. C. Sacchi, and E. Rigamonti, *Macromolecules,* **13,** 798 (1980). (c) J. Soto, M. L. Steigerwald, and R. H. Grubbs, *J. Am. Chem. Soc.*, **104,** 4479 (1982); (d) R. H. Grubbs, Plenary Lecture given at the 4th International Symposium on Homogeneous Catalysis, Leningrad, September 1984.
17. S. Carra, F. Parisi, I. Pasquon, and P. Pino, eds., *Giulio Natta, Present Significance of His Scientific Contribution,* Editrice di Chimica Srl., Milan, 1982.
18. B. Bogdanovic, *Adv. Organometal. Chem.*, **17,** 105 (1979).
19. P. W. Jolly and G. Wilke, *The Organic Chemistry of Nickel,* Vol. 2, Academic Press, New York, 1975.
20. B. Bogdanovic, H. Biserka, H. G. Karmann, H. G. Nüssel, D. Walter, and G. Wilke, *Ind. Eng. Chem.*, **62,** 34 (1970).
21. R. H. Grubbs and A. Miyashita, *J. Am. Chem. Soc.*, **100,** 1300, 2418, 7416 (1978).
22. (a) S. J. McLain and R. R. Schrock, *J. Am. Chem. Soc.*, **100,** 1315 (1978); (b) S. J. McLain, J. Sancho, and R. R. Schrock, *J. Am. Chem. Soc.*, **102,** 5610 (1980).
23. G. Natta, G. Dall'asta, and G. Mazzanti, *Angew. Chem.*, **76,** 765 (1964).
24. H. S. Eleuterio, U. S. Patent 3,074,918 (1963).
25. N. Calderon, *Acc. Chem. Res.*, **5,** 127 (1972).
26. (a) J. L. Herrison and Y. Chauvin, *Makromol. Chem.*, **141,** 161 (1970); (b) T. J. Katz, *Adv. Organometal. Chem.*, **16,** 283 (1977); (c) N. Calderon, E. A. Ofstead, and W. A. Judy, *Angew. Chem. Int. Ed. Engl.*, **15,** 401 (1976); (d) N. Calderon, J. P. Lawrence, and E. A. Ofstead, *Adv. Organometal. Chem.*, **17,** 449 (1979); (e) R. H. Grubbs, *Prog. Inorg. Chem.*, **24,** 1 (1978); (f) K. J. Ivin, *Olefin Metathesis,* Academic Press, London, 1983.
27. T. J. Katz, S. J. Lee, and N. Acton, *Tetrahedron Lett.*, 4247 (1976).
28. W. Reppe, O. Schlichting, K. Klager, and T. Topel, *Ann.*, **560,** 1 (1948).
29. (a) K. P. C. Vollhardt, *Acc. Chem. Res.*, **10,** 1 (1977); (b) R. L. Funk and K. P. C. Vollhardt, *J. Am. Chem. Soc.*, **101,** 215 (1979).
30. N. Hagihara, *Nippon Kagaku Zasshi,* **73,** 237 (1952).
31. (a) G. N. Schrauzer, *Chem. Ber.*, **94,** 1403 (1961); (b) *Adv. Organometal. Chem.*, **2,** 94 (1965).
32. J. P. Collman, *Acc. Chem. Res.*, **1,** 136 (1968).
33. G. Wilke, *Pure Appl. Chem.*, **50,** 677 (1978).
34. (a) H. Yamazaki and Y. Wakatsuki, *J. Organometal. Chem.*, **139,** 157, 169 (1977); (b) Y. Wakatsuki, K. Aoki, and H. Yamazaki, *J. Am. Chem. Soc.*, **101,** 1123 (1979).
35. (a) Y. Wakatsuki and H. Yamazaki, *Synthesis,* 26 (1976); (b) *J. Chem. Soc. Dalton Trans.*, 1278 (1978).

36. H. Bönnemann, *Angew. Chem. Int. Ed., Engl.*, **17,** 505 (1978); (b) H. Bönnemann and W. Brijoux, *Aspects of Homogeneous Catalysis,* Vol. 5, D. Reidel, Dordrecht, 1984.

37. H. Shirakawa and S. Ikeda, *J. Polymer Sci., Polymer Chem. Ed.*, **12,** 1924 (1974).

38. S. Ikeda, *Kogyo Kagaku Zasshi,* **70,** 1880 (1967).

39. P. M. Maitlis, *Acc. Chem. Res.*, **9,** 93 (1976).

40. (a) P. Heimbach, P. Jolly, and G. Wilke, *Adv. Organometal. Chem.*, **8,** 29 (1970); (b) J. Kiji, K. Yamamoto, S. Mitani, Y. Soshikawa, and J. Furukawa, *Bull. Chem. Soc. Jpn.*, **46,** 179 (1973).

41. G. Wilke, *J. Organometal. Chem.*, **200,** 349 (1980).

42. (a) E. J. Smutny, *J. Am. Chem. Soc.*, **89,** 6793 (1967); (b) S. Takahashi, T. Shibano, and N. Hagihara, *Bull. Chem. Soc. Jpn.*, **41,** 254 (1968); (c) *Kogyo Kagaku Zasshi,* **72,** 184 (1969).

43. (a) J. Tsuji, *Acc. Chem. Res.*, **6,** 8 (1973); (b) *Adv. Organometal. Chem.*, **17,** 141 (1979).

44. R. Baker, *Chem. Rev.*, **73,** 487 (1973).

45. (a) T. Saito, Y. Uchida, A. Misono, A. Yamamoto, K. Morifuji, and S. Ikeda, *J. Organometal. Chem.*, **6,** 572 (1966); (b) S. Otsuka and K. Taketomi, *Europ. Polymer J.*, **2,** 289 (1966).

46. (a) G. Allegra, F. L. Guidice, G. Natta, U. Giannini, G. Fagherazzi, and P. Pino, *J. Chem. Soc. Chem. Commun.*, 1263 (1967); (b) G. Natta, V. Giannini, P. Pino, and A. Cassasa, *Chem. e Ind.*, **47,** 524 (1965).

47. A. C. L. Su, *Adv. Organometal. Chem.*, **17,** 269 (1978).

48. T. Matsumoto and J. Furukawa, *J. Polymer Sci.*, **B6,** 896 (1968); **B7,** 541 (1969).

49. V. I. Klepikova, G. P. Kondratenkov, V. A. Kormer, M. I. Lobach, and L. A. Churlyaeva, *J. Polymer Sci., Polymer Lett. Ed.*, **11,** 193 (1973); V. A. Kormer and M. I. Lobach, *Macromolecules,* **10,** 572 (1977); R. Warrin, P. Teyssiée, P. Bourdaudurq, and F. Dawans, *J. Polymer Sci., Polymer Lett. Ed.*, **11,** 177 (1973).

50. (a) Ph. Teyssiée, *Fund. Res. Homog. Catal.*, **3,** 107 (1979); (b) J. Furukawa, *Acc. Chem. Res.*, **13,** 1 (1980).

51. S. M. Atlas and H. M. Mark, *Cat. Rev.*, **13,** 1 (1976).

52. S. Otsuka and A. Nakamura, *Polymer Lett.*, **5,** 973 (1967).

53. (a) J. Falbe, ed., *New Syntheses with Carbon Monoxide,* Springer-Verlag, Berlin, 1980; (b) I. Wender and P. Pino, *Organic Syntheses Via Metal Carbonyls,* Wiley-Interscience, New York, Vol. 1, 1968, Vol. 2, 1977; (c) H. Falbe, *Carbon Monoxide in Organic Synthesis,* Springer-Verlag, 1970; (d) C. Masters, *Adv. Organometal. Chem.*, **17,** 61 (1979); (e) R. P. A. Sneeden, in *Comprehensive Organometallic Chemistry,* Vol. 8, Pergamon Press, Oxford, 1982, p. 19; (f) I. Tkatchenko, in *Comprehensive Organometallic Chemistry,* Vol. 8, Pergamon Press, Oxford, 1982, p. 101; (g) W. Keim, ed., *Catalysis in C_1 Chemistry,* D. Reidel Publishing Co., Dordrecht, 1983; (h) R. L. Pruett, *Adv. Organometal. Chem.*, **17,** 1 (1979); (i) P. C. Ford, ed., *ACS Symposium Series,* **152,** *Catalytic Activation of Carbon Monoxide,* 1981; (j) M. A. Vanice, *Catal. Rev. Sci. Eng.*, **14,** 153 (1976); (k) V. Ponec, *Catal. Rev. Sci. Eng.*, **18,** 151 (1978); (l) G. Henrici-Olivé and S. Olivé, *Angew. Chem., Int. Ed. Engl.*, **15,** 136 (1976); (m) E. L. Muetterties and J. Stein, *Chem. Rev.*, **79,** 479 (1979).

54. O. Roelen, Ger. Pat. 949548 (1938); O. Roelen, *Angew. Chem. A.*, **60,** 62 (1948).

55. R. F. Heck and D. S. Breslow, *J. Am. Chem. Soc.*, **83,** 4023 (1961).

56. N. H. Alemaroğlu, J. M. L. Renninger, and E. Oltay, *Monatsh. Chem.*, **107,** 1043 (1976).

57. R. L. Pruett and J. A. Smith, *J. Org. Chem.*, **83,** 327 (1961).

58. C. Tolman, *Chem. Rev.*, **77,** 313 (1977).

59. C. K. Brown and G. Wilkinson, *J. Chem. Soc. (A),* 2753 (1970) and references cited therein.

60. A. Stefani, G. Consiglio, C. Botteghi, and P. Pino, *J. Am. Chem. Soc.*, **95**, 6504 (1973).
61. (a) P. Pino and G. Consiglio, *Fund. Res. Homog. Catal.*, **3**, 519 (1979); (b) A. Stefani, D. Tatone, and P. Pino, *Helv. Chim. Acta.*, **59**, 1639 (1977); (c) P. Pino, G. Consiglio, C. Botteghi, and C. Salomon, *Adv. Chem. Ser.*, **132**, 295 (1974).
62. B. Fell and H. Bahrmann, *J. Mol. Catal.*, **2**, 211 (1977).
63. W. Reppe, *Liebigs Ann. Chem.*, **582**, 1 (1953).
64. R. J. Angelici, *Acc. Chem. Res.*, **5**, 535 (1972).
65. D. Milstein and J. L. Huckaby, *J. Am. Chem. Soc.*, **104**, 6150 (1982).
66. (a) H. Koch and W. Giefert, *Brennstoff Chem.*, **36**, 321 (1955); (b) Y. Souma, J. Iyoda, and H. Sano, *J. Org. Chem.*, **38**, 2016 (1973); (c) Y. Souma and H. Sano, *Bull. Chem. Soc. Jpn.*, **46**, 3237 (1973); (d) Y. Souma and H. Sano, *Shokubai (Catalyst)*, **23**, 48 (1981).
67. (a) D. Forster, *Adv. Organometal. Chem.*, **17**, 255 (1979); (b) D. Forster, *J. Chem. Soc. Dalton Trans.*, 1639 (1979); (c) T. Mizoroki, T. Matsumoto, and A. Ozaki, *Bull. Chem. Soc. Jpn.*, **52**, 479 (1979).
68. (a) J. F. Roth, J. H. Craddock, A. Hershman, and F. F. Paulik, *Chemtech*, **1**, 600 (1971); (b) J. F. Roth, from address given at Utah University, November 6, 1980; (c) J. F. Roth, *J. Organometal. Chem.*, **279**, 1 (1985).
69. R. L. Pruett, *Ann. N. Y. Acad. Sci.*, **295**, 239 (1977).
70. P. T. Wolczanski and J. E. Bercaw, *Acc. Chem. Res.*, **13**, 121 (1980).
71. (a) M. R. Churchill and H. J. Wasserman, *J. Chem. Soc. Chem. Commun.*, 274 (1981); (b) P. Belmonte, R. R. Schrock, M. R. Churchill, and W. J. Young, *J. Am. Chem. Soc.*, **102**, 2858 (1980); (c) P. A. Belmonte, F. G. N. Cloke, and R. R. Schrock, *J. Am. Chem. Soc.*, **105**, 2643 (1983); (d) M. R. Churchill and H. J. Wasserman, *Inorg. Chem.*, **21**, 226 (1982); M. R. Churchill and W. J. Young, *Inorg. Chem.*, **20**, 382 (1981).
72. (a) R. C. Brady and R. Pettit, *J. Am. Chem. Soc.*, **102**, 618 (1980); (b) **103**, 1287 (1981).
73. G. Henrici-Olivé and S. Olivé, *J. Mol. Catal.*, **16**, 111 (1982) raised a question regarding the Pettit mechanism.
74. W. A. Herrmann, *Angew. Chem. Int. Ed. Engl.*, **21**, 117 (1982).
75. Ch. K. Rofer-DePoorter, *Chem. Rev.*, **81**, 447 (1981).
76. D. R. Fahey, *J. Am. Chem. Soc.*, **103**, 136 (1981).
77. J. A. Roth and M. Orchin, *J. Organometal. Chem.*, **172**, C27 (1979).
78. M. Röper and H. Loevenich, in *Catalysis in C_1 Chemistry*, W. Keim, ed., D. Reidel, Dordrecht, 1983.
79. (a) N. Rizkalla and C. N. Winnick, *Ger. Offenleg.*, 2,610,035 (1976); (b) J. L. Ehrler and B. Juran, *Hydrocarbon Proc.*, February, 109 (1982).
80. (a) R. M. Laine, R. G. Rinker, and P. C. Ford, *J. Am. Chem. Soc.*, **99**, 252 (1977); (b) P. C. Ford, ed., *ACS Symposium Series*, **152**, 79, 95, 107, 123 (1981).
81. (a) D. M. Fenton and P. J. Steinwand, *J. Org. Chem.*, **37**, 2034 (1972); (b) **39**, 701 (1974); (c) D. M. Fenton and K. L. Olivier, *Chemtech.*, 220 (1972).
82. (a) F. Rivetti and U. Romano, *J. Organometal. Chem.*, **174**, 221 (1979); (b) *Chimia e Industria (Milan)*, **62**, 7 (1980); (c) *J. Organometal. Chem.*, **154**, 323 (1978).
83. (a) S. Uchiumi, *Shokubai*, **23**, 477 (1981); (b) G. Nawata and H. Miyazaki, *Yuki Gosei Kagaku Kyokaishi (J. Synth. Org. Chem. Jpn)*, **41**, 545 (1983); (c) S. Uchiumi and M. Yamashita, *J. Japan Petrol. Inst.*, **25**, 197 (1982).
84. (a) R. A. Sheldon and J. K. Kochi, *Metal-Catalyzed Oxidation of Organic Compounds*, Academic Press, New York, 1981; (b) *Adv. Catal.*, **25**, 272 (1976); (c) R. A. Sheldon, *J. Mol. Catal.*, **20**, 1 (1983).
85. F. C. Phillips, *Am. Chem. J.*, **16**, 255 (1894).

86. (a) R. Jira and W. Freiesleben, *Organometal. React.*, **3,** 1 (1972); (b) R. Jira, W. Blau, and D. Grimm, *Hydrocarbon Proc.*, March, 97 (1976); (c) J. Smidt, W. Hafner, R. Jira, J. Seldmeier, R. Sieber, R. Rütlinger, and H. Kojer, *Angew. Chem.*, **71,** 176 (1959); **74,** 93 (1962); (d) J. Smidt, *Chem. Ind. (London),* 54 (1962).

87. J. Hillis, J. Francis, M. Ori, and M. Tsutsui, *J. Am. Chem. Soc.*, **96,** 4800 (1974).

88. (a) P. M. Henry, *J. Am. Chem. Soc.*, **86,** 3246 (1964); (b) P. M. Henry, *J. Org. Chem.*, **38,** 2415 (1973); (c) M. Kasaki, M. Isemura, Y. Kitaura, S. Shinoda, and Y. Saito, *J. Mol. Cata.*, **2,** 351 (1977).

89. J. Tsuji, *Organic Syntheses with Palladium Compounds,* Springer-Verlag, Berlin, 1980.

90. J. Tsuji, M. Kaito, T. Yamada, and T. Mandai, *Bull. Chem. Soc. Jpn.*, **51,** 1915 (1978).

91. I. I. Moiseev, M. N. Vargaftik, and Y. K. Syrkin, *Dokl. Akad. Nauk SSSR,* **133,** 377 (1960).

92. (a) T. Majima and H. Kurosawa, *J. Chem. Soc. Chem. Commun.*, 610 (1977); (b) H. Kurosawa, T. Majima, and N. Asada, *J. Am. Chem. Soc.*, **102,** 6996 (1980).

93. (a) J. Tsuji, *Acc. Chem. Res.*, **6,** 8 (1973); *Adv. Organometal. Chem.*, **17,** 141 (1979).

94. K. L. Olivier, D. M. Fenton, and J. Biale, *Hydrocarb. Process,* **51,** 95 (1972).

95. *Hydrocarb. Proc.*, **46,** 141 (April 1967); U.S. Patent, 3 360 584 (1967).

96. A. O. Chong and K. B. Sharpless, *J. Org. Chem.*, **42,** 1587 (1977).

97. (a) K. B. Sharpless, S. S. Woodard, and M. G. Finn, *Pure Appl. Chem.*, **55,** 1823 (1983). (b) K. G. Sharpless and T. R. Verhoeven, *Aldrichimica Acta,* **12,** 63 (1979); (c) K. B. Sharpless, C. H. Behrens, T. Katsuki, A. W. M. Lee, V. S. Martin, M. Takatani, S. M. Viti, F. J. Walker, and S. S. Woodard, *Pure Appl. Chem.*, **55,** 589 (1983).

98. (a) H. Mimoun, I. Sérée de Roch, and L. Sajus, *Tetrahedron,* **26,** 37 (1970); H. Mimoun, M. M. P. Machirant, and I. Sérée de Roch, *J. Am. Chem. Soc.*, **100,** 5437 (1978); (b) H. Mimoun, L. Saussine, E. Daire, M. Postel, J. Fischer, and R. Weiss, *J. Am. Chem. Soc.*, **105,** 3101 (1983); (c) H. Mimoun, *J. Mol. Chem.*, **7,** 1 (1980); (d) *Pure Appl. Chem.*, **53,** 2389 (1981); (e) *Angew. Chem. Int. Ed. Engl.*, **21,** 734 (1982).

99. (a) K. B. Sharpless, C. H. Behrens, T. Katsuki, A. W. M. Lee, V. S. Martin, M. Takatani, S. M. Viti, F. J. Walker, and S. S. Woodard, *Pure Appl. Chem.*, **55,** 589 (1983); (b) R. C. Michaelson, R. E. Palermo, and K. B. Sharpless, *J. Am. Chem. Soc.*, **99,** 1990 (1977). (c) T. Katsuki and K. B. Sharpless, *J. Am. Chem. Soc.*, **102,** 5974 (1980); (d) B. Rossiter, T. Katsuki, and K. B. Sharpless, *J. Am. Chem. Soc.*, **103,** 464 (1981); (e) S. Yamada, T. Mashiko, and S. Terashima, *J. Am. Chem. Soc.*, **99,** 1988 (1977).

100. (a) R. H. Grubbs, *Prog. Inorg. Chem.*, **24,** 1 (1979); (b) R. H. Grubbs, *Comprehensive Organometallic Chemistry,* Vol. 8, Pergamon Press, Oxford 1982, p. 499; (c) T. J. Katz, *Adv. Organometal. Chem.*, **16,** 283 (1977); (d) J. Halpern, *Organic Synthesis via Metal Carbonyls,* Vol. II, I. Wender and P. Pino, eds., Wiley, New York, 1977, p. 705; (e) N. Calderon, J. P. Lawrence, and E. A. Ofstead, *Adv. Organometal. Chem.*, **17,** 449 (1979); (f) C. P. Casey, *Chemtech,* 378 (1978); (g) K. J. Ivin, *Olefin Metathesis,* Academic Press, London, 1983.

101. *Hydrocarbon Proc.*, **46,** 232 (Nov. 1967).

102. (a) J. L. Herrison and Y. Chauvin, *Makromol. Chem.*, **141,** 161 (1970); (b) J. P. Soufflet, D. Commereuc, and Y. Chauvin, *Compt. Rend. Sec. C,* **276,** 169 (1973).

103. (a) C. P. Casey and T. J. Burkhardt, *J. Am. Chem. Soc.*, **95,** 5833 (1973); (b) **96,** 7808 (1974).

104. (a) F. N. Tebbe, G. W. Parshall, and G. S. Reddy, *J. Am. Chem. Soc.*, **100,** 3611 (1978); (b) F. N. Tebbe, G. W. Parshall, and D. W. Ovenall, *J. Am. Chem. Soc.*, **101,** 5074 (1979).

105. T. R. Howard, J. B. Lee, and R. H. Grubbs, *J. Am. Chem. Soc.*, **102,** 6876 (1980).

106. (a) S. Devarajan, D. R. M. Walton, and G. J. Leigh, *J. Organometal. Chem.*, **181,** 88

(1979); (b) A. Mortreux, J. C. Delgrange, M. Blanchard, and B. Lubochinsky, *J. Mol. Catal.*, **2,** 73 (1977).

107. S. F. Pedersen, R. R. Schrock, M. R. Churchill, and H. J. Wasserman, *J. Am. Chem. Soc.*, **104,** 6808 (1982).

108. (a) K. H. Dötz, *Angew. Chem. Int. Ed. Engl.*, **23,** 587 (1984); (b) K. A. Brown-Wensley, S. L. Buchwald, L. Canizzaro, L. Clawson, S. Ho, D. Meinhardt, J. R. Stille, D. Strauss, and R. H. Grubbs, *Pure Appl. Chem.*, **55,** 1733 (1983).

CHAPTER 8

Organic Synthesis Using Transition Metal Complexes

This chapter is principally concerned with laboratory scale organic synthesis. Larger-scale synthetic processes were discussed in Chapter 7. The distinction, however, is arbitrary, since it is often difficult to predict when a process that is useful in the laboratory may acquire industrial significance due to a demand in society. Conversely, certain industrial processes can often be applied conveniently to small-scale laboratory syntheses.

As discussed in Chapter 7, synthetic reactions utilizing transition metal complexes are characterized by high selectivity and mild reaction conditions. Until quite recently textbooks of organic chemistry only covered organometallic compounds of main-group elements, such as alkyllithium and Grignard reagents. Nowadays every synthetic chemist is aware of the usefulness of transition metal compounds. Synthetic methods involving transition metal compounds have advanced remarkably by keeping pace with fundamental research in organotransition metal chemistry. A compound that was once produced in low yield by conventional techniques with use of a multistep sequence of reactions can now often be prepared by a shorter route with high selectivity. The application of transition metal complexes to an organic synthesis can produce a "streamlining" of the overall process.

Reactions involving transition metal complexes consist of the elementary processes discussed in Chapter 6. Most of these processes, such as oxidative addition, reductive elimination, insertion, deinsertion, and external attack on coordinated ligands, proceed under mild reaction conditions. The reactions are often specific to the transition metal atom and can be controlled by the ligands associated with it. Thus, by the proper choice of a transition metal complex for a particular substrate, it is now possible to design novel synthetic reactions. Consequently, there has been a proliferation of new methods in the modern organic literature.

8.1. HYDROGENATION AND RELATED REACTIONS

Of all the reactions promoted by transition metal complexes, the hydrogenation of unsaturated compounds has attracted most attention.[1] It could even be argued that the attention is disproportionately high. A monograph containing almost 2000 references specifically related to hydrogenation appeared in 1973,[1a] and an updated review by the same author five years later quoted an additional 500 papers.[1b]

The earliest observation of the homogeneous activation of molecular hydrogen was reported by Calvin in 1938[2] and concerned the reduction of benzoquinone to hydroquinone. In the following year Iguchi discussed the hydrogenation of fumarate with rhodium complexes, but the significance of this work was not recognized until many years later.[3] Perhaps the most influential article, after Halpern and other researchers[1e] set the stage through their pioneering work for understanding the mechanism of olefin hydrogenation, is that of Wilkinson and his co-workers.[4] It appeared in 1965 after the phenomenon of reversible binding of molecular hydrogen and ethylene by Vaska's complex had been established in 1962.[5] Wilkinson's elegant work combined kinetic studies with physicochemical observations (especially NMR studies) of the interaction of small molecules with complexes, thus enabling fundamental steps such as hydrogen activation and olefin coordination to be studied. In contrast, mechanistic studies of the mode of action of heterogeneous catalysts are severely limited by the lack of such direct experimental evidence.

a. Basic Concept of Olefin Hydrogenation

Figure 8.1 shows the mechanism of olefin hydrogenation by Wilkinson's catalyst, $RhCl(PPh_3)_3$.[6] It is somewhat different in detail from that originally proposed by Wilkinson; side reactions in which dinuclear complexes are formed have been omitted for simplicity.

The complex $RhCl(PPh_3)_3$ dissociates one of its PPh_3 ligands L in solution to give complex [A], which has a coordinated solvent S. Oxidative addition of H_2 to [A] gives a dihydride complex [B]. The same dihydride can also be produced, though much less rapidly, by direct oxidative addition of H_2 to undissociated $RhCl(PPh_3)_3$, giving [E] and subsequent dissociation of triphenylphosphine. The octahedral complex [C] is formed by displacement of S from [B] by an olefin. Wilkinson originally proposed that in the next step the two hydride ligands attack the coordinated olefin in a concerted manner.[1c] However, it is now generally accepted that there is a two-step process in which the olefin inserts into the Rh–H bond to give the hydride–alkyl species [D], and alkane is reductively eliminated to regenerate [A]. The coordinatively unsaturated complex [A] reacts further with H_2 to continue the catalytic cycle. It should be noted that both the initial complex $RhCl(PPh_3)_3$ and [A], containing solvent S as a 2-electron donor, have a 16-electron configuration,

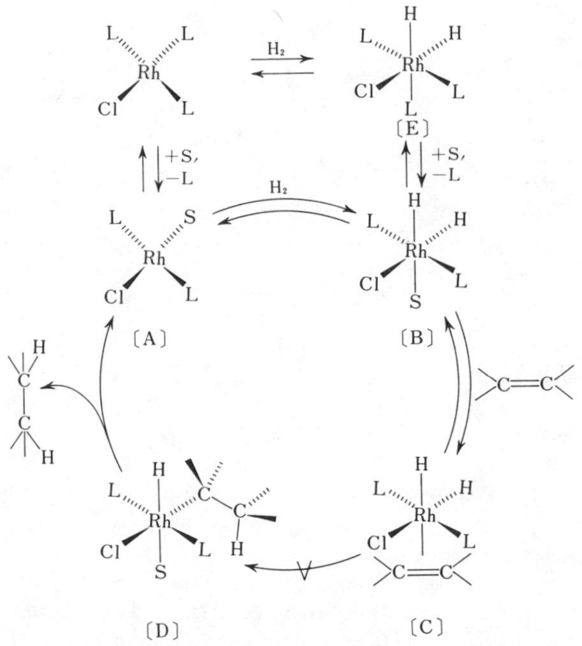

FIGURE 8.1. Mechanism of olefin hydrogenation by Wilkinson's complex.

whereas the octahedral complexes have an 18-electron configuration. The catalytic cycle thus involves coordinatively unsaturated d^8 square planar Rh(I) and coordinatively saturated d^6 octahedral Rh(III) species. In none of the intermediate species is the 18-electron configuration exceeded.

In this mechanism the first step is assumed to be addition of H_2 to [A]. However, the Rh(I) complex may first form a π complex with the olefin and then undergoes oxidative addition of H_2 to give the hydrido-olefin complex. This process does occur in some catalytic reactions. Which alternative is favored depends on the complex and its affinity for the olefin and H_2 and on the relative concentrations of the reactants. Figure 8.2 illustrates the generalized mechanisms of olefin hydrogenation taking both routes into consideration.

In cycle A, H_2 first oxidatively adds to a coordinatively unsaturated species L_nM, step (a). Complexation of an olefin in step (b) gives the dihydrido-olefin complex [M]. The ensuing olefin insertion into one of the two M–H bonds [step (e)] gives a hydrido-alkyl complex that on reductive elimination in step (f) liberates alkane with regeneration of the coordinatively unsaturated L_nM. This reacts further with H_2 to drive the catalytic cycle A in an anticlockwise direction. Figure 8.2 also illustrates cycle B, involving initial olefin complexation [step (c)] to L_nM. Molecular hydrogen oxidatively adds to the olefin complex in step (d) to give the dihydrido-olefin species [M] that undergoes

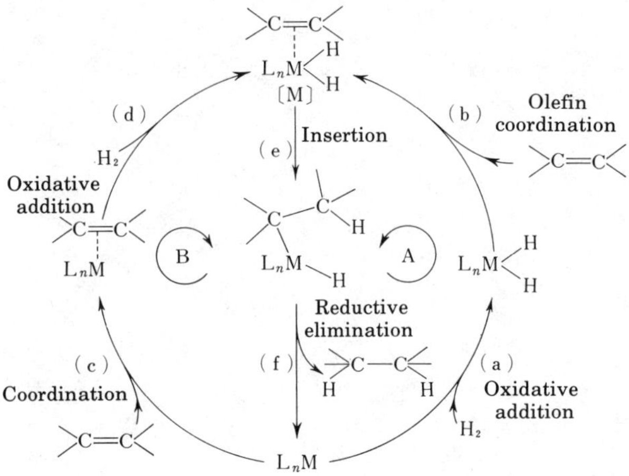

FIGURE 8.2. Two possible catalytic cycles in olefin hydrogenation with a transition metal complex catalyst.

olefin insertion and reductive elimination of alkane. This regenerates L_nM and thus drives the catalytic cycle B in a clockwise direction.

It is not easy in real catalytic systems to establish which cycle is operating, but pertinent information can often be obtained by carrying out kinetic studies on the individual elementary steps of the cycle and by examining the rates of addition of olefin and H_2.

The stereochemistry of the hydrogenation products of various olefins has been established as cis. This is consistent with a process in which the olefin coordinates initially in a side-on manner and undergoes migratory insertion into the M–H bond; the resulting hydrido-alkyl then eliminates alkane in a concerted manner.

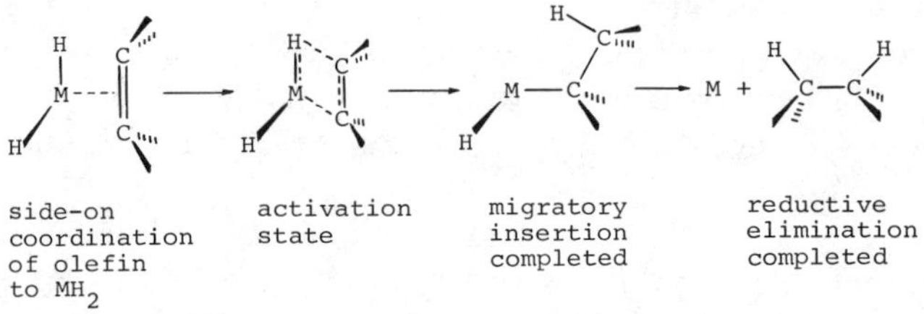

(8.1)

In addition to Wilkinson's catalyst, a variety of other hydrogenation catalysts have been reported, including $[Co(CN)_5]^{3-}$, a platinum–tin(II) chloride complex $[Pt(SnCl_3)_5]^{3-}$, combinations of various transition metal compounds with alkylaluminum compounds (Ziegler type catalysts), and various polynuclear transition metal complexes.

b. Asymmetric Hydrogenation

Many naturally occurring organic compounds, such as amino acids and carbohydrates, display optical activity with only one of the enantiomers that show physiological activity. For example, L-glutamic acid markedly enhances the taste of food, whereas D-glutamic acid does not. Proteins are composed of sequential combinations of L-amino acids only. The importance of asymmetric synthesis should be obvious from these few examples alone. Although extensively used in industry, heterogeneous catalysts are not as effective as homogeneous ones for asymmetric synthesis where optical yields of almost 100% have been achieved. An asymmetric hydrogenation with a rhodium complex having chiral phosphine ligands led to the first industrial process for the production of L-DOPA, L-dihydroxyphenylalanine, a compound that is active against Parkinson's disease.

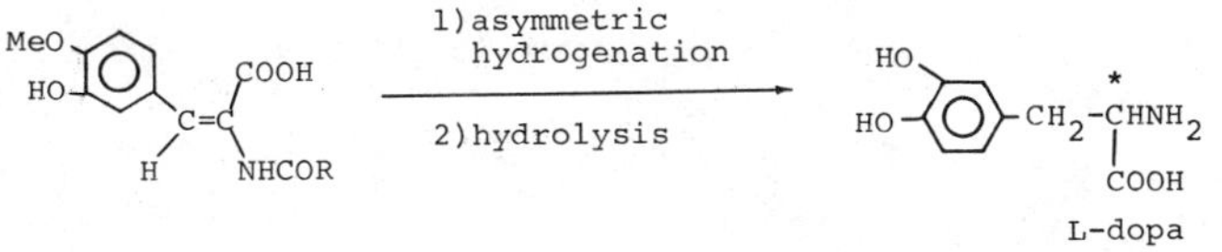

1) asymmetric hydrogenation
2) hydrolysis

L-dopa

A variety of chiral tertiary phosphine ligands have been prepared and examined for effectiveness in this reaction. Representative examples are shown in Figure 8.3.

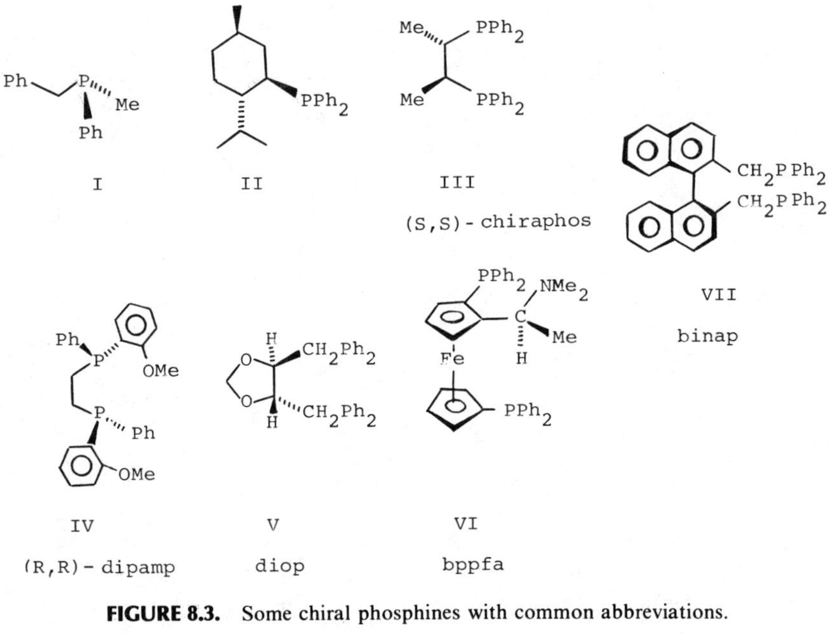

I

II

III

(S,S)-chiraphos

VII

binap

IV

(R,R)-dipamp

V

diop

VI

bppfa

FIGURE 8.3. Some chiral phosphines with common abbreviations.

The asymmetry of the monophosphines may be associated with the phosphorus atom, as in I, or with a substituent, as in II, or both. The bidentate phosphines may form metal chelates with five-membered rings (III and IV), seven-membered rings (V and VI), or even a nine-membered ring, as is the case with binap (VII), which has an axial chirality element.

In general, the unidentate chiral phosphines used in earlier work gave low and variable optical yields. The bidentate ditertiary phosphines are more rigidly coordinated to the metal and result in a larger constraint on the coordinated substrate.

The nature of the substrate also affects the stereoselectivity. Substrates in which a coordinating polar group is adjacent to the double bond usually give higher optical yields than do simple olefins. The α-*N*-acylaminoacrylic acids and their derivatives give particularly high optical yields because they combine with the metal through both the double bond and the carbonyl group.[7] Figure 8.4 illustrates coordination of (*Z*)-α-acetamidocinnamate with a rhodium complex containing a chiral bidentate diphosphine ligand (P-*-P).

On coordination of the olefin asymmetry is induced at the α carbon atom of the olefin so that two diastereomers (A and B) are formed. In the former the rhodium atom is bound to the si face and in the latter to the re face (for the conventions used in stereochemistry see ref. 8a). When (*S,S*)-chiraphos (see Fig. 8.3) was used in combination with ethyl (*Z*)-α-acetamidocinnamate, a rhodium complex corresponding to B in Figure 8.4 was isolated and its structure was determined by X-ray analysis. It was also shown by NMR spectroscopy to be the predominant species in solution.[9]

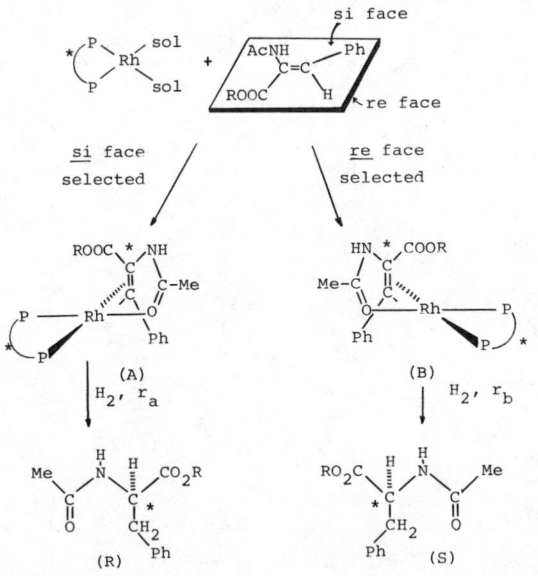

FIGURE 8.4. Two pathways in asymmetric hydrogenation of (*Z*)-α-acetamidocinnamate catalyzed by a rhodium complex having a chiral bidentate ligand.

It is expected that cis addition of H_2 to the re face of the olefin coordinated to the metal would yield *N*-acetyl-(*S*)-phenylalanine ester, as shown in Figure 8.4. However, it was found that the predominant product of hydrogenation with the rhodium complex having the (*S,S*)-chiraphos was the *R* isomer with over 95% enantiomeric excess. [The enantiomeric excess is a currently used measure of optical yield and is defined as the quantity $(X_R - X_S)/(X_R + X_S) \times 100$, where X_R and X_S are the relative quantities of *R* and *S* enantiomers. If 98% of one isomer and 2% of the other are produced, the enantiomeric excess (e.e.) is 96%.]

The result was interpreted by assuming that H_2 adds more rapidly to the minor diastereomer A than to the major diastereomer B ($r_a > r_b$). A similar conclusion was reached for the asymmetric hydrogenation of the same substrate using a rhodium complex containing the (*R,R*)-dipamp ligand, although in this case the chirality of the principal product was opposite to the product obtained with (*S,S*)-chiraphos. NMR investigations revealed that the minor diastereomer was hydrogenated at a much faster rate than the predominant diastereomer.[8i,10]

These results are schematically illustrated in Figure 8.5.

The figure shows that despite its smaller concentration the minor diastereomer will be hydrogenated more rapidly, provided the activation energy, $\Delta G^{\ddagger}_{minor}$, for H_2 addition to the minor diastereomer is much smaller than that for the major diastereomer.

The crucial question in asymmetric hydrogenation is the stereochemical arrangement of the chiral phosphine in the transition state for hydrogenation. Although this question is difficult to answer, the arrangement of thc diphosphine ligands in the substrate–catalyst adduct is known with considerable certainty. Figure 8.6 illustrates the preferred arrangement of the four phenyl rings in [Rh(*S,S*-chiraphos)]$^+$ and [Rh((*R,R*)-dipamp)]$^+$ viewed from the side of the rhodium atom, looking toward the chiral phosphine ligands, that is, the substrate approaches from the viewer's side.

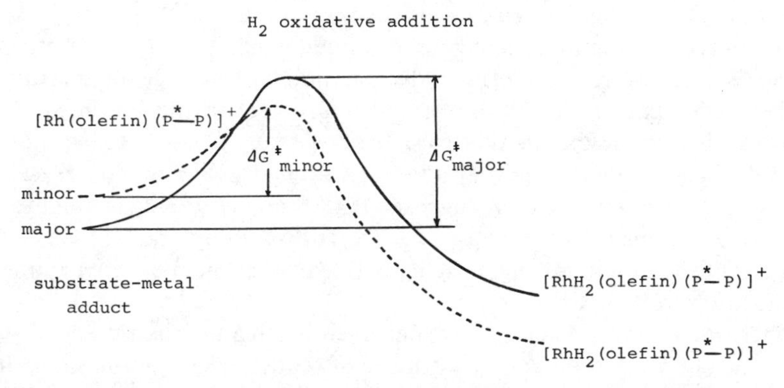

FIGURE 8.5. Schematic representation of control of asymmetric hydrogenation by a chiral rhodium catalyst. (From J. Halpern, *Science,* **217,** p. 406, copyright 1982 by the AAAS.)

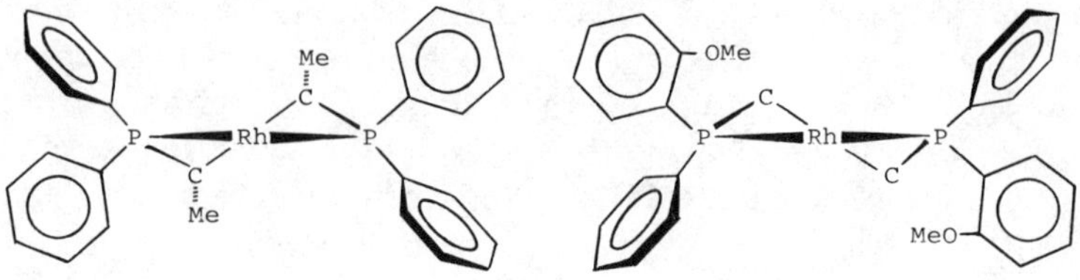

[Rh(*S*, *S*-chiraphos)]$^+$ [Rh(*R*,*R*-dipamp)]$^+$

FIGURE 8.6. View of coordination of chiral diphosphine ligands to a rhodium complex. Substrate approaches rhodium and is influenced by stereochemical arrangements of the phenyl and methoxyphenyl rings.

The figure shows that one of the two phenyl rings attached to each phosphorus atom has its edge pointing toward the coordinated substrate, whereas the other phenyl ring has a flat face directed toward the substrate. Thus, a chiral template will have more room where the flat face of the phenyl ring is oriented, whereas the edge arrangement will cause stronger steric repulsion. In addition to this steric effect, the flat face of the phenyl ring may have some electronic interaction with the substrate through its π-electron cloud. In each of the templates of the two chiral phosphines coordinated to the metal, an array of alternating "edge-face" phenyl groups can be seen. The difference in the orientation of the phenyl rings will lead to a preference in matching an enantioface of the substrate with the template provided by the particular chiral complex and will also affect the energy of the transition state. That [Rh((*S*,*S*)-chiraphos)]$^+$ gives predominantly the *R* amino acid ester under ambient conditions whereas [Rh((*R*,*R*)-dipamp)]$^+$ yields mainly the *S* enantiomer may reflect the opposite environments provided by the two chiral ligands.

The remarkable findings obtained in these studies warn us that the predominant species identified in a catalyst system may not be the one that carries the catalytic cycle but may represent a blind alley. This caution is particularly appropriate when the catalyst system is very active.

The more rigid chiral ditertiary phosphine ligands usually give better optical yields. However, for olefin hydrogenation an increase of coordination number accompanies the oxidative addition of H_2 to the intermediate substrate–catalyst complex. In this case ligands that are too rigid may not be desirable. The observation that the rate of hydrogenation decreases with increasing phosphine rigidity suggests that there is a balance between an acceptable hydrogenation rate and ligand rigidity, if high optical yields are to be accomplished by imposing a highly defined stereochemical environment on the substrate to be hydrogenated.[11]

Other functional groups are hydrogenated less readily than the C–C double bond, and there are relatively few studies of reductions of ketones, aldehydes, esters, amides, nitriles, and anhydrides. Aldehydes and ketones are reduced to alcohols under oxo conditions, that is, by catalytic reactions involving metal carbonyl hydrides. A cationic catalyst precursor of the type $[Rh(diene)L_2]^+$ has proved effective for the reduction of ketones. Part of the reason for the low activity of these substrates may be their weaker coordinating abilities to transition metals. The mechanism of reduction of these substrates is not clear, but it is likely that in the hydrogenation of a ketone the reaction proceeds by the insertion of the carbonyl group into the M–H bond that is generated by oxidative addition of H_2. The alkoxo–hydrido complex thus formed may then react with H_2 and reductively eliminate the product alcohol. (Fig. 8.7).

Alkoxides of early transition metals are well known, but they are not reactive. On the other hand, alkoxides of group 8–10 transition metals are limited to methoxides and phenoxides, which do not have β-hydrogen atoms.[12] Very recently, however, some stable alkoxo complexes of group 8 and 9 transition metals that have fluoroalkyl groups have been isolated by treatment of certain Co, Rh, and Ru hydrides with hexafluoroacetone, and their properties have been examined.[13] For fluorinated ketones insertion into the M–H bond to form alkoxo–hydrido complexes has been established. It was further demonstrated that treatment of the alkoxo–hydrido complex with H_2 caused the liberation of $(CF_3)_2CHOH$. Thus, the mechanism shown in Figure 8.7 seems quite plausible. In fact, $RuH_2(PPh_3)_4$ was shown to catalyze the hydrogenation of $CF_3COC_6H_5$.

Alcohols or tetrahydrofuran can also be used as hydrogen donors.[14] This method is sometimes more convenient experimentally than direct hydrogenation. Isopropyl alcohol is the reagent of choice, and the asymmetric hydrogenation of ketones in moderate optical yields has been achieved using chiral phosphine–rhodium or phosphine–iridium complexes.[15]

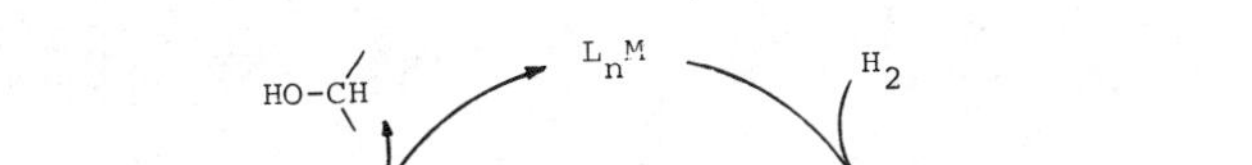

$$Me_2CHOH + Ph(R)C{=}O \xrightarrow[M\ =\ Rh,\ Ir]{[M(diene)P^*_2]^+} Ph(R)\overset{*}{C}H{-}OH + Me_2C{=}O$$

A reactive anionic ruthenium complex has been reported to reduce ketones and esters.[16] It is expected that more useful hydrogenation catalysts of ketones and other carbonyl compounds will be developed.

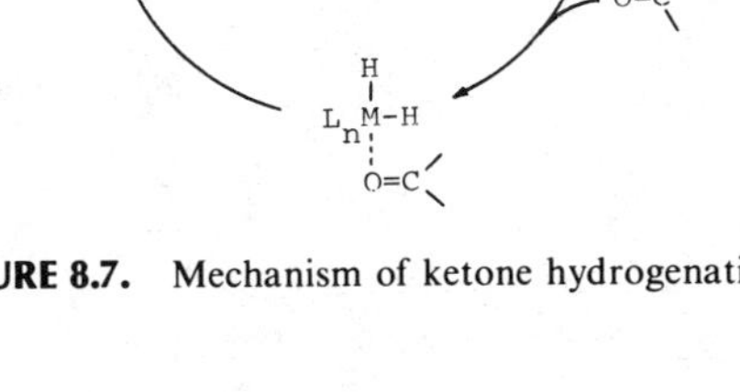

FIGURE 8.7. Mechanism of ketone hydrogenation.

For the hydrogenation of arenes, heterogeneous catalysts are much more active than homogeneous catalysts. Some homogeneous catalysts have been reported to be capable of reducing aromatic compounds, albeit with low catalytic activity. Ziegler catalysts composed of transition metal compounds and organoaluminum compounds,[17] a π-allylcobalt complex,[18] π-arene-rhodium[19] and π-arene-ruthenium complexes,[19,20] and a Rh(I) complex under phase transfer conditions[21] have been shown to be moderately active catalysts.

c. Hydrosilylation

The addition of a compound containing an Si–H bond to olefins, acetylenes, aldehydes, and ketones is called hydrosilylation,[22-27] and many transition metal complexes are known to catalyze this reaction.[22] Schematically, hydrosilylation can be regarded as analogous to hydrogenation, with one of the hydrogen atoms in H_2 being replaced by the SiR_3 group.

A variety of organosilicon compounds can be prepared by use of hydrosilylation. A representative reaction is shown below.

$$R_3SiH + \gt C{=}C\lt \xrightarrow{RhCl(PPh_3)_3} R_3Si{-}\overset{|}{\underset{|}{C}}{-}\overset{|}{\underset{|}{C}}{-}H \quad (8.2)$$

The mechanism of hydrosilylation probably resembles that of hydrogenation, as shown in Figure 8.8.

In this mechanism an olefin coordinates to a low-valent transition metal complex. The subsequent oxidative addition of R_3SiH to the resulting π complex with cleavage of the Si–H bond has a precedent in the reaction of $[IrCl(CO)(PPh_3)_2]$ with silanes to give a silyliridium(III) hydride [Eq. (8.3)].[23] This reaction is analogous to the well-known oxidative addition of H_2, which gives a dihydride.

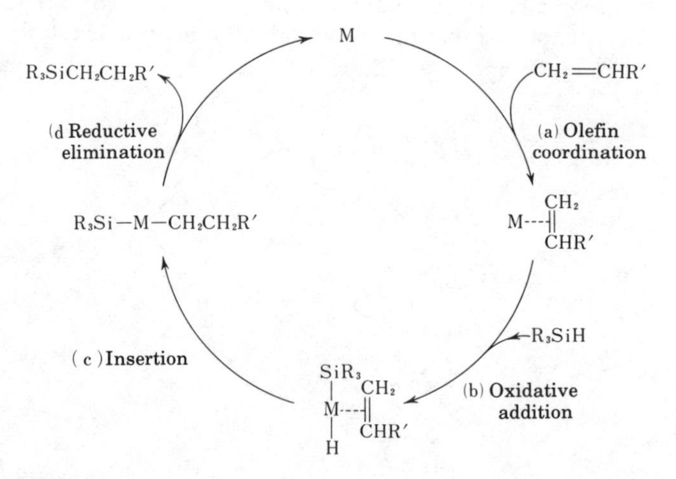

FIGURE 8.8. Mechanism of olefin hydrosilylation (ligands are omitted).

$$\text{trans-IrCl(CO)(PPh}_3)_2 + \text{R}_3\text{SiH} \longrightarrow \text{Ir(H)Cl(SiR}_3\text{)(CO)(PPh}_3)_2 \quad (8.3)$$

Insertion of the coordinated olefin into the M–H bond [step (c)] gives an alkyl–alkylsilyl complex that reductively eliminates the hydrosilylated derivative of the olefin. As an alternative to the mechanism shown in Figure 8.8, a catalytic cycle involving initial oxidative addition of R_3SiH to the metal followed by olefin insertion is also possible, as discussed previously for olefin hydrogenation.

In addition to its use in laboratory syntheses the hydrosilylation reaction has broad applications in the manufacture of silicone polymers. One practical application is the "curing" of a liquid silicone polymer that has an unsaturated functionality to a gum rubber or to a hard material suitable for dental use. In this toughening process polymer chains are cross-linked by reaction of the Si–H function of one chain with the vinyl group of another chain.

$$(\sim O)_2Si(R)H + H_2C{=}CH\text{–}Si(R)(O\sim)_2 \longrightarrow (\sim O)_2Si(R)\text{–}CH_2\text{–}CH_2\text{–}Si(R)(O\sim)_2$$

Various transition metal complexes can be used for hydrosilylation, but platinum compounds, such as chloroplatinic acid, $H_2PtCl_6 \cdot 6H_2O$ (sometimes called Speier's catalyst), are usually the catalysts of choice. In a comparative study of the activities of various catalysts for the hydrosilylation of ethylene with aryldimethylsilane, the rates using H_2PtCl_6, $RhCl(PPh_3)_3$, and $Co_2(CO)_8$ decreased in the ratio of 4000 : 200 : 1.[26]

An induction period is usually observed when H_2PtCl_6 is used. This period may correspond to the time required to reduce Pt(IV) to Pt(II) or Pt(0), the silane being the likely reducing agent. The induction period is reduced or eliminated altogether when lower-valent platinum complexes such as $[Pt(C_2H_4)Cl_2]_2$ and $Pt(C_2H_4)(PPh_3)_2$ are used.

Use of acrylonitrile as the olefin substrate in hydrosilylation gives an organosilicon compound that has a cyanoethyl group. Copolymerization of this monomer with more normal silicone monomers affords an oil-resistant silicone polymer with polar cyanoethyl groups on the backbone.

Ketones and aldehydes are readily hydrosilylated to give silyl ethers or alkoxysilanes. Since silyl ethers are readily hydrolyzed to alcohols, this route can be regarded as an alternative to the catalytic hydrogenation of ketones and aldehydes, which are more resistant to hydrogenation than are olefins.

$$R_3SiH + {>}C{=}O \xrightarrow{RhCl(PPh_3)_3} R_3Si\text{–}O\text{–}CH{<}$$
$$\xrightarrow{MeOH} R_3SiOMe + HOCH{<} \quad (8.4)$$

By using chiral tertiary phosphines, asymmetric hydrosilylation can be achieved.[27]

d. Hydrocyanation

Hydrogen cyanide does not add to unactivated olefins, but the addition is catalyzed by certain transition metal complexes. Oxidative addition of HCN to low-valent transition metal complexes is known.[28]

$$L_nM + HCN \rightleftharpoons L_nM\langle^{H}_{CN} \quad (8.5)$$

Olefin insertion into the resulting M–H bond gives an alkyl–cyano complex that reductively eliminates alkyl cyanide. This regenerates a coordinatively unsaturated species that again undergoes reaction (8.5), thus carrying the catalytic cycle. Reductive elimination of an organo nitrile from a phenylnickel cyanide complex has been demonstrated.[29a] It is assisted by triethyl phosphite, which may form a five-coordinate intermediate from which elimination can occur.[29b]

An important industrial application of this catalytic process is the two-step hydrocyanation of butadiene to adiponitrile, which is converted to hexamethylene diamine for use in the manufacture of Nylon 66.

HCN → CN; double bond migration ⇌ CN; HCN → NC–CN

(8.6)

This adiponitrile synthesis was commercialized by Du Pont. The HCN addition to the butadiene to give adiponitrile must be regiospecific, and the factors controlling the additions and any subsequent isomerization have been recently discussed.[30]

e. Isomerization of Olefins

Transition metal complexes can catalyze the migration of an olefinic double bond. Such double-bond migration is important in laboratory syntheses as well as in certain industrial processes. For example, some unsaturated steroids that are not readily accessible by other synthetic paths may be prepared from more readily available unsaturated steroids by double-bond migration. Double-bond isomerization may be one of key processes in the synthesis of adiponitrile, according to Eq. (8.6).

Two main mechanisms for double-bond migration have been recognized. One involves olefin insertion into a metal hydride with subsequent β-hydrogen elimination (Fig. 8.9*a*) and the other involves a 1,3-hydrogen shift as shown in Figure 8.9*b*.

The first mechanism is believed to operate in double-bond isomerizations catalyzed either by metal hydrides such as $[NiHL_4]^+$ and $RhHCl(PPh_3)_3$ or by metal hydrides that are generated in some way in the catalytic system.

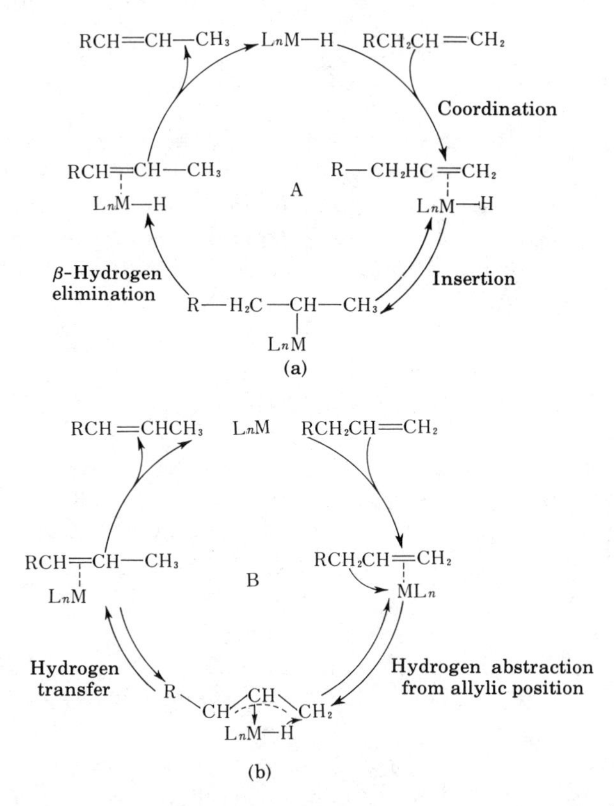

FIGURE 8.9. Mechanism of olefin isomerization. a: pathway through σ-alkyl intermediate, b: through a π-allyl intermediate.

In the first type of isomerization catalysis the coordinatively unsaturated species at 12 o'clock in Figure 8.9*a* may coordinate an olefin that then inserts into the M–H bond to give an alkyl complex at 6 o'clock. If β-hydrogen elimination from the methyl group occurs, the complex reverts to the starting complex (if the olefin or the metal hydride is deuterated, this would lead to H–D exchange). If the β-hydrogen at the methylene group is abstracted, however, an olefin whose terminal double bond has been moved internally is produced. Insertion of a new olefin molecule into the regenerated metal hydride drives the catalytic cycle forward.

In isomerizations involving a 1,3-hydrogen shift, it is believed that a π-allyl–hydride intermediate is formed, as shown in Figure 8.9*b*. Such a mechanism is applicable to catalysts that are not metal hydrides and that contain metals that readily form π-allyl complexes. This mechanism has been proposed for the double-bond isomerization of olefins catalyzed by $Fe_3(CO)_{12}$ and $PdCl_2(PhCN)_2$.[31,32] In this mechanism the coordinatively unsaturated metal complex abstracts a hydrogen at the allylic position to form the π-allyl–hydrido complex shown at 6 o'clock in Figure 8.9*b*. Transfer of the hydride ligand to the sterically less hindered terminal allylic carbon gives an olefin whose double bond has been shifted from the terminal to an internal position.

As discussed in Section 6.3, model systems demonstrating both types of behavior are known. The hydrido–ethylene complex $[CpRh(H)(C_2H_4)(PMe_3)]^+$ has been shown to be in equilibrium with the corresponding ethyl complex $[CpRh(C_2H_5)(PMe_3)]^+$,[33] and the presence of an equilibrium between a π-allyl–hydrido and a π-olefin complex has been observed spectroscopically for the following system.[34]

$$\pi\text{-}C_3H_5\text{Ni(H)}(PF_3) \underset{< -50^\circ C}{\overset{> -40^\circ C}{\rightleftharpoons}} (CH_2{=}CHCH_3)\text{Ni--}PF_3 \tag{8.7}$$

Olefins can also undergo skeletal isomerization, for example, the nickel-catalyzed conversion of *cis*-1,4-hexadiene to *trans*-2-methyl-1,3-pentadiene.[35] A complicated mechanism involving cleavage of the C(2)–C(3) bond has been proposed. An olefin metathesis reaction could also account for the observed reaction products.

8.2. C–C BOND FORMATION CATALYZED BY TRANSITION METAL COMPLEXES

Organic synthesis involves constructing target molecules by putting together building blocks and introducing functional groups. Thus, the development of methodology to form C–C bonds selectively and efficiently constitutes a pivotal problem in organic synthesis.[36,69]

To do this by means of a transition metal complex, it is necessary to prepare an active species with an M–C bond and to induce it to react as required. We discussed the methods of preparing transition metal alkyls in Chapter 4. For synthetic applications, however, it is not always necessary to isolate the transition metal alkyl; generation of an active species *in situ* is enough for most purposes.

The following are the main methods for preparing active species containing a σ metal–carbon bond.

1. Alkylation (including arylation) of a transition metal compound with a suitable alkylating agent.
2. Oxidative addition.
3. Insertion of an unsaturated compound such as an olefin into an M–H bond.
4. External attack on a π-bonded ligand.

In method 1 various alkylating reagents such as organolithium compounds, Grignard reagents, organomercury compounds, or other organotransition metal compounds are employed. Organic chemists sometimes call this type of reaction *transmetallation*. However, the term transmetallation is better reserved for the transfer of a metal-containing group such as Bu_3Sn and Me_3Si. The term alkylation is sufficient to describe the introduction of an alkyl group (including aryl and vinyl groups) into a metal entity.

The second approach, oxidative addition, is quite important as a method of generating an active species with a metal–carbon bond. In the oxidative addition of an organic halide a compound with an M–C and an M–halogen bond is generated. (In this category we also include reactions of anionic metal complexes with organic halides to give metal alkyls.) The combination of oxidative addition of an organic halide to a low-valent transition metal complex with subsequent alkylation of the resulting metal–halogen bond gives a complex with two M–C bonds in the same molecule; the complex is then ready for C–C bond formation by reductive elimination.

An active species with a metal–carbon bond can also be created by approaches 3 and 4, as discussed in Chapter 6. Insertion of an olefin or CO into a metal–hydride bond gives a new metal alkyl or acyl species. Attack on the coordinated CO, olefin, or π-allyl ligand by a carbon nucleophile causes C–C bond formation or generation of a new transition metal complex with a reactive σ metal–carbon bond. The latter is susceptible to further reactions, leading to a new C–C bond. Ingenious combinations of these synthetic methods can yield a variety of preparative strategies.

a. C-C Coupling Combining Oxidative Addition, Alkylation, and Reductive Elimination

(M) —R-X→ (M)(R)(X) —R'm, mX→ (M)(R)(R') ——→ M + R-R

oxidative addition — alkylation — reductive elimination

Dialkylnickel complexes containing a bipyridine ligand undergo C–C coupling reactions on treatment with aryl halides such as chlorobenzene.[37] The chlorobenzene oxidatively adds to the dialkylnickel compound, eliminating the C–C coupled product and affording a phenylnickel complex.

$$(\text{bipy})\text{NiR}_2 + \text{C}_6\text{H}_5\text{X} \longrightarrow (\text{bipy})\text{Ni(Ph)X} + \text{R–R} \quad (8.8)$$

Based on this reaction, a new coupling reaction of Grignard reagents with aryl halides catalyzed by nickel complexes was devised.[38]

$$\text{Ar–X} + \text{RMgX} \xrightarrow{[\text{Cat}]} \text{Ar–R} + \text{MgX}_2 \quad (8.9)$$

Other transition metals, such as palladium, iron, cobalt, and copper, in combination with suitable ligands can also be employed as catalysts.[38,39]

For the nickel-catalyzed cross-coupling of the Grignard reagent with an aryl halide, the following mechanism has been proposed (Fig. 8.10).

The catalyst precursor $NiCl_2L_2$ is first alkylated by the Grignard reagent to give a dialkylnickel complex. This complex is then converted by reaction with aryl halide into the arylnickel halide [A], as shown in Eq. (8.8). Alkylation of this complex with an alkyl Grignard reagent produces a nickel complex with both alkyl and aryl groups (complex [B]). Reductive elimination of this complex yields the cross-coupled product R–Ar and a Ni(0) complex that immediately adds Ar–X to reform [A], which thus carries the catalytic cycle.

Although the scheme shown probably represents what is happening in the catalytic cycle, uncertainties remain regarding the detailed mechanism. A complex corresponding to [B] that has both alkyl and aryl groups has been isolated by Morrel and Kochi[40,41] and was shown to undergo reductive elimination and liberation of Ar–R when it was treated with aryl halide. However, the aryl–alkyl complex isolated by Kochi has a configuration in which the triethylphosphine ligands (L) are mutually trans. This arrangement is not suitable for direct reductive elimination, and the alkyl and aryl groups must somehow be brought to adjacent positions. These could be either mutually cis sites in a square planar or tetrahedral geometry or adjacent apical and equatorial positions in a trigonal bipyramidal complex formed by interaction with another ligand. The reductive elimination of Ar–R from NiAr(R)(dmpe) (dmpe = $Me_2PCH_2CH_2PMe_2$) was observed to be markedly enhanced by the addition of a tertiary phosphine, suggesting that reductive elimination takes place from a pentacoordinated species.[29b] This is yet another example of the difficulties encountered in identifying what is actually taking place in a catalytic process composed of a series of elementary steps.

Markedly differing catalytic activities for C–C cross-coupling within the platinum group triad are evident, nickel being the most suitable, palladium

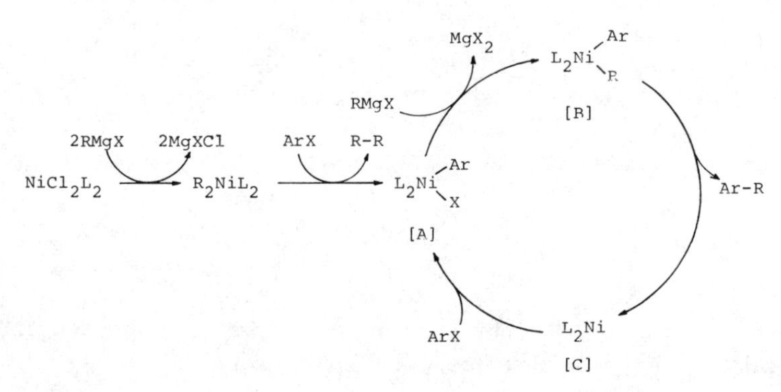

FIGURE 8.10. Mechanism of nickel-catalyzed cross-coupling of Grignard reagents with aryl halides.

being somewhat inferior, and platinum being ineffective. The variation in the catalytic activities is intimately associated with the basic properties of organonickel, organopalladium, and organoplatinum complexes. Nickel(II) complexes are known to readily interconvert between tetrahedral and square planar geometries and to be transformed into pentacoordinated species on interaction with a fifth ligand. The behavior of diorganopalladium compounds, on the other hand, is mostly confined to the square planar geometry, except in transition states, and the reductive elimination of R–R from *cis*-PdR_2L_2 complexes occurs by a dissociative mechanism.[42] The behavior of the platinum analogs further diverges from that of the palladium dialkyls, since thermolysis studies of *cis*-PtR_2L_2 complexes show that β-hydrogen elimination is much favored over reductive elimination, even when the two alkyl groups are in cis positions.

The nature of the alkylating agent may also affect the course of the reaction. For example, alkyllithium reagents, which have strongly carbanionic alkyl groups, are better alkylating reagent for transition metal compounds than are Grignard reagents, but they also provide the possibility of further alkylation of the MR_2L_2 species to yield ionic complexes of the type $[MR_3L]^-Li^+$ and $Li_2[MR_4]$ (M = Pd, Pt).[43,44] The behavior of these ionic complexes is considerably different from that of the neutral complexes so that the catalytic process may be seriously affected by an excess of the carbanionic alkylating reagent.

Vinyl halides can also be induced to couple with the alkyls of main-group elements.

$$\text{PhCH=CHBr} + \text{MeLi} \xrightarrow{PdL_4} \text{PhCH=CHMe} \tag{8.10}$$

This reaction is regiospecific: *E*-β-bromostyrene affords *E*-β-methyl-styrene exclusively, and the corresponding *Z* isomer yields the *Z*-β-methyl-styrene with 99% selectivity. The reaction can also be applied to asymmetric C–C coupling. From a racemic Grignard reagent an optically active hydrocarbon can be produced by cross-coupling with vinyl bromide or chlorobenzene.[45]

$$\text{Ph(Me)CH–MgCl} + \text{CH}_2\text{=CHBr} \xrightarrow[L^*]{[Ni]} \text{CH}_2\text{=CH–C(Me)(H)Ph} \tag{8.11}$$

Use of chiral aminoalkyl phosphines is particularly effective for this purpose. The effect is thought to arise from diastereomeric selection of one of the enantiomers in the racemic alkyl magnesium chloride assisted by complexation with the amino group. The chiral alkyl group is transferred to the nickel atom to give a complex that also contains a vinyl group from vinyl bromide. The chiral alkyl group and the vinyl group are then reductively eliminated to give an optically active olefin, according to Eq. (8.11). This reaction may be regarded as a kinetic resolution of a racemic Grignard reagent, that is, one enantiomer of the Grignard reagent is consumed faster than the other enantiomer, resulting in an optical induction. Since the Grignard reagent rapidly undergoes inversion, the remaining enantiomer is quickly epimerized, and the consumed enantiomer is supplied at a faster rate than the consumption rate for the coupling reaction. Thus, high yields of a single enantiomer can be obtained.

$$\left(R^1R^2R^3C\text{-}MgX \underset{\text{racemic}}{\overset{\text{fast}}{\rightleftharpoons}} X\text{-}Mg\text{-}CR^1R^2R^3 \right) \xrightarrow[{[ML^*]}]{R^4X} R^1R^2R^3C\text{—}R^4$$

In some cases a skeletal isomerization of the alkyl group accompanies the C–C coupling reaction. This reaction probably arises from the isomerization of a branched alkyl group to a sterically less demanding linear alkyl group by reversible β-hydrogen elimination under the steric influence of a tertiary phosphine ligand.

$$(CH_3)_2CH\text{-}ML_n \rightleftharpoons CH_3CH{=}CH_2 \cdots ML_n(H) \rightleftharpoons CH_3CH_2CH_2\text{-}ML_n \quad (8.12)$$

Such skeletal isomerization has also been observed in isolated transition metal alkyls.[46]

An application of the metal-promoted C–C coupling is the one-step synthesis of cyclophanes using dihaloaromatics and di-Grignard reagents.[47]

$$\text{2,6-}Cl_2C_4H_3Y + XMg(CH_2)_nMgX \xrightarrow[\text{dppe}]{NiCl_2} \text{[}C_4H_3Y(CH_2)_n\text{] cyclophane} \quad (8.13)$$

Y = N, CH

Application of the same principle to polycondensation reactions affords polymers from organic dihalides and magnesium.[48]

$$X\text{-}C_6H_4\text{-}X + Mg \xrightarrow{NiCl_2L_2} \text{-}(C_6H_4)\text{-}(C_6H_4)\text{-}(C_6H_4)\text{-}(C_6H_4)\text{-}\cdots + MgX_2$$

$$X\text{-}C_4H_2S\text{-}X + Mg \xrightarrow{NiCl_2L_2} \text{-}(C_4H_2S)\text{-}(C_4H_2S)\text{-}(C_4H_2S)\text{-}(C_4H_2S)\text{-}\cdots + MgX_2 \quad (8.14)$$

$$X\text{-}(CH_2)_6\text{-}X + Mg \xrightarrow{CuBrL_2} \text{-}(CH_2)_6\text{-}\cdots + MgX_2$$

Polyaromatics prepared in this way have more regular structures than those of polyaromatics prepared by means involving radical reactions. They are thermally stable, insensitive to air, and can be converted into electric conductors by doping with iodine or other reagents.

The perfluoroalkyls of main-group elements can be prepared *in situ* and are employed in subsequent cross-coupling reactions. Perfluoroalkyls of zinc can be prepared by the reaction of the perfluoroalkyl iodide with zinc under ultrasonic irradiation. Cross-coupling of the perfluoroalkyl group with aryl iodides is catalyzed by a palladium–phosphine complex.[49]

$$\begin{gathered} R_fI + Zn \xrightarrow[\text{DMF}]{\text{ultrasonic irradiation}} R_fZnI \\ R_fZnI + ArI \xrightarrow{PdCl_2L_2} R_f{-}Ar + ZnI_2 \end{gathered} \qquad (8.15)$$

Acyl halides also oxidatively add to low-valent transition metal complexes, and the acyl complexes formed can be utilized for the synthesis of ketones. For example, $RhCl(CO)(PPh_3)_2$ is alkylated by an alkyllithium. Oxidative addition of an acyl chloride to the resulting alkylrhodium complex gives an acyl–alkylrhodium(III) complex that reductively eliminates a ketone.

$$trans\text{-}RhCl(CO)L_2 \xrightarrow[-\,LiCl]{RLi} trans\text{-}RhR(CO)L_2 \xrightarrow{R'COCl} [Rh(COR')(R)(Cl)(CO)L_2] \xrightarrow{-\,RCOR'} RhCl(CO)L_2 \qquad (8.16)$$

The rhodium complex [recovered as $RhCl(CO)(PPh_3)_2$] can be reused.[50]

This reaction cannot be accomplished in one pot since the ketone produced reacts further with the alkyllithium. The reaction can be made catalytic, however, by use of an alkylating reagent of weak carbanionic character. Alkyltin compounds are convenient for this purpose, and a catalytic ketone synthesis can then be accomplished.[51,52]

$$RCOCl + SnR'_4 \xrightarrow{[Pd]} RCOR' + R'_3SnCl \qquad (8.17)$$

As catalysts, $Pd(PPh_3)_4$, $PhCH_2PdCl(PPh_3)_2$, and other palladium compounds have been used. A probable reaction scheme is shown in Figure 8.11.

In this reaction the acyl chloride oxidatively adds to a Pd(0) complex. The fact that added PPh_3 inhibits the reaction suggests that a coordinatively

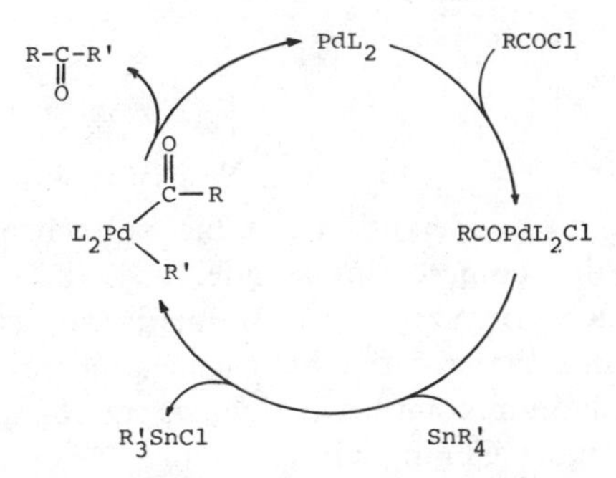

FIGURE 8.11. Mechanism of palladium-catalyzed ketone synthesis from acyl chlorides and alkyltin compounds.

unsaturated species must be generated. Alkylation of the acylpalladium chloride species with the alkyltin compound gives an acyl–alkylpalladium species that reductively eliminates an unsymmetric ketone with regeneration of the coordinatively unsaturated Pd(0) species, which carries the catalytic cycle. The alkylation by the alkyltin compound constitutes the rate-determining step, and the rate decreases in the order PhC≡C > PrC≡C > PhCH=CH > Ph > $PhCH_2$ > CH_3OCH_2 > CH_3 > Bu. The stereochemistry of the vinyl group of the vinyltin compound is retained, but inversion at the α carbon of $PhCHDSnBu_3$ was observed.[52c]

b. C–C Coupling Utilizing Oxidative Addition Combined with CO Insertion

$$\text{(M)} \xrightarrow[\text{oxidative addition}]{\text{R–X}} \text{(M)}\begin{smallmatrix}\text{R}\\ \text{X}\end{smallmatrix} \xrightarrow[\text{insertion}]{\text{CO}} \text{(M)}\begin{smallmatrix}\text{C(=O)–R}\\ \text{X}\end{smallmatrix} \xrightarrow[\text{+Base}]{\text{HY}} \text{RCOY} + \text{HX·Base}$$

Oxidative addition of organic halide RX to a low-valent transition metal complex and the subsequent CO insertion into the metal–carbon bond affords an acylmetal halide complex. The overall process is thus equivalent to the oxidative addition of an acyl halide.

$$\text{RX} + \text{PdL}_n \longrightarrow \text{R–Pd(L)}_2\text{–X} \xrightarrow{\text{CO}} \text{RCO–Pd(L)}_2\text{–X}$$

The acyl complexes yield amides on treatment with primary or secondary amines, esters on reaction with alcohols, and aldehydes with H_2.

$$\text{L}_2\text{Pd}\begin{smallmatrix}\text{C(=O)–R}\\ \text{X}\end{smallmatrix} \begin{cases} \xrightarrow{\text{R'NH}_2} \text{RCONHR'} + \text{HX} + \text{PdL}_2 \\ \xrightarrow{\text{R'OH}} \text{RCOOR'} + \text{HX} + \text{PdL}_2 \\ \xrightarrow{\text{H}_2} \text{RCHO} + \text{HX} + \text{PdCl}_2 \end{cases} \qquad (8.18)$$

By trapping hydrogen halide with a suitable base, such as a tertiary amine, and converting the Pd(0) complex immediately into the alkylpalladium halide by interaction with RX, the system can be made catalytic. For this type of reaction an alkyl halide having a β hydrogen is not appropriate, since the intermediate alkylpalladium complex is then susceptible to β-hydrogen elimination before it undergoes CO insertion.

A similar synthesis of carboxylic esters can be accomplished with a cobalt carbonyl. For this purpose, the anionic cobalt carbonyl, $[Co(CO)_4]^-$ (prepared by treatment of $Co_2(CO)_8$ with sodium amalgam), can be employed. The reaction proceeds as follows:

$$[Co(CO)_4]^- + RX \longrightarrow RCo(CO)_4 \xrightarrow{CO} RCOCo(CO)_4 \xrightarrow{R'OH} RCOOR' + HCo(CO)_4$$

By using an amine to react with the $HCo(CO)_4$, the process can be made catalytic.[53]

The combination of carbonylation and oxidative addition of an organic halide can be applied to cyclization reactions.[54]

$$C_6H_4(Br)(CH_2)_nNHR \xrightarrow[Pd(OAc)_2,\ PPh_3,\ Et_3N]{CO\ 100°C} C_6H_4(CH_2)_n N(R)C(=O) \quad (8.19)$$

Recently, catalytic double carbonylation has been reported.[55-57] In this process two CO molecules are introduced in one step to give α-keto amides from aryl halides and amines.

$$RX + 2CO + HNR'_2 \xrightarrow{[Pd]} RCOCONR'_2 \quad (8.20)$$

Since two reactive carbonyl groups are introduced into the product, this process may find considerable synthetic utility. One application is the preparation of α-amino acids from α-keto acids.

$$RCOCONR'_2 \xrightarrow{H_2O} RCOCOOH \xrightarrow{PhNHNH_2} R\text{-}C(=N\text{-}NHPh)\text{-}COOH \xrightarrow{[H]} R\text{-}CH(NH_2)COOH \quad (8.21)$$

Another application is the preparation of heterocycles.

$$C_6H_4(COCONR_2)(NHAc) \xrightarrow{HCl} \text{isatin (indoline-2,3-dione, N-H)} \quad (8.22)$$

The mechanism of the double-carbonylation process can be represented as shown in Figure 8.12.

The process is catalyzed by palladium complexes that have appropriate tertiary phosphine ligands. When a Pd(II) halide is used, it is converted under the reaction conditions into a Pd(0) complex that oxidatively adds aryl halides. Aryl bromides and iodides can be used as substrates, but the process is not applicable to aryl chlorides because these do not readily add to the Pd(0) complex. The arylpalladium(II) complex formed undergoes insertion of only one CO molecule, not two, as established by detailed mechanistic studies. It was found that the other CO coordinates to the palladium by producing an ionic carbonyl intermediate. Attack of an amine at the coordinated CO generates a carbamoyl–acyl species that reductively eliminates the α-keto amide and regenerates the Pd(0) species, thus carrying the catalytic cycle.

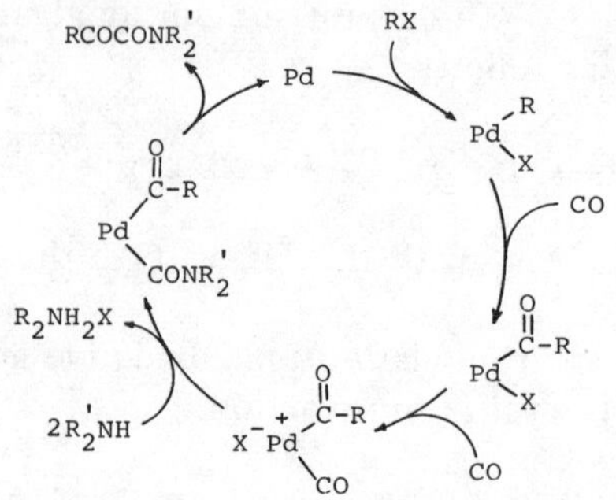

$$\mathrm{RX} + \mathrm{Pd} \rightarrow \mathrm{R{-}Pd{-}X} \xrightarrow{\mathrm{CO}} \mathrm{RC(O){-}Pd{-}X} \xrightarrow{\mathrm{CO}} \mathrm{X^{-}\,Pd^{+}(CO)(C(O)R)} \xrightarrow{2R'_2NH,\ -R'_2NH_2X} \mathrm{Pd(C(O)R)(CONR'_2)} \rightarrow \mathrm{RCOCONR'_2} + \mathrm{Pd}$$

FIGURE 8.12. Mechanism of double carbonylation of aryl halides catalyzed by palladium complexes.

Another special type of synthetic method merits attention. This reaction exploits the characteristic reactivity of a hydrosilane and the ability of cobalt carbonyl to catalyze carbonylation of an olefin. The process gives silyl enol ethers that have one more carbon atom than the olefins.[58]

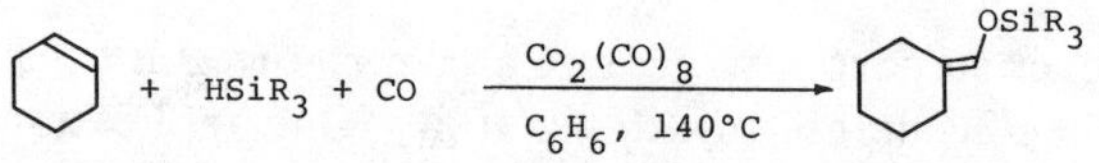

$$\text{cyclohexene} + HSiR_3 + CO \xrightarrow[C_6H_6,\ 140°C]{Co_2(CO)_8} \text{cyclohexylidene-}CH{-}OSiR_3 \qquad (8.23)$$

The active catalyst for this system is produced by the following reaction.

$$HSiR_3 + Co_2(CO)_8 \longrightarrow HCo(CO)_4 + R_3SiCo(CO)_4$$

The catalytic cycle may start by insertion of olefin into the Co–H bond. Further reaction with CO and $HSiR_3$, producing aldehyde, is related to the hydroformylation. The aldehyde formed, however, further reacts with a silylcobalt complex, $R_3SiCo(CO)_3$, to give an alkylcobalt complex that liberates the enol silyl ether by β-hydrogen elimination.

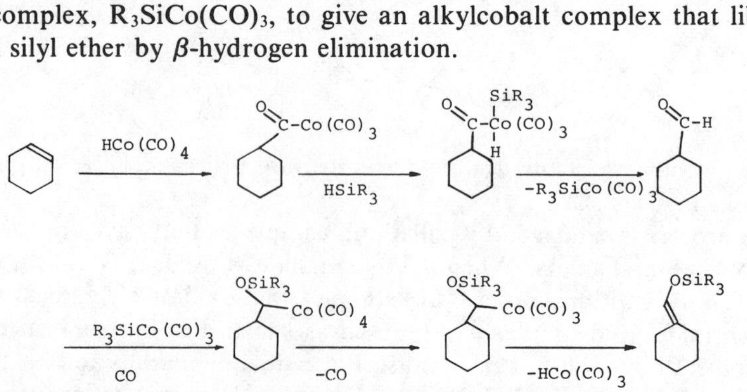

$$\text{cyclohexene} \xrightarrow{HCo(CO)_4} C_6H_{11}C(O){-}Co(CO)_3 \xrightarrow{HSiR_3} C_6H_{11}C(O){-}Co(CO)_3(SiR_3)(H) \xrightarrow{-R_3SiCo(CO)_3} C_6H_{11}CHO$$

$$\xrightarrow[CO]{R_3SiCo(CO)_3} C_6H_{11}CH(OSiR_3){-}Co(CO)_4 \xrightarrow{-CO} C_6H_{11}CH(OSiR_3){-}Co(CO)_3 \xrightarrow{-HCo(CO)_3} C_6H_{10}{=}CH{-}OSiR_3$$

Aldehydes are further carbonylated to α-siloxyaldehyde having one more carbon atom by a similar catalyst system.

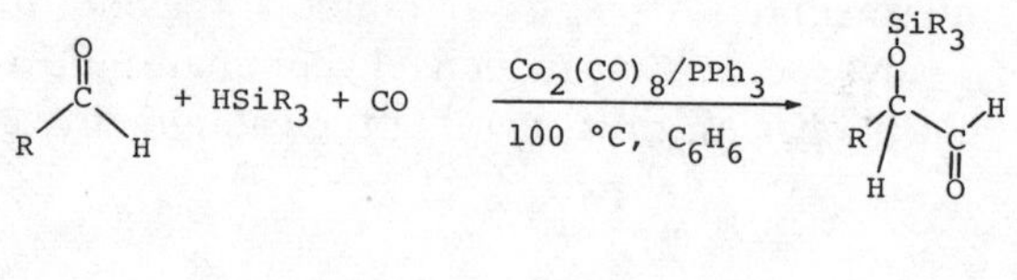

$$RCHO + HSiR_3 + CO \xrightarrow[100\ °C,\ C_6H_6]{Co_2(CO)_8/PPh_3} RCH(OSiR_3){-}CHO$$

c. Combination of Olefin Insertion and β-Hydrogen Elimination with Oxidative Addition of Organic Halides

$$\text{(M)} \xrightarrow[\text{oxidative addition}]{\text{R-X}} \text{(M)}(\text{R})(\text{X}) \xrightarrow[\text{insertion}]{=\!\!\!/\text{R}'} \text{(M)}(\text{CH}_2\text{CHR}'\text{R})(\text{X}) \xrightarrow{\beta\text{-H elimination}} \text{(M)} + \text{RCH=CHR}'$$

The best-studied preparative methods of this type are syntheses by means of palladium compounds.[59] Since the synthetic reactions have been most extensively studied by Heck, they are sometimes called Heck reactions. These reactions may be regarded as the substitution of the hydrogen atom of an olefin by a vinyl or aryl group, but mechanistically they involve olefin insertion into a Pd–C bond formed by oxidative addition and subsequent β-hydrogen elimination.

$$RCH{=}CH_2 + ArX + NEt_3 \xrightarrow{[Pd]} RCH{=}CHAr + Et_3NHX \tag{8.24}$$

$$RCH{=}CH_2 + CH_2{=}CHX + NEt_3 \xrightarrow{[Pd]} RCH{=}CH{-}CH{=}CH_2 + Et_3NHX \tag{8.25}$$

Reaction (8.24) has been developed on the basis of previous observation of the stoichiometric arylation of an olefin,[60] the reaction being made catalytic with the use of an amine as a dehydrohalogenation agent.

In this type of reaction $Pd(PPh_3)_4$ is the most frequently employed catalyst. In the catalytic arylation of an olefin, as in Eq. (8.24), aryl halide oxidatively adds to Pd(0) to give an arylpalladium halide intermediate. Olefin insertion into the Pd–aryl bond and subsequent β-hydrogen abstraction yields the arylated olefin and generates a hydridopalladium halide species. This in turn is dehydrohalogenated by amine to regenerate the Pd(0) species that carries the catalytic process.

$$PdL_n + ArX \longrightarrow ArPdXL_2 \xrightarrow{RCH=CH_2} X\text{-}PdL_2\text{-}CH(R)CH_2Ar$$
$$\xrightarrow{-RCH=CHAr} [PdHXL_2] \xrightarrow{L,\ Et_3N} PdL_n + Et_3NH^+X^- \tag{8.26}$$

Reaction (8.25) involving vinyl halide that give diene derivatives proceed similarly, starting with oxidative addition of vinyl halide to Pd(0).

$$RCH{=}CHBr + PdL_n \longrightarrow RCH{=}CH\text{-}PdL_2\text{-}Br \xrightarrow{CH_2=CHCO_2Me} RCH{=}CHCH_2CH(PdL_2Br)CO_2Me$$
$$\rightleftharpoons (RCH{=}CHCH{=}CHCO_2Me)\,H\text{-}PdL\text{-}Br \xrightarrow[-Et_3NHBr]{Et_3N} RCH{=}CHCH{=}CHCO_2Me + PdL_n \tag{8.27}$$

In place of zero-valent complexes such as $Pd(PPh_3)_4$, catalyst precursors like $Pd(OAc)_2$–tertiary phosphine mixtures can also be employed for these reactions. The divalent palladium compound is readily reduced in the reaction system to generate the active catalyst.

The stereochemistry of these reactions can be explained by cis olefin insertion into the Pd–C bond, followed by cis Pd–H elimination to liberate the product olefin. Treatment of *E* and *Z* isomers of β-methylstyrene with bromobenzene in the presence of a $Pd(OAc)_2$–PPh_3 mixture and triethylamine produces *E*- and *Z*-phenylated products, respectively, with high regiospecificities.

$$PhBr + PhCH{=}CHMe\ (E\text{-isomer}) + Et_3N \xrightarrow[PPh_3]{Pd(OAc)_2} Ph(Ph)C{=}CHMe\ (E\text{-isomer}\ (79\%)) + Et_3NH^+Br^-$$

$$PhBr + PhCH{=}CHMe\ (Z\text{-isomer}) + Et_3N \xrightarrow[PPh_3]{Pd(OAc)_2} Ph(Ph)C{=}CHMe\ (Z\text{-isomer}\ (73\%)) + Et_3NH^+Br^- \qquad (8.28)$$

The regiospecificities of the reaction products are accounted for as shown below by assuming that the cis addition of Pd–Ph to the coordinated olefin is followed by abstraction of the β-hydrogen atom from the palladium-bonded alkyl group in an eclipsed conformation.

Reaction pathway for the *E* isomer:

(H)(Ph)C=C(Me)(H) with Pd-Ph —cis addition→ Ph(H)(Pd)C–C(Me)(H)(Ph) —rotation→ Ph(H)(Pd)C–C(Ph)(Me)(H) —cis elimination, -Pd-H→ (H)(Ph)C=C(Ph)(Me) (E-isomer)

Reaction pathway for the *Z* isomer:

(H)(Ph)C=C(H)(Me) with Pd-Ph —cis addition→ Ph(H)(Pd)C–C(H)(Me)(Ph) —rotation→ Ph(H)(Pd)C–C(Me)(Ph)(H) —cis elimination, -Pd-H→ (H)(Ph)C=C(Me)(Ph) Z-isomer

Organomercurials can also be used to arylate olefins in the presence of palladium salts. The organomercuric compound reacts with the Pd(II) compound to give an organopalladium compound that undergoes the olefin insertion–elimination process to give alkyl(aryl)-substituted olefins.

$$RHgX + PdX_2 \longrightarrow RPdX + HgX_2$$

$$RPdX + CH_2{=}CHY \longrightarrow RCH_2CH(Y){-}PdX \longrightarrow RCH{=}CHY + PdHX \longrightarrow Pd(0) + HX \qquad (8.29)$$

The relative rates of reaction and the regioselectivity depend on both steric and electronic factors.[59] The less sterically hindered is the olefin, the greater is the rate; the new carbon–carbon bond is formed at the least sterically hindered or at the most electron-deficient carbon atom. The relative rates for olefinic substrates were found to be as follows:

olefin	: C_2H_4	> $CH_2{=}CH{-}OAc$	> propene	> $CH_2{=}C(Me)Ph$	> $CH_2{=}CPh_2$
relative rate	: 14000	970	220	42	1

The regiospecificity of the substitution is highest when Y in Eq. (8.29) is an electron-withdrawing group such as CN, CO_2Me, or Ph.

In contrast to the syntheses summarized in Eqs. (8.24) and (8.25), which are based on oxidative addition, the synthesis using organomercurials is not catalytic, although the Pd(0) formed in the reactions can be reused, if it is separately reoxidized. Because of the ease of mercuriation of a pyrimidine ring at the 5 position, the reaction has found an extensive application in nucleoside chemistry.[61]

5-ClHg-uridine —[1) Li_2PdCl_4, C_2H_4; 2) Pd/C, H_2]→ 5-ethyluridine

Alkylmagnesium and alkyllithium compounds can be employed in place of organomercurials for olefin substitution.

$$\text{MeMgX or MeLi} + \text{PhCH=CH}_2 \xrightarrow{\text{Pd(II)}} \text{MeCH=CHPh} \tag{8.30}$$

The yield in the reaction decreases in the order $Pd(acac)_2 > Pd(OAc)_2 > PdCl_2$ and is increased by substitution of an electron-withdrawing substituent in the phenyl ring. Studies employing specifically deuterated styrenes have shown that a Pd–Me species generated *in situ* undergoes cis addition to the double bond and that the resulting σ-bonded group decomposes by cis elimination of Pd–H.[62]

$$\text{(H)(Ph)C=C(D)(H)} \xrightarrow[\text{Pd(acac)}_2]{\text{MeLi}} \text{(H)(Ph)C=C(Me)(D)}$$

$$\text{(H)(Ph)C=C(H)(D)} \xrightarrow[\text{Pd(acac)}_2]{\text{MeLi}} \text{(H)(Ph)C=C(Me)(H)}$$

The formation of β-d_1-E-methylstyrene from E-β-d_1-styrene can be accounted for as follows.

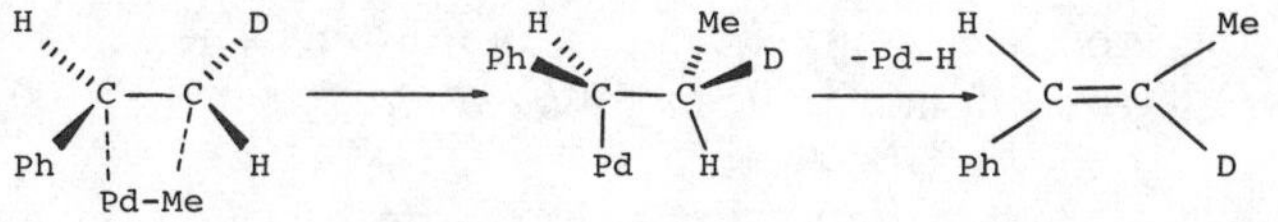

The nondeuterated product is Z-β-methylstyrene. By analogy, the formation of E-β-methylstyrene and β-d_1-Z-methylstyrene can be explained by a cis addition and cis(syn) elimination mechanism. The substitution of olefins using methyllithium is in contrast to the substitution with soft nucleophiles such as $NaCH(CO_2Et)_2$, which attack the coordinated olefin from an anti direction as described in the following section.

d. Synthesis Utilizing Nucleophilic Attack on the Coordinated Ligand

Unsaturated hydrocarbons such as ethylene, butadiene, and benzene are normally not susceptible to nucleophilic addition or substitution reaction. On coordination to electron-withdrawing transition metal centers, however, they are rendered susceptible to attack by a wide variety of nucleophiles, such as H^-, R^-, CN^-, RO^-, or amines.[63] The reactions are usually stoichiometric but they become catalytic if the reduced transition metal product can be reoxidized. We have already dealt with one industrial application of this concept, namely, the Wacker process described in Section 7.4. In this section we are primarily concerned with synthetic applications.

(1) Reactions of Coordinated Olefins

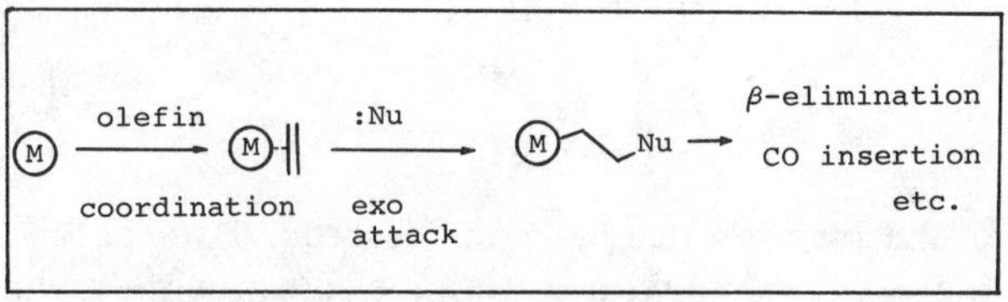

As discussed in Section 7.4, when an olefin coordinated to a transition metal compound in a fairly high oxidation state is attacked externally (exo) by a nucleophile, it undergoes a π–σ rearrangement to give an alkyl group in which the substituent is attached at the β position. Some synthetic applications have been devised, palladium compounds being especially useful. Either stoichiometric addition or substitution can be achieved using Pd(II) compounds. Olefins can be alkylated by carbanions such as diethyl methylmalonate in the presence of triethylamine. Exposure of the reaction mixture to an atmosphere of hydrogen yields saturated products, whereas β-hydrogen elimination produces disubstituted olefins.[64]

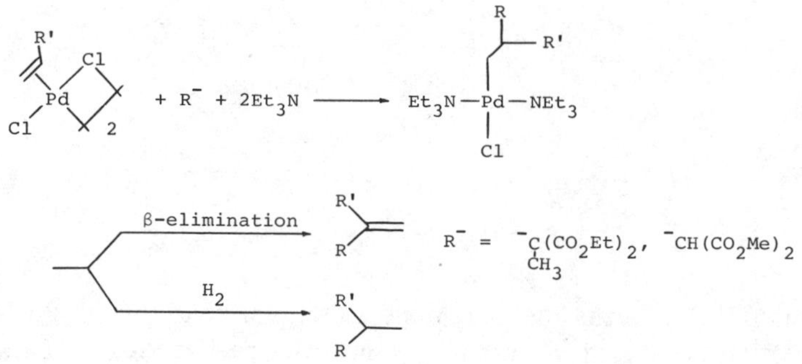

It has been shown that substitutions at the α carbon of styrene by carbanions proceed by external attack.[62,63]

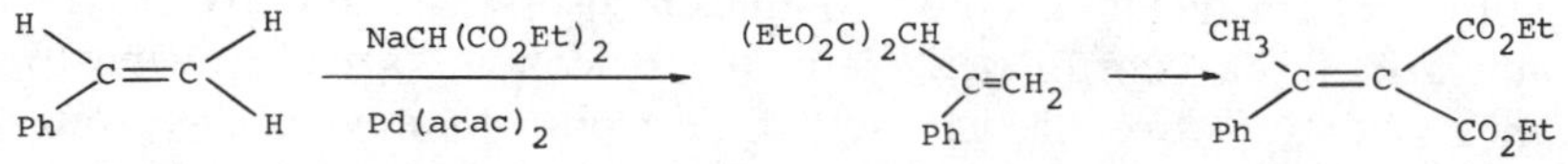

Complexes of the type $PdCl_2$ (olefin) undergo nucleophilic attack by amines at low temperatures, the stereochemistry of the amination being trans.[65]

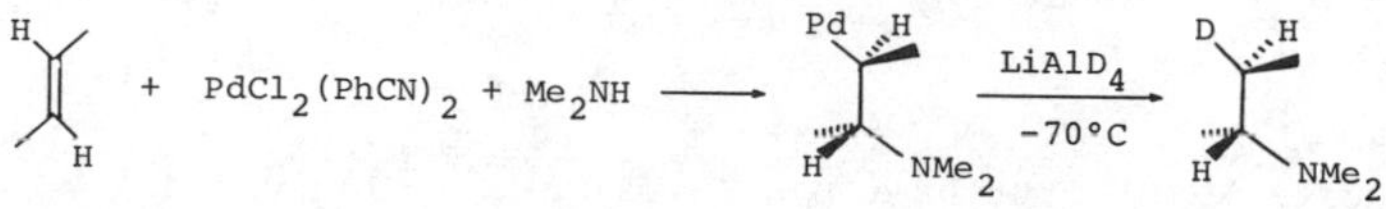

In the case of allylamine a five-membered intermediate is generated by nucleophilic attack at the coordinated double bond of the allylic moiety whose interaction with the Pd(II) is enhanced by coordination of the dimethylamino group.[64d]

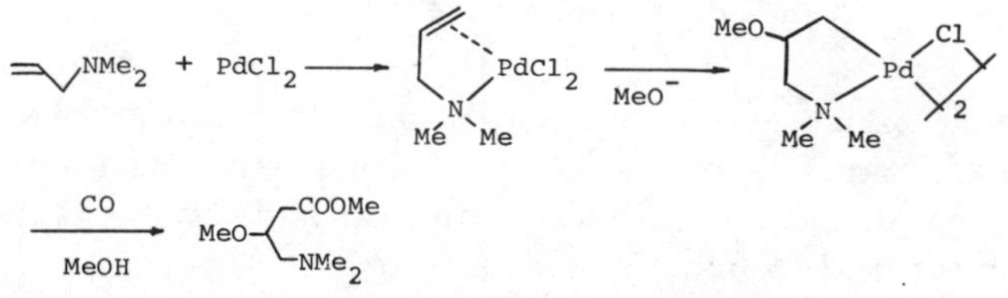

The palladium-assisted attack of a nucleophile at the allylic carbon has been applied to the synthesis of prostaglandins.[66]

Some examples of the application of the principle of nucleophilic attack at an allylic group coordinated to Pd(II) to a catalytic reaction may be found in the synthesis of heterocycles by intramolecular nucleophilic attack.[67,68] The generated Pd(0) species can be reoxidized by $CuCl_2$[67] or benzoquinone.[68]

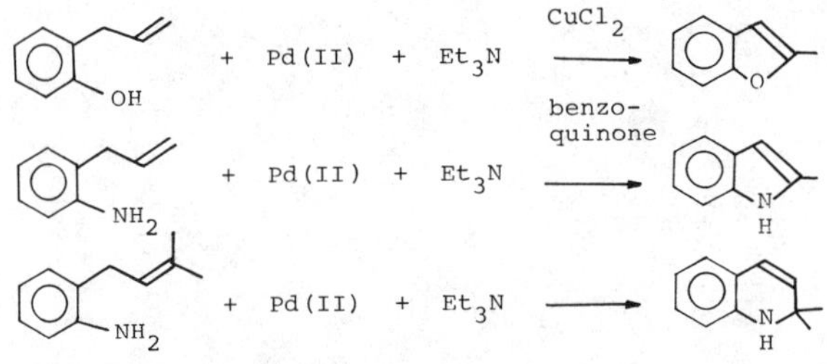

Although Pt(II) forms olefin complexes similar to those of Pd(II), they have not often been used in organic synthesis, in part because the alkylplatinum(II) intermediates are so stable. The only exception is platinum-catalyzed hydrosilylation. However, the inertness of alkylplatinum complexes makes them ideally suited for mechanistic studies, particularly for identifying intermediates.

Cationic olefin metal carbonyl complexes also serve as active centers for the activation of a coordinated olefin to nucleophilic attack, owing to the presence of the positive charge and the electron-withdrawing CO groups. A prime example is the species $[C_5H_5Fe(CO)_2(olefin)]^+$, which undergoes nucleophilic attack at the olefin and produces a stable σ-alkyl complex.[69,70]

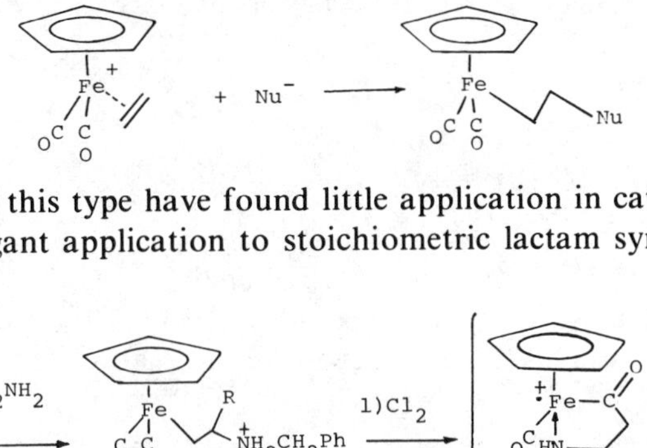

Complexes of this type have found little application in catalytic synthesis, although an elegant application to stoichiometric lactam synthesis has been reported.[71]

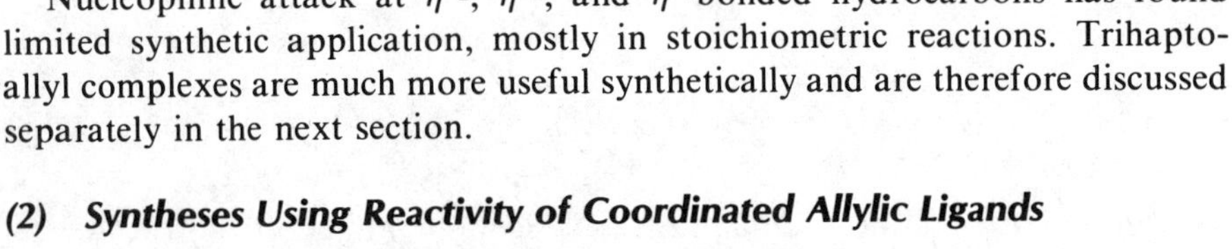

Nucleophilic attack at η^4-, η^5-, and η^6-bonded hydrocarbons has found limited synthetic application, mostly in stoichiometric reactions. Trihapto-allyl complexes are much more useful synthetically and are therefore discussed separately in the next section.

(2) *Syntheses Using Reactivity of Coordinated Allylic Ligands*

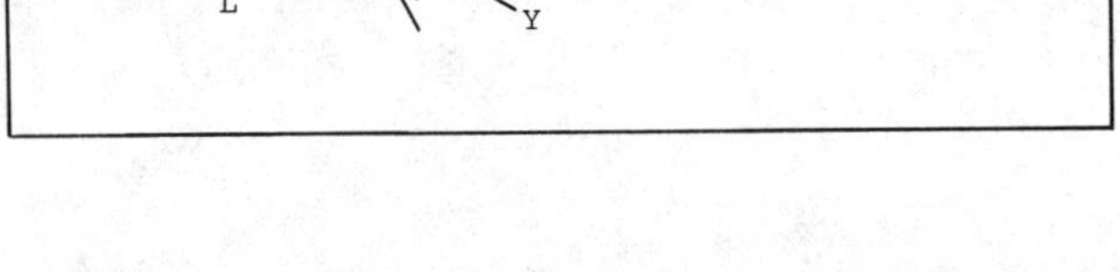

Reactions of η^3-allyl complexes with respect to the polymerization and oligomerization of dienes were discussed in Section 7.2. The reactivity of η^3-allyl complexes, particularly palladium derivatives, has been increasingly utilized for organic syntheses. η^3-Allyl complexes can be prepared in various ways *in situ.*[72-75]

An allylic hydrogen atom adjacent to a coordinated double bond is susceptible to abstraction by a transition metal, and an η^3-allyl complex is formed.

(8.31)

8-1

When X = halide or acetate ion, the η^3-allyl complex produced is bridged as in **8-1**. The bridges can be broken by ligands such as tertiary phosphines or by polar solvents, giving neutral mononuclear complexes like **8-2** or ionic complexes such as **8-3**.

(8.32)

8-1 **8-2** **8-3**

Nucleophiles, for example, stabilized carbanions[76] and enamines,[77] attack the allylic ligand, resulting in allylic alkylation.

$$(\eta^3\text{-allyl})Pd(PPh_3)Cl + {}^{-}CH(COOR)_2 \longrightarrow CH_2{=}CHCH_2CH(COOR)_2 \quad (8.33)$$

Attack on the η^3-allyl ligand usually occurs from the exo side (trans attack), but recently cis attack has also been reported (cf. Section 6.4).

These are useful reactions, but they have the disadvantage of being stoichiometric in palladium. If the η^3-allyl complexes are generated by oxidative additions of allylic compounds to Pd(0), however, catalytic systems can be constructed that are useful in organic syntheses. For example, allyl acetate is easily cleaved at the carbon–oxygen bond on reaction with Pd(0) complexes in the presence of a suitable tertiary phosphine ligand to give an η^3-allylpalladium complex of type **8-2**. This may react further to form cationic η^3-allylpalladium complexes of type **8-3** that are attacked by nucleophiles at the η^3-allyl center, producing alkylation products and regenerating a Pd(0) species. The Pd(0) complex further oxidatively adds allyl acetate to carry the catalytic cycle. The η^3-allyl complex has been isolated and characterized in the case of L = PCy_3.[78]

$$CH_2{=}CHCH_2OAc + PdL_n \longrightarrow (\eta^3\text{-allyl})Pd(L)(OAc) \xrightarrow{L} [(\eta^3\text{-allyl})PdL_2]^+ OAc^- \quad (8.34)$$

$$\xrightarrow{R^-M^+} CH_2{=}CHCH_2R + MOAc + PdL_n$$

The process has been extensively employed by Tsuji, Trost, and their coworkers for the synthesis of natural products.

The oxidative addition of allylic acetate to Pd(0) and subsequent nucleophilic attack at the allylic ligand both proceed with inversions resulting in net retention of configuration as confirmed by using optically active allylic acetate[79] (cf. Section 6.4b).

The stereochemistry of the reaction can be controlled by means of a chiral ligand, and asymmetric C–C bond coupling has been achieved.[80,81,88]

$$\text{(cyclopentenyl)CH(CH}_3\text{)OAc} + {}^-CH(CO_2Me)_2 \xrightarrow[(+)\text{-diop}]{Pd(PPh_3)_4} \text{(cyclopentenyl)C}^*\text{H(CH}_3\text{)CH(CO}_2\text{Me)}_2$$

Allyl–oxygen bond cleavage can be induced by allowing palladium complexes to react with allyl ethers and allyl carbonates.[82] For example, the allyl–oxygen bond in allylic ethers is broken as shown below.[83]

$$R'CH{=}CHCH_2OR + HCO_2^- \xrightarrow[-CO_2,\ -ROH]{[Pd]} R'CH_2CH{=}CH_2 + R'CH{=}CHCH_3 \qquad (8.35)$$

In this process an η^3-allyl alkoxide complex is formed by oxidative addition of the allylic ether, and the alkoxide ligand is then replaced by formate ion. Decarboxylation of the formate ligand is followed by reductive elimination of the allyl and hydrido ligands to liberate the hydrogenation product of the allylic entity. A Pd(0) species is regenerated in the process to carry the catalytic cycle further.

$$\left\{\begin{matrix} R'CH{=}CHCH_2OR \\ \text{or} \\ R'CH(OR)CH{=}CH_2 \end{matrix}\right\} + PdL_n \longrightarrow \left[(\eta^3\text{-}R'C_3H_4)Pd(L)(OR) \xrightarrow{HCO_2^-} (\eta^3\text{-}R'C_3H_4)Pd(L)(O_2CH) \xrightarrow{-CO_2} (\eta^3\text{-}R'C_3H_4)Pd(L)(H) \right]$$

$$\longrightarrow R'CH_2CH{=}CH_2 + R'CH{=}CHCH_3 + PdL_n \qquad (8.36)$$

By combining allyl–oxygen bond cleavage in diallyl ethers with CO insertion into the allyl-metal bond, allyl esters of β,γ-unsaturated acids can be synthesized.[84]

$$CH_2{=}CHCH_2OCH_2CH{=}CH_2 \xrightarrow[[Pd]]{CO} CH_2{=}CHCH_2C(=O)OCH_2CH{=}CH_2 \qquad (8.37)$$

Cleavage of the allyl–oxygen bond in allyl carbonate followed by decarboxylation and nucleophilic attack at the allylic ligand provides a useful means for the allylation of nucleophiles.

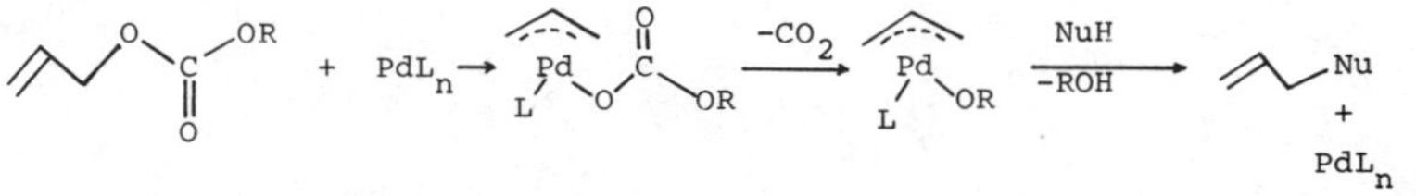

(8.38)

By combining the allyl–oxygen bond cleavage in allylic cabonates with CO insertion, synthetic routes to carboxylic esters of unsaturated acids have been found.[84]

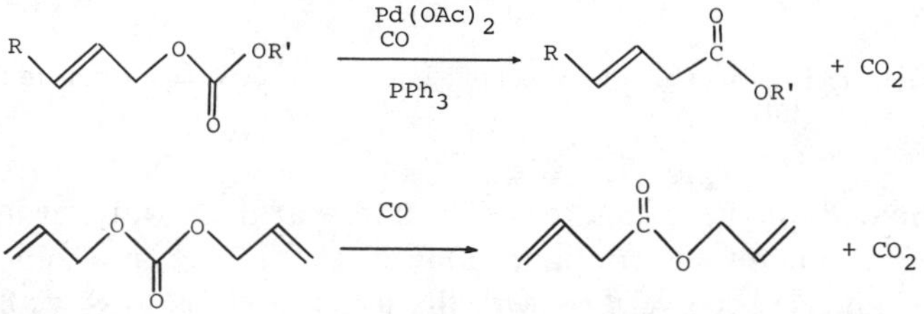

Another application is C–C bond formation between allylic ethers, or their analogs, and Grignard reagents catalyzed by nickel and palladium complexes.[85]

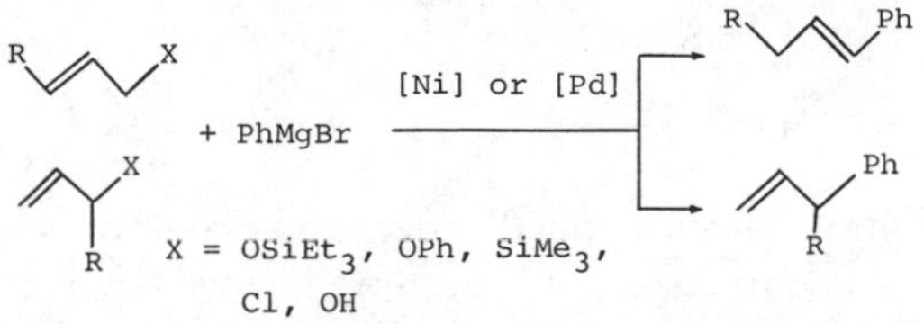

(8.39)

Felkin[86] has shown that in coupling reactions between allyl alcohols and Grignard reagents C–O bond cleavage takes place to give η^3-allyl nickel complexes that are attacked by the alkyl group of the Grignard reagent. For this type of reaction Grignard reagents having β-hydrogen atoms cannot be used since they reduce the allyl alcohols.

(8.40)

In the above reaction the allyl alcohol is considered to react with the Grignard reagent to yield an allyloxymagnesium derivative that subsequently undergoes allyl–oxygen bond cleavage on complexation with the nickel, as shown in Figure 8.13.

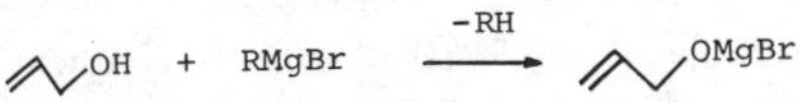

The catalyst precursor NiL_2Cl_2 is assumed to be reduced by the Grignard reagent and enters the catalytic cycle as L_2Ni (Fig. 8.13). If the allyl–oxygen bond is broken by expulsion of the OMgX entity from the exo position and

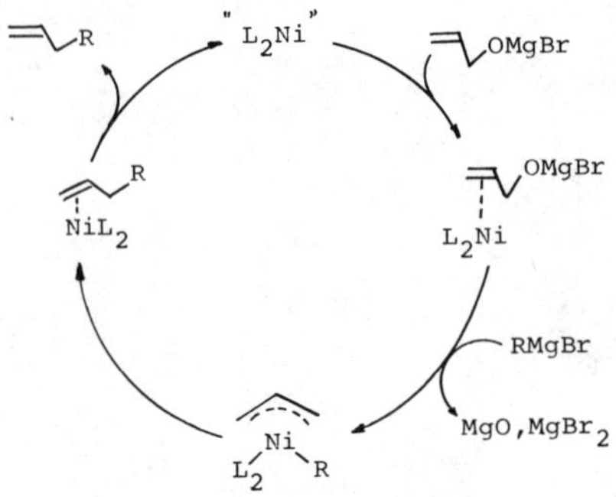

"L_2Ni" ; OMgBr ; L_2Ni ; RMgBr ; MgO, $MgBr_2$; Ni ; L_2 ; R ; NiL_2

FIGURE 8.13. Mechanism of allyl–alkyl coupling in reaction between a Grignard reagent and an allyl alcohol.

the coordinated alkyl group internally attacks at the η^3-allyl group from the endo side, inversion of configuration at the α-carbon atom should be observed. This has been verified with the use of a chiral allyl alcohol in combination with PhMgBr.[69,87]

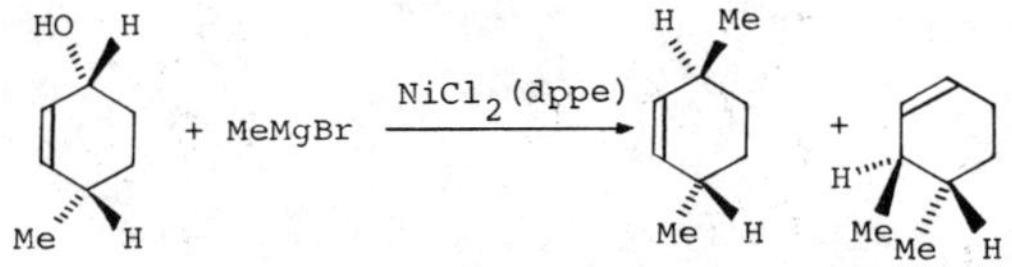

HO H ; Me H + MeMgBr $\xrightarrow{NiCl_2(dppe)}$ H Me ; Me H + H ; Me Me H

The Felkin reaction has been applied to terpenoid synthesis.[86d] With the use of chiral phosphines, asymmetric C–C coupling has been achieved: chiral olefins are thereby produced in the coupling of allyl alcohols or allyl ethers with Grignard reagents.[88]

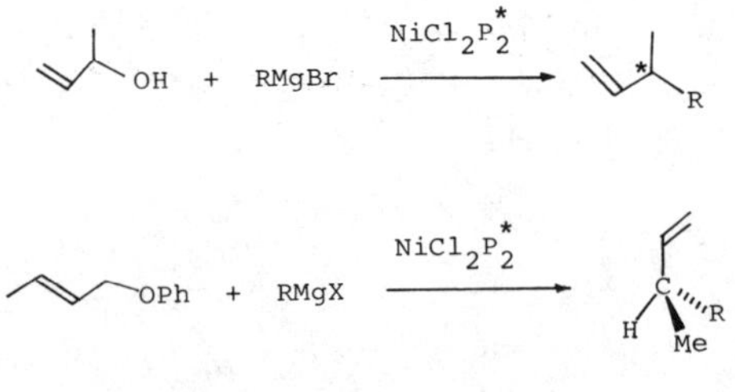

OH + RMgBr $\xrightarrow{NiCl_2P^*_2}$ R

OPh + RMgX $\xrightarrow{NiCl_2P^*_2}$ C, R, H, Me

More applications will undoubtedly be developed in order to exploit the reactivity of the intermediate allyl complexes. Preparative methods utilizing the reactivity of a related trimethylenemethane complex have also been developed.[89] The trimethylenemethane complex of palladium can be generated by treatment of (2-(acetoxymethyl)-3-allyl)trimethylsilane with $Pd(PPh_3)_4$. Reaction of the trimethylenemethane complex with olefins that have electron-withdrawing groups (EWG) selectively affords cyclopentanoid compounds, which are important natural products.

Me_3Si–CH₂–C(=CH₂)–CH₂–OAc + PdL_n $\xrightarrow{-AcOSiMe_3}$ (trimethylenemethane)Pd(L_2) $\xrightarrow[-PdL_n]{\text{CH}_2\text{=CH–EWG}}$ 3-methylene-cyclopentyl–EWG

EWG = electron-withdrowing group

e. Syntheses Utilizing C–H Bond Activation

$$\text{(M)–Y} + \text{H–R} \xrightarrow{-HY} \text{(M)–R} \longrightarrow \text{CO insertion etc}$$

Recent developments in the activation of C–H bonds in alkanes show great potential in organic synthesis, although there are no applications yet. However, C–H bond activation at sp^2 carbon atoms has already found some preparative applications.

Palladium compounds such as $Pd(OAc)_2$ are known to promote olefin insertion into a C–H bond in aromatic compounds; for example, *trans*-stilbene can be prepared from benzene and styrene. Styrene can also be made from benzene and ethylene, but styrene is further arylated to *trans*-stilbene.[90]

$$C_6H_5CH{=}CH_2 + C_6H_6 \xrightarrow[\text{NaOAc, HOAc}]{PdCl_2} C_6H_5CH{=}CHC_6H_5 + Pd + 2NaCl \quad (8.41)$$

$$C_2H_4 + C_6H_6 \xrightarrow[\text{HOAc}]{Pd(OAc)_2} C_6H_5CH{=}CH_2 \quad (8.42)$$

These stoichiometric reactions can be made catalytic when they are coupled with a redox system that reoxidizes the reduced palladium.[90c]

$$C_6H_6 + C_2H_4 + 1/2O_2 \xrightarrow{[Pd]} C_6H_5CH{=}CH_2 + H_2O \quad (8.43)$$

The mechanism of these processes presumably involves olefin insertion into the Pd–aryl bond formed by cleavage of the arene C–H bond and subsequent β-hydrogen elimination to give the olefin.

$$Ar{-}PdX + CH_2{=}CH(Y) \longrightarrow Ar{-}CH_2{-}CH(Y){-}PdX \xrightarrow{-HPdX} ArCH{=}CHY \quad (8.44)$$

The C–H bond activation in aromatics is enhanced by a chelating interaction with the neighboring substituent[91,92] and palladium-promoted syntheses have been carried out on this basis.

$$\text{C}_6\text{H}_5\text{NHCOR} \xrightarrow{Pd(OAc)_2} \mathbf{8\text{-}4} \xrightarrow{CO/EtOH} o\text{-}C_6H_4(NHCOR)(CO_2Et) \quad (8.45)$$

8-4

Complex **8-4**, which has a Pd–C bond, can be treated with CO and ethanol to introduce an ethoxycarbonyl group at the ortho position of acetanilide.[93]

f. Syntheses Utilizing Oxidative Addition and Decarbonylation

$$M \xrightarrow[\text{oxidative addition}]{RCOY} M(C(O)R)(Y) \xrightarrow{\text{decarbonylation}} M(CO)(R)(Y) \xrightarrow[\text{reductive elimination}]{-RY} M + CO$$

Y = H, halogen

To activate a neighboring group and enable a specific reaction to be carried out, it is sometimes necessary to first introduce a carbonyl group and then subsequently to remove it. This can be a very powerful procedure, but it has not often been used. The most frequently used reagent is Wilkinson's complex, $RhCl(PPh_3)_3$. Aldehydes are decarbonylated on treatment with Wilkinson's complex to afford alkanes.[94-96]

$$RCHO + RhCl(PPh_3)_3 \longrightarrow RH + RhCl(CO)(PPh_3)_2 + PPh_3 \quad (8.46)$$

In this reaction the aldehyde oxidatively adds to the Rh(I) complex with cleavage of the aldehyde C–H bond. The acyl complex thus formed undergoes deinsertion of the CO group to provide an alkyl–hydrido complex that reductively eliminates the alkane.

$$RhClL_3 \underset{}{\overset{-L}{\rightleftharpoons}} RhClL_2 \overset{RCHO}{\rightleftharpoons} R\text{-}C(O)\text{-}RhCl(H)L_2 \rightleftharpoons R\text{-}RhCl(CO)(H)L_2 \longrightarrow RH + RhCl(CO)L_2 \quad (8.47)$$

L = PPh_3

Oxidative addition of the C–H bond of an aldehyde has a precedent in the reaction of 8-quinoline carboxaldehyde with a rhodium complex to form an acyl–hydride complex that is stabilized by chelation, which aids isolation and characterization.[97]

$$\text{8-formylquinoline} + RhClL_3 \longrightarrow \textbf{8-5} \quad (8.48)$$

8-5

The decarbonylation of an unsaturated aldehyde occurs with retention of the olefin geometry.

$$(H)(Ph)C{=}C(CHO)(R) + RhCl(PPh_3)_3 \longrightarrow (H)(Ph)C{=}C(H)(R) + RhCl(CO)(PPh_3)_2 + PPh_3 \quad (8.49)$$

Decarbonylation is also stereospecific, with retention of configuration at the α-carbon atom attached to the aldehyde group.

$$\text{Ph}_2\text{C}_3\text{H}(R)(CHO) + RhClL_3 \longrightarrow \text{Ph}_2\text{C}_3\text{H}(R)(H) + RhCl(CO)L_2 \quad (8.50)$$

These decarbonylation reactions have been applied to more complex systems, such as disaccharides, steroid derivatives, and macrolides. The reactions are usually stoichiometric under mild conditions, since $RhCl(CO)(PPh_3)_2$ is less reactive than $RhClL_3$ (the carbonyl ligand acts as a π acceptor, withdrawing electron density from the metal). At temperatures exceeding 200 °C catalytic decarbonylation can be performed.

Acyl halides can also be decarbonylated. The reaction proceeds with cleavage of the carbon–halogen bond.[98]

$$RhClL_3 \underset{}{\overset{-L}{\rightleftharpoons}} RhClL_2 \overset{PhCH_2COCl}{\rightleftharpoons} PhCH_2C(O){-}RhL_2Cl_2 \rightleftharpoons (OC)(PhCH_2)RhL_2Cl_2 \longrightarrow RhCl(CO)L_2,\ PhCH_2Cl \qquad L = PPh_3 \quad (8.51)$$

Use of an acyl halide having a β-hydrogen atom leads to the formation of an olefin by β-hydrogen elimination from the intermediate σ-alkyl intermediate.

$$RCH_2CH_2COCl + RhClL_3 \longrightarrow RCH{=}CH_2 + RhCl(CO)L_2 + HCl$$

The decarbonylation of acyl chlorides is stoichiometric if carried out in aromatic solvents at between 20 and 100 °C. These reactions can be made catalytic in Rh complex at higher temperatures, although some sensitive organic compounds may not tolerate the severe conditions required.

Certain carboxylic esters react with Ni(0) complexes by undergoing cleavage of an acyl–oxygen bond [cf. Eq. (6.53)]. The combination of the reaction with decarbonylation provides a synthetic route to olefins from thiol esters.[99]

$$RCH_2-C(R^1)(R^2)-CO-S-C_5H_4N \xrightarrow[DMF]{NiCl_2/Zn} RCH=C(R^1)(R^2) + CO + C_5H_5NS \qquad (8.52)$$

g. Syntheses Utilizing Insertion and Alkyl Transfer

$$\text{(M)}-H \xrightarrow[\text{insertion}]{>C=C<\ \text{etc.}} \text{(M)}\text{-}CH_2CH_2\text{-}H \xrightarrow{M'X \to MX} \text{(M')}\text{-}CH_2CH_2\text{-}H$$

The transfer of an alkyl group from an organometallic compound of a main-group element to a transition metal provides a useful means of preparing organotransition metal complexes, as discussed in Chapters 4 and 6. The transfer of an alkyl group from a transition metal to a main-group element or to other transition metal elements may also afford a useful synthetic approach. Particularly well studied is the use of early transition metal complexes, such as those of zirconium and titanium.

As shown in Eq. (6.117), $Cp_2ZrH(Cl)$ readily undergoes insertion of olefins or acetylenes to give alkyl- or alkenylzirconium complexes. Although these organozirconium complexes are rather stable and not very reactive, the alkyl or alkenyl group bound to the zirconium can be readily transferred to other metals, thus generating more reactive organometallic compounds.[100-103]

$$RC\equiv CR' + Cp_2Zr(Cl)H \longrightarrow Cp_2Zr(Cl)-C(R)=C(H)R'$$

$$\xrightarrow{AlCl_3} \left[AlCl_2-C(R)=C(H)R' \right] \xrightarrow{CH_3COCl} CH_3CO-C(R)=C(H)R' \qquad (8.53)$$

In this reaction the alkenyl group formed by insertion of the alkyne into the Zr–H bond of $Cp_2Zr(H)Cl$ is transferred to aluminum, and the resulting alkenylaluminum compound rapidly reacts with acyl halides to produce ketones. The high reactivity of the alkenylaluminum is in contrast to the low reactivity of the alkenylzirconium complex toward direct attack of acyl halides.[101]

The alkyl transfer reaction can be performed with metals other than aluminum (e.g., Zn, Cu, Pd, and Ni), and various other synthetic procedures can be realized. A catalytic process based on alkenylzirconium complexes is the cross-coupling reaction of an alkenyl group with the aryl group of an aryl halide catalyzed by $Ni(PPh_3)_4$. The reaction probably proceeds by oxidative addition of an aryl halide to the Ni(0) complex to afford an arylnickel halide, which is subsequently alkylated by transfer of the alkenyl group from the alkenylzirconium complex. Reductive elimination of the alkenyl and aryl groups gives the arylated olefin and regenerates a Ni(0) complex, which further carries the catalytic cycle.

$$(R^1)(H)C{=}C(R^2)(ZrCp_2Cl) + ArX \xrightarrow{Ni(PPh_3)_4} (R^1)(H)C{=}C(R^2)(Ar) \quad (8.54)$$

A recent application of alkyl transfer from a transition metal to a main-group element is catalytic olefin hydroalumination, which converts aluminum hydrides into aluminum alkyls.[104-107]

$$LiAlH_4 + 4CH_2{=}CHR \xrightarrow{[Cat]} LiAl(CH_2CH_2R)_4 \quad (8.55)$$

The insertion of olefins into Al–H bonds was discovered in 1949 by Ziegler, but the process required more severe conditions than those for hydroboration. Use of early transition metal catalysts, such as $TiCl_4$, $CpTiCl_2$, $Cp_2Ti(AlH_3)_2$, and $ZrCl_4$, allows the hydroalumination reaction to proceed under much milder conditions. In this type of reaction hydride transfer from $LiAlH_4$ to a zirconium or titanium complex gives a hydrido complex having either Zr–H or Ti–H bonds. Insertion of the olefin into the transition metal–hydride bond gives a transition metal alkyl. Exchange of the alkyl group on the transition metal with the hydrido ligand attached to the aluminum provides the alkylaluminum compound and regenerates the transition metal hydride, which carries the catalytic cycle. Hydrolysis of the aluminum alkyls produced in this way gives alkanes, halogenation provides alkyl halides, and oxidation followed by hydrolysis yields alcohols.

A typical reaction of Grignard reagents with organic carbonyl compounds is 1,2-addition, in which an alkyl group undergoes nucleophilic attack on the carbonyl carbon atom. The reaction can be modified by addition of a transition metal compound. Thus, reaction of isobutyl magnesium bromide with a ketone in the presence of Cp_2TiCl_2 leads to reduction of the ketone to an alcohol.

$$Me_2CHCH_2MgBr + R^1C(=O)R^2 \xrightarrow{Cp_2TiCl_2} BrMgOCHR^1R^2 \xrightarrow{H_2O} R^1R^2CHOH \quad (8.56)$$

In this reaction isobutylmagnesium bromide alkylates Cp_2TiCl_2 to give an isobutyltitanium complex, which on β-hydrogen elimination is believed to generate a titanium hydride species. Reaction of the titanium hydride with the ketone gives an alkoxotitanium complex that is further alkylated by the isobutylmagnesium bromide, thus regenerating the isobutyltitanium complex to close the catalytic cycle (Fig. 8.14).[108]

h. Cyclopropanation

The fundamental reactions of metal–carbene complexes were discussed in Section 6.4, and olefin metathesis reactions involving metal–carbene com-

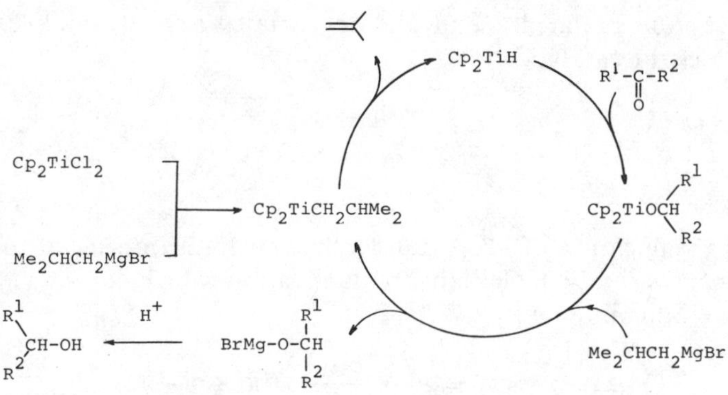

Cp_2TiCl_2 + Me_2CHCH_2MgBr → $Cp_2TiCH_2CHMe_2$ → Cp_2TiH (+ isobutene); Cp_2TiH + R^1-CO-R^2 → $Cp_2TiOCH(R^1)(R^2)$; + Me_2CHCH_2MgBr → $BrMg$-O-$CH(R^1)(R^2)$ + $Cp_2TiCH_2CHMe_2$; $BrMg$-O-$CH(R^1)(R^2)$ →(H^+) R^1R^2CH-OH

FIGURE 8.14. Proposed mechanism of reduction of a ketone with a Grignard reagent catalyzed by Cp_2TiCl_2.

plexes were treated in Section 7.5. In the reactions of carbene complexes with olefins, cyclopropane derivatives are sometimes formed by reductive elimination from metallacyclobutanes.

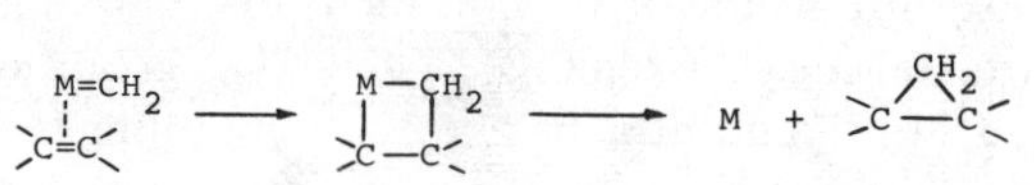

$$M{=}CH_2 + {>}C{=}C{<} \longrightarrow \text{metallacyclobutane } (M{-}CH_2{-}C{-}C) \longrightarrow M + \text{cyclopropane } ({>}C{-}CH_2{-}C{<}) \quad (8.57)$$

Systems in which carbene intermediates are formed can therefore be used to prepare cyclopropane derivatives. A mixture of $PhWCl_3$ and $AlCl_3$ acts as a catalyst for the metathesis of *cis*-2-butene. Addition of ethyl acrylate to the system yields a cyclopropane derivative, the intermediate carbene complex presumably being trapped by the electronegative olefin.[109]

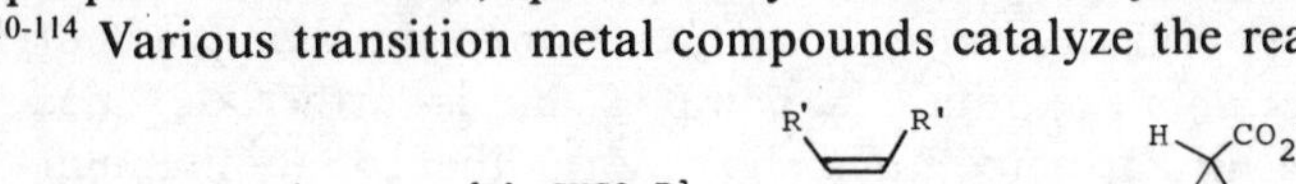

$$CH_3CH{=}CHCH_3\ (cis) + CH_2{=}CHCO_2Et \xrightarrow[AlCl_3]{PhWCl_3} \text{cyclopropane}(CH_3, H; H, CO_2Et) \quad (8.58)$$

Carbenes can also be formed from diazo compounds, and when diazo compounds react with transition metal complexes, transition metal–carbene complexes may be formed. These carbene complexes can react with olefins to give cyclopropane derivatives, presumably via metallacyclobutane intermediates.[110-114] Various transition metal compounds catalyze the reaction.

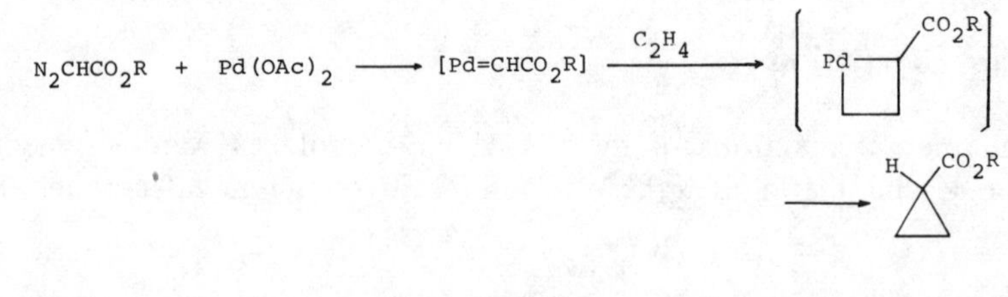

$$N_2CHCO_2R + Rh_2(OAc)_4 \longrightarrow [Rh{=}CHCO_2R] \xrightarrow{R'CH{=}CHR'} \text{cyclopropane}(H, CO_2R; R', R')$$

$$N_2CHCO_2R + Pd(OAc)_2 \longrightarrow [Pd{=}CHCO_2R] \xrightarrow{C_2H_4} [\text{Pd-cyclobutane}(CO_2R)] \longrightarrow \text{cyclopropane}(H, CO_2R)$$

The synthesis of cyclopropane derivatives from methylene bromide and electronegative olefins is catalyzed by a cobalt compound. The reaction is believed to proceed by formation of metal–carbene and metallacyclobutane intermediates and subsequent reductive elimination.[114] The asymmetric synthesis of cyclopropane derivatives can also be achieved with use of transition metal complexes that have chiral ligands. Cyclopropanation of styrene with diazoacetate catalyzed by the cobalt complex of a camphor-modified dioxime ligand proceeds with high stereospecificity.

$$PhCH{=}CH_2 + N_2CHCO_2R \xrightarrow{[Co]}$$ (cyclopropanes: H, CO_2R, Ph, H) + (Ph, CO_2R, H, H)

The asymmetric synthesis of optically active chrysanthemic acid with a chiral copper catalyst has also been achieved.[115]

$$+ N_2CHCO_2R \xrightarrow{[CuL_n^*]}$$ (CO_2R, *, *) + (CO_2R, *, *)

8.3. STOICHIOMETRIC REACTIONS USING TRANSITION METAL COMPOUNDS

a. Synthesis Using Organocopper Reagents

Organocopper reagents are probably the most frequently used transition metal compounds in organic synthesis.[116-120] At least 100 papers are published each year dealing with synthesis using organocopper compounds, showing how popular these reagents are among synthetic organic chemists. Many well-known reactions that are promoted by metallic copper or inorganic copper compounds, such as the Ullman reaction, the Sandmeyer reaction, and the Gatterman reaction, involving the coupling of aryl groups and the synthesis of aryl halides from diazonium salts are likely to proceed through organocopper intermediates.

Alkylcopper compounds can be prepared by treating copper salts with alkylating agents.[116,121] They are rarely isolated in pure state, and isolated solid methyl copper is known to decompose explosively.[122b] Alkylcopper complexes, however, are stabilized by coordination with tertiary phosphines.[122,123]

When copper salts, usually halides, are treated at low temperature with more than two equivalents of an alkylating agent such as an alkyllithium or less often a Grignard reagent, ionic organocopper complexes called organocuprates are formed [Eq. (8.59)]. These ate complexes are more stable than simple copper alkyls. When more than three equivalents of an alkyllithium is employed, complex adducts of the type R_nCuLi_{n-1} ($n > 2$) are obtained.

$$2RLi + CuX \xrightarrow{< 0°C} [R_2Cu]Li + LiX \qquad (8.59)$$

The organocuprates (Gilman reagents) prepared *in situ* have found enormous application in organic synthesis, but detailed structural studies are limited. Based on molecular weight measurements and NMR studies, however, the following dimeric structure for $LiCuMe_2$ has been proposed.[124]

$$\begin{array}{ccc} CH_3 - Cu - CH_3 \\ | \quad\quad\quad | \\ Li \quad\quad\quad Li \\ | \quad\quad\quad | \\ CH_3 - Cu - CH_3 \end{array}$$

8-6

Cuprates of the type R_2CuLi are called homocuprates, and R^1R^1CuLi complexes are called mixed homocuprates, whereas RZCuLi (Z = OR′, SR′, CN, Cl, Br) are known as heterocuprates.

The most important reactions of alkyl cuprates are cross-coupling reactions [Eq. (8.60)] and 1,4-additions to conjugated enones [Eq. (8.61)].

$$R_2CuLi + R'X \longrightarrow R\text{-}R' \tag{8.60}$$

$$R_2CuLi + CH_2{=}CH\text{-}\underset{\underset{O}{\|}}{C}\text{-}CH_3 \longrightarrow RCH_2CH{=}\underset{\underset{OLi}{|}}{C}CH_3 \xrightarrow{H^+} RCH_2CH_2\underset{\underset{O}{\|}}{C}CH_3 \tag{8.61}$$

The reaction of an alkyl cuprate with an alkyl halide is first order with respect to each reactant. Stereochemical inversion at the alkyl carbon atom of the alkyl halide has been observed indicating an S_N2 reaction mechanism.[125] The reactivity of alkyl halides is in the order primary > secondary >> tertiary and I > Br > Cl as in straightforward S_N2 reactions. As the reaction mechanism for the cross-coupling reaction, a simple S_N2 process in which the copper-bonded alkyl group attacks the alkyl halides [as in Eq. (8.62)] and oxidative addition are conceivable.

$$R\text{-}\overset{R}{\overset{|}{Cu}}(I)^- + \;\rangle C\text{-}X \xrightarrow[\text{inversion}]{-X^-} R\text{-}C\langle \; + \; RCu(I) \tag{8.62}$$

However, oxidative addition to form a simple $R_2R'Cu(III)$ species is not compatible with the highly selective formation of the cross-coupling product R–R′ [Eq. (8.60)]. To circumvent these difficulties, a mechanism in which R′X adds to a dimeric R_2CuLi has been proposed.

$$\begin{array}{ccc} R\text{-}Cu\text{-}R \\ | \quad\quad | \\ Li \quad\quad Li \\ | \quad\quad | \\ R\text{-}Cu\text{-}R \end{array} \xrightarrow{R'X} \begin{array}{ccc} R' \\ | \\ R\text{-}Cu\text{-}R \\ | \quad\quad | \\ Li \quad\quad Li \\ | \quad\quad | \\ R\text{-}Cu\text{-}R \\ | \\ X \end{array} \longrightarrow R\text{-}R'$$

In this mechanism the preferential reductive elimination of R–R′, rather than R–R, can be accounted for by the arrangement of the R and R′ groups in the dimeric structure. Obviously, more detailed studies on the structure and nature of alkyl cuprates are required.

The cross-coupling reaction proceeds smoothly with alkenyl halides under mild conditions. The reactivity order of the alkenyl halides is again I > Br > Cl. The cross-coupling reaction proceeds with retention of stereochemistry at the vinylic carbon atom.

$$PhCH{=}CHBr + LiCuR_2 \longrightarrow PhCH{=}CHR \quad (8.63)$$

The advantages of alkyl cuprates for organic synthesis are (1) the reactions occur under very mild conditions; (2) the stereospecificity is high with retention of the stereochemistry of the alkyl group in the alkyl cuprate; (3) substitution of alkenyl halides can be achieved, a difficult process with conventional methods. A disadvantage of the process is the low reactivity arising from insolubility of the cuprate in solvents suitable for the reactions. Another disadvantage based on homocuprates is that one alkyl group attached to copper is wasted. This can be costly if the R group is difficult to synthesize. This problem can be avoided by using heterocuprates, although these are usually less thermally stable than homocuprates and reactions must be performed at low temperatures. Acetylenic mixed cuprates RCu(C≡CR′)Li are also used to avoid the loss of an expensive R group, but these are much less reactive than the corresponding homocuprates.

Since acyl halides also react with alkyl cuprates, unsymmetrical ketones can easily be synthesized.

$$\text{(alkenyl)COCl} + Et_2CuLi \longrightarrow \text{(alkenyl)COEt} \quad (8.64)$$

An example of the application of a heterocuprate is shown below.[126a]

$$Br(CH_2)_{10}COCl + t\text{-}Bu(t\text{-}BuO)CuLi \longrightarrow Br(CH_2)_{10}CO(t\text{-}Bu)$$

This process makes use of the higher reactivity of the acyl halide compared to the alkyl halide.

In addition to the lithium cuprates, magnesium-containing alkyl cuprates can be produced by the reaction of copper halides with Grignard reagents and can be used analogously to the lithium cuprates. When mixed homocuprates R^1R^2CuMgX are used, the alkyl group of longer chain length reacts preferentially.

$$CH_2{=}CH(CH_2)_8CH_2MgCl + CH_3Cu \longrightarrow ClMg^+[CH_2{=}CH(CH_2)_8CH_2CuCH_3]^-$$

$$\xrightarrow{I(CH_2)_{10}COOEt} CH_2{=}CH(CH_2)_{19}COOEt \quad (8.65)$$

The reagent $RCuMgX_2$, prepared from RMgX and CuX, can be conveniently used to add an alkyl group to a terminal acetylene functionality.[126b]

$$RCuMgX_2 + R'C{\equiv}CH \longrightarrow \underset{R}{\overset{R'}{}}C{=}C\underset{CuMgX_2}{\overset{H}{}} \begin{cases} \xrightarrow{R''X} \underset{R}{\overset{R'}{}}C{=}C\underset{R''}{\overset{H}{}} \\ \xrightarrow{I_2} \underset{R}{\overset{R'}{}}C{=}C\underset{I}{\overset{H}{}} \end{cases} \quad (8.66)$$

The process is useful for the preparation of trisubstituted olefins. Alkylcuprates, R_2CuLi, cannot be used in this case because they only abstract the acidic terminal proton from the alkyne. The syn addition of $RCuMgX_2$ to a terminal alkyne has a high regio- and stereospecificity.[120]

Epoxides undergo nucleophilic ring opening reactions with alkyl cuprates. The alkyl group of R_2CuLi usually attacks the less substituted site in the epoxide ring.[127]

(8.67)

OH, O + $LiCuMe_2$ ⟶ OH, OH, Me (83%) + OH, Me, OH (14%)

1,4-Addition of the type shown in Eq. (8.61) has been used for the synthesis of certain natural products. The following is a typical example of the application of a mixed cuprate to the synthesis of a prostaglandin derivative.[128]

(8.68)

$Me(CH_2{=}CH)CuLi$ + cyclopentenone ⟶ $[\;]^- M^+$ —(I–CH₂CH=CH(CH₂)₃COOMe)⟶ product with COOMe

b. Synthesis Using Metal Carbonyls

Carbonyls of nickel, iron, and cobalt are not expensive and are often utilized in organic synthesis.[129] Characteristics of the processes that use metal carbonyls are (1) utilization of the reactivity of the low-valent, electron-rich metal carbonyls; (2) introduction of the migratory CO insertion step into the synthetic sequence; and (3) *Umpolung* (charge affinity inversion) of the acyl group from the usual RCO^+ nature of an acyl halide, for example, to the RCO^- character, as in the metal carbonyl anions.

The old observation[130] of allyl–allyl coupling in the reaction of allyl bromide with $Ni(CO)_4$ in polar solvents such as DMF [Eq. (8.69)] has been developed into a useful organic synthetic procedure and has yielded a variety of terpenoids.

$$\text{CH}_2\text{=CHCH}_2\text{Br} + \text{Ni(CO)}_4 \xrightarrow[\text{DMF}]{-2\text{CO}} [\text{8-7}] \xrightarrow[-2\text{CO}]{\text{CH}_2\text{=CHCH}_2\text{Br}} \text{1,5-hexadiene} + \text{NiBr}_2 \tag{8.69}$$

8-7

Under conditions of high dilution, α,ω-dibromoalkenes gives macrocyclic compounds containing 12-, 14-, and 18-carbon chains.[131] Macrolides can also be synthesized in this way.

$$(\text{CH}_2)_n(\text{CH=CHCH}_2\text{Br})_2 + \text{Ni(CO)}_4 \longrightarrow \text{cyclo-}(\text{CH}_2)_n(\text{CH=CH})_2 \tag{8.70}$$

$$\xrightarrow{\text{Ni(CO)}_4} \tag{8.71}$$

In nonpolar solvents complex **8-7** is transformed into a dimeric η^3-allylnickel μ-bromide, and this reacts with alkyl halides to give C–C coupled products.[132]

$$\textbf{8-7} \xrightarrow[\text{benzene}]{-\text{CO}} [(\eta^3\text{-C}_3\text{H}_5)\text{Ni}(\mu\text{-Br})]_2 \xrightarrow{\text{RX}} \text{RCH}_2\text{CH=CH}_2 + \text{NiBrX}$$

Iron carbonyls have been extensively used to synthesize compounds that are difficult to prepare by other methods.[133,134] Dimeric $Fe_2(CO)_9$ has been employed as a source of coordinatively unsaturated $Fe(CO)_4$. The reaction of $Fe(CO)_4$ with dibromoketones is believed to proceed via cationic oxallyl complexes such as **8-8**.

$$\text{dibromoketone} + \text{Fe(CO)}_4 \longrightarrow [\text{Br}\ \ \text{Fe}^+\text{(CO)}_4]\ \text{Br}^- \longrightarrow [\text{OFe(CO)}_n\text{Br}\ \text{allyl}^+]\ \text{Br}^-$$

8-8

The cationic allyl complex **8-8** undergoes 3+4-cycloaddition reactions with dienes to give 5-cycloheptenones, which are useful as precursors of various biologically active compounds. An elegant stereocontrolled approach to *C*-nucleosides has been developed.[135]

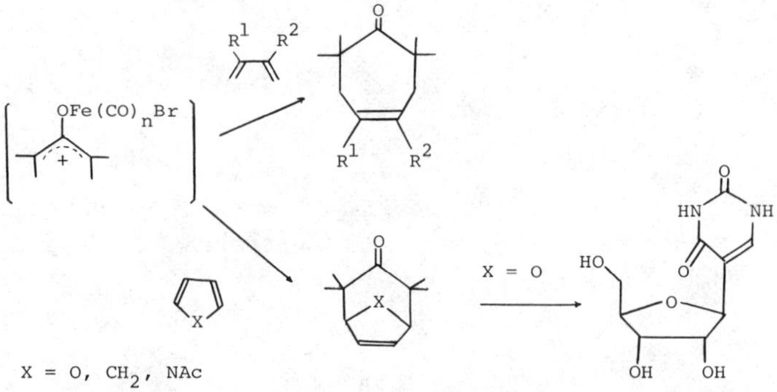

In the above examples the CO ligands stabilize the intermediates but do not participate in the synthesis. However, as we have seen in Section 7.3, migratory insertions of carbon monoxide into metal–carbon bonds take place readily and can form a basis for stoichiometric synthetic approaches to useful organic carbonyl compounds.

Reaction (8.69), for example, which gives biallyl, can be modified to produce an α,γ-unsaturated acyl halide by carrying out the reaction under CO in solvents of moderate polarity, such as ether.

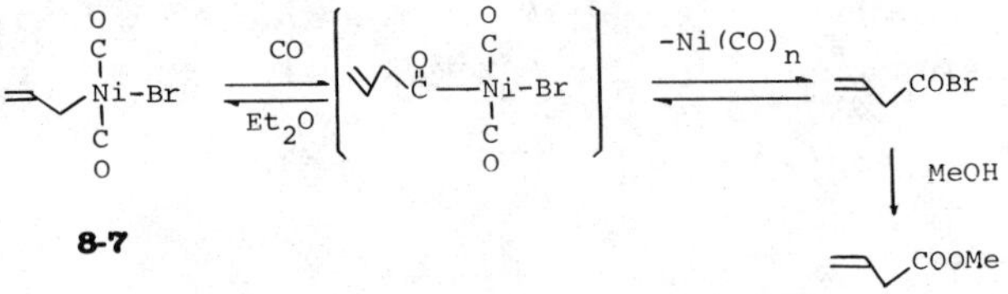

(8.72)

Other unsaturated compounds, such as alkenes, dienes, and alkynes, can be induced to insert into the allyl–Ni bond.

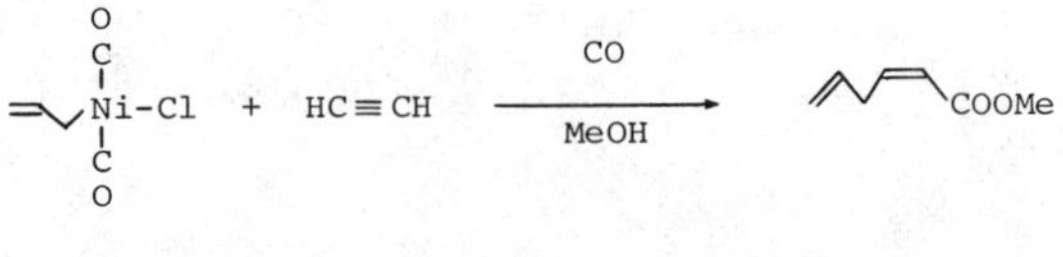

(8.73)

Anionic metal carbonyls are stronger nucleophiles than their neutral counterparts.[136-138] Anionic metal carbonyls can be prepared by treating neutral metal carbonyls with alkyllithium reagents or alkali metals (cf. Section 4.5).

$$Ni(CO)_4 + RLi \xrightarrow{-70^\circ} Li[RCONi(CO)_3]^{129}$$

$$Fe(CO)_5 + RLi \longrightarrow Li[RCOFe(CO)_4]^{129}$$

$$Fe(CO)_5 + Na(Hg) \xrightarrow{THF} Na_2[Fe(CO)_4]^{138}$$

$$Co_2(CO)_8 + Na(Hg) \longrightarrow Na[Co(CO)_4]^{53}$$

Reactions of anionic acyliron complexes with electrophiles such as allyl halides and vinyl halides give the organic carbonyl compounds expected from the *Umpolung* of the acyl group.

R^2X → R^1CR^2 (C=O)
$[R^1C^-]$ (C=O)
$CH_2=CHC(=O)R^2$ → $R^1CCH_2CH_2CR^2$ (C=O, C=O)
R^2COR^3 → $R^1C(=O)-C(OH)R^2R^3$

The reactivity of arenes is considerably altered when they are coordinated to transition metals having π-acceptor ligands such as CO; in particular, nucleophilic reactions can be carried out on the arenes. The approach has potential utility in organic synthesis, but details will not be treated here since we discussed the basic reactions of this type in Section 6.4.

The reactivity of dienes are also modified when they are bound to transition metals.[139] A useful synthetic method is to attach cyclohexadiene to the $Fe(CO)_3$ entity and convert it into a cationic cyclohexadienyl complex by abstracting hydride.

$$\text{cyclohexadiene (1,3- or 1,4-)} \xrightarrow{Fe(CO)_5} \text{(diene)}Fe(CO)_3 \xrightarrow[-H^-]{Ph_3C^+BF_4^-} [(C_6H_7)Fe(CO)_3]^+BF_4^-$$

The η^5-cyclohexadienyl complex reacts rapidly with nucleophiles. When the dienyl complex has a methoxy group, it exerts a directing effect such that nucleophiles add to the para terminus.

$$[(\text{OMe},\text{Me-}C_6H_5)Fe(CO)_3]^+BF_4^- \xrightarrow{NaR} (\text{OMe-diene, Me R})Fe(CO)_3 \xrightarrow[2)\ H_3O^+]{1)\ Me_3NO} \text{4-Me-4-R-cyclohex-2-enone}$$

An application of the above reaction is synthesis of spirocyclic compounds.

$$MeO_2C(CH_2)_n,\ CH(CO_2Me)_2\text{-substituted (OMe-diene)}Fe(CO)_3 \xrightarrow[\substack{2)\ Me_4N^+OAc^- \\ 3)\ NaH}]{1)\ Me_3NO} \text{spiro keto ester } (n = 3) + \text{spiro keto ester } (n = 2)$$

(products: OMe, O, CO_2Me; n = 3 and n = 2)

The reader is referred to references 1 and 53 in Chapter 7 for general reviews.

REFERENCES

1. Reviews on hydrogenation: (a) B. R. James, *Homogeneous Hydrogenation,* Wiley, New York, 1973; (b) *Adv. Organometal Chem.,* **17,** 319 (1979); (c) in *Comprehensive Organometallic Chemistry,* Vol. 8, G. Wilkinson, F. G. A. Stone, and E. W. Abel, eds., Pergamon Press, Oxford, 1982, p. 285; (d) R. S. Coffey, *Aspects of Homogeneous Catalysis,* **1,** 3 (1970); (e) J. Halpern, *J. Organometal. Chem.,* **200,** 133 (1980); (f) P. N. Rylander, *Cata-*

lytic Hydrogenation in Organic Syntheses, Academic Press, New York, 1979; (g) D. Valentine and J. W. Scott, *Synthesis,* 329 (1978); (h) P. J. Brothers, *Prog. Inorg. Chem.,* **28,** 1 (1981); (i) A. J. Birch and D. H. Williamson, *Org. Reactions,* **24,** 1 (1976); (j) F. J. McQuillin, *Homogeneous Hydrogenation in Organic Chemistry,* D. Reidel, Dordrecht, 1976; (k) F. H. Jardine, *Prog. Inorg. Chem.,* **28,** 63 (1981).

2. (a) M. Calvin, *J. Am. Chem. Soc.,* **61,** 2230 (1973); (b) *Trans. Faraday Soc.,* **34,** 1181 (1938).
3. (a) M. Iguchi, *Nippon Kagaku Kaishi* (*J. Chem. Soc. Jpn.*), **60,** 1287 (1939); (b) **63,** 634, 1752 (1942).
4. (a) J. F. Young, J. A. Osborn, F. H. Jardine, and G. Wilkinson, *Chem. Commun.,* 131 (1965); (b) D. Evans, J. A. Osborn, F. H. Jardine, and G. Wilkinson, *Nature,* **208,** 1203 (1965); (c) J. A. Osborn, F. H. Jardine, J. F. Young, and G. Wilkinson, *J. Chem. Soc., (A),* 1711 (1966).
5. L. Vaska and J. W. DiLuzio, *J. Am. Chem. Soc.,* **84,** 679 (1962).
6. J. Halpern, in *Organotransition Metal Chemistry,* Y. Ishii and M. Tsutsui, eds., Plenum Press, 1975, p. 109.
7. J. Halpern, A. S. C. Chan, P. P. Reley, and J. J. Pluth, *Adv. Chem. Ser.,* **173,** 16 (1979).
8. Reviews and books on asymmetric syntheses: (a) J. D. Morrison and H. S. Mosher, *Asymmetric Organic Reactions,* Prentice Hall, Englewood Cliffs, N. J., 1972; (b) Y. Izumi and A. Tai, *Stereodifferentiating Reactions,* Academic Press, New York, 1977; (c) P. Pino and G. Consiglio in *Fundamental Research in Homogeneous Catalysis,* M. Tsutsui and R. Ugo, eds., Plenum Press, New York, 1977; (d) D. Valentine and J. W. Scott, *Synthesis,* 329 (1978); (e) H. B. Kagan and J. C. Fiand, *Top. Stereochem.* **10,** 175 (1978); (f) J. Halpern, *Science,* **217,** 401 (1982); (g) W. S. Knowles, W. C. Christopfel, K. E. Koenig, and C. F. Hobbs, *Adv. Chem. Ser.,* **196,** 325 (1982); (h) B. Bosnich and N. K. Roberts, *Adv. Chem. Ser.,* **196,** 337 (1982); (i) J. M. Brown, P. A. Chaloner, and D. Parker, *Adv. Chem. Ser.,* **196,** 355 (1982); (j) W. S. Knowles, M. J. Sabacky, and B. D. Vineyard, *Chemtech,* 590 (1972); (k) W. S. Knowles, M. J. Sabacky, and B. D. Vineyard, *Ann. N. Y. Acad. Sci.,* **295,** 274 (1977); (l) B. D. Vineyard, W. S. Knowles, and M. J. Sabacky, *J. Mol. Cat.,* **19,** 159 (1983); (m) B. Bosnich and M. D. Fryzuk, *Top. Stereo Chem.,* **12,** 119 (1981); (n) J. D. Morrison, W. F. Masler, and M. K. Neuberg, *Adv. Catal.,* **25,** 81 (1976); (o) L. Horner, *Pure Appl. Chem.,* **52,** 843 (1983); (p) J. M. Brown, P. A. Chaloner, R. Glaser, and S. Geresh, *Tetrahedron,* **36,** 815 (1980).
9. (a) A. C. S. Chan and J. Halpern, *J. Am. Chem. Soc.,* **102,** 838 (1980); (b) A. C. S. Chan, J. J. Pluth, and J. Halpern, *J. Am. Chem. Soc.,* **102,** 5952 (1980); (c) *Inorg. Chim. Acta,* **37,** 2477 (1979).
10. (a) J. M. Brown and P. A. Chaloner, *Tetrahedron Lett.,* 1877 (1977); (b) *J. Chem. Soc. Chem. Commun.,* 321, 344 (1978).
11. J. D. Oliver and D. P. Riley, *Organometallics,* **2,** 1032 (1983).
12. (a) D. C. Bradley, R. C. Mehrotra, and D. P. Gaur, *Metal Alkoxides,* Academic Press, London, 1978; (b) M. A. Bennett and T. Yoshida, *J. Am. Chem. Soc.,* **95,** 3028 (1973); (c) J. Arthur, D. R. Robertson, D. A. Tocher, and T. A. Stephenson, *J. Organometal Chem.* **208,** 389 (1981); (d) S. Komiya, S. Tane-ichi, A. Yamamoto, and T. Yamamoto, *Bull. Chem. Soc. Jpn.* **53,** 673 (1980).
13. Y. Hayashi, S. Komiya, T. Yamamoto, and A. Yamamoto, *Chem. Lett.,* 1363 (1984).
14. (a) H. Imai, T. Nishiguchi, and K. Fukuzumi, *J. Org. Chem.,* **41,** 665 (1976); (b) R. Spogliarich, A. Tencich, J. Kasper, and M. Graziani, *J. Organometal. Chem.,* **240,** 453 (1982).
15. R. Spogliarich, G. Zassinovich, I. Kaspar, and M. Graziani, *J. Mol. Catal.,* **16,** 359 (1982) and references therein.
16. (a) G. P. Pez and R. A. Grey, *Fund. Res. Homog. Catal.,* **4,** 97 (1984); (b) G. P. Pez, R. A. Grey, and J. Corsi, *J. Am. Chem. Soc.,* **103,** 7528 (1981); (c) R. A. Grey, G. P. Pez, and A.

Wallo, *J. Am. Chem. Soc.*, **103,** 7536 (1981); (d) R. Wilczynski, W. A. Fordyce, and J. Halpern, *J. Am. Chem. Soc.*, **105,** 2066 (1983).

17. M. F. Sloan, A. S. Matlock, and D. S. Breslow, *J. Am. Chem. Soc.*, **95,** 4014 (1963).
18. E. L. Muetterties and J. R. Bleeke, *Acc. Chem. Res.*, **12,** 324 (1979).
19. P. M. Maitlis, *Acc. Chem. Res.*, **11,** 301 (1978).
20. M. A. Bennett, *Chemtech,* **10,** 444 (1980); M. A. Bennett, T.-N. Huang, and T. W. Turney, *J. Chem. Soc. Chem. Commun.*, 312 (1979).
21. K. R. Januszkiewicz and H. Alper, *Organometallics,* **2,** 1055 (1983).
22. J. L. Speier, *Adv. Organometal. Chem.*, **17,** 407 (1979).
23. C. S. Cundy, B. M. Kingston, and M. F. Lappert, *Adv. Organometal. Chem.*, **11,** 253 (1973).
24. J. F. Harrod and A. J. Chalk, in *Organic Syntheses via Metal Carbonyls,* Vol. 2, I. Wender and P. Pino, eds., Wiley-Interscience, New York, 1977, p. 673.
25. R. M. Meals, *Pure Appl. Chem.*, **13,** 141 (1966).
26. P. Svoboda, M. Capka, J. Hetflejš, and V. Chvalovsky, *Coll. Czech. Chem. Commun.*, **37,** 1585 (1972).
27. I. Ojima, K. Yamamoto, and M. Kumada, *Aspects Homog. Catal.*, **3,** 186 (1977).
28. E. S. Brown, *Aspects Homog. Catal.*, **2,** 57 (1974).
29. (a) G. Favero, M. Gaddi, A. Morvillo, and A. Turco, *J. Organometal. Chem.*, **149,** 395 (1978); (b) S. Komiya, Y. Abe, A. Yamamoto, and T. Yamamoto, *Organometallics,* **2,** 1466 (1983).
30. C. A. Tolman, W. C. Seidel, J. D. Druliner, and P. J. Domaille, *Organometallics,* **3,** 33 (1984).
31. D. Bingham, B. Hudson, D. E. Webster, and P. B. Wells, *J. Chem. Soc., Dalton Trans.*, 1521 (1974).
32. C. P. Casey and C. R. Cyr, *J. Am. Chem. Soc.*, **95,** 2248 (1973).
33. H. Werner and R. Freser, *Angew. Chem. Int. Ed. Engl.*, **18,** 157 (1979).
34. H. Bönnemann, *Angew. Chem.*, **82,** 699 (1970).
35. R. G. Miller, P. A. Pinke, R. D. Stauffer, H. J. Golden, and D. J. Baker, *J. Am. Chem. Soc.*, **96,** 4211, 4221, 4229 (1974).
36. (a) R. Noyori, in *Transition Metal Organometallics in Organic Synthesis,* Vol. 1, H. Alper, ed., 1976, p. 83; (b) J. P. Collman and L. S. Hegedus, *Principles and Applications of Organotransition Metal Chemistry,* University Science Book, Mill Valley, Calif., 1980; (c) J. K. Stille and D. E. James, in *The Chemistry of Double-Bonded Functional Groups,* Part 2, S. Patai, ed., Wiley, London, 1977, p. 1099. (d) H. M. Colquhoun, J. Holton, D. J. Thompson, and M. V. Twigg, *New Pathways for Organic Synthesis,* Plenum Press, New York, 1984.
37. M. Uchino, A. Yamamoto, and S. Ikeda, *J. Organometal. Chem.*, **24,** C64 (1970); **84,** 93 (1975).
38. (a) K. Tamao, K. Sumitani, and M. Kumada, *J. Am. Chem. Soc.*, **94,** 9268 (1972); (b) K. Tamao, K. Sumitani, Y. Kiso, M. Zembayashi, A. Fujioka, S. Kodama, I. Nakajima, A. Minato, and M. Kumada, *Bull. Chem. Soc. Jpn.*, **49,** 1958 (1976); (c) R. J. Corriu and J. P. Masse, *J. Chem. Soc. Chem. Commun.*, 144 (1972).
39. (a) A. Sekiya and N. Ishikawa, *J. Org. Chem.*, **41,** 2661 (1976); *J. Organometal. Chem.*, **118,** 349 (1976); (b) S.-I. Murahasi, M. Yamamura, K. Yanagisawa, N. Mita, and K. Kondo, *J. Org. Chem.*, **44,** 2408 (1979).
40. D. G. Morrel and J. K. Kochi, *J. Am. Chem. Soc.*, **97,** 7262 (1975).
41. G. Smith and J. K. Kochi, *J. Organometal. Chem.*, **198,** 199 (1980).

42. (a) A. Gillie and J. K. Stille, *J. Am. Chem. Soc.*, **102,** 4933 (1980); (b) F. Ozawa, T. Ito, Y. Nakamura, and A. Yamamoto, *Bull. Chem. Soc. Jpn.*, **54,** 1868 (1981).
43. H. Nakazawa, F. Ozawa, and A. Yamamoto, *Organometallics*, **2,** 241 (1983).
44. R. Uson, J. Fornies, P. Espinet, F. Martinez, and M. Tomas, *J. Chem. Soc. Dalton Trans.*, 463 (1981).
45. (a) T. Hayashi, M. Fukushima, M. Konishi, and M. Kumada, *Tetrahedron Lett.*, **21,** 79 (1980); (b) T. Hayashi, M. Konishi, M. Fukushima, T. Mise, M. Kagotani, M. Tajika, and M. Kumada, *J. Am. Chem. Soc.*, **104,** 180 (1982); (c) T. Hayashi, M. Tajika, K. Tamao, and M. Kumada, *J. Am. Chem. Soc.*, **98,** 3718 (1976).
46. (a) M. A. Bennett and R. Charles, *J. Am. Chem. Soc.*, **94,** 666 (1972); (b) D. P. Arnold, M. A. Bennett, G. T. Crisp, and J. C. Jeffery, *Adv. Chem. Ser.*, **196,** 195 (1982); (c) S. Komiya, Y. Morimoto, A. Yamamoto, and T. Yamamoto, *Organometallics*, **1,** 1528 (1982).
47. K. Tamao, S. Kodama, T. Nakatsu, Y. Kiso, and M. Kumada, *J. Am. Chem. Soc.*, **97,** 4405 (1975).
48. (a) T. Yamamoto, Y. Hayashi, and A. Yamamoto, *Bull. Chem. Soc. Jpn.*, **51,** 2091 (1978); (b) T. Yamamoto, K. Sanechika, and A. Yamamoto, *J. Polymer Sci., Polymer Lett. Ed.*, **18,** 9 (1980); (c) *Bull. Chem. Soc. Jpn*, **56,** 1497, 1503 (1983).
49. (a) W. Ishikawa and T. Kitazume, *Yuki Gosei Kagaku Kyokaishi* (*J. Synth. Org. Chem. Jpn.*), **41,** 432 (1983); (b) T. Kitazume and N. Ishikawa, *Chem. Lett.*, 137 (1982).
50. L. S. Hegedus, P. M. Kendall, S. M. Lo, and J. R. Sheats, *J. Am. Chem. Soc.*, **97,** 5448 (1975).
51. (a) K. Takagi, T. Okamoto, Y. Sakakibara, A. Ohno, S. Oka, and N. Hayama, *Chem. Lett.*, 951 (1975); (b) M. Kosugi, Y. Shimizu, and T. Migita, *J. Organometal. Chem.*, **129,** C36 (1977); (c) *Chem. Lett.*, 1423 (1977).
52. (a) D. Milstein and J. K. Stille, *J. Am. Chem. Soc.*, **100,** 3636 (1978); (b) *J. Org. Chem.*, **44,** 1613 (1979); (c) J. W. Labadie and J. K. Stille, *J. Am. Chem. Soc.*, **105,** 6129 (1983).
53. R. F. Heck, in *Organic Synthesis via Metal Carbonyls*, Vol. I, I. Wender and P. Pino, eds, Wiley, New York, 1968, p. 374.
54. M. Mori, K. Chiba, and Y. Ban. *J. Org. Chem.*, **43,** 1684 (1978).
55. F. Ozawa, H. Soyama, T. Yamamoto, and A. Yamamoto, *Tetrahedron Lett.*, **23,** 3383 (1982).
56. T. Kobayashi and M. Tanaka, *J. Organometal. Chem.*, **233,** C64 (1982).
57. (a) F. Ozawa, T. Sugimoto, Y. Yuasa, M. Santra, T. Yamamoto, and A. Yamamoto, *Organometallics*, **3,** 683 (1984); (b) F. Ozawa, T. Sugimoto, T. Yamamoto, and A. Yamamoto, *Organometallics*, **3,** 692 (1984); (c) J. Chen and A. Sen, *J. Am. Chem. Soc.*, **106,** 1506 (1984); (d) F. Ozawa, H. Soyama, H. Yanagihara, I. Aoyama, H. Takino, K. Izawa, T. Yamamoto, and A. Yamamoto, *J. Am. Chem. Soc.*, **107,** 3235 (1985).
58. S. Murai and N. Sonoda, *Angew, Chem., Int. Ed. Engl.*, **18,** 837 (1979); *Yuki Gosei Kagaku Kyokaishi* (*J. Synth. Org. Chem. Jpn.*), **39,** 301 (1981).
59. (a) R. F. Heck, *Acc. Chem. Res.*, **12,** 146 (1979); (b) *Pure Appl. Chem.*, **50,** 69 (1978); (c) *Org. React.* **27,** 345 (1982).
60. K. Mori, T. Mizoroki, and A. Ozaki, *Bull. Chem. Soc. Jpn.*, **46,** 1505 (1973).
61. (a) D. E. Bergstrom and J. L. Ruth, *J. Am. Chem. Soc.*, **98,** 1587 (1976); (b) D. E. Bergstrom and M. K. Ogawa, *J. Am. Chem. Soc.*, **100,** 8106 (1978); (c) C. F. Bigge, P. Kalaritis, and M. P. Mertes, *Tetrahedron Lett.*, 1653 (1979).
62. S.-I. Murahashi, M. Yamamura, and N. Mita, *J. Org. Chem.*, **42,** 2870 (1977).
63. A. J. Birch and I. D. Jenkins, in *Transition Metal Organometallics in Organic Synthesis*, Vol. I, H. Alper, ed., 1976, p. 1.
64. (a) T. Hayashi and L. Hegedus, *J. Am. Chem. Soc.*, **99,** 7093 (1977); (b) L. S. Hegedus, *J. Mol. Catal.* **19,** 201 (1983); (c) L. S. Hegedus and K. Siirala-Hansén, *J. Am. Chem. Soc.*,

97, 1184 (1975); (d) D. Madema, R. van Helden, and C. F. Kohl, *Inorg. Chim. Acta,* **3,** 255 (1969).

65. B. Åkermark, J. E. Bäckvall, K. Siirala-Hansén, K. Sjöberg, and K. Zetterberg, *Tetrahedron Lett.,* **15,** 1363 (1974).
66. R. A. Holton, *J. Am. Chem. Soc.,* **99,** 8083 (1977).
67. T. Hosokawa, S. Miyagi, S.-I. Murahashi, and A. Sonoda, *J. Org. Chem.,* **43,** 2752 (1978).
68. (a) L. S. Hegedus, G. F. Allen, and E. L. Waterman, *J. Am. Chem. Soc.,* **98,** 2674 (1976); (b) L. S. Hegedus, G. F. Allen, J. J. Bozell, and E. L. Waterman, *J. Am. Chem. Soc.,* **100,** 5800 (1978).
69. (a) S. G. Davies, *Organotransition Metal Chemistry: Application to Organic Synthesis,* Pergamon Press, Oxford, 1982; (b) R. Scheffold, *Modern Synthetic Methods,* Vol. 3, *Transition Metals in Organic Synthesis,* Otto Salle Verlag, Frankfurt, 1983.
70. P. Lennon, A. M. Rosan, and M. Rosenblum, *J. Am. Chem. Soc.,* **99,** 8426 (1977) and references cited therein.
71. (a) P. K. Wong, M. Madhavarao, E. F. Marten, and M. Rosenblum, *J. Am. Chem. Soc.,* **99,** 2823 (1977); (b) S. R. Berryhill and M. Rosenblum, *J. Org. Chem.,* **45,** 1984 (1980).
72. (a) B. M. Trost, *Tetrahedron,* **33,** 2615 (1977); (b) B. M. Trost and T. R. Verhoven in *Comprehensive Organometallic Chemistry,* Vol. 8, G. Wilkinson, F. G. A. Stone, and E. W. Abel, eds., Pergamon Press, Oxford, 1982, p. 299.
73. J. Tsuji, *Organic Synthesis by Means of Transition Metal Complexes,* Springer-Verlag, Berlin, 1975.
74. J. Tsuji, *Organic Synthesis with Palladium Compounds,* Springer-Verlag, Berlin, 1980.
75. G. P. Chiusoli, *Pure Appl. Chem.,* **52,** 635 (1980).
76. B. M. Trost, L. Weber, P. E. Strege, T. J. Fullerton, and T. J. Dietsche, *J. Am. Chem. Soc.,* **100,** 3416 (1978).
77. (a) J. Tsuji, H. Takahashi, and M. Morikawa, *Tetrahedron Lett.,* 4387 (1965); (b) J. Tsuji, *Bull. Chem. Soc. Jpn.,* **46,** 1896 (1973).
78. T. Yamamoto, O. Saito, and A. Yamamoto, *J. Am. Soc.,* **103,** 5600 (1981).
79. T. Hayashi, T. Hagihara, M. Konishi, and M. Kumada, *J. Am. Chem. Soc.,* **105,** 7767 (1983).
80. B. M. Trost and P. E. Strege, *J. Am. Chem. Soc.,* **99,** 1649 (1974).
81. (a) T. Hayashi, in *Asymmetric Reactions and Processes in Chemistry,* E. L. Eliel and S. Otsuka, eds., ACS Symposium Series No. 185, American Chemical Society, Washington, D.C., 1982; (b) T. Hayashi, M. Konishi, M. Fukushima, K. Kanehira, T. Hioki, and M. Kumada, *J. Org. Chem.,* **48,** 2195 (1983).
82. J. Tsuji, *Yuki Gosei Kagaku Kyokaishi* (*J. Synth. Org. Chem. Jpn.*), **41,** 619 (1983).
83. (a) J. Tsuji and T. Yamakawa, *Tetrahedron Lett.,* **20,** 613 (1979); (b) J. Tsuji, I. Shimizu, and I. Minami, *Chem. Lett.,* 1017 (1984).
84. (a) J. Tsuji, K. Sato, and H. Okumoto, *Tetrahedron Lett.,* **23,** 5189 (1982); (b) *J. Org. Chem.* **49,** 1341 (1984).
85. T. Hayashi, M. Konishi, K. Yokota, and M. Kumada, *J. Chem. Soc. Chem. Commun.,* 313 (1981).
86. (a) H. Felkin and G. Swierczewski, *Tetrahedron,* **31,** 2735 (1975); (b) C. Chuit, H. Felkin, C. Frajerman, G. Roussi, and G. Swierczewski, *J. Organometal. Chem.,* **127,** 371 (1977); (c) H. Felkin, E. Jampel-Costa, and G. Swierczewski, *J. Organometal. Chem.,* **134,** 265 (1977); (d) B. L. Buckwalter, I. R. Burfitt, H. Felkin, M. Joly-Goudket, K. Naemura, M. F. Solomon, E. Wenkert, and P. M. Wovkulich, *J. Am. Chem. Soc.,* **100,** 6445 (1978).
87. G. Consiglio, F. Morandini, and O. Piccolo, *J. Am. Chem. Soc.,* **103,** 1846 (1981).
88. (a) G. Consiglio, F. Morandini, and O. Piccolo, *Helv. Chim. Acta,* **63,** 987 (1980); (b) *J. Chem. Soc. Chem. Commun.,* 112 (1983).

89. B. M. Trost and D. M. T. Chan, *J. Am. Chem. Soc.*, **105,** 2315, 2326 (1983).
90. (a) Y. Fujiwara, R. Asano, I. Moritani, and S. Teranishi, *J. Org. Chem.*, **41,** 1681 (1976); (b) I. Moritani and Y. Fujiwara, *Tetrahedron Lett.*, **12,** 1119 (1967); (c) Y. Fujiwara, I. Moritani, S. Danno, R. Asano, and S. Teranishi, *J. Am. Chem. Soc.*, **91,** 7166 (1969).
91. H.-P. Abicht and K. Issleib, *Z. Chem.*, **17,** 1 (1977).
92. M. I. Bruce, *Angew. Chem. Int. Ed. Engl.*, **16,** 73 (1977).
93. H. Horino and N. Inoue, *J. Org. Chem.*, **46,** 4416 (1981).
94. (a) K. Ohno and J. Tsuji, *J. Am. Chem. Soc.*, **90,** 99 (1968); (b) *Tetrahedron Lett.*, 2173 (1967).
95. H. H. Walborsky and L. E. Allen, *J. Am. Chem. Soc.*, **93,** 5465 (1971).
96. (a) J. Tsuji, *Organic Synthesis via Metal Carbonyls,* Vol. II, I. Wender and P. Pino, eds., Wiley-Interscience, New York, 1977, p. 595; (b) M. C. Baird, in *The Chemistry of Acid Derivatives,* Suppl. B, Part 2, S. Patai, ed., Wiley, Chichester, 1979, p. 825.
97. J. W. Suggs, *J. Am. Chem. Soc.*, **100,** 630 (1978).
98. K. S. Y. Lau, Y. Becker, F. Huang, N. Baenziger, and J. K. Stille, *J. Am. Chem. Soc.*, **99,** 5664 (1977).
99. T. Goto, M. Onaka, and T. Mukaiyama, *Chem. Lett.*, 709 (1980).
100. J. Schwarz and J. A. Labinger, *Angew. Chem. Int. Ed. Engl.*, **15,** 333 (1976).
101. D. B. Carr, M. Yoshifuji, L. I. Shoer, K. I. Gall, and J. Schwarz, *Ann. N. Y. Acad. Sci.*, **295,** 127 (1977).
102. (a) E. Negishi, *Acc. Chem. Res.*, **15,** 340 (1982); (b) in *New Applications of Organometallic Reagents in Organic Synthesis,* D. Seyferth, ed., Elsevier, New York, 1976, p. 93.
103. D. B. Carr and J. Schwarz, *J. Am. Chem. Soc.*, **99,** 638 (1977).
104. F. Sato, in *Fundamental Research in Homogeneous Catalysis,* Vol. 2, Y. Ishii and M. Tsutsui, eds., Plenum Press, New York, 1978, p. 81.
105. F. Sato, S. Sato, and M. Sato, *J. Organometal. Chem.*, **131,** C26 (1977).
106. K. Isagawa, K. Tatsumi, and Y. Otsuji, *Chem. Lett.*, 1145 (1976).
107. E. C. Ashby, J. J. Lin, and A. B. Goel, *J. Org. Chem.*, **43,** 2567 (1978).
108. F. Sato, T. Jinbo, and M. Sato, *Tetrahedron Lett.*, **21,** 2171, 2175 (1980).
109. P. G. Gassman and T. H. Johnson, *J. Am. Chem. Soc.*, **98,** 6055 (1976).
110. R. G. Salomon and J. K. Kochi, *J. Am. Chem. Soc.*, **95,** 3330 (1973).
111. (a) A. Nakamura, A. Konishi, Y. Tatsuno, and S. Otsuka, *J. Am. Chem. Soc.*, **100,** 3443 (1978); (b) A. Nakamura, A. Konishi, R. Tsujitani, and S. Otsuka, *J. Am. Chem. Soc.*, **100,** 3449 (1978).
112. A. J. Anciaux, A. J. Hubert, A. F. Noels, N. Periniot, and P. Teyssié, *J. Org. Chem.*, **45,** 695 (1980).
113. M. P. Doyle and J. G. Davidson, *J. Org. Chem.*, **45,** 1538 (1980).
114. (a) H. Kanai, Y. Nishiguchi, and H. Matsuda, *Bull. Chem. Soc. Jpn.*, **56,** 1592 (1983); (b) H. Kanai, N. Hiraki, and S. Ikeda, *Bull. Chem. Soc. Jpn.*, **56,** 1025 (1983).
115. T. Aratani, Y. Yoneyoshi, and T. Nagase, *Tetrahedron Lett.*, 2599 (1977).
116. G. H. Posner, *An Introduction to Synthesis Using Organocopper Reagents,* Wiley, New York, 1980.
117. G. H. Posner, *Org. React.*, **22,** 253 (1975); **19,** 1 (1972).
118. H. O. House, *Acc. Chem. Res.*, **9,** 59 (1976).
119. R. G. R. Bacon and H. A. O. Hill, *Quart. Rev.*, **19,** 95 (1965).
120. (a) J. F. Normant, *Synthesis,* 63 (1972); (b) *Pure Appl. Chem.*, **50,** 709 (1978).
121. (a) H. Gilman and L. A. Wood, *J. Am. Chem. Soc.*, **65,** 435 (1943); (b) C. E. H. Bawn and

F. J. Whitby, *Disc. Faraday Soc.*, **7,** 228 (1947); (c) G. Costa, A. Camus, and E. Pauluzzi, *Gazz. Chim. Ital.*, **86,** 997 (1956).

122. (a) A. Yamamoto, A. Miyashita, T. Yamamoto, and S. Ikeda, *Bull. Chem. Soc. Jpn.*, **45,** 1583 (1972); (b) T. Ikariya and A. Yamamoto, *J. Organometal. Chem.*, **72,** 145 (1974); (c) A. Miyashita and A. Yamamoto, *Bull. Chem. Soc. Jpn.*, **50,** 1102 (1977).
123. (a) G. M. Whitesides, E. R. Stedronsky, C. P. Casey, and S. Filippo, Jr., *J. Am. Chem. Soc.*, **92,** 1426 (1970); (b) G. M. Whitesides, E. J. Panek, and E. R. Stedronsky, *J. Am. Chem. Soc.*, **94,** 232 (1972); (c) M. Tamura and J. K. Kochi, *J. Am. Chem. Soc.*, **93,** 1485 (1971).
124. (a) R. G. Pearson and C. D. Gregory, *J. Am. Chem. Soc.*, **98,** 4098 (1976); (b) J. S. Filippo, Jr., *Inorg. Chem.*, **17,** 275 (1978).
125. C. R. Johnson and G. A. Dutra, *J. Am. Chem. Soc.*, **95,** 7783 (1973).
126. (a) G. H. Posner, C. E. Whitten, and J. J. Sterling, *J. Am. Chem. Soc.*, **95,** 7788 (1973); (b) H. Westmijze, J. Meijer, and D. Vermeer, *Receuil,* **96,** 168 (1977).
127. B. C. Hartman, T. Livinghouse, and B. Rickborn, *J. Org. Chem.*, **38,** 4346 (1973).
128. G. H. Posner, J. J. Sterling, C. E. Whitten, C. M. Lenz, and D. J. Brunnelle, *J. Am. Chem. Soc.*, **97,** 101 (1975).
129. (a) M. Ryang, *Organometal. Chem. Rev.*, **A5,** 67 (1970); (b) M. Ryang and S. Tsutsumi, *Synthesis,* 55 (1971); (c) L. Cassar, G. P. Chiusoli, and F. Guerrieri, *Synthesis,* 509 (1973); (d) H. Alper, *J. Organometal. Chem. Libr.*, **1,** 305 (1976).
130. I. D. Webb and G. T. Borcherdt, *J. Am. Chem. Soc.*, **73,** 2654 (1951).
131. (a) E. J. Corey and E. Hamanaka, *J. Am. Chem. Soc.*, **89,** 2758 (1967); (b) E. J. Corey and E. Wat, *J. Am. Chem. Soc.*, **89,** 2757 (1967); (c) E. J. Corey and J. A. Kirst, *J. Am. Chem. Soc.*, **94,** 667 (1972).
132. (a) M. F. Semmelhack, *Org. React,* **19,** 115 (1972); (b) E. J. Corey and M. F. Semmelhack, *J. Am. Chem. Soc.*, **89,** 2755 (1967).
133. R. Noyori, *Acc. Chem. Res.*, **12,** 61 (1979).
134. R. Noyori, *Transition Metal Organometallics in Organic Synthesis,* H. Alper, ed., Academic Press, New York, 1976.
135. (a) R. Noyori, T. Sato, and Y. Hayakawa, *J. Am. Chem. Soc.*, **100,** 2561 (1978); (b) R. Noyori and Y. Hayakawa, *Org. React.*, **29,** 163 (1983).
136. R. F. Heck, *Acc. Chem. Res.*, **2,** 10 (1967).
137. F. Guerrieri and G. P. Chiusoli, *J. Organometal. Chem.*, **15,** 209 (1968).
138. J. P. Collman, *Acc. Chem. Res.*, **8,** 342 (1975).
139. (a) A. J. Pearson, *Acc. Chem. Res.*, **13,** 463 (1980); (b) A. J. Birch, *Tetrahedron,* **37,** Supplement 1, 289 (1981).

Chapter 9

Topics Related to Catalysis and Organometallics

In this last chapter we shall deal with topics that are related to organometallics but have not been discussed in previous chapters. The main topics include biologically active catalyst systems, supported transition metal catalysts, and transition metal clusters.

9.1. ORGANOMETALLICS IN BIOINORGANIC CHEMISTRY

Although a variety of metals is known to serve as catalysts in biological systems,[1] only coenzyme B_{12} has been unequivocally established to have a metal–carbon σ bond. The structure of vitamin B_{12} coenzyme is shown in Figure 9.1.

In coenzyme B_{12}, cobalt is situated at the center of a conjugated macrocyclic ring composed of four pyrrole derivatives. The cobalt atom is directly bonded to the 5′ carbon atom of the deoxyribose moiety, which is further connected to adenine. The combination of ribose with adenine is called adenosine. On the other side of the planar macrocyclic ring cobalt is attached to a dimethylbenzimidazole base that is linked with the side chain of a pyrrole ring through ribose and phosphate entities. Other biologically important metal complexes containing tetrapyrrole rings include the iron-containing family of hemoglobin, myoglobin, and cytochromes and the magnesium-containing chlorophyll. These complexes all possess the porphin ring, as shown in Figure 9.2*a,* composed of four pyrrole rings connected with methine bridges.[2]

Coenzyme B_{12} differs from these biologically active metal complexes in one respect. Instead of the porphin ring, it contains a *corrin* ring, which is composed of four pyrrole rings joined by three methine bridges and one direct,

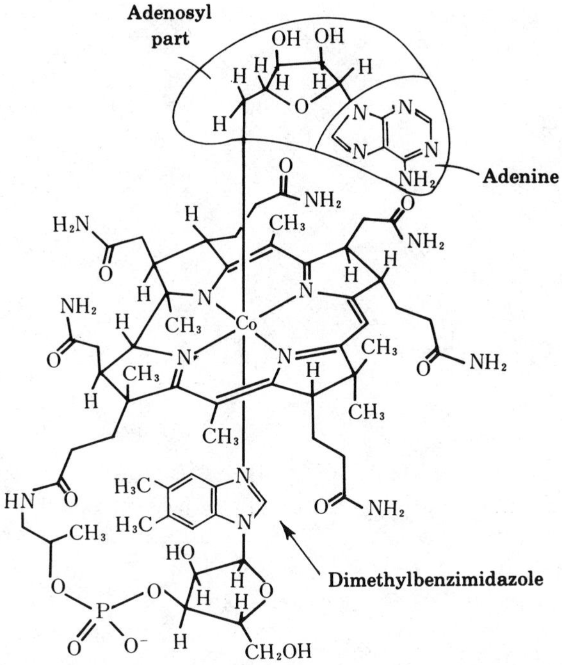

Adenosyl part
Adenine
Dimethylbenzimidazole

9-1

FIGURE 9.1. The structure of vitamin B_{12} coenzyme.

(a) porphin
(b) corrin
(c) phthalocyanine (tetrazaporphyrin)

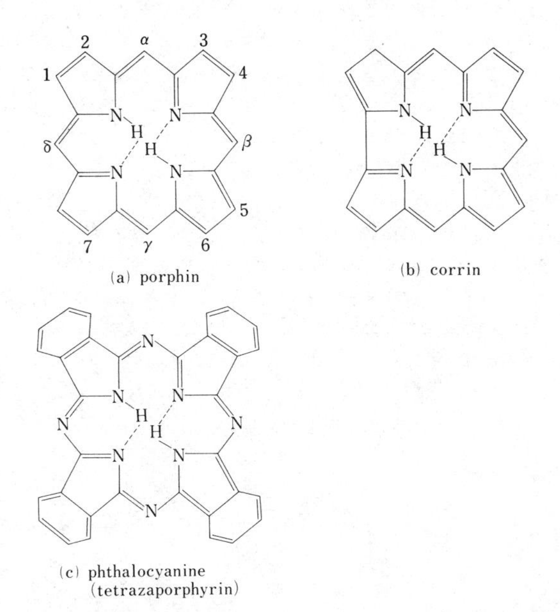

FIGURE 9.2. Porphyrin and its relatives.

ring-to-ring bond. Various groups such as methyl, ethyl, and vinyl are attached to the four pyrrole rings at the 1–8 positions and at the methine bridges, α, β, γ, and δ in the porphin or corrin ring. The substituted porphins are called porphyrins. The porphyrins and corrin are constructed by complicated biosynthetic sequences from building blocks such as 5-aminolevulinic acid.[2i]

```
                                              COOH
                                              |
                                         COOH CH2
                                         |    |
 COOH        COOH                        CH2  CH2
 CH2         CH2                           \  /
 |           CH2                           ====
 CH2------>  C=O       -2H2O              /    \
 |           |         ------->          \      /     ---------> porphyrin
 O=C <-.     CH2                          \ N  /
 CH2    `--- NH2                       CH2  H
 |                                     |
 NH2                                   NH2

 5-aminolevulinic acid                 porphobilinogen
```

In contrast to naturally occurring porphyrins, which are biosynthesized as shown above, a man-made pigment called phthalocyanine (Fig. 9.2*c*, tetrazaporphyrin) can be prepared in a one-pot reaction, for example, by treating phthalonitrile (*o*-dicyanobenzene) with metal ions.[3]

The two protons in the tetrapyrrole ring systems shown in Figure 9.2 can be replaced by divalent or trivalent metal ions to form metal complexes. The metal-containing tetrapyrrole ring systems exhibit specific colors arising from the metal-containing conjugated macrocycles. Note that the metal-containing conjugated systems in porphyrins extend over a wider range than that of the corrin system. The difference in the ring structures between the porphyrin and the corrin probably affects the rigidity of the molecular plane formed by the tetrapyrrole rings as well as the electronic state of the molecule and the Co–C bond strength. However, the precise reason for the requirement of the corrin ring for coenzyme B_{12} activity still remains to be clarified.

The groups attached to the cobalt atom above and below the corrin ring can be replaced by other groups. In the structure shown in Figure 9.1 the entity without the adenosine is called cobalamin. Replacement of the adenosyl group with the methyl group leads to methylcobalamin, and substitution by a CN group gives cyanocobalamin. The adult human body contains 2–5 mg of coenzyme B_{12} and its derivatives, mainly in the liver. Coenzyme B_{12} is effective in the treatment of pernicious anemia. Vitamin B_{12} was first isolated in 1958 in the form of cyanocobalamin.[4] Thus, cyanocobalamin is called vitamin B_{12}, and the adenosyl-bonded cobalamin is called coenzyme B_{12}, even though cyanocobalamin itself is not present in biological systems and is not catalytically active.

Challenge and Conquest

The structure of vitamin B_{12} was determined by a British X-ray crystallographer Dorothy Crowfoot-Hodgkin.[5] The work was accomplished by means of a primitive computer, before the advent of modern computer-assisted X-ray crystallography. At that time it must have been desperately time-consuming and laborious work to solve the structure of such a complicated large molecule. Nevertheless, she was equal to this formidable challenge and clarified the structure after eight years. She was awarded the Nobel prize for her work in 1964.

An equally difficult task is the synthesis of complicated molecule such as coenzyme B_{12}, which has many chiral centers. For some synthetic chemists, however, this is a challenge that must be accepted. The structure of vitamin B_{12}, which to most chemists looks terribly complicated, appealed to Woodward as a "lovely structure." Whereas Hilary climbed Mt. Everest because "it is there," the established structure of vitamin B_{12} must have motivated an ambitious synthetic chemist as Woodward to start its total synthesis. However, even for Woodward, who succeeded in the total syntheses of many extremely complicated compounds such as chlorophyll, quinine, cholesterol, strychinine, and reserpine, the total synthesis of vitamin B_{12} was a tough job. Eventually, however, the total synthesis was achieved in 1972 with the cooperation of Eschenmoser in Switzerland. It required 65–70 synthetic steps resulting from the effort of 99 chemists in 11 years (J. H. Krieger, *Chem. Eng. News,* March 12, 1973).

Coenzyme B_{12} catalyzes several types of important biological reactions. The most remarkable are rearrangements represented by the following generalized process.

$$\mathrm{-\overset{|}{\underset{R}{C}}-\overset{H}{\underset{|}{C}}-} \rightleftharpoons \mathrm{-\overset{H}{\underset{|}{C}}-\overset{|}{\underset{R}{C}}-} \tag{9.1}$$

The group R may be a carbon atom with substituents, the oxygen atom of an alcohol, or an amino group. An example of the skeletal rearrangement reaction is

$$\mathrm{HOOC{-}\underset{CH_3}{CH}{-}\underset{O}{\overset{}{C}}{-}S{-}R \longrightarrow HOOC{-}CH_2{-}CH_2{-}\underset{O}{C}{-}S{-}R} \tag{9.2}$$

In this process C–C bond breaking and making as well as C–H bond breaking and making take place simultaneously. Such chemical reactions are difficult to carry out by conventional synthetic means. Although the detailed mechanism has not been established, it is quite likely that facile Co–C breaking and making are responsible for the rearrangement reaction. It is also probable that a change of the valence state of cobalt is important in the process.

Coenzyme B_{12}, as shown in Figure 9.1, contains an adenosyl moiety bound to cobalt, and the metal atom is formally in the +3 oxidation state. One electron reduction gives a Co(II) species called coenzyme B_{12r}, and further reduction gives B_{12s}, which contains Co(I). Such reductions can be achieved in biological systems by FAD (flavion adenine dinucleotide).[1e] The reduced cobalamin B_{12s} is a powerful nucleophile. Like the low-valent transition metal complexes discussed in Section 6.2, it reacts with CH_3I to give methylcobalamin.

$$[B_{12}(Co^{I})] + CH_3I \longrightarrow [B_{12}(Co^{III})\text{—}CH_3]^+ + I^- \tag{9.3}$$

Similar nucleophilic attack of B_{12s} at the C-5 of the ribose moiety of ATP (adenosine triphosphate) is believed to expel a triphosphate anion as the leaving group and to form a C–Co bond, giving coenzyme $B_{12}(Co^{III})$, thus achieving an unusual adenosyl transfer reaction.

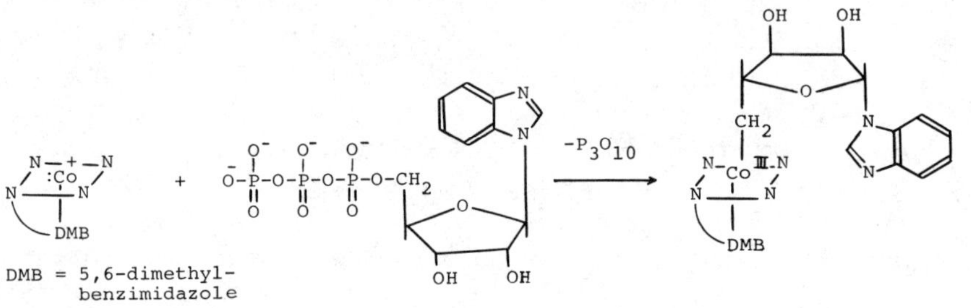

$-P_3O_{10}$

DMB = 5,6-dimethylbenzimidazole

(9.4)

Methylcobalamin is also formed by reaction of a B_{12s}-bound enzyme with methyltetrahydrofolate. An important role of methylcobalamin in the biological system is the methyl transfer reaction to give methionine.[1e]

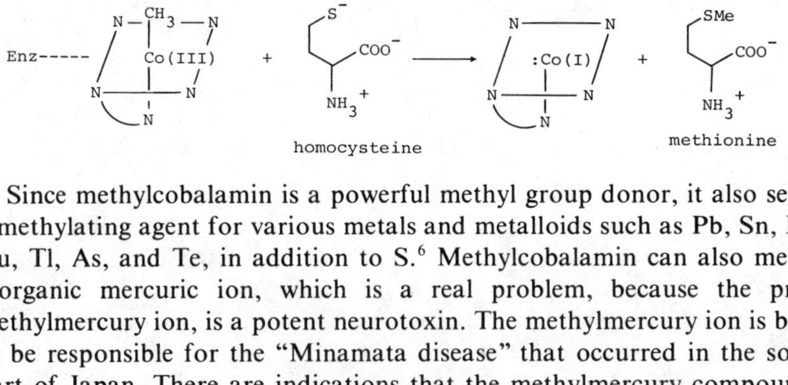

homocysteine methionine (9.5)

Since methylcobalamin is a powerful methyl group donor, it also serves as a methylating agent for various metals and metalloids such as Pb, Sn, Pd, Pt, Au, Tl, As, and Te, in addition to S.[6] Methylcobalamin can also methylate inorganic mercuric ion, which is a real problem, because the product, methylmercury ion, is a potent neurotoxin. The methylmercury ion is believed to be responsible for the "Minamata disease" that occurred in the southern part of Japan. There are indications that the methylmercury compound was formed in a factory then manufacturing acetaldehyde by an old process using a Hg(II) compound. The waste of the process, which contained the methylmercury compound, was discharged into a river where it subsequently entered the fish. Many residents in the city who ate the fish containing the methylmercury compound were killed or disabled. This incident triggered the movement to clean up the environment and to tighten restrictions against the discharge of waste products. Fundamental studies on coenzyme B_{12} point to the danger of conversion of inorganic mercurials into the most toxic methylmercury compound. Thus, the methylmercury compound may be formed without being discharged from a factory. In the biological system, however, there is another kind of bacterium that decomposes methylmercury to mercury, and the balance is somehow maintained. This, of course, does not exempt mankind from the responsibility of not discharging mercury and its compounds into the environment.

Although coenzyme B_{12} is a very complicated complex, as is evident from Figure 9.1, a seemingly simple cobalt complex containing dimethylglyoxime is capable of mimicking much of the chemical behavior of coenzyme B_{12}, as was demonstrated by Schrauzer.[7] Dimethylglyoxime (DMG) is a chelating agent used in the analysis of nickel, but it reacts with Co(III) as well to give a complex (**9-2**) where the two glyoxime units form a planar ring around cobalt.

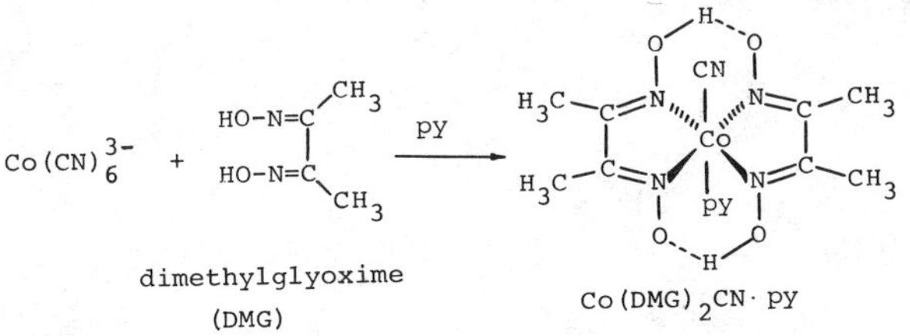

$Co(CN)_6^{3-}$ + dimethylglyoxime (DMG) $\xrightarrow{py}$ $Co(DMG)_2CN \cdot py$

9-2

Complex **9-2** has been named *cobaloxime* by analogy with cobalamin. On treatment with sodium amalgam, $Co(III)(DMG)_2CN$ is reduced to a Co(I)–DMG complex, which is a potent nucleophile. Like cobalamin B_{12s}, it undergoes oxidative addition with CH_3I to produce methylcobaloxime, which exhibits similar behavior to methylcobalamin. The crystal structure of a substituted alkylcobaloxime shows that the Co–N (in-plane) and Co–C bond lengths are very similar to those found for coenzyme B_{12}.[8] In addition to cobaloxime, other cobalt complexes, such as **9-3** and **9-4**, that have planar ligands have been synthesized and their chemical behavior studied.

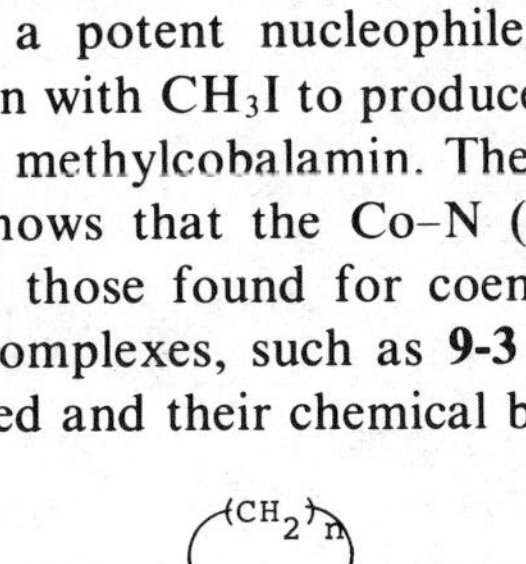

9-3

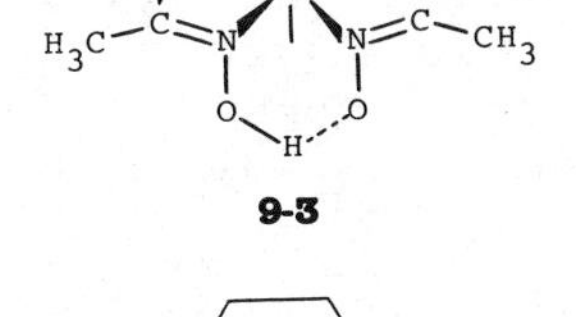

Co(salen)

9-4

The Co–C bond strengths in cobalamins[9] and cobaloxime containing various nitrogen bases[10] have been recently estimated to be in the range of 70–100 kJ mol^{-1}. These values are smaller than the M–C bond dissociation energies of the other transition metal alkyls listed in Table 3.2. The smaller Co–C bond dissociation energy may be associated with the ability of alkylcobalamins and alkylcobaloximes to undergo facile homolytic Co–C bond cleavage, either thermally or photochemically. Although the precise mechanisms for the skeletal rearrangements and alkyl transfer reactions promoted by the coenzyme B_{12} have not been established, proposals based on homolytic cleavage of the Co–C bond have been made,[2c,d] and these are consistent with the weak Co–C bond strength. Cleavage of an α-cyanoethyl group from cobalt can be induced by light or even by x rays used for crystallographic analysis.[11] This homolysis causes racemization of *S*-1-cyanoethylcobaloxime and isomerization of β-cyanoethylcobaloxime to the α-cyano isomer, a process that may have implications for the activity of coenzyme B_{12}, as represented by Eq. (9.1).

$$\underset{(\text{Co})}{CH_2}\text{—}CH_2CN \longrightarrow H\text{—}\underset{(\text{Co})}{\overset{CH_3}{C}}\text{—}CN$$

An interesting and biologically significant aspect of alkylcobalt complexes is that an alkylcobalt complex having a porphyrin ring system shows considerably different behavior from that of coenzyme B_{12}. The Co(II)-containing hemoglobin, called coboglobin, takes up O_2 reversibly as Fe(II)-containing hemoglobin does,[12] but coboglobin shows resistance to reduction to the Co(I) state. Thus, the difference between the porphyrin and corrin systems considerably affects their biological activity.[13]

Although coenzyme B_{12} has long been considered as the only organotransition metal complex that occurs in biological systems and has definite biological activity, recent studies show that other transition metal complexes with metal–carbon bonds may be involved in biological reactions.

Cytochrome P-450 is a hemoprotein that shows an absorption maximum at 450 nm on complexation with carbon monoxide. The hemoprotein has monooxygenase activity and also promotes epoxidation of olefins.[1e] Recently it was found that cytochrome P-450 is destroyed after it has catalyzed epoxidation and that the iron porphyrin system is converted into an *N*-alkylated compound with a 2-hydroxyethyl group attached to a nitrogen atom in one of the tetrapyrrole rings. Acetylene also serves to destroy the cytochrome P-450 and induces *N*-alkylation with the formylmethyl group. Furthermore, an alkyl group bound to an Fe(III) or Co(III) porphyrin was shown to be transferred reversibly to one of the nitrogen atoms in the tetrapyrrole ligand.[14-16]

$$\text{(N}_4\text{)Fe–R} \underset{S_2O_4Na_2}{\overset{FeCl_3,\ -20^\circ}{\rightleftharpoons}} \text{(N}_3\text{N–R)Fe–Cl} \qquad (9.6)$$

R = Me, Ph, $-CH{=}CH_2$

9-5 **9-6**

Thus, if a 2-hydroxyethyl group were somehow generated on the iron atom of an iron porphyrin, it is likely that the group would be transferred to one of the nitrogen atoms in the porphyrin ring. Ethylene is a possible starting material for the formation of a 2-hydroxyethyl moiety. Although it may sound surprising for those who regard ethylene only as a petrochemical feedstock, ethylene is known as a ubiquitous hormone that promotes germination, growth, fruit ripening, senescence, and abscission. For example, bananas are transported to and stored in big cities while they are green. Ripening can be triggered by treatment with ethylene. Ethylene released from a piece of fruit may also accelerate the ripening of other fruits; an ethylene absorber is used to prevent this contagious effect during transportation. The receptor site for ethylene in plants has not been identified, but a transition metal complex is thought to be a likely candidate. Recently a copper complex bound with ethylene in the side-on manner has been characterized, and copper has been proposed as the binding site.[17] However, the known process of 2-hydroxyethyl group transfer from iron to nitrogen in the porphyrin system suggests the possibility of the following process in which OH^- nucleophilically attacks ethylene η^2-bonded to iron, thus generating a coordinated 2-hydroxyethyl moiety.

^-OH $H_2C{=}CH_2$ N N M N N X   →   OH CH$_2$ CH$_2$ N N M N N   →   OH CH$_2$ CH$_2$ N N M N N   (9.7)

The hydroxyethyl group can then be transferred to nitrogen in the porphyrin ring. The process of external nucleophilic attack on ethylene is reminiscent of the Wacker process for the Pd(II)-promoted conversion of ethylene into acetaldehyde. The proposal here is admittedly speculative, and we must await further experimental corroboration to see if any organometallic intermediates are involved.† However, it seems likely that coenzyme B_{12} does not represent the only biologically active organometallic compound and that other biologically active organometallic compounds will be discovered.

9.2. SUPPORTED TRANSITION METAL CATALYSTS

Homogeneous transition metal complex catalysts have the advantage of performing selective catalytic reactions under usually mild conditions. The high selectivities arise from the discrete, well-defined structures of transition

† In a currently accepted proposal oxoiron (V) species is considered responsible for the epoxidation of olefins.

metal complexes, which are amenable to modern physicochemical means of identification. A serious disadvantage of homogeneous transition metal complex catalysts lies in the difficulty in separating them from the products. In some cases, for example, the oxo process, the products are easily separated by simple means, such as distillation. However, it is difficult to separate high boiling point products from thermally unstable catalysts. Heterogenization of soluble catalysts by supporting them on organic polymers or inorganic solids could, in principle, remove these disadvantages. Such catalysts have been called variously, hybrid, supported, anchored, immobilized, fixed, grafted, or linked catalysts.[19]

In addition to the improvement in separation ability, supporting of homogeneous catalysts can be expected to improve their efficiency because the aggregation of catalytically active species that can occur in homogeneous catalysts can be blocked in the heterogenized system. Catalyst properties also can be modified by putting the transition metal catalyst on solid supports, although examples of successful application of the technique are still limited.

The commonly employed techniques of supporting the transition metal complexes are (1) anchoring them to cross-linked polystyrene and (2) fixing them on a solid surface of silica or alumina.

Cross-linked polystyrenes are prepared by copolymerizing styrene with a limited amount of divinylbenzene. Tertiary phosphines acting as ligands for transition metal complexes can be introduced into the monomeric styrene or alternatively into the phenyl ring of polystyrene as shown below.

$$\text{P}\!-\!C_6H_5 \xrightarrow[SnCl_4]{MeOCH_2Cl} \text{P}\!-\!C_6H_4\!-\!CH_2Cl \xrightarrow{LiPPh_2} \text{P}\!-\!C_6H_4\!-\!CH_2PPh_2 \quad (9.8)$$

$$\text{P}\!-\!C_6H_5 \xrightarrow[FeBr_3]{Br_2} \text{P}\!-\!C_6H_4\!-\!Br \xrightarrow{LiBu} \text{P}\!-\!C_6H_4\!-\!Li \xrightarrow{PPh_2Cl} \text{P}\!-\!C_6H_4\!-\!PPh_2 \quad (9.9)$$

polystyrene backbone

A transition metal complex can be introduced into these ligand-substituted polymers by ligand exchange reactions:

$$\text{P}\!-\!C_6H_4\!-\!CH_2PPh_2 + RhCl(PPh_3)_3 \longrightarrow \text{P}\!-\!C_6H_4\!-\!CH_2PPh_2\!-\!RhCl(PPh_3)_2 \quad (9.10)$$

9-7

The amount of divinylbenzene added in copolymerization with styrene influences the properties of the base polymers. The copolymers range from a type that readily swells in solvents to a hard and insoluble solid. In hard copolymers a transition metal complex can be anchored only on the surface, whereas a considerable amount of the metal complex can be attached to a low cross-linking copolymer, which swells better. However, in this type of polymer resin the flexibility of the polymer backbones may cause complexation of a single metal atom to two polymer chains or dimerization due to formation of chlorine bridges, as shown in Figure 9.3.

Such dimerization can lower the activity of the polymer-supported catalyst by scavenging the coordinative unsaturation required for catalytic activity.

By employing a similar technique, chiral ligands such as that shown below can also be introduced into the copolymer of styrene.

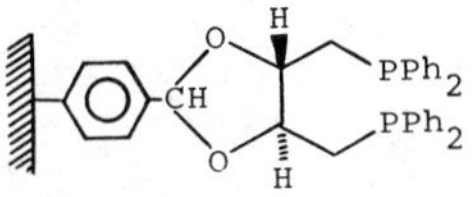

9-8 (A diop-containing polystyrene)

Polymer-anchored rhodium-diop catalysts have proved to be as effective for some asymmetric hydrogenation reactions as their soluble analogs.[20,21]

A problem associated with the phosphine-anchored polystyrene system is leaching of the active catalyst from the support. To avoid this problem, the ligand can be more firmly fixed on the polymer chain by forming a covalent metal–carbon bond.[22]

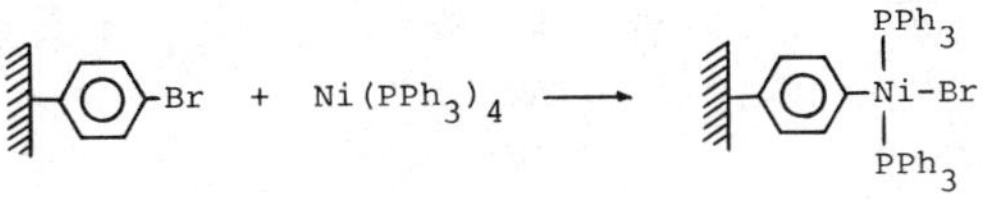

(9.11)

Another approach to anchoring the metal on a polymer is to bind a metal complex on the side chain of a polymer by π-complex formation[23]:

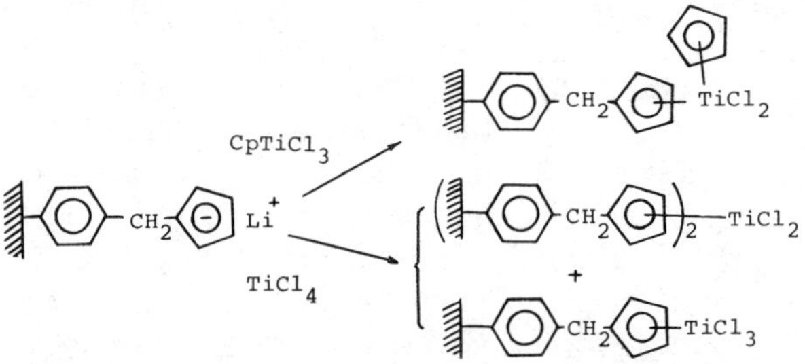

(9.12)

Improved activities for olefin hydrogenation using the reduced type of these polymeric catalysts over the corresponding soluble catalyst derived from Cp_2TiCl_2 have been reported.

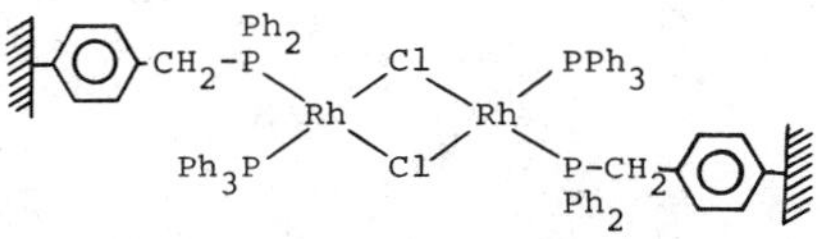

FIGURE 9.3. Dimerization of a polymer-supported rhodium complex.

Transition metal complexes supported on polymers have the following advantages: (1) The nature of the active site is much better understood than in heterogeneous catalysts, and most active sites in the supported transition metal complexes are expected to participate in the catalysis; (2) since coordinatively unsaturated metal complexes are more readily formed in the supported catalysts than in the homogeneous system, the catalyst activity per metal atom is usually higher than in the homogeneous catalysts; and (3) they have less tendency to aggregate and be deactivated than the homogeneous catalysts.

However, preparation of the transition metal complexes supported on organic polymers is a rather delicate task. Slight differences in the degree of cross-linking of polymers may profoundly affect the total amount of metal complex supported on the polymer. Quite often it is difficult to prepare a supported transition metal complex that is catalytically more active than the corresponding homogeneous catalyst. One of the reasons is the slow rate of the diffusion of the reaction substrate into the swollen organic polymer.

In contrast, attachment of transition metal complexes to the surface of inorganic solids does not have the problem associated with swelling of polymers, but it is a difficult process for many transition metal complexes. A successfully developed industrial application of a transition metal complex anchored to a solid support is the silica-supported chromocene used for ethylene polymerization by Union Carbide.[24]

As shown in Figure 9.4, chromocene reacts with OH groups on the surface of silica, and one of the cyclopentadienyl groups is released to give a singly Cp-bonded species (A). Oxidative addition of another adjacent OH group to Cr(II) gives a doubly anchored active species (B) that has a Cr–H bond. Repeated insertion of ethylene molecules into the Cr–H bond leads to polyethylene.

This catalyst is highly efficient, and the catalyst residue in the polyethylene produced is so innocuous that it may be left in the polymer, thus allowing the costly deashing process discussed in Section 7.2 to be eliminated.

By allowing organometallic complexes to react with the hydroxyl groups on inorganic supports, reasonably well-defined anchored catalysts can be prepared. Various transition metal complexes including $Zr(\eta^3\text{-}C_3H_5)_4$, $Cr(\eta^3\text{-}C_3H_5)_3$, $Cr_2(\eta^3\text{-}C_3H_5)_4$, $Mo(\eta^3\text{-}C_3H_5)_4$, $Mo_2(\eta^3\text{-}C_3H_5)_4$, $Fe_3(CO)_{12}$, $Co_2(CO)_8$, and $Rh(\eta^3\text{-}C_3H_5)_3$ have been fixed in this way on the surface of silica and

-Si-OH + $CrCp_2$ $\xrightarrow{-C_5H_6}$ (A) $\longrightarrow$ (B)

FIGURE 9.4. Anchoring of chromocene to silica surface.

alumina.[25-29] An advantage of anchoring η^3-allyl complexes to the inorganic support is the relatively easy preparation of highly dispersed and active catalyst species by utilizing the high reactivity of the η^3-allyl groups toward the hydroxyl group. Furthermore, modification of the catalyst species by treating it with hydrogen or oxygen can be accomplished keeping the dispersed state of the active sites on the support surface. The reactions that may occur on the surface of silica when it is treated first with a dinuclear η^3-allyl molybdenum complex[27] and then with H_2 and O_2 are illustrated in Figure 9.5.

The binding of the dinuclear molybdenum complex to the solid surface of γ-alumina takes place readily at 0 °C. Surface species A is formed, accompanied by release of H_2. On H_2 reduction of A at 595 °C coordinatively unsaturated species B is formed, which can be oxidized to C and further to D depending on the reaction temperature. These surface species (A–D) catalyze various reactions such as ethylene hydrogenation, dehydrooxidation, and olefin metathesis. The catalyst prepared this way proved to be much more active than that prepared by the conventional impregnation method.

The nature of supported transition metal catalysts can be examined by various spectroscopic approaches including IR, UV diffuse reflectance, ESR, XPS (X-ray photoelectron spectroscopy), and EXAFS (extended X-ray absorption fine structure) as well as by chemical means. The most revealing information supporting the di-μ-oxo-Mo_2(VI) structure (D) on the alumina surface was obtained from EXAFS data.[27e] By an analogous approach, an η^3-allyl rhodium entity has been anchored on the surface of zeolite.[30]

The attachment of a transition metal complex to a solid surface is straightforward in the above case, where the organotransition metal complex itself is reactive with the solid surface to form covalent bonds. Otherwise, special methods have to be devised for anchoring ligands on the surface of silica or alumina. A frequently employed method is to introduce into a ligand a readily hydrolyzable group that reacts with the hydroxyl group on the silica surface. For example, a tertiary phosphine ligand, which has an organosilicon moiety

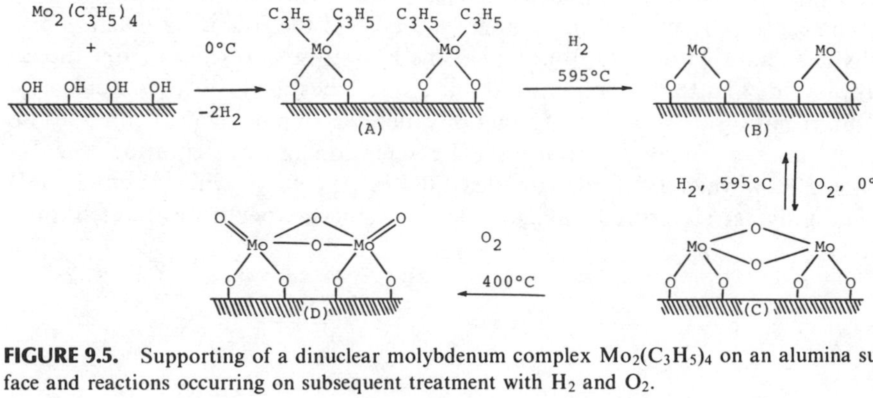

FIGURE 9.5. Supporting of a dinuclear molybdenum complex $Mo_2(C_3H_5)_4$ on an alumina surface and reactions occurring on subsequent treatment with H_2 and O_2.

with a readily hydrolyzable alkoxy group, can be prepared by the following reaction.[31]

$$EtO\text{-}Si(Me)_2\text{-}CH{=}CH_2 + HPPh_2 \xrightarrow{h\nu} EtO\text{-}Si(Me)_2\text{-}CH_2CH_2PPh_2 \quad (9.13)$$

9-9

The ethoxy group in **9-9** reacts with the OH group on the silica surface to give a tertiary phosphine ligand firmly bound to the solid surface by a Si–O–Si bond.

$$\text{Si-OH} + EtO\text{-}Si(Me)_2\text{-}CH_2CH_2PPh_2 \xrightarrow{-EtOH} \text{Si-O-}Si(Me)_2\text{-}CH_2CH_2PPh_2 \quad (9.14)$$

9-9

Metal complexes may be anchored on functionalized silica surfaces prepared as in Eq. (9.14). An alternative approach is to prepare the alkoxysilane-containing ligand (**9-9**) first and then treat it with a transition metal complex to give complex **9-10**, which has an alkoxysilyl group. Subsequent attachment of **9-10** to the solid surface by reaction of **9-10** with the OH groups provides the silica-supported transition metal complex **9-11**.

$$EtO\text{-}Si(Me)_2\text{-}CH_2CH_2PPh_2 + [RhCl(cod)]_2 \longrightarrow EtO\text{-}Si(Me)_2\text{-}CH_2CH_2PPh_2RhCl(cod)$$

9-10

$$\downarrow \text{Si-OH}$$

$$\text{Si-O-}Si(Me)_2\text{-}CH_2CH_2PPh_2RhCl(cod)$$

9-11

Although successful industrial applications of supported transition metal catalysts are still limited, the advantages of the surface-anchored catalyst have been successfully utilized in organic synthesis. In the following reaction of an allylic acetate with a Pd(0) complex the acetato group leaves from the anti (trans) side and the nucleophilic amine approaches the η^3-allyl complex again from the anti side to give a product in which the stereochemical configuration at the allylic carbon is retained. The retention of configuration does not exceed 65% when a soluble complex $Pd(PPh_3)_4$ is used, whereas employment of a polymer-supported catalyst increases the stereochemical retention to 100%.[32]

$$MeO_2C\text{-}C_6H_8\text{-}OAc \xrightarrow{Pd(PPh_3)_4} [MeO_2C\text{-}(\eta^3\text{-}C_6H_8)Pd^+(PPh_3)_2]\ OAc^- \xrightarrow{HNEt_2} MeO_2C\text{-}C_6H_8\text{-}NEt_2 \quad (9.15)$$

In another ingenious application the yield of a macrocyclization product is reported to be increased by use of a polymer-supported palladium complex.[33] Undoubtedly, we shall see more applications of polymer-supported transition metal complex to the catalysis of specific reactions.

Although they are not polymer anchored, discussion of one-dimensional polymers containing transition metals seems to be pertinent here. A series of polymers containing platinum, palladium, and nickel in the backbones has been prepared by Hagihara.[34] For example, a polymer containing linearly joined platinum and acetylene units can be synthesized by dehydrochlorination or oxidative coupling reaction.

$$Cl\text{-}\underset{PBu_3}{\overset{PBu_3}{Pt}}\text{-}Cl + HC{\equiv}C\text{-}C{\equiv}C\text{-}\underset{PBu_3}{\overset{PBu_3}{Pt}}\text{-}C{\equiv}C\text{-}C{\equiv}CH \xrightarrow[HNEt_2]{Cu} \left(\underset{PBu_3}{\overset{PBu_3}{Pt}}\text{-}C{\equiv}C\text{-}C{\equiv}C\right)_n$$

$$n\ HC{\equiv}C\text{-}C{\equiv}C\text{-}\underset{PBu_3}{\overset{PBu_3}{Pt}}\text{-}C{\equiv}C\text{-}C{\equiv}CH \xrightarrow{[CuCl\text{-}O_2]} \left(\underset{PBu_3}{\overset{PBu_3}{Pt}}\text{-}C{\equiv}C\text{-}C{\equiv}C\text{-}C{\equiv}C\text{-}C{\equiv}C\right)_n$$

The square planar geometry of platinum is maintained in these linear polymers. Despite the rigid rodlike shape of the linear molecules, the PBu_3-coordinated platinum–polyacetylene polymers are soluble in organic solvents and form liquid crystals. Polymers containing both nickel and platinum atoms connected by acetylene linkages also can be prepared by alkynyl exchange reactions.[35]

$$n\ HC{\equiv}C\text{-}\underset{PBu_3}{\overset{PBu_3}{Ni}}\text{-}C{\equiv}CH + n\ HC{\equiv}C\text{-}C{\equiv}C\text{-}\underset{PBu_3}{\overset{PBu_3}{Pt}}\text{-}C{\equiv}C\text{-}C{\equiv}CH$$

$$\longrightarrow \left(\underset{PBu_3}{\overset{PBu_3}{Ni}}\text{-}C{\equiv}C\text{-}C{\equiv}C\text{-}\underset{PBu_3}{\overset{PBu_3}{Pt}}\text{-}C{\equiv}C\text{-}C{\equiv}C\right)_n + n\ HC{\equiv}CH$$

Another type of one-dimensional polymer is the family of partially oxidized tetracyanoplatinate(II) salts, which have long been known to conduct electricity.[36-38] Pure potassium tetracyanoplatinate(II), $K_2Pt(CN)_4 \cdot 3H_2O$, is a white, nonconducting solid. Partial oxidation of this compound causes a change in color to bronze and makes it an electric conductor. In crystals of the partially oxidized product planar platinum complexes are arranged in infinite stacks as illustrated in Figure 9.6.

The Pt–Pt distances decrease from 348 pm in the starting, nonconducting complex $K_2Pt(CN)_4 \cdot 3H_2O$ to 287 pm in the stacked structure, a value close to the Pt–Pt separation of 277.5 pm in metallic platinum. In this stacked molecule a kind of one-dimensional bond structure is set up by overlap of the d_{z^2} orbitals resulting in the enhanced electric conductivity along the Pt–Pt axis, as

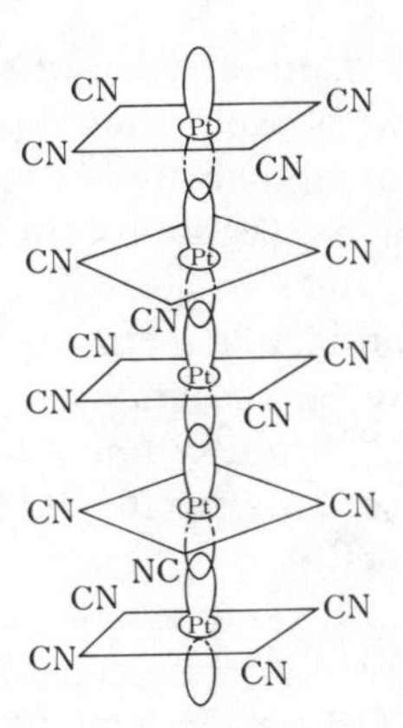

FIGURE 9.6. Diagram of the stacking of $[Pt(CN)_4]^{n-}$ ions showing how d_{z^2} orbitals overlap. (Reproduced by permission from F. A. Cotton and G. Wilkinson, *Advanced Inorganic Chemistry,* 4th ed., Wiley, New York, 1980, p. 1111.)

supported by MO calculations.[39] Future development in this area leading to one-dimensional polymers with organometallic ligands is of interest.

9.3. METAL CLUSTER COMPLEXES AND CATALYSIS

In some transition metal complexes more than three metal atoms are assembled to form metal cluster complexes[40] containing metal–metal bonds.† Particularly well studied are cluster complexes of metal carbonyls that form large cluster molecules. Other ligands such as cyclopentadienyl groups, tertiary phosphines, hydrides, and olefins are also used for stabilizing metal clusters. Heterogeneous catalysts containing transition metals are thought to owe their specific catalytic activities to the particular structures on the metal surfaces. However, presently available means for characterizing the nature of metal surfaces still remain rather limited and are not sufficiently well developed to allow detailed understanding of what is happening on the metal surfaces. On the other hand, metal cluster complexes can be treated as discrete molecules, and their structures and behavior are amenable to detailed examination by various physical methods. In particular, X-ray and neutron diffraction studies have established the structures of many metal cluster complexes, and these can be used as models for studies of reaction mechanisms catalyzed by heterogeneous catalysts. Most metal clusters are soluble in organic solvents and are diamagnetic, so their behavior in solution can be studied by NMR techniques. Thus, metal clusters are expected to fill the information gap between the mononuclear transition metal complexes and heterogeneous catalysts and may provide a clue to the behavior of substrates adsorbed on metal surfaces.

† At times dinuclear complexes with metal–metal bonds are also included in metal cluster complexes.[40p]

The similarity between an assembly of metal atoms and metal clusters should be evident in the examples shown in Figure 9.7.

The central metal atom in the environment of hexagonal close packing is surrounded by 12 neighboring metal atoms, as illustrated in Figure 9.7*a–c*. A metal cluster of composition $[Rh_{13}(CO)_{24}H_3]^{2-}$ has an analogous structure where the central rhodium atom is surrounded by 12 rhodium atoms coordinated with 24 carbonyl ligands, some of which are terminal and some bridging.[41] The modes of bonding of CO ligands to metals are believed to reflect the modes of CO adsorption on heterogeneous catalyst surfaces.

Although metal clusters are important in relation to catalysis, the chemistry of transition metal clusters has now developed enough to merit their separate treatment. Thus, we first discuss briefly the recent development in chemistry of metal cluster complexes and then treat their catalytic aspects.

a. Isolobal Analogy in Metal Clusters

Before discussing the chemistry of transition metal clusters, it may be appropriate to treat the concept of isolobal analogy.[42] The basic concept of isolobal analogy proposed by Hoffmann was briefly discussed in Chapter 2. The isolobal analogy is particularly useful in understanding complicated structures and the behavior of metal clusters. Furthermore, it can be used in certain cases as a predictive tool for the synthesis of novel compounds.[40o]

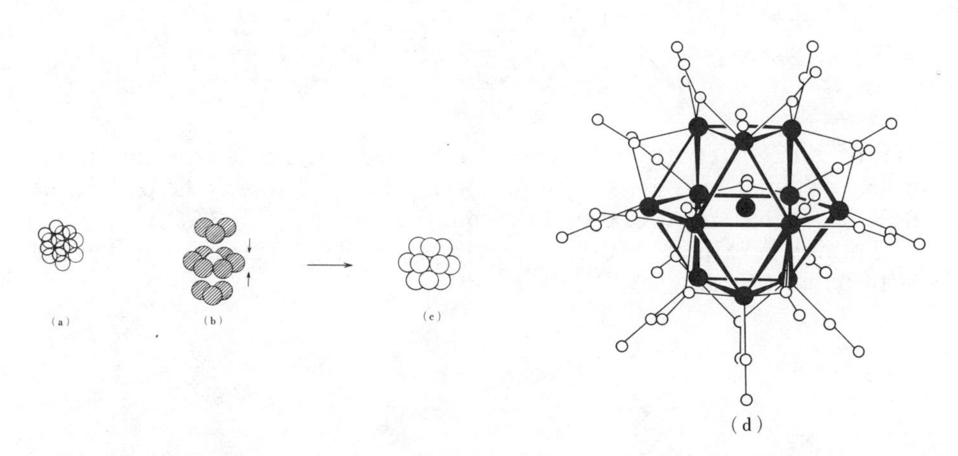

FIGURE 9.7. Comparison of the structures of a metal cluster $Rh_{13}(CO)_{24}H_3^{2-}$ with a metal assembly of hexagonal close packing. (*a*) Two layers of hexagonal close packing. (*b*) Side view of hexagonal close packing where three layers are taken apart. The white sphere represents the central metal atom. (*c*) Reconstruction of the assembly of 13 rhodium atoms in hexagonal close packing. The central metal atom is surrounded by 12 rhodium atoms. (*d*) Structure of $Rh_{13}(CO)_{24}H_3^{2-}$. The filled circles represent Rh atoms and open circles carbonyl groups. (From E. L. Muetterties, *Science,* **196,** p. 843, copyright 1977 by the AAAS.)

A d^8 ML_4 species such as $Fe(CO)_4$ is isolobal with methylene (or carbene) CH_2.

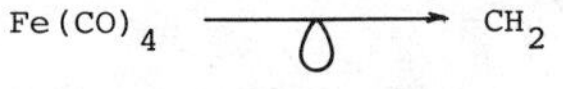

$Fe(CO)_4$ ⟷ CH_2

Both fragments have two electrons in delocalized a_1 and b_2 orbitals that are equivalent to two localized hybrids, as shown below.

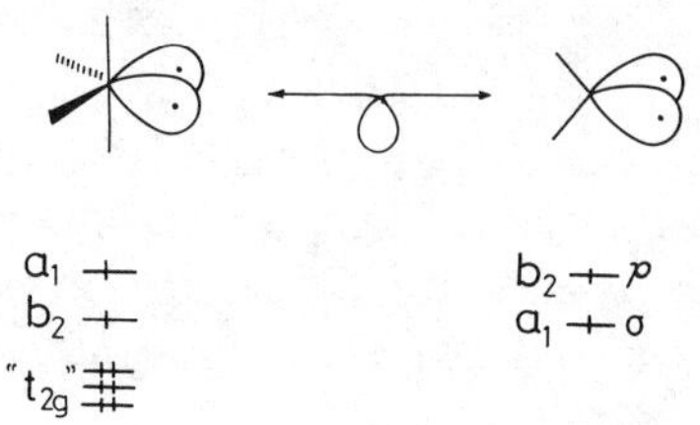

a_1
b_2
"t_{2g}"

b_2 — p
a_1 — σ

On combination of the two isolobal fragments, CH_2 and $Fe(CO)_4$, three known compounds can be obtained: ethylene, a tetracarbonyliron carbene complex, and $Fe_2(CO)_8$. The last molecule is unstable and has been observed only in a matrix.

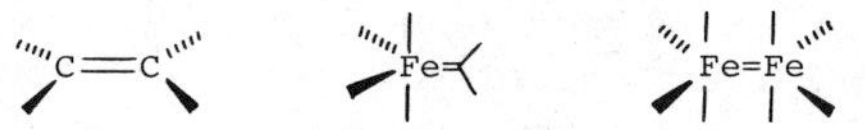

C=C Fe= Fe=Fe

Whereas CH_2 may be trimerized to cyclopropane, the d^8 $M(CO)_4$ species may be trimerized with methylene to give metallacyclopropanes:

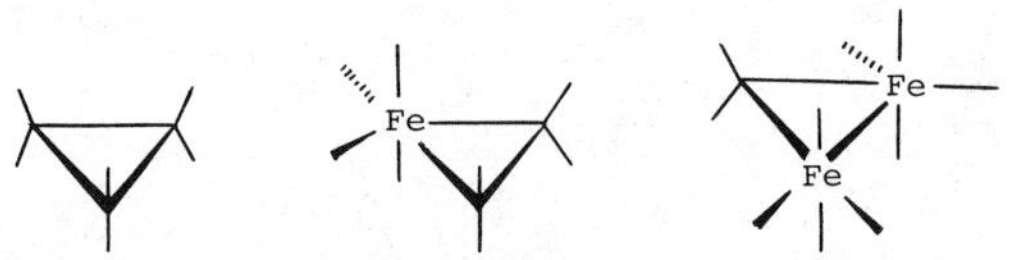

Fe Fe Fe

The d^8 $M(CO)_4$ entity itself can be trimerized, but the all-metallic three-membered complexes are known for Os and Ru, whereas for Fe part of the three-membered ring is bridged by carbonyl groups (cf. Fig. 3.25).

Consider next the d^9 ML_3 fragment [e.g., $Co(CO)_3$]. This is isolobal with a methylidyne (or carbyne) CH, as shown below.

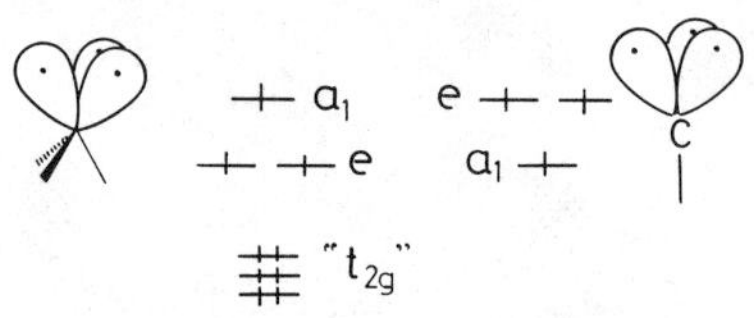

— a_1 e — —
— — e a_1 —
C
"t_{2g}"

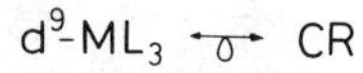

d^9-ML_3 ⟷ CR

Their similarity can be seen in the following entire series of mixed organic and inorganic tetrahedranes.

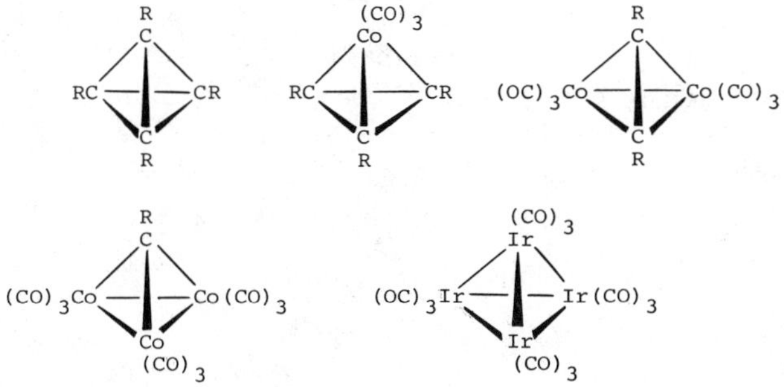

A merit of the isolobal analogy exists in recognition of a seemingly very complicated structure, such as $Os_5(CO)_{19}$, as comprised of a typical trigonal bipyramidal $Os(CO)_5$ derivative, with two of the equatorial carbonyls substituted by olefins.

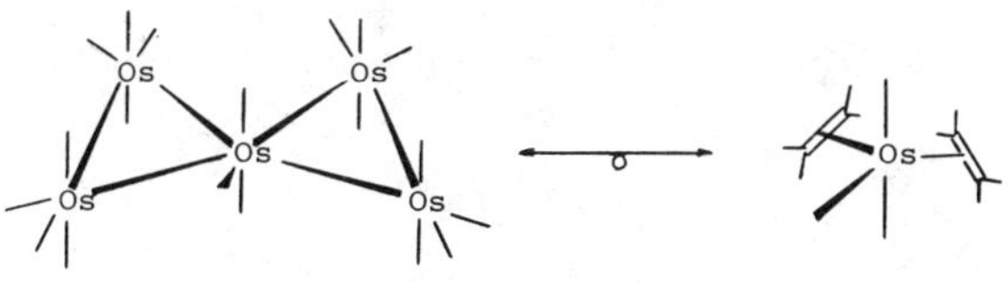

b. Synthesis and Properties of Metal Clusters

Previously, metal clusters have been prepared haphazardly, but systematic synthesis methods are gradually being developed. Simple metal carbonyl clusters can be prepared by reducing metal salts in an atmosphere of carbon monoxide. Bigger metal clusters are synthesized from smaller metal cluster units by various approaches. One important approach is called the redox condensation method, which allows the formation of polynuclear metal carbonyls under mild conditions. An example is the synthesis of an iron carbonyl cluster shown below.

$$[Fe_3(CO)_{11}]^{2-} + Fe(CO)_5 \xrightarrow[\text{THF}]{25\ ^\circ\text{C}} [Fe_4(CO)_{13}]^{2-} + 3\ CO \tag{9.16}$$

The first step in the reaction is electron transfer between $[Fe_3(CO)_{11}]^{2-}$ and $Fe(CO)_5$ to give the reactive intermediates $[Fe_3(CO)_{11}]^-$ and $[Fe(CO)_5]^-$. This process is followed by liberation of CO and metal–metal bond formation to give aggregated species. The reaction is accelerated by heat, light, or oxidation. The process of building up a bigger cluster is endothermic at room temperature because M–CO bonds, which have dissociation energies ranging from 120 to 190 kJ mol^{-1}, have to be broken in the process; this is only partly compensated by formation of M–M bonds that have bond energies of 80–130 kJ mol^{-1}. However, at higher temperatures the enthalpy increase is compensated by the entropy increase due to the liberation of gaseous CO, so the equilibrium shifts to favor the formation of bigger clusters. However, these become increasingly difficult to prepare as their size increases because of the difficulty in controlling the delicate preparative conditions.

Another approach to build up metal clusters is the addition of a coordinatively unsaturated species to other complexes having metal–metal or metal–carbon multiple bonds. The method is suitable for the preparation of clusters containing different metal atoms. Typical examples are shown below:

M=M —M'→ M'(M—M)

ex.

$Cp'Rh(\mu\text{-}CO)_2RhCp'$ + $M(CO)_5(thf)$ ⟶ $Cp'Rh(\mu\text{-}CO)_2RhCp'\cdot M(CO)_5$

$Cp' = C_5Me_5$ M = Cr, Mo

M(M=M) —M'→ M_3M' (tetrahedron)

ex.

$H_2Os_3(CO)_{10}$ + $CpCo(CO)_2$ ⟶ $(OC)_3Os$–$Os(CO)_3$, Os, H, H, CpCo(CO) cluster

M=C< —M'→ M—M' bridged by C

ex.

$(OC)_5M{=}CR_2$ + $Pt(C_2H_4)(PR_3)_2$ ⟶ $(OC)_5M(\mu\text{-}CR_2)Pt(PR_3)_2$

M = Cr, Mo, W

M≡CR —M'→ M'(M≡CR) —M"→ M, M', M", CR cluster

ex.

$Cp(CO)_2W{\equiv}CR$ —$Pt(C_2H_4)L_2$→ $L_2Pt(\mu\text{-}CR)W(CO)_2Cp$ —$Fe_2(CO)_9$→ $L(CO)Pt(\mu_3\text{-}CR)(Fe)W(CO)_2Cp$

L = PMe_3 or PMe_2Ph

The number of combinations of heteroatom metal clusters prepared in this way can be enormous.

An intriguing type of metal cluster is the one that encapsulates nonmetal atoms such as C, H, N, P, and S. The most well-studied clusters of this type are carbide clusters in which a carbide ion is encapsulated in the center of a polyhedron. The carbide clusters can be prepared by thermolysis of metal carbonyls, the source of the carbide carbon usually being carbon monoxide.

$$[Ru_6(CO)_{18}]^{2-} \xrightarrow[\text{diglyme}]{162\,^\circ C} [Ru_6C(CO)_{16}]^{2-} \xrightarrow[\text{tetraglyme}]{210\text{-}230\,^\circ C} [Ru_{10}C_2(CO)_{24}]^{2-}$$

Sometimes organic halides or CS_2 are employed as the carbon source.

$$[Rh(CO)_4]^- + CCl_4 \longrightarrow [Rh_6C(CO)_{15}]^{2-}$$

$$Co_2(CO)_8 + CS_2 \longrightarrow Co_6C(CO)_{12}S_2$$

Assemblage and rearrangement of these metal clusters exhibit very rich chemistry.[40l] Since most metal–metal bonds are weaker than metal–ligand bonds, addition of nucleophiles often leads to metal–metal bond breaking accompanied by a change in the cluster shapes. Single metal–metal bonds may be broken by adding donor ligands with concomitant opening of the cluster framework. Conversely, elimination of ligands can lead to formation of metal–metal bonds and assemblage of metal clusters. To avoid total degradation, strong metal–metal bonds must be present or the cluster core must be stabilized by bridging ligands such as phosphides.

The following example illustrates interconversions among μ_3-ligand bridged trinuclear metal clusters.[43]

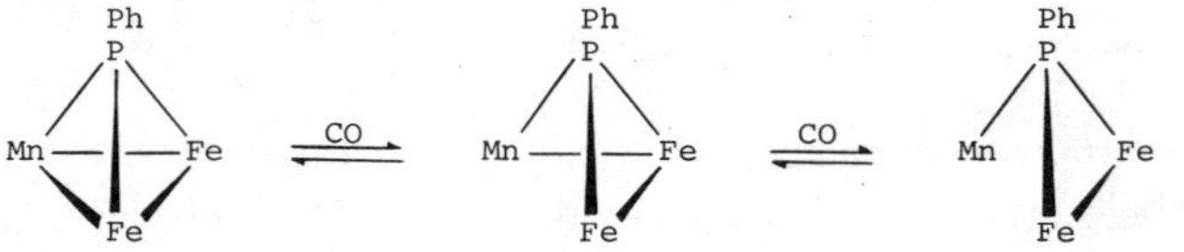

Ph P Mn Fe Fe ⇌(CO) Ph P Mn Fe Fe ⇌(CO) Ph P Mn Fe Fe

$PhPFe_2MnCp(CO)_8$ $PhPFe_2MnCp(CO)_9$ $PhPFe_2MnCp(CO)_{10}$

Another example of the phosphido group R_2P that stabilizes cluster skeletons by bridging metal atoms is shown below.[44]

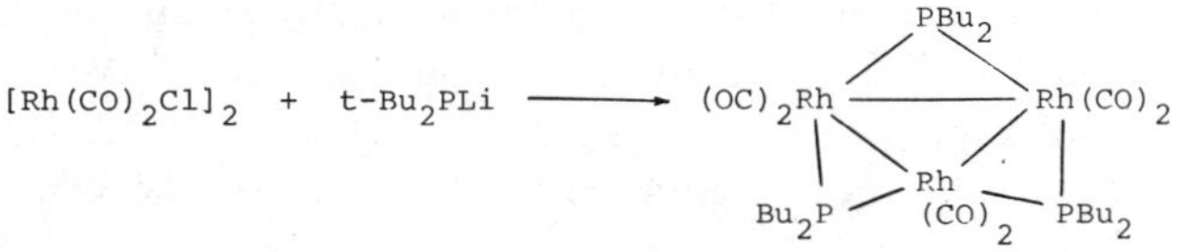

$$[Rh(CO)_2Cl]_2 + t\text{-}Bu_2PLi \longrightarrow$$

$(OC)_2Rh$, $Rh(CO)_2$, $Rh(CO)_2$, PBu_2, Bu_2P, PBu_2

Carbynes and alkynes also stabilize clusters by preventing fragmentation. A well-known example is $RCCo_3(CO)_9$, which is synthesized from RCX_3 (X = halogen) and $Co_2(CO)_8$.[45]

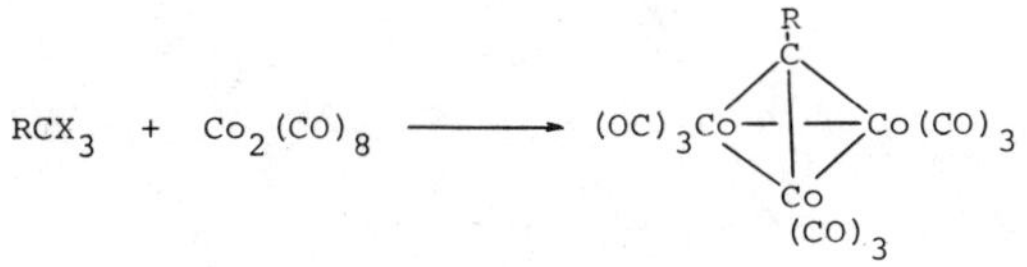

$$RCX_3 + Co_2(CO)_8 \longrightarrow$$

R C $(OC)_3Co$ $Co(CO)_3$ Co $(CO)_3$

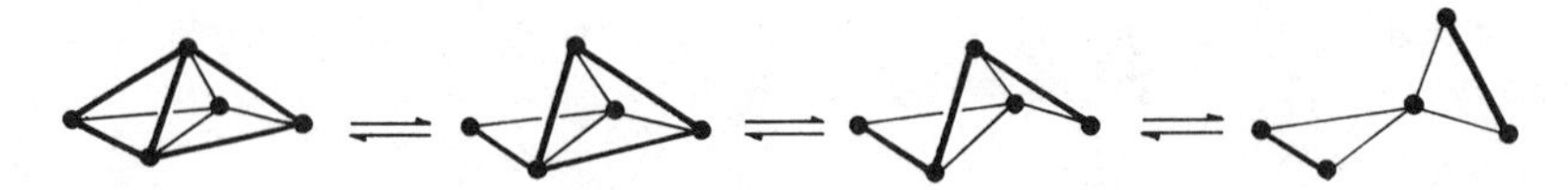

FIGURE 9.8. Opening and reconstruction modes of a trigonal bipyramidal cluster skeleton.

Increasing the nuclearity of the clusters augments the number of possible opening modes. Figure 9.8 shows the opening of a trigonal bipyramidal cluster skeleton.[40l]

As the number of metal atoms increases, possible structures of metal clusters increase as well. At least four different types of closed metal polyhedra are known for hexanuclear clusters[40l], as shown in Figure 9.9.

Opening of a hexanuclear metal cluster $Os_6(CO)_{18}$ by stepwise treatment with CO and $P(OMe)_3$ has been reported to give a complex of planar framework, as shown in Figure 9.10.[46]

Some metal clusters show interesting fluxional behavior.[47] The complex $[Pt_3(CO)_3(\mu_2\text{-}CO)_3]_5^{2-}$, which looks like a five-story pagoda as shown in Figure 9.11, is quite nonrigid in solution and shows a swivelling movement as revealed by NMR spectroscopy. It has also been observed that the trigonal platinum metal unit in the cluster exchanges with another metal cluster.[48]

The 18-electron rule generally governs the bonding in clusters with up to five metal atoms. The breakdown of the 18-electron rule beyond nuclearity 5 means that from here on a metal–metal bond no longer corresponds to a 2-electron interaction and localized bonding no longer exists. In electron counting of these polynuclear clusters one has to consider the cluster complexes with delocalized metal–metal bonds.

There exist certain analogies between the shapes of metal clusters and of polyhedral boron hydrides.[49] Based on the analogies, empirical rules predicting shapes of metal clusters have been proposed by Wade[49] and Mingos.[50] The electron configurations in these metal clusters have also been accounted for by Lauher[51] on the basis of extented Hückel MO calculations. Although these rules are of considerable use for clusters of relatively small nuclearities, clusters of higher nuclearities do not show obvious relations between their structures and the electron count. In these cases packing possibilities and ligand stereochemistries are optimized in a yet unpredictable way.[40l,52]

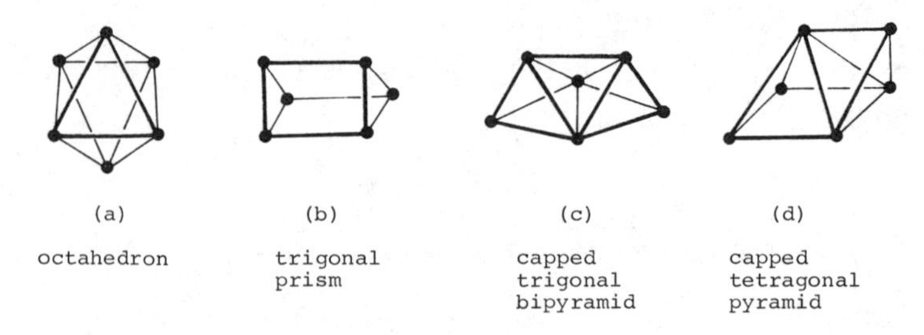

FIGURE 9.9. Different types of closed metal polyhedra of hexanuclear clusters.

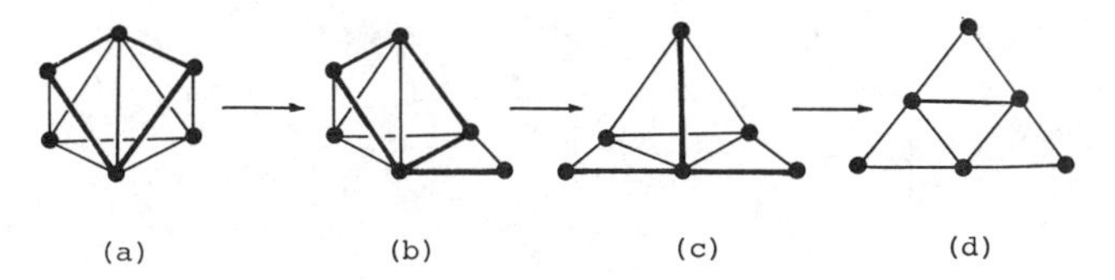

FIGURE 9.10. Unfolding of a hexanuclear cluster.

It may be appropriate to briefly discuss the chemistry of particularly intriguing types of complexes that have not yet found the place for treatise. These are transition metal complexes of carboranes.[53,54]

Decaborane $B_{10}H_{14}$ is a compound having a molecular shape like a nest (*nido* in Greek) that can accommodate two carbon atoms to form an icosahedron. Treatment of the decaborane with acetylene in diethyl sulfide induces closure of the *nido* compound to give a closed cage (*closo* in Greek).

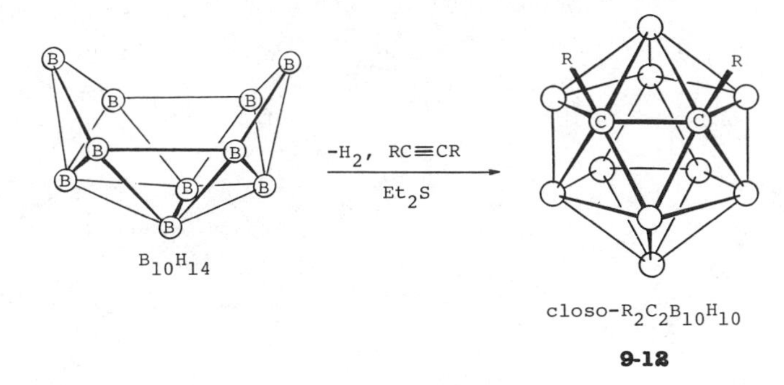

9-12

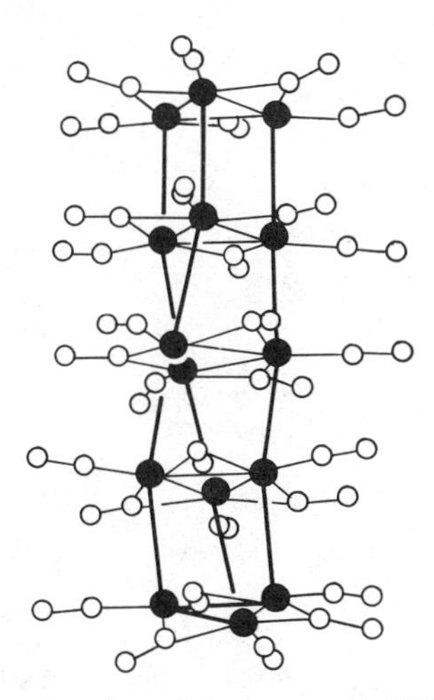

FIGURE 9.11. Schematic representation of the structure of $[Pt_{15}(CO)_{30}]^{2-}$. The Pt–Pt atomic distances in the trigonal planes are 266 pm. The Pt–Pt distances between the planes are 308 pm. (Reprinted with permission from J. C. Calabrese et al., *J. Am. Chem. Soc.*, **96**, 2614. Copyright 1974 American Chemical Society.)

Such a carbon-containing borane is called carborane. Heating the carborane (**9-12**) with an alkoxide ion causes removal of a boron atom from the icosahedron to give a *nido*-type carborane anion.

$$\text{closo-}B_{10}C_2H_{12} + EtO^- + 2EtOH \longrightarrow [B_9C_2H_{12}]^- + H_2 + B(OEt)_3$$

9-12

Further removal of a proton on treatment with NaH gives a dianion (**9-13**).

$$[B_9C_2H_{12}]^- + NaH \xrightarrow{THF} [B_9C_2H_{11}]^{2-} + H_2 + Na^+$$

9-13

Complex **9-13** (Fig. 9.12) is a strong base and takes up protons and metal ions. The dianion has a shape resembling a vase and is called dicarbollide

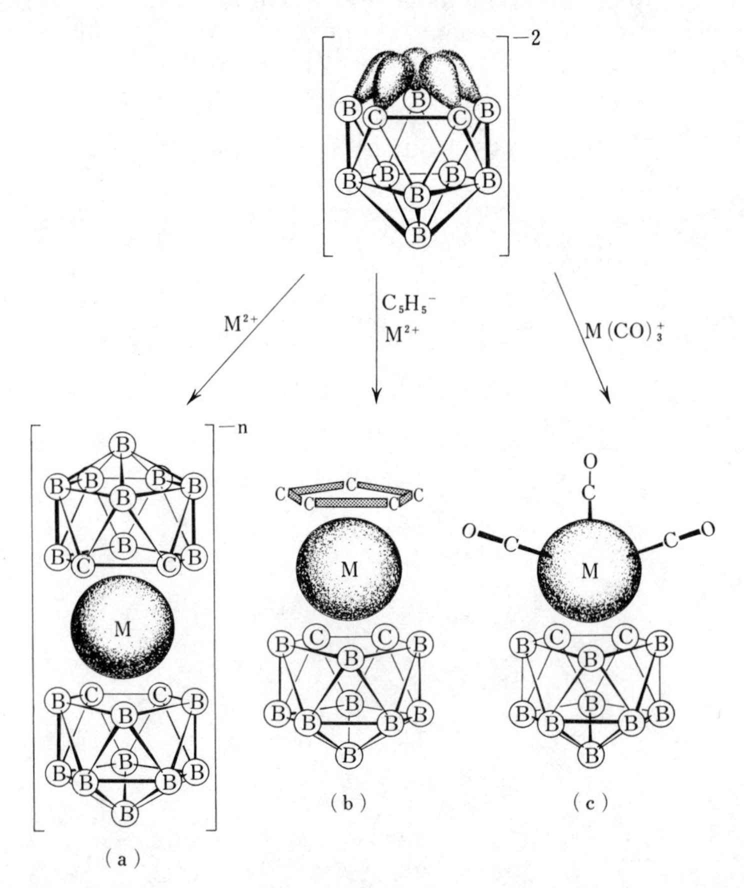

FIGURE 9.12. Formation of carbollyl transition metal complexes from dicarbollide anion. (Adapted from J. E. Huheey, *Inorganic Chemistry,* 2nd ed., Harper & Row, New York, 1978.) **9-13**, Probable structure of $B_9C_2H_{11}^{-2}$ anion. Each of the three boron atoms and the two carbon atoms on the open face of the cage directs an orbital toward the apical position occupied formerly by the twelfth boron atom in *closo*-$B_{10}C_2H_{12}$ (**9-12**). (*a*) Dicarbollyl species analogous to ferrocene. (*b*) Mixed carbollylcyclopentadienyl complex. (*c*) Mixed carbollyl–carbonyl complex.

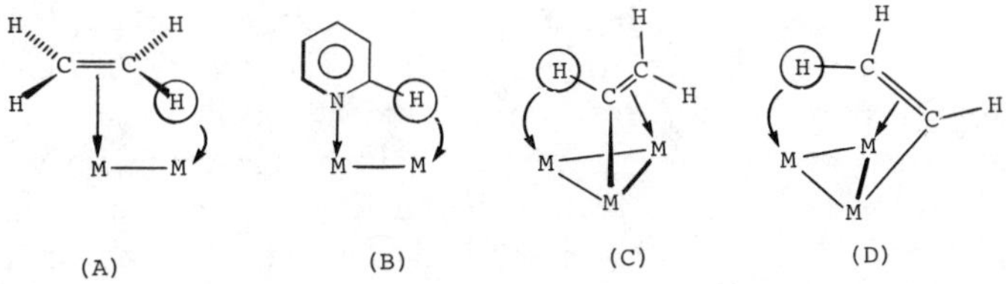

(A) (B) (C) (D)

FIGURE 9.13. Modes of C–H activation by metal clusters. (Adapted from ref. 55.)

(after *olla*, Spanish for vase). The pentagonal face of the dicarbollide resembles the cyclopentadienyl ion $C_5H_5^-$ and serves as a ligand to form metallocenelike complexes.

Catalytic activities of group 8–10 transition metal complexes that have the carborane ligand for olefin hydrogenation and isomerization have been demonstrated.[54]

In addition to rearrangement, transition metal clusters exhibit a variety of reactivities toward organic compounds.[55] One example of a particular mode of C–H activation has been discussed in Section 6.2 [Eq. (6.42)]. Examples of the C–H activation affected by assemblies of metal atoms are shown in Figure 9.13.

These are just a few examples of actions of metal clusters on organic compounds. Obviously, many more such reactions uniquely characteristic of metal clusters are expected to be revealed in the future.

c. Relevance of Metal Clusters as Models for Heterogeneous Catalysts

As nuclearity of metal clusters increases, the molecular size approaches that of particles of heterogeneous catalysts. Figure 9.14 shows an example of a plati-

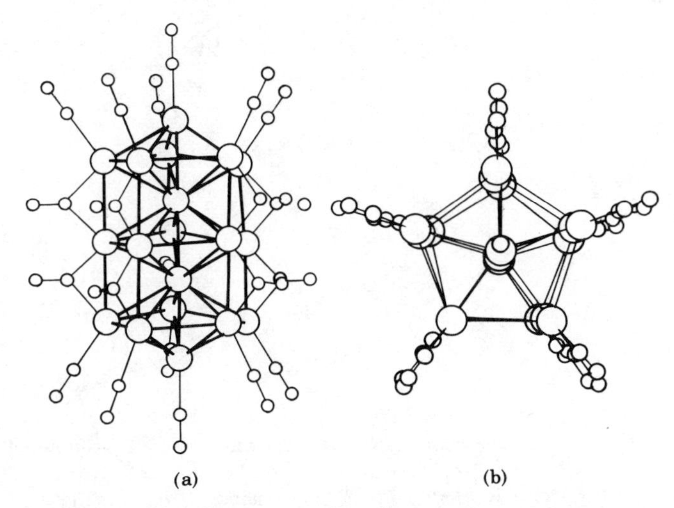

FIGURE 9.14. Structure of $[Pt_{19}(CO)_{12}(\mu_2\text{-}CO)_{10}]^{4-}$ (*a*) Side view. (*b*) Top view. (Reprinted with permission from D. M. Washecheck et al., *J. Am. Chem. Soc.*, **101,** 6110. Copyright 1979 American Chemical Society.)

num carbonyl cluster.[56] This beautiful structure can be regarded as being composed of three pentagonal bipyramids joined together through two common platinum atoms. At present in the petrochemical industry, a platinum catalyst supported on alumina is used for reforming petroleum in a process called platforming (*plat* from platinum and *forming* from reforming). The platinum cluster in Figure 9.14 has about the same size as the smallest particles used in the process. Thus, the metal cluster can be regarded as a metal particle that has been solubilized by putting ligands onto it.

Interestingly, the structure shown in Figure 9.14 is analogous with Ni, Pd, and Pt whiskers, which are attracting special attention because of their extraordinary strength as a construction material.

However, a symmetrical metal cluster such as that shown in Figure 9.14 does not provide information regarding modes of interaction of substrates with active sites on solid catalysts. Metal clusters containing alkynes, olefins, and nitriles are known, and they may be regarded as models, suggesting the modes of interactions of these substrates on the active sites.

Figure 9.15 illustrates the structures of cobalt and iron clusters that contain coordinated alkyne molecules.

Alkyne molecules are able to coordinate with two metal atoms utilizing their orthogonal two π orbitals, as shown in Figure 9.15*a*.[57a] The alkyne molecules can further form electron-deficient bonds with three and four metal atoms as in (*b*)[57b] and (*c*).[57c] As the number of metal atoms interacting with the alkyne increases from 2 to 3 to 4, as shown in (*a*), (*b*), and (*c*), the C≡C distance in the alkyne molecule increases from 136 to 141 to 144 pm. (Cf. the

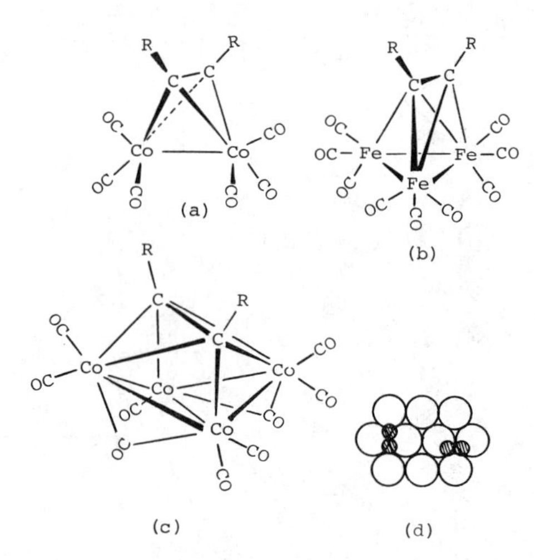

FIGURE 9.15. Modes of alkyne coordination to metals in cobalt and iron clusters (*a*–*c*). (*a*) $Co_2(CO)_6(RC{\equiv}CR)$; C≡C bond distance, 135 pm. (*b*) $Fe_3(CO)_9(RC{\equiv}CR)$; C≡C bond distance, 141 pm. (*c*) $Co_4(CO)_{10}(RC{\equiv}CR)$; C≡C bond distance, 144 pm. (*d*) Presumed modes of coordination of acetylene molecules adsorbed on Pt(111) plane. (From E. L. Muetterties, *Science*, **196**, p. 845, copyright 1977 by the AAAS.)

C≡C distance in free alkyne, alkene, and alkane, which is 120, 134, and 154 pm, respectively.) The increase in the interatomic distance reflects the increase in back bonding from the metal atoms to the π^* orbitals of the alkyne.

Thus, the triple bond in an alkyne molecule in a metal cluster complex is similar in nature to a single bond, and the coordinated alkyne can thus be regarded as activated and ready to be reduced. For comparison, presumed modes of alkyne coordination to three and four metal atoms in the Pt(1 1 1) plane are shown in Figure 9.15*d*.

We have discussed the mechanism of olefin hydrogenation by mononuclear transition metal complexes in Section 8.1. In that case an olefin-coordinated metal dihydride species ([M] in Fig. 8.2) was assumed to be an active species. In the hydrogenation of an olefin on the surface of a heterogeneous catalyst, however, simultaneous coordination of two hydrogen atoms and one olefin molecule to a single metal atom exposed on the surface of the catalyst is sterically somewhat unlikely. It is more reasonable to assume that a hydrogen molecule is adsorbed on two metal atoms and the olefin to another atom (Fig. 9.16).

The mechanism represented in Figure 9.16 differs from that in Figure 8.2 in that elementary reactions such as olefin insertion into the M–H bond and reductive elimination of the alkyl–hydrido ligand proceed not on a single metal atom but between multiple metal atoms. For such reactions to occur, a species adsorbed on a metal atom is required to be transferred onto another metal atom. Thus, on the catalyst species adsorbed species may be "wandering around" from one metal atom to another.

It is difficult to prove whether dynamic behavior of this sort occurs on heterogeneous catalysts. In the case of metal clusters, however, NMR studies have confirmed that both hydride ligands and carbonyl ligands can migrate between the metal atoms.

Although mononuclear transition metal complexes catalyze a variety of reactions, there are certain types of reactions that can be promoted only by

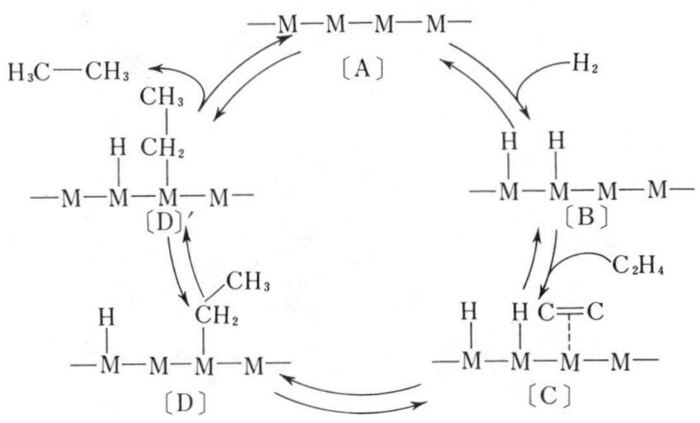

—M—M—M—M—
〔A〕
H_2
$H_3C—CH_3$
CH_3
H CH_2
—M—M—M—M—
〔D〕′
H H
—M—M—M—M—
〔B〕
C_2H_4
CH_3
H CH_2
—M—M—M—M—
〔D〕
H H C=C
—M—M—M—M—
〔C〕

FIGURE 9.16. Conceptual scheme of the mechanism of olefin hydrogenation by a heterogeneous transition metal catalyst. —M—M—M—M— represents the metal surface. Note, however, that it is not one-dimensional. Compare with Figure 8.2. (From E. L. Muetterties, *Science,* **196,** p. 841, copyright 1977 by the AAAS.)

solid catalysts. Fischer–Tropsch synthesis of hydrocarbons from synthesis gas is one of those processes.[58] To cleave the C–O bond in carbon monoxide, considerable activation of the molecule is required, and assemblies of metal atoms serve the purpose. In many cases, however, unequivocal evidence for the active participation of metal clusters in the catalytic processes is quite difficult to obtain. The following example[59] is particularly intriguing in that it shows how a metal cluster could take part in catalytic reactions (Fig. 9.17).

Treatment of a carbide-encapsulated octahedral cluster complex, $[Fe_6C(CO)_{16}](Et_4N)_2$[A], in methanol with tropylium bromide, a mild one-electron oxidizing agent, yields a tetranuclear iron cluster, $[Fe_4C(CO)_{12}CO_2CH_3]$[D], that has a methoxycarbonyl group attached to the central carbon atom. The reaction probably proceeds through intermediates [B] and [C], as shown in Figure 9.17. The oxidation is regarded as having caused the removal of two vertices from the octahedral dianion [A], thus exposing the central atom in [B].

One molecule of liberated CO then reacts with the exposed, apparently electrophilic carbon to give a metal ketenylidene specis [C], which then undergoes reaction with solvent methanol, yielding the methoxycarbonyl complex [D], whose structure has been established by X-ray analysis.[59b]

The μ_4-carbide intermediate [B] can also be derived by treatment of the methoxycarbonyl complex [D] with trifluoromethanesulfonic acid, and its molecular structure has been established as well. The molecule comprises a "butterfly" configuration of four iron atoms with the exposed carbon atom sitting on top of the butterfly.[59d] Further reaction of [D] with H_2 was shown to afford methyl acetate.

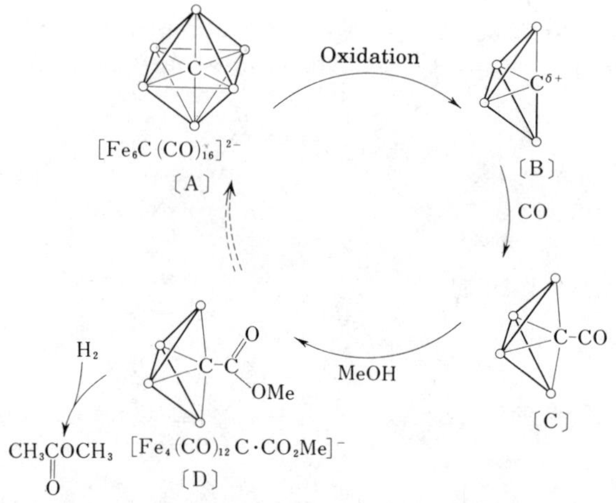

FIGURE 9.17. Reactions of a carbon atom encapsulated in an iron cluster. (Reprinted with permission from Bradley et al., *J. Am. Chem. Soc.*, **101,** 7417. Copyright 1979 American Chemical Society.)

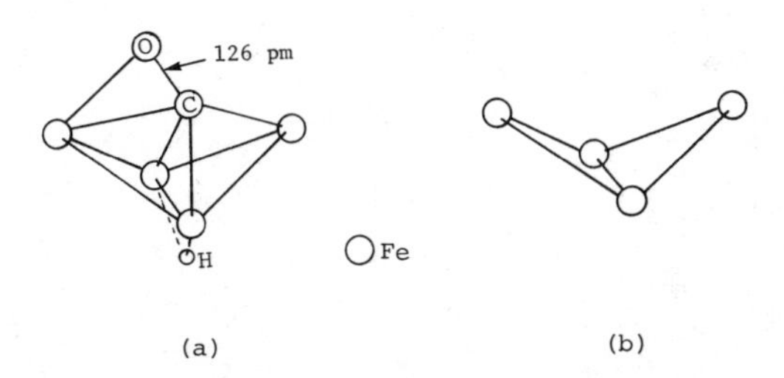

FIGURE 9.18. Interaction of a CO molecule with four metal atoms comprising a butterfly configuration in $[Fe_4(CO)_{13}H]^-$. (*a*) Skeletal structure with omission of part of CO ligands. (*b*) Butterfly core of four metal atoms. (Reproduced with modification from ref. 60).

Although the reactions illustrated in Figure 9.17 do not form a catalytic cycle, the reactions show the possibility of catalytically converting CO and H_2 to methyl acetate. In the process of activation of carbon monoxide on heterogeneous catalysts, it is believed that surface carbon atoms are formed via dissociative adsorption of CO. Since the carbide-encapsulated complex [A] is derived from $Fe(CO)_5$, carbon–oxygen bond cleavage must be involved in the reaction. This reaction step and the subsequent reactions shown in Figure 9.17 are considered to be relevant to the mechanisms of hydrogenation of carbon monoxide.

Figure 9.18*a* shows the central part of the molecular structure of a tetranuclear iron cluster complex $[Fe_4(CO)_{13}H]^-$.[60]

In this molecule 12 of the 13 carbonyl groups are terminally bound to four iron atoms, forming a butterfly arrangement with a dihedral angle of 117°, in which all five Fe–Fe bond distances are equivalent. A unique carbonyl group is trapped between the two butterfly wings, the C–O bond distance (126 pm) being considerably longer than that of free CO. The central CO groups may be regarded as having been activated on interaction with the four iron atoms in the butterfly arrangement. Further activation of the CO group may well lead to the C–O bond cleavage to form a carbide-containing complex. Current ideas in surface chemistry include models such as that shown in Figure 9.19 as sites for adsorption of substrates. On the metal surfaces (as represented in the figure) the presence of "steps" with a height of monoatomic layer forming a "terrace" and of "kinks" is assumed. The butterfly framework as shown in Figure 9.18*b* may represent the kink on the surface of a catalyst.

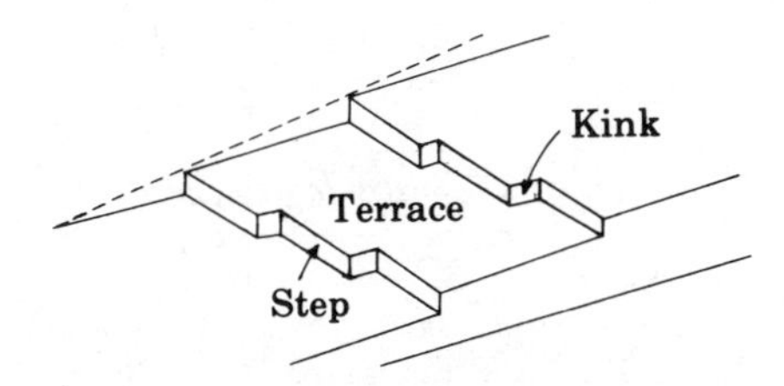

FIGURE 9.19. Schematic representation of adsorption active sites on metal surfaces.

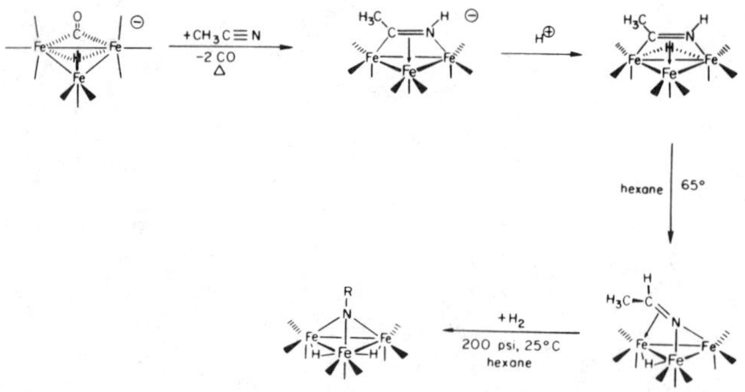

FIGURE 9.20. Stepwise reduction of acetonitrile by a triiron cluster. (Reprinted with permission from M. A. Andrews and H. D. Kalsz, *J. Am. Chem. Soc.*, **99,** 6763. Copyright 1977 American Chemical Society.)

Another interesting example related to the cooperative function of an assembly of transition metal atoms in promoting specific reactions is stoichiometric stepwise reduction pathway for acetonitrile over a tri-iron carbonyl cluster.[61] The sequence in Figure 9.20 has been identified (with carbonyl groups omitted for clarity). This scheme provides a clear example of the sequence of events that can occur on a cluster leading to the extensive reduction of the C≡N bond of acetonitrile. Such a reduction is usually difficult to achieve by a mononuclear complex.

In certain cases, metal clusters are actually added and their presence in the catalytic systems can be established by physical methods; for example, ethylene glycol synthesis from CO and H_2 is promoted by addition of rhodium clusters at 1000 atm and 200 °C. However, uncertainties always remain regarding the identity of the true active species under severe working conditions. Whether the intact cluster or a highly reactive mononuclear species produced by disintegration of the cluster is actually operating as the catalyst is an extremely difficult problem to solve. Thus, an example of a real cluster catalyst system where the metal cluster has been unequivocally established as the actual working catalyst species has yet to be seen. Meanwhile, the chemistry of metal clusters will flourish and keep fascinated chemists involved in the study of a variety of metal clusters.

REFERENCES

1. (a) M. N. Hughes, *The Inorganic Chemistry of Biological Processes,* 2nd ed., Wiley, Chichester, 1981; (b) J. M. Pratt, *Inorganic Chemistry of Vitamin B_{12},* Academic Press, New York, 1972; (c) J. M. Wood and D. G. Brown, *Struct. Bond.,* **11,** 47 (1972); (d) H. P. C. Hogenkamp and G. N. Sands, *Struct. Bond.,* **20,** 23 (1974); (e) C. Walsh, *Enzymatic Reaction Mechanisms,* W. H. Freeman, San Francisco, 1979; (f) R. P. Hanzlik, *Inorganic Aspects of Biological and Organic Chemistry,* Academic Press, New York, 1976; (g) A. W. Addison, W. R. Cullen, D. Dolphin, and B. R. James, eds., *Biological Aspects of Inorganic*

Chemistry, Wiley, New York, 1977; (h) R. F. Gould, ed., *Bioinorganic Chemistry* (Adv. in Chem. Ser., Vol. 100, American Chemical Society, Washington, D.C., 1971; (i) F. E. Brinckman and J. M. Bellama, *Organometals and Organometalloids. Occurrence and Fate in the Environment* (Adv. in Chem. Ser., Vol. 82), American Chemical Society, Washington, D.C., 1978.

2. (a) J. E. Falk, in *Porphyrins and Metalloporphyrins,* 1964; J. G. Furhop and K. M. Smith, eds., Elsevier, Amsterdam, 1975; (b) G. N. Schrauzer, *Angew. Chem.,* **88,** 465 (1976); (c) B. M. Barbior, *Acc. Chem. Res.,* **8,** 376 (1975); (d) R. H. Abeles and D. Dolphin, *Acc. Chem. Res.,* **9,** 114 (1976); (e) A. W. Johnson, *Chem. Soc. Rev.,* **9,** 125 (1980); (f) D. Dolphin, ed., *The Porphyrins,* 7 vols., Academic Press, New York, 1978, 1979; (g) R. H. Abeles in *Bioinorganic Chemistry,* (Adv. in Chem. Ser., Vol. 100), American Chemical Society, Washington, D.C., 1971; (h) R. V. Stevens, in *Vitamin B-12,* D. Dolphin, ed., Wiley, New York, 1982; (i) A. I. Scott, *Acc. Chem. Res.,* **11,** 29 (1978); (j) T. S. Thayer, *Organometallic Compounds and Living Organisms,* Academic Press, Orlando, 1984.
3. K. Kasuga and M. Tsutsui, *Coord. Chem. Rev.,* **32,** 67 (1980).
4. H. A. Barker, H. Weissbach, and R. D. Smyth, *Proc. Natl. Acad. Sci. U.S.,* **44,** 1093 (1958).
5. (a) P. Senhard and D. Hodgkin, *Nature,* **192,** 937 (1961); (b) D. Crowfoot-Hodgkin, *Proc. Roy. Soc. (London),* **A288,** 294 (1965); (c) D. Crowfoot-Hodgkin, J. Kamper, M. Mackay, J. Pickworth, K. N. Trueblood, and J. G. White, *Nature,* **178,** 64 (1956).
6. (a) W. P. Ridley, L. J. Dizikes, and J. M. Wood, *Science,* **197,** 329 (1977); (b) J. S. Thayer and F. E. Brinckman, *Adv. Organometal. Chem.,* **20,** 313 (1982).
7. (a) G. N. Schrauzer and J. Kohnle, *Chem. Ber.,* **97,** 3056 (1964); (b) G. N. Schrauzer, *Acc. Chem. Res.,* **1,** 97 (1968); (c) G. N. Schrauzer, *Angew. Chem. Int. Ed. Engl.,* **15,** 417 (1976).
8. P. G. Lenhert, *Chem. Commun.,* 980 (1967).
9. G. N. Schrauzer and J. Grate, *J. Am. Chem. Soc.,* **103,** 541 (1981).
10. J. Halpern, *Acc. Chem. Res.,* **15,** 238 (1982).
11. (a) Y. Ohashi and Y. Sasada, *Nature (London),* **267,** 142 (1977); (b) Y. Ohashi, Y. Sasada, and Y. Ohgo, *Chem. Lett.,* 457 (1978); (c) Y. Ohashi, K. Yanagi, T. Kurihara, Y. Sasada, and Y. Ohgo, *J. Am. Chem. Soc.,* **104,** 6353 (1982); (d) Y. Ohashi, *Yuki Gosei Kagaku Kyokaishi (J. Synth. Org. Chem., Jpn),* **41,** 1120 (1983).
12. B. M. Hoffman and D. H. Petering, *Proc. Nat. Acad. Sci. U.S.,* **67,** 637 (1970); **69,** 2122 (1972).
13. Y. Murakami, *Biomimetic Chemistry,* (Adv. in Chem. Ser. Vol. 191), American Chemical Society, Washington, D.C., 1980.
14. (a) P. R. Ortiz de Montellano, H. S. Beilan, K. L. Kunze, and B. A. Mico, *J. Biol. Chem.,* **256,** 4395 (1981); (b) O. Augusto, K. L. Kunze, and P. R. Ortiz de Montellano, *J. Biol. Chem.,* **257,** 6231 (1982); (c) P. R. Ortiz de Montellano, K. L. Kunze, and O. Augusto, *J. Am. Chem. Soc.,* **104,** 3545 (1982).
15. (a) D. Mansuy, *Pure Appl. Chem.,* **52,** 698 (1980); (b) D. Mansuy, *Rev. Biochem. Toxicol.,* **3,** 283 (1981); (c) D. Mansuy, J.-P. Battioni, D. Dupré, E. Sartori, and G. Chottard, *J. Am. Chem. Soc.,* **104,** 6159 (1982).
16. (a) D. Dolphin, D. Halks, and E. Johnson, *Inorg. Chem.,* **20,** 4348 (1981); (b) H. J. Callot and F. Metz, *J. Chem. Soc. Chem. Commun.,* 947 (1982).
17. J. S. Tompson, R. L. Harlow, and J. F. Whitney, *J. Am. Chem. Soc.,* **105,** 3522 (1983).
18. (a) J. T. Groves and T. E. Nemo, *J. Am. Chem. Soc.,* **105,** 5786, 5791, 6283 (1983); (b) R. A. Sheldon, *J. Mol. Catal.,* **20,** 1 (1983).
19. (a) C. U. Pittman, Jr., in *Polymer-Supported Reactions in Organic Synthesis,* P. Hodge and D. C. Sherrington, eds., Wiley, New York, 1980; (b) F. R. Hartley and P. N. Vezey, *Adv. Organometal. Chem.,* **15,** 189 (1977); (c) R. H. Grubbs, *Chemtech,* 512 (1977); (d) Y. Chauvin, D. Commereuc, and F. Dawans, *Prog. Polym. Sci.,* **5,** 95 (1977); (e) J. C. Bailar, Jr.,

Cat. Rev., Sci. Eng., **10,** 17 (1974); (f) V. A. Zakharov and Yu. I. Yermakov, *Cat. Rev., Sci. Eng.*, **15,** 67 (1979); (g) E. M. Cernia and M. Graziani, *J. Appl. Polym. Sci.*, **18,** 2725 (1974); (h) C. U. Pittman, Jr., and G. O. Evans, *Chemtech,* 560 (1973); (i) P. Hodge, *Chem. Brit.*, **14,** 237 (1978); (j) A. L. Robinson, *Science,* **194,** 1261 (1976); (k) Z. M. Michalska and D. E. Webster, *Chemtech,* 117 (1975); (l) L. L. Murrel, in *Advances Materials in Catalysis,* J. J. Burton and R. L. Garten, eds., Academic Press, New York, 1977. (m) C. U. Pittman, Jr., A. Hirao, C. Jones, R. M. Hanes, and Q. Ng, *Ann. N. Y. Acad. Sci.*, **295,** 15 (1977); (n) Yu. I. Yermakov, B. N. Kuznetsov, and V. A. Zakharov, *Catalysis by Supported Complexes,* Elsevier, Amsterdam, 1981; (o) J. W. Coenen, R. S. Downing, and J. J. F. Scholten, *J. Mol. Catal.*, **11,** 213 (1981); (p) P. Hodge and D. Sherrington, eds., *Polymer-Supported Reactions in Organic Synthesis,* Wiley, Chichester, 1980.

20. W. Dumont, J.-C. Poulin, T.-P. Dang, and H. B. Kagan, *J. Am. Chem. Soc.*, **95,** 8295 (1973).
21. T. Masuda and J. K. Stille, *J. Am. Chem. Soc.*, **100,** 268 (1978); N. Takaishi, H. Imai, C. A. Bertelo, and J. K. Stille, *J. Am. Chem. Soc.*, **100,** 264 (1978).
22. N. Kawata, T. Mizoroki, A. Ozaki, and M. Okawara, *Chem. Lett.*, 1165 (1973).
23. W. D. Bonds, Jr., C. H. Brubaker, Jr., E. S. Chandrasekanen, C. Gibbons, R. H. Grubbs, and L. C. Kroll, *J. Am. Chem. Soc.*, **97,** 2128 (1975).
24. (a) D. T. Laverty, J. J. Rooney, and A. Stewart, *J. Catal.*, **45,** 110 (1976); (b) F. J. Karol, C. Wn, W. T. Reichle, and N. J. Maraschin, *J. Catal.*, **60,** 68 (1979).
25. D. G. H. Ballard, *Adv. Catal.*, **23,** 263 (1973).
26. (a) Yu. I. Yermakov and V. Zakharov, *Adv. Catal.*, **24,** 173 (1975); (b) Yu. I. Yermakov, *Fundamental Research in Homogeneous Catalysis,* Plenum Press, New York, 1984.
27. (a) Y. Iwasawa, *J. Mol. Catal.*, **17,** 93 (1982) and references cited therein; (b) Y. Iwasawa, M. Yamagishi, and S. Ogasawara, *J. Chem. Soc. Chem. Commun.*, 246 (1982); (c) 871 (1980); (d) Y. Iwasawa and H. Hamamura, *J. Chem. Soc. Chem. Commun.*, 130 (1983); (e) Y. Iwasawa, Y. Sato, and H. Kuroda, *J. Catal.*, **82,** 289 (1983).
28. J. P. Candlin and H. Thomas, *Adv. Chem. Ser.*, **132,** 212 (1974).
29. J. Schwartz and M. D. Ward, *J. Mol. Cat.*, **8,** 465 (1980).
30. T. N. Huang and J. Schwartz, *J. Am. Chem. Soc.*, **104,** 5244 (1982).
31. (a) A. K. Smith, J. M. Basset, and P. M. Maitlis, *J. Mol. Cat.*, **2,** 223 (1977); (b) Z. M. Michalska, *J. Mol. Cat.*, **3,** 125 (1977/1978).
32. B. M. Trost and E. Keinan, *J. Am. Chem. Soc.*, **100,** 7779 (1978).
33. B. M. Trost and R. W. Warner, *J. Am. Chem. Soc.*, **104,** 6112 (1982).
34. (a) Y. Fujikura, K. Sonogashira, and N. Hagihara, *Chem. Lett.*, 1067 (1975); (b) S. Takahashi, M. Kariya, Y. Yatake, K. Sonogashira, and N. Hagihara, *Macromolecules,* **10,** 879 (1977); **11,** 1063 (1978).
35. K. Sonogashira, K. Ohga, S. Takahashi, and N. Hagihara, *J. Organometal. Chem.*, **188,** 237 (1980).
36. J. S. Miller and A. J. Epstein, *Prog. Inorg. Chem.*, **20,** 1 (1976).
37. J. S. Miller, *Adv. Chem. Ser.*, **50,** 1 (1976).
38. J. R. Ferraro and K. B. Merteo, *Coord. Chem. Rev.*, **36,** 357 (1981).
39. M.-H. Whangbo and R. Hoffmann, *J. Am. Chem. Soc.*, **101,** 6093 (1978).
40. (a) B. F. Johnson, ed., *Transition Metal Clusters,* Wiley, New York, 1980; (b) E. L. Muetterties, T. Rhodin, E. Band, C. Bruker, and W. P. Pretzer, *Chem. Rev.*, **79,** 91 (1979); (c) E. L. Muetterties, *Science,* 839 (1977); (d) A. K. Smith and J. M. Bassett, *J. Mol. Catal.*, **2,** 229 (1977); (e) P. Chini, *J. Organometal. Chem.*, **200,** 37 (1980); (f) P. Chini, G. Longoni, and V. G. Albano, *Adv. Organometal. Chem.*, **14,** 285 (1976); (g) M. Ichikawa, *Chemtech,* 674 (1982); (h) F. G. A. Stone, *Acc. Chem. Res.*, **14,** 318 (1981); (i) E. L. Muetterties, *Bull. Soc. Chim. Belg.*, **84,** 959 (1975); (j) E. L. Muetterties, *Chem. Rev.*, **79,** 479 (1979); (k) M.

Tachikawa and E. L. Muetterties, *Prog. Inorg. Chem.*, **28,** 203 (1981); (l) H. Vahrenkamp, *Adv. Organometal. Chem.*, **22,** 169 (1983); (m) E. L. Muetterties, *J. Organometal. Chem.*, **200,** 177 (1980); (n) H. Vahrenkamp, *Struct. Bond. (Berlin),* **32,** 1 (1977); (o) F. G. A. Stone, *Angew. Chem. Int. Ed. Engl.*, **23,** 89 (1984); (p) F. A. Cotton, *Chem. Soc. Rev.*, **12,** 35 (1983).

41. V. G. Albano, G. Ciani, S. Martinengo, and S. Sirone, *J. Chem. Soc. Dalton Trans.*, 978 (1979).
42. For detailed discussion see R. Hoffmann, *Angew. Chem. Int. Ed. Engl.*, **21,** 711 (1982); *Science,* **211,** 995 (1982).
43. G. Huttner, J. Schneider, H. D. Müller, G. Mohr, J. von Seyerl, and L. Wohlfahrt, *Angew. Chem. Int. Ed. Engl.*, **18,** 76 (1979).
44. J. L. Atwood, W. E. Hunter, R. A. Jones, and T. C. Wright, *Inorg. Chem.*, **22,** 993 (1983).
45. D. Seyferth, *Adv. Organometal. Chem.*, **14,** 97 (1976).
46. R. J. Goudsmit, B. F. G. Johnson, J. Lewis, P. R. Raithby, and K. H. Whitmire, *J. Chem. Soc. Chem. Commun.*, 640 (1982).
47. J. C. Calabrese, L. F. Dahl, P. Chini, G. Longoni, and S. Martinengo, *J. Am. Chem. Soc.*, **96,** 2614 (1974).
48. C. Brown, B. T. Heaton, A. D. C. Towl, P. Chini, A. Fumagalli, and G. Longoni, *J. Organometal. Chem.*, **181,** 233 (1979).
49. (a) K. Wade, *Chem. Brit.*, **11,** 177 (1975); (b) *Adv. Inorg. Chem. Radiochem.*, **18,** 1 (1976).
50. (a) D. M. P. Mingos, *Nature (London), Phys. Sci.*, **236,** 99 (1972); (b) R. Mason and D. M. P. Mingos, *MTP Int. Rev. Sci.: Phys. Chem. Ser. Two,* **11,** 121 (1975).
51. (a) J. W. Lauher, *J. Am. Chem. Soc.*, **100,** 5305 (1978); (b) **101,** 2604 (1979).
52. P. R. Raithby, in *Transition Metal Clusters,* B. F. Johnson, ed., Wiley, New York, 1980, p. 5.
53. (a) E. L. Muetterties, ed., *Boron Hydride Chemistry,* Academic Press, New York, 1975; (b) T. Onak, *Adv. Organometal. Chem.*, **3,** 263 (1965); (c) M. F. Hawthorne, *Acc. Chem. Res.*, **6,** 124 (1973); (d) R. G. Adler and M. F. Hawthorne, *J. Am. Chem. Soc.*, **92,** 6174 (1970); (e) M. F. Hawthorne and T. D. Andrews, *J. Am. Chem. Soc.*, **87,** 2496 (1965).
54. (a) T. E. Paxon, K. P. Callahan, E. L. Hoel, and M. F. Hawthorne, in *Organotransition-Metal Chemistry,* Y. Ishii, M. Tsutsui, eds., Plenum Press, New York, 1975, p. 1; (b) T. E. Paxson and M. F. Hawthorne, *J. Am. Chem. Soc.*, **96,** 4674 (1974); (c) R. T. Baker, M. S. Delaney, R. E. King III, C. B. Knobler, J. A. Long, T. B. Marder, T. E. Paxson, R. G. Teller, and M. F. Hawthorne, *J. Am. Chem. Soc.*, **106,** 2965 (1984).
55. A. J. Deeming, in *Transition Metal Clusters,* B. F. Johnson, ed., Wiley, New York, 1980, p. 391.
56. D. M. Washecheck, E. J. Wucherer, L. F. Dahl, A. Ceriotti, G. Longoni, M. Manassero, M. Sansoni, and P. Chini, *J. Am. Chem. Soc.*, **101,** 6110 (1979).
57. (a) E. L. Muetterties, *Angew. Chem. Int. Ed. Engl.*, **17,** 545 (1978); (b) J. F. Blount, L. F. Dahl, C. Hoogzand, and W. Hübel, *J. Am. Chem. Soc.*, **88,** 292 (1966); (c) L. F. Dahl and D. L. Smith, *J. Am. Chem. Soc.*, **84,** 2450 (1962).
58. (a) W. A. Hermann, *Angew. Chem. Int. Ed. Engl.*, **21,** 117 (1982); (b) V. Ponec, *Catal. Rev. Sci. Eng.*, **18,** 151 (1978); (c) J. Falbe, ed., *New Syntheses with Carbon Monoxide,* Springer-Verlag, Berlin, 1980.
59. (a) J. S. Bradley, *Adv. Organometal. Chem.*, **22,** 1 (1983); (b) J. S. Bradley, G. B. Ansell, and E. W. Hill, *J. Am. Chem. Soc.*, **101,** 7417 (1979); (c) J. S. Bradley, *Phil. Trans. Roy. Soc. (London),* **A308,** 103 (1982); (d) J. S. Bradley, G. B. Ansell, M. E. Lepnowicz, and E. W. Hill, *J. Am. Chem. Soc.*, **103,** 4968 (1981).
60. M. Manassero, M. Sansoni, and G. Longoni, *J. Chem. Soc. Chem. Commun.*, 919 (1976).
61. M. A. Andrews and H. D. Kaesz, *J. Am. Chem. Soc.*, **99,** 6763 (1977).

Index